Microbiology of the Everglades Ecosystem

Microbiology of the Everglades Ecosystem

Editors

James A. Entry
Nutrigrown LLC. 7389
Washington Boulevard
Suite 102
Elkridge Maryland 21075

Andrew D. Gottlieb
Harvard Dr.
Lake Worth, FL
USA

Krish Jayachandran
Department of Environmental Studies
Florida International University
Miami, FL
USA

Andrew Ogram
Soil Microbiology, Soil and Water Science Department
University of Florida
Gainesville, FL
USA

CRC Press
Taylor & Francis Group
Boca Raton London New York

CRC Press is an imprint of the
Taylor & Francis Group, an **informa** business
A SCIENCE PUBLISHERS BOOK

Cover painting reproduced by permission of Patricia Littlechild.

CRC Press
Taylor & Francis Group
6000 Broken Sound Parkway NW, Suite 300
Boca Raton, FL 33487-2742

First issued in paperback 2020

© 2015 by Taylor & Francis Group, LLC
CRC Press is an imprint of Taylor & Francis Group, an Informa business

No claim to original U.S. Government works

ISBN-13: 978-1-4987-1183-8 (hbk)
ISBN-13: 978-0-367-73841-9 (pbk)

Library of Congress Cataloging-in-Publication Data

Microbiology of the everglades ecosystem / James A. Entry, Andrew Gottlieb, Krish Jayachandran, Andrew Ogram, editors.
    pages cm
"A CRC title."
Includes bibliographical references and index.
ISBN 978-1-4987-1183-8 (hardcover : alk. paper) 1. Swamp ecology--Florida--Everglades. 2. Soil microbial ecology--Florida--Everglades. 3. Soils--Organic compound content--Florida--Everglades. 4. Everglades (Fla.) I. Entry, James A., editor. II. Gottlieb, Andrew (Andrew D.) editor. III. Jayachandran, K. V., editor. IV. Ogram, Andrew, editor.

QH541.5.S9.M53 2015
577.6809759'39--dc23
                                             2014044900

**Visit the Taylor & Francis Web site at
http://www.taylorandfrancis.com**

**and the CRC Press Web site at
http://www.crcpress.com**

Preface

Given the importance of microbiology to the structure and function of the Everglades ecosystem, the information available is limited and often scattered throughout varying sources. Microbiological information is buried in journals devoted to water quality, biology, and hydrology, as well as other sources. Our goal is to provide students, researchers and managers the first comprehensive source where the science of Everglades microbiology is organized into a more detailed and holistic form. The book characterizes the tight coupling between primary producers and bacterial communities and how their interactions help to shape the Everglades landscape. Although the examples provided are primarily from the Florida Everglades, the concepts and theory discussed are relevant to wetland systems globally.

The book is divided into three sections. The first section is devoted to abiotic factors that affect microbiological interactions including water nutrient concentration and trends, soil nutrient distribution, and mercury-sulfate interactions. A chapter concerning the implications of global climate change and sea level rise is also included. The second section contains information on periphyton structure and function in the Everglades landscape including the importance of species based assessments of water quality, the impact of hydrology, nutrient loading, and conductivity on periphyton communities, the role of periphyton mats in consumer community structure, patterns and regulators of nitrogenase activity, the structure and function of cyanobacterial mats in Belize, biological indicators of water quality and habitat, and distributions of epiphytic diatoms along phosphorus and salinity gradients in Florida Bay. Additionally, developing methods including pigment-based chemotaxonomy and the use of passive remote sensing to detect calcareous periphyton mats in the Everglades are also discussed. Although there is a lack of molecular information on bacterial and algal communities, the third section presents molecular based information on microbial communities in the Everglades ecosystem. This includes microbial ecology of mercury methylation and demethylation, methanogens in sawgrass communities of the Everglades and Biscayne Bay, syntrophic bacteria, methanogen and methanotroph associations, and biological control of invasive plants using native microorganisms, and degradation of microcystins by indigenous bacterial communities in the Everglades.

This book is intended to serve as a detailed Everglades and Florida Bay microbiology reference source. Given that environmental pressures such as nutrient enrichment, drying and salinization are wide spread phenomenon, the ideas presented in this book are also relevant to tropical and subtropical systems throughout the globe. The book provides recommendations regarding future research needs that will help foster collaborative research and a greater understand. The editors would like to thank the authors who all contributed their expertise, time and energy to this book. We would also like to thank Robin Woodward-Entry for her work on the graphics in this book and Pat Littlechild for the painting used in the front cover.

James A. Entry
Andrew D. Gottlieb
Krish Jayachandran
Andrew Ogram

Contents

Section III: Microbiology of the Everglades

CHAPTER

1

Importance of Microorganisms to the Everglades Ecosystem

Andrew D. Gottlieb,[1,*] *James A. Entry,*[2] *Andrew Ogram*[3]
and *Krish Jayachandran*[4]

Importance of the Everglades Ecosystem

Significant scientific advances made in recent years elucidate the importance of microbial ecology to virtually all aspects of ecology of the greater Everglades ecosystem. This volume assembles much of the recent Everglades microbial ecology research and includes chapters on a variety of topics ranging from basic science to potential implications for management strategies for this unique ecosystem. Included in this volume are discussions of such diverse topics as environmental concerns (nutrient impacts, mercury cycling, and the potential impacts of sea level rise), the function of periphyton communities across gradients in nutrients and hydrology, and the fundamental ecology of methane (CH_4), nitrogen fixation, and algal toxins.

The Everglades ecosystem occupies approximately 9000 km² in South Florida and represents the largest subtropical wetland in the United States. The historic Everglades extended from the headwaters north of Lake Okeechobee to the mangrove estuaries of Florida Bay and the Gulf of Mexico. More than half of the original system has been lost to drainage and development. The present Everglades ecosystem is comprised of

---
[1] Harvard Drive, Lake Worth, FL 33460-6332.
 Email: adgottlieb71@gmail.com
[2] Nutrigrown LLC., 7389 Washington Boulevard, Suite 102, Elkridge Maryland 21075.
 Email: jim.entry@nutrigrown.com
[3] Soil and Water Science Department, University of Florida, Gainesville, Florida 32611-0290.
 Email: aogram@ufl.edu
[4] Environmental Studies Department, Florida International University, Miami, FL 33199.
 Email: jayachan@fiu.edu
* Corresponding author

three Water Conservation Areas (WCAs) (WCA1, WCA2, and WCA3) and Everglades National Park. Over the last century the construction of more than 2500 km of canals and levees and hundreds of water control structures has disrupted the natural sheet flow and altered the hydroperiod in large areas of the Everglades (Harvey et al. 2005). In addition to the loss of area and changes to hydrology, the biotic integrity of the remaining Everglades is also threatened due to undesirable changes in water quality.

The fauna and flora of the pre-development Everglades were adapted to extremely low phosphorus (P) concentrations; phosphorus concentrations in the water column of the more pristine areas of the Everglades ecosystem are typically below 10 µg total phosphorus (TP) L^{-1}. Ecosystem form and function can change with very small increases of P availability (Hecky and Kilham 1988). Water containing elevated nutrient concentrations, originating from urban and agricultural sources, ultimately flows into the EPA (Daroub et al. 2009; Lang et al. 2010; Daroub et al. 2011) and is responsible for altered ecosystem structure and function (DeBusk et al. 1994; 2001; Davis et al. 2003; Noe et al. 2003; Childers et al. 2003; King et al. 2004; Liston and Trexler 2005; Hagerthey et al. 2008) including conversion of sawgrass (*Cladium jamaicense* Crantz.) stands to cattail (*Typha domingensis* Pers.) (DeBusk et al. 1994; 2001; Doren et al. 1997; Lorenzen et al. 2000; Miao et al. 2000; 2001; McCormick et al. 2009) and dissolution of periphyton mats (Gaiser et al. 2006; Gottlieb et al. 2006; McCormick et al. 2001). Changes in composition of primary producer communities affect higher trophic levels, as well as the nature of the organic substrate available to bacteria.

Importance of Everglades Microbiology

The Everglades ecosystem supports a large community of microorganisms which serves diverse and complex functions. Microbes are responsible for the mineralization of nutrients including carbon (C), nitrogen (N), phosphorus (P), and sulfur (S). It is predominantly through the actions of microbes that the nutrients contained in organic detritus are mineralized to forms available to plants and other microbes. It is perhaps the single-most important function of microbes; recycling nutrients tied up in organic material back into forms useable by plants, microbes, and small invertebrates. This process, known as the microbial loop, helps to explain how oligotrophic systems such as the Everglades can support such abundant species diversity and niche specialization across the landscape (Fenchel et al. 2008; Anderson and Ducklow 2001; Azam et al. 1992; Porter 1988). Microorganisms can also degrade pesticides and other hazardous materials deposited in the Everglades, often rendering them less or even non-toxic, potentially lowering or eliminating ground and surface water contamination.

Microbial processes control trace gas flux to the atmosphere. The flux of trace gases between Everglade's detritus and atmosphere is the result of the simultaneous production and consumption processes in soil (Conrad 1996; Laanbroek 2010; Madsen 2011). When soils are inundated, complex interactions between groups of anaerobic microorganisms are responsible for production of trace gases H_2, CO_2, CH_4, N_2, N_2O, and NO. N_2, N_2O, CH_4, and CO_2 are the dominant gaseous products of anaerobic degradation of organic material in most regions of the Everglades, and are released into the atmosphere, in contrast to the other trace gases (Conrad 1996; Laanbroek 2010; Madsen 2011). Although rates are highly variable, freshwater wetlands such as the

Everglades are significant sources of green house gases, such as N_2O, CO_2, and CH_4. Nitrous oxide, a major product of respiratory denitrification, is much more efficient in trapping heat than either CO_2 or CH_4, and may be produced by a very broad range of microorganisms in the Everglades (Chaunhan et al. 2004; 2006).

A significant percentage of the produced CH_4 is oxidized or consumed by methanotrophic bacteria at anoxic-oxic interfaces (i.e., at boundaries between benthic floc and aerobic water and at the root surfaces of aquatic plants). Many wetland macrophytes are adapted to O_2 limited conditions such that they prevail in anaerobic soils (Laanbroek 2010; Madsen 2011). Wetland plants transport O_2 to roots, creating an aerobic rhizosphere which sustains a number of beneficial oxidation processes. Through the aerenchyma pores that transport O_2 by the plant into the root zone, CH_4 can enter the plant aerenchyma system and subsequently be emitted into the atmosphere (Sorrell et al. 2000; Miao et al. 2000; Bricks et al. 1992). Much of the O_2 released into the root zone can be used to oxidize CH_4 before it enters the atmosphere (Laanbroek 2010; Madsen 2011). The O_2 can also be used to regenerate alternative electron acceptors. The continuous supply of alternative electron acceptors will diminish the role of methanogenesis in the anaerobic mineralization processes in the root zone and repress the production and emission of CH_4 (Laanbroek 2010; Madsen 2011).

When water levels are low, large areas of the Everglades can become aerobic. Aerobic conditions combined with large amounts of easily degradable organic material increase heterotrophic microbial activity resulting in increased oxidation and decomposition (Laanbroek 2010; Madsen 2011). Aerobic conditions also favor CO co-oxidation by the ammonium monooxygenase of nitrifying bacteria and NO consumption by either reduction to N_2O in denitrifiers or oxidation to NO_2 in heterotrophic bacteria (Conrad 1996; Laanbroek 2010; Madsen 2011). In addition, shallow surface water environments often contain low CO_2, thereby promoting the use of bicarbonate as a terminal electron acceptor by certain species of cyanobacteria (during photosynthesis). This process results in precipitation of $CaCO_3$, as well as P and other coprecipitates (Merz 1992).

Microorganisms

The importance of microorganisms to the function, stability and resilience of the Everglades ecosystem cannot be overstated. Microbial communities are highly diverse and interact with each other and with their environment through complex networks. Microorganisms exhibit different structural and functional characteristics that both define the organism and its role in the microbial community (Sylvia et al. 1998; Rao 1999; Pepper et al. 2014).

The composition and activities of microbial communities are primarily controlled by three factors: 1) the availability of electron donors; 2) the availability of electron acceptors; and 3) the limiting nutrient. In aerobic environments of the Everglades, particularly in soils during the dry season, organic carbon is plentiful as an electron donor, as is the availability of O_2 as a terminal electron acceptor. Depending on the location, the soil may be limited in nitrogen or in phosphorus, which would control much of the activities of the community. When the Everglades is inundated during the rainy season, O_2 concentrations decrease sharply in the water column and into the soil, such that the availability of alternative terminal electron acceptors or metabolisms is critical.

The wide range of environmental conditions therefore supports a great diversity of prokaryotic types, including aerobic and anaerobic heterotrophs, photosynthetic autotrophs, anaerobic autotrophs, and lithotrophs, each group functioning under conditions that may be either spatially or temporally separated.

In addition to prokaryotes, fungi are abundant in soil population compared to other microorganisms (Sylvia et al. 1998; Rao 1999; Pepper et al. 2014). Fungi serve critical functions as primary decomposers of plant matter, thereby providing nutrients for bacteria. Fungi may also serve as important in the soil as food sources for other, larger organisms. They form beneficial symbiotic relationships with plants, including mycorrhizal relationships with wetland plants (Sylvia et al. 1998; Rao 1999; Pepper et al. 2014).

Algae are a group of photosynthetic organisms ranging from unicellular to multicellular (i.e., colonial and filamentous) in form and generally contain chlorophyll, yet lack true roots, stems and leaves found in vascular plants (Round 1973). Algae utilize solar energy for cell maintenance, growth, and reproduction through photosynthesis. Algal growth requires sufficient sunlight and nutrients, especially N and P (Vymazal 1995). Algae not only grow on soil and plant surfaces directly exposed to sunlight, but also live below the surface of the water. Some algal groups (cyanobacteria) can fix N_2 adding large amounts of N to the ecosystem (Fernandez-Valiente et al. 2001; Rejmankova et al. 2000; 2014; Inglett et al. 2004; Sylvia et al. 1998). Algae are further divided into phyla, traditionally based on pigments and form (including but not limited to): Chlorophytes (green algae), Rhodophytes (red algae), Phaeophytes (brown algae), Chrysophytes (golden algae), and Cyanophytes (blue-green algae). Chlorophytes predominantly make chlorophyll, beta carotene, and lutein. The Cyanophytes contain chlorophyll and other pigments (predominantly carotene) that tint the algae a blue-green color. The Bacillariophyaceae contain chlorophyll as well as carotene and xanothophyll that make the algae golden brown in color (Round 1973). The distribution and abundance of algal taxa found in the Everglades and other wetland systems are directly related to nutrient status, hydrology and salinity (Gaiser et al. 2006; Gottlieb et al. 2006; Wachnicka and Wingard 2014). By understanding current patterns and historical trends in algal community structure, we can monitor and assess changes in system function.

Microbiology of the Everglades Ecosystem

Microorganisms are essential to nearly all processes in the Everglades and other wetland ecosystems, including nutrient cycling and nutrient availability, water quality, organic soil formation and mineralization. A book or compendium on microbiology in this important ecosystem has not been written. This book is intended to provide the reader with the latest information on microbiology in the Everglades ecosystem. The structural, functional and methodological information presented here is relevant not only to the Everglades but to wetlands globally.

The first section of the book characterizes abiotic factors that affect water and soil microorganisms, including discussion of nutrients, mercury, and global climate change mediated sea level rise. In general, nutrient input to the WCAs and the Everglades National Park is decreasing. Diverse groups of anaerobic prokaryotes are responsible for conversion of inorganic mercury to the more toxic methyl mercury which bioaccumulates in the Everglades ecosystem. Mass storage calculations show that the Everglades is a sink for Hg, where most of the Hg input to the system, primarily from atmospheric deposition, will be accumulated in soil and floc. Although, in the southern Everglades, the signature of sea level rise has been blurred with significant hydrologic modifications that are decreasing freshwater flows and accelerating saltwater intrusion (Ross et al. 2000), overall there is ample evidence to support the belief that coastal ecology in the Everglades is changing in concert with rising sea levels. Sea level rise is predicted to shift ecotones and elevate salinities in certain areas of the Everglades ecosystem. Changes in vegetation communities are often the most apparent manifestation of migrating environmental gradients, yet microbial communities and functions will likely respond before vegetation shifts are seen.

Section two of Microbiology of the Everglades Ecosystem focuses on the structural and functional relationships algae play in the Everglades landscape. Algae is nearly ubiquitous throughout the Everglades and is frequently seen as a thick periphyton mat (a complex microbiological community of algae, bacteria, fungi, and detritus) growing on soil surfaces, plants, and free floating depending on the local physical and chemical conditions in the landscape. Understanding the structure and function of algal communities is important as algae are the base of the food web, are important, often dominant primary producers in the system, provide shelter and structure for fish and invertebrates (Liston and Trexler 2005; Sargeant et al. 2011), and play an important role in biogeochemical cycling affecting N and P dynamics and soil formation processes (Gleason and Spackman 1974). In addition to affecting ecosystem structure and function, algae are also used as environmental indicators (Gaiser 2009). Algal communities respond to environmental change rapidly and are used reliably as indicators of nutrient status, salinity, and hydroperiod. By comparing modern day algal assemblages to dated, cored materials, we can gain a longer-term understanding of environmental conditions and landscape structure, ultimately helping to better understand variance and change through time. Additionally, marine and estuarine paleo-assemblages are used to hindcast hydrologic conditions in upstream, freshwater marshes.

The characterization and experimental methods chapters presented provide a snapshot of the status of the system, current and emerging monitoring methods (including remote mapping approaches and chemotaxonomy), and new information regarding periphyton taxonomy and mat functionality (i.e., nutrient cycling and food web dynamics). A discussion of comparable Caribbean flora and function helps to recognize the global distribution of periphyton mats and their significance in the landscape, particularly the role cyanobacteria play in nitrogen cycling. Although there are differences in community structure between wetlands and across landscapes, responses to stressors such as nutrient enrichment, drying and salinization share similarities. Much of the historic periphyton research focused on the photosynthetic fractions of the algal mat. The functional role of bacteria in food webs is also critical, making nitrogen and phosphorus available for uptake by higher organisms. The tight coupling between these two assemblages makes it crucial to understand the community as a whole.

The third section of Microbiology of the Everglades Ecosystem discusses bacterial associations of methanogens and methanotrophs carrying out mercury methylation and demethylation within the sawgrass areas of both the Everglades and Biscayne Bay watersheds. The use of N fixation measurements as an early microbial indicator of nutrient enrichment, as well as algal toxin degradation by indigenous bacterial communities, is also discussed. Although molecular information is presented in section three of this book, very little molecular work has been conducted to characterize the effects of nutrient enrichment on Everglades bacterial communities. Molecular work is necessary to fully understand how microbial communities respond to nutrient inputs, as well as how microbial communities interact with different components of their environment. In addition to providing a better understanding of the impacts of nutrients to microbial communities and the biogeochemical cycles they control, study of microbial communities can provide economic, rapid, and reliable bioindicators of anthropogenic change.

From nutrient cycling to carbon flux, microbial communities help to shape the landscape. By understanding the relationships of physical and chemical drivers on microbiological communities, we can better manage and restore the existing Everglades landscape and related functionality. Ongoing research that will produce new information and allow us to better understand the effects of changes in conductivity and salinity, nutrients, and other physciochemical drivers on microbial structure and function is needed. Changes at the microbial level may be populated through the food web and ultimately effect larger organisms such as fish and birds. Understanding the range and rate of historical changes using paleoecological techniques provides insight into how human populations will need to adapt to or manage change. The studies and theory presented are relevant not only to South Florida audiences but are relevant globally. Declining water quality, decreased freshwater availability, increasing salinity are nearly ubiquitous stressors affecting natural systems. For those wetlands and other ecosystems neighboring large urban areas, the pressures are even greater. Microbiological research, along with other scientific approaches (varying disciplines of chemistry, biology and engineering) must be integrated into regional planning to inform future decision making with holistic and realistic approaches.

References

Anderson, Thomas R. and Hugh W. Ducklow. 2001. Microbial loop carbon cycling in ocean environments studied using a simple steady-state model. Aquat. Microb. Ecol. 26: 37–49.

Brix, H., B.K. Sorrell and P.T. Orr. 1992. Internal pressurization and convective gas flow in some emergent freshwater macrophytes. Limnol. Oceanogr. 37: 1420–1433.

Chauhan, A., A. Ogram and K.R. Reddy. 2004. Syntrophic-methanogenic associations along a nutrient gradient in the Florida Everglades. Appl. Environ. Microbiol. 70: 3475–3484.

Chauhan, A., A. Ogram and K.R. Reddy. 2006. Syntrophic archaeal associations differ with nutrient impact in a freshwater marsh. J. Appl. Microbiol. 100: 73–84.

Childers, D.L., R.F. Doren, R. Jones, G.B. Noe, M. Rugge and L.J. Scinto. 2003. Decadal change in vegetation and soil phosphorus pattern across the Everglades landscape. J. Environ. Qual. 32: 344–362.

Conrad, R. 1996. Soil microorganisms as controllers of atmospheric trace gases (H_2, CO, CH_4, OCS, N_2O, and NO). Microbiol. Rev. 60: 609–640.

Daroub, S.H., T.A. Lang, O.A. Diaz and S. Grunwald. 2009. Long-term water quality trends after implementing best management practices in south Florida. J. Environ. Qual. 38: 1683–1693.

Daroub, S.H., S. Van Horn, T.A. Lang and O.A. Diaz. 2011. Best management practices and long-term water quality trends in the Everglades Agricultural Area. Crit. Rev. Environ. Sci. Technol. 41:(S1) 608–632.

Davis, S.E., C. Corronado-Molina, D.L. Childers and J.W. Day. 2003. Temporarily dependant C, N, and P dynamics associated with the decay of *Rhizophora mangle* L. Leaf litter in oligotrophic mangrove wetlands of the Southern Everglades. Aquatic Bot. 75: 199–215.

DeBusk, W.F., K.R. Reddy, M.S. Kochand and Y. Wang. 1994. Spatial distribution of nutrients in a northern Everglades marsh: Water Conservation Area 2A. Soil Sci. Soc. Am. J. 58: 543–552.

DeBusk, W.F., S. Newman and K.R. Reddy. 2001. Spatio–temporal patterns of soil phosphorus enrichment in Everglades Water Conservation Area 2A. J. Environ. Qual. 30: 1438–1446.

Doren, R.F., T.V. Armentano, L.D. Whiteaker and R.D. Jones. 1997. Marsh vegetation patterns and soil phosphorus gradients in the Everglades ecosystem. Aquatic Bot. 56: 145–163.

Entry, J.A. and A. Gottlieb. 2014. The impact of stormwater treatment areas on water quality in the Everglades Protection Area. Environ. Monit. Assess. 186: 1021–1037.

Fenchel, T. 2008. The Microbial Loop—25 years later. J. Exp. Mar. Biol. Ecol. 366: 99–103.

Gaiser, E.E. 2009. Periphyton as an early indicator of restoration in the Florida Everglades. Ecol. Indic. 9s: 537–545.

Gaiser, E.E., D.L. Childers, R.D. Jones, J.H. Richards, L.J. Scinto and J.C. Trexler. 2006. Periphyton responses to eutrophication in the Florida Everglades: cross-system patterns of structural and compositional change. Limnol. Oceanogr. 51: 617–630.

Gaiser, E.G., P. McCormick, S.E. Hagerthey and A.D. Gottlieb. 2011. Landscape patterns of periphyton in the Florida Everglades. Crit. Rev. Environ. Sci. Technol. 41(S1): 92–120.

Gleason, P.J. and W. Spackman. 1974. Calcareous Periphyton and Water Chemistry in the Everglades. Miami Geological Society, Coral Gables, Florida.

Gottlieb, A., J.H. Richards and E.E. Gaiser. 2005. Effects of desiccation duration on the community structure and nutrient retention of short and long hydroperiod Everglades periphyton mats. Aquatic Bot. 82: 99–112.

Gottlieb, A., J.H. Richards and E.E. Gaiser. 2006. Comparative study of periphyton community structure in long and short hydroperiod Everglades marshes. Hydrobiologia 569: 195–207.

Harvey, J.W., J.E. Saiers and J.T. Newlin. 2005. Solute transport and storage mechanisms in wetlands of the Everglades, south Florida. Water Resour. Res. 41: W05009, doi:10.1029/2004WR003507.

Hecky, R.E. and P. Kilham. 1988. Nutrient limitation of phytoplankton in fresh-water and marine environments—a review of recent-evidence on the effects of enrichment. Limnol. Oceanogr. 33: 796–822.

Inglett, P.W., K.R. Reddy and P.V. McCormick. 2004. Periphyton chemistry and nitrogenase activity in a northern Everglades ecosystem. Biogeochemistry 67: 213–233.

King, R.S., C.J. Richardson, D.L. Urban and E.A. Romanowicz. 2004. Spatial dependency of vegetation-environment linkages in an anthropogenically influenced ecosystem. Ecosystems 7: 75–97.

Laanbroek, H.J. 2010. Methane emission from natural wetlands: interplay between emergent macrophytes and soil microbial processes. A mini-review. Ann. Bot. 105: 141–153.

Lambeti, G.A. 1996. The role of periphyton in benthic foodwebs. Algal Ecol. 17: 533–572.

Lang, T.A., O. Oladeji, M. Josan and S. Daroub. 2010. Environmental and management factors that influence drainage water P loads from Everglades Agricultural Area farms of South Florida. Agric. Ecosys. Environ. 138: 170–180.

Liston, S.E. and J.C. Trexler. 2005. Spatial and temporal scaling of macroinvertebrate communities inhabiting floating periphyton mats in the Florida Everglades. J. North Am. Benthol. Soc. 24: 832–844.

Madsen, E.L. 2011. Microorganisms and their roles in fundamental biogeochemical cycles. Curr. Opin. Biotechnol. 22(3): 456–464.

McCormick, P.V., M.B. O'Dell, R.B.E. Shuford, J.G. Backus and W.C. Kennedy. 2001. Periphyton response to experimental phosphorus enrichment in a subtropical wetland. Aquatic Bot. 71: 119–139.

McCormick, P.V., S. Newman and L.W. Vilchek. 2009. Landscape responses to wetland eutrophication: loss of slough habitat in the Florida Everglades, USA. Hydrobiologia 621: 105–114.

Merz, Martina U.E. 1992. The biology of carbonate precipitation by cyanobacteria. Facies 26: 81–101.

Miao, S.L., S. Newman and F.H. Sklar. 2000. Effects of habitat nutrients and seed sources on growth and expansion of *Typha domingensis*. Aquatic Bot. 68: 297–311.

Miao, S.L., P.V. McCormick, S. Newman and S. Rajagopalan. 2001. Interactive effects of seed availability, water depth, and phosphorus enrichment on cattail colonization in an Everglades wetland. Wetlands Ecol. Manage. 9: 39–47.

Noe, G.B., L.J. Scinto, J. Taylor, D. Childers and R.D. Jones. 2003. Phosphorus cycling and partitioning in an oligotrophic Everglades wetland ecosystem: a radioisotope tracing study. Freshwater Biol. 48: 1993–2008.

Pepper, I.L., C.P. Gerba and T.G. Gentry. 2014. Environmental Microbiology, 3rd Edition. Academic Press, location.

Rao, S. 1999. Soil Microbiology, Fourth edition. Science Publishers, Enfield.

Rejmankova, E. and Jaroslava Kristonos. 2000. A function of cyanobacterial mats in phosphorus-limited tropical wetlands. Hydrobiologia 431: 135–153.

Round, F.E. 1973. The Biology of the Algae. Edward Arnold Publishers.

Sargeant, B.L., E.E. Gaiser and J.C. Trexler. 2011. Indirect and direct controls of macroinvertebrates and small fish by abiotic factors and trophic interactions in the Florida Everglades. Fresh. Biol. 56: 2334–2346.

Sorrell, B.K., I.A. Mendelssohn, K.L. McKee and R.A. Woods. 2000. Ecophysiology of wetland plant roots: a modeling comparison of aeration in relation to species distribution. Ann. Bot. 86: 675–685.

Sylvia, D., M., J.J. Fuhrmann, P.G. Hartel and D.A. Zuberer. 1998. Principles and Applications of Soil Microbiology. Prentice Hall. Upper Saddle River, New Jersey.

Valiente-Fernandez, E., A. Quesada, C. Howard-Williams and I. Hawes. 2001. N_2-Fixation in cyanobacterial mats from ponds on the McMurdo Ice Shelf, Antarctica. Microb. Ecol. 42: 338–249.

Vymazal, J. 1995. Algae and element cycling in wetlands. CRC press, Chelsea, Michigan: Lewis, Publishers 698 pp.

Vymazal, J. 2007. Removal of nutrients in various types of constructed wetlands. Sci. Total Environ. 380: 48–65.

SECTION I

Abiotic Factors that Influence Water and Soil Microorganisms

Water Quality in the Everglades Protection Area

James A. Entry[1,*] and *Andrew D. Gottlieb*[2]

Introduction

The Everglades Protection Area (EPA), comprised of three Water Conservation Areas (WCAs) (WCA1, WCA2, and WCA3) and Everglades National Park (Park), is the largest subtropical wetland in the United States (Fig. 1). The historic Everglades extended from the wet prairies, sloughs, and tree islands located south of Lake Okeechobee to the mangroves of Florida Bay and the Gulf of Mexico (Davis 1943; Egler 1952; Loveless 1959). Over half of the original Everglades ecosystem has been lost to drainage and development (Harvey et al. 2005). More than 2500 km of canals and levees and hundreds of water control structures, have altered the natural sheet flow, water depth and hydroperiod in large areas of the Everglades (Harvey et al. 2005). The remaining Everglades is also threatened due to undesirable changes in water quality.

Water containing elevated nutrient concentrations, originating from urban and agricultural sources, ultimately flowing into the EPA (Daroub et al. 2009; Lang et al. 2010; Daroub et al. 2011) is associated with altered ecosystem structure and function (DeBusk et al. 1994; 2001; Davis et al. 2003; Noe et al. 2003; Childers et al. 2003; King et al. 2004; Liston and Trexler 2005; Hagerthey et al. 2008) including conversion of sawgrass (*Cladium jamaicense* Crantz.) stands to cattail (*Typha domingensis* Pers.) (DeBusk et al. 1994; 2001; Doren et al. 1997; Lorenzen et al. 2000; Miao et al. 2000; 2001; McCormick et al. 2001) and dissolution of periphyton mats (McCormick et al.

[1] Nutrigrown LLC, 7389 Washington Boulevard, Suite 102, Elkridge Maryland 21075.
Email: jim.entry@nutrigrown.com
[2] Harvard Drive, Lake Worth, FL 33460.
Email: adgottlieb71@gmail.com
* Corresponding author

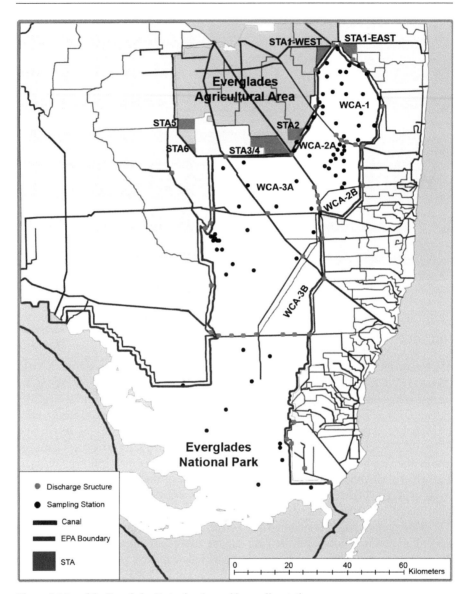

Figure 1. Map of the Everglades Protection Area with sampling stations.

Color image of this figure appears in the color plate section at the end of the book.

2009; Gaiser et al. 2006). Pre-development Everglades ecosystem phosphorus (P) concentrations in the water column are typically below 10 μg total phosphorus (TP). Therefore, Everglades fauna and flora are adapted to extremely low P concentrations; (TP) L^{-1} (Noe et al. 2001; McCormick et al. 2001; Gottlieb et al. 2006).

Small increases in P availability can result in large changes in ecosystem form and function. In response to past adverse changes in the Everglades ecosystem, the federal government sued the State of Florida in 1988 for violation of state water quality standards. The lawsuit was settled in 1992. The 1992 Consent Decree established a long-term limit that now requires all structures discharging into the Park, and marsh water samples at selected interior monitoring stations in the Refuge and the Park to meet 10 µg TP L⁻¹ (Case No. 88-1886-CIV-MORENO).

Six large constructed wetlands called stormwater treatment areas (STAs) were built to remove nutrients and improve water quality flowing into the EPA. Water flows from agricultural and urban areas and is diverted to one or more of the STAs. The STAs are designed to remove nutrients by cycling stormwater first through cells containing emergent vegetation and then through cells containing submerged aquatic vegetation. Emergent vegetation, usually cattail, removes nutrients, especially P from stormwater (primarily via root uptake). Nutrients are then cycled through the system and eventually redeposited into the soil via litterfall. Since water levels are high enough to keep the soil anaerobic, organic matter decomposition and nutrient mineralization are minimized and resuspention into the water column is limited (Noe et al. 2010; Harvey et al. 2011). Water is then conveyed to submerged aquatic vegetation (SAV) cells where additional nutrients are removed by plants (and related epiphytes) such as *Hydrilla verticillata* (L.F.) Royle and *Najas guadalupensis* Morong., in addition to the macroalgae, *Chara* sp. After nutrients are removed stormwater is pumped via canal into one of the WCAs.

The efficacy of nutrient removal from stormwater varies temporally and among STAs and depends on several factors including: 1) antecedent land use, 2) nutrient and hydraulic loading and residence time, 3) vegetation condition, 4) soil type, 5) cell topography, 6) cell size and shape, and 7) construction activities to improve P removal. Nearly all of the STAs have received stormwater in excess of their design capacity. Excess loading as well as design issues associated with STA1-East have resulted in increased mineral and nutrient inflow to WCA1 (Ivanoff and Chen. 2012).

There is a two tiered approach to removing nutrients from water flowing into the Everglades ecosystem. First, best management practices (BMPs) preclude or reduce nutrient application to farm fields, as well as remove nutrients from farm field and urban runoff and leachate flowing into canals which flow into the STAs. If BMPs are properly working, they are able to reduce water flowing from agricultural and urban lands to 80–150 µg TP L⁻¹ (Pietro et al. 2008; Ivanoff and Chen 2012). Historically canal water originating from farm fields more typically contained 200–300 µg TP L⁻¹ (Baker et al. 2012). The second tier (of the nutrient removal system) is STAs. If STA inflow contains 100–150 µg TP L⁻¹ and if working optimally, the STAs can reduce TP concentrations in outflow to 12–25 µg L⁻¹ (Ivanoff and Chen 2012; Dierberg et al. 2002), and potentially further using periphyton storm water treatment technologies as polishing cells (Dodds 2003).

The 58,320 ha⁻² WCA1 receives pumped inflows that are first treated in STA1-West and STA1-East, located adjacent to WCA1's northern boundary. Untreated water has historically been discharged to WCA1, but at a much lower frequency, rate, and volume than treated flows (USFWS 2007a; USFWS 2007b). Stormwater originating from urban and agricultural sources is treated in STA1-East whereas stormwater treated in STA1-West originates in the 280,000 ha⁻² Everglades Agricultural Area (EAA). STA1-West receives water draining primarily from sugarcane (*Saccharum* spp.) and vegetable production. Water Conservation Area 2 is 54,400 ha^2 and is located south of WCA1. Similar to the other WCAs, WCA 2 is an impounded marsh. It is surrounded by canals and levees constructed in the 1950s. This WCA is divided into the 44,800 ha^2 WCA-2A and the 9,600 ha^2 WCA-2B by the L-35B canal (Fig. 1). The northwest section of WCA2 is dominated by a sawgrass plain that extends southeast. The central area of WCA2 is comprised of sawgrass, wet prairies, sloughs, and tree islands. Slough and wet prairie were the most dominant vegetative forms of the south central and southeastern areas, respectively (Davis 1943; Hanan and Ross 2010; Larsen and Harvey 2010).

Water management operations reduced water depths and hydroperiods in northern WCA2-A resulting in increased oxidation of organic soil, and areas of upland terrestrial vegetation subjected to increased fire frequency and intensity (Osbourne et al. 2013; Qian et al. 2009a; 2009b). In addition to drying, increased water depths and prolonged hydroperiods from water management structure and operations resulted in the loss of vast areas of tree islands and wet prairie vegetation in southern WCA2 (Larsen and Harvey 2010; Larsen et al. 2011; Wang et al. 2011). Altered stage effects, as well as decades of inflow water containing high nutrient concentrations created a soil nutrient gradient from the northern part of the WCA to the southern interior (Scheidt and Kalla 2007; Osbourne et al. 2011). This nutrient loading also resulted in a plant species shift from sawgrass to stands of cattail (Childers et al. 2003; King et al. 2004; Hagerthey et al. 2008).

Water Conservation Area 3 is bounded to the north by the L-5 levee, the Rotenberger and Holeyland wildlife management areas, and the EAA and in the south by the L-29 levee, Tamiami Trail, and Everglades National Park. To the west lies the Big Cypress National Preserve, and to the east lies WCA2 and the greater Miami Dade-Broward metropolitan area (Fig. 1). Water Conservation Areas 3 is divided into WCA3-A 202,019 ha^2 and 35,223 ha^2 WCA3-B by interior levees to reduce water losses and seepage. Approximately 60% of the water input to WCA3 is from precipitation while 17% enters the area from WCA2 via the S-11A, S-11B, and S-11C structures (David 1996). Water exits WCA3 and flows into the Everglades National Park through the four S12 structures and the S333 located along the (western and central regions of) Tamiami Trail. The natural hydroperiod of WCA3 has been modified because of impoundment and due to water supply and flood control demands (Walters et al. 1992). Although most inflow is from rainfall, inflow distribution along the northern boundary of WCA3A is skewed (high Miami Canal and S-11 structure flows) leading to excessive drainage in northern WCA3-A and prolonged flooding in WCA-3A South (David 1996).

Everglades National Park encompasses 556,900 ha^2 of freshwater sloughs, sawgrass prairies, marl-forming wet prairies, mangrove forests and saline tidal areas (Ross et al. 2000; Davis 1943). Sloughs comprise much of the central drainage of the Park. Southern marl-forming prairies have shorter hydroperiods and are characterized by high periphyton mat cover and the formation of marl soils (Gottlieb et al. 2006; Gleason and Spackman 1974). The mangrove ecosystem occupies the southern and western borders of the Park where freshwater ecosystems merge with the brackish estuaries of Florida Bay and the Gulf of Mexico (Egler 1952). The objective of this chapter is to characterize water column macro and micronutrient concentrations and seasonal trends in the WCAs and Everglades National Park prior to and after full start-up of the STAs.

Material and Methods

Experimental Design

The experiment was a block design with four blocks (WCA1, WCA2, WCA3, and the Park) and two STA treatments (prior to and after STA's became fully operational for both wet and dry seasons). The length of time samples were collected and the number of samples taken at each monitoring site each month depended on water depth, weather conditions, and sample storage limitations (SFWMD 2006).

Sample Collection and Analysis

All samples collected from November, 1986 through December 2011 were collected and analyzed by the SFWMD chemistry laboratory in West Palm Beach, Florida and were downloaded from the DBHYDRO website at http://www.sfwmd.gov/portal/ page/portal/xweb% 20environmental %20monitoring/dbhydro%20application. Marsh sites (Table 1) were accessed by float helicopter and sampled by wading out into the marsh to collect 3 L of undisturbed water. Samples were stored on ice at 4°C, filtered and preserved within 4 hr of collection (SFWMD 2010).

Table 1. Water Conservation Areas (WCAs) and Everglades National Park (ENP), stormwater treatment pre and post sampling periods, and the number of sites sampled in the Everglades Protection Area (EPA).

EPA	ha[-1]	STA[a]	Pre STA Period	STA Operational[c]
WCA1	58,320	STA1-East	1-4-1999 to 31-12-2005	1-1-2006 to 31-12-2011
		STA1-West	1-4-1986 to 31-12-1999	1-1-1999 to 31-12-2011
WCA2	54,400	STA2	1-1-1994 to 31-12-1999	1-1-2000 to 31-12-2011
WCA3	237,242	STA3/4	1-1-1994 to 31-12-2004	1-1-2005 to 31-12-2011
ENP[c]	556,900	All STAs	1-1-1986 to 31-12-1998	1-1-1999 to 31-12-2011

[a] STA size and efficiency of TP removal information is in Germain and Pietro (2011) and Ivanoff and Chen (2012).
[b] STA post operational sampling period started when each STA was fully operational.
[c] Everglades National Park post operational sampling period started when STA-West was fully operational.

Laboratory Analysis

All samples collected from November 1986 through December 2011 were analyzed using standard methods (APHA 2005; Sharp and Solorzano 1980). Ammonium (NH_4), NO_3, and NO_x were analyzed using Standard Method (SM) 4500 PF using a 4500-P-F TrAAcs 800 continuous flow analyzer. Soluble reactive phosphorus (SRP) and TP were analyzed using SM 4500-PF and SM 4500-PFNO3F, using a Lachat QuikChem 8000 PIA (flow injection analysis) respectively (APHA 2005). Total Kjeldahl Nitrogen (TKN) was analyzed using the EPA 351.2 method using a Lachat QuikChem 8000

PIA. Sulfate and Cl was determined by the EPA 300.1 method using a Dionex DX 500 ion chromatograph (APHA 2005). After water samples were filtered through a 0.45 µm filter, a 2.0 ml sub-sample was analyzed for Na, K, Ca, Mg and Fe using a Perkin Elmer Optima 4300 DV inductively coupled plasma emission spectrometer (SM 3120.B method; APHA 2005). Iron was determined using EPA 210.1 1 and 245.1, respectively (APHA 2005).

Statistical Analysis

Data to determine differences among means (WCAs and ENP and prior to and after STAs became operational) were based on the dry season months (December 1 through March 31) and wet season months (June 1 through September 30) using all sites. Data were subjected to a Wilcoxon and Kruskal Wallis test (Snedecor and Cochran 1994) to determine differences among treatment means with a NPAR1WAY analysis using Statistical Analysis Software programs (SAS Institute Inc. 2010). Plots of surface water TP over time are from un-impacted sites only (those sites with soils less than 500 mg/kg TP) and are based on data from all months.

Kendall tau or Tobit trend analyses for the entire year were based on the center dry season months (December 1 through March 31) and wet season months (June 1 through September 30) (Helsel and Hirsch 2002). Transition months may distort trend analysis and were hence discarded (Helsel and Hirsch 2002). Seasonal Kendall tau trend analysis was performed with residuals from locally weighted scatterplot smoothing (LOWESS) curves (Helsel and Hirsch 2002). Tobit trends were analyzed using SAS programs QLIM procedure (SAS 2010).

When Kendall tau or Tobit slopes are negative the analyses indicates that nutrient concentration in water was decreasing and when slopes are positive the analyses indicates that nutrient concentration was increasing. The more negative or more positive the slope, the more rapid the concentration of that nutrient in water was decreasing or increasing.

If the data reported was below the detection limit, we reported that number as one half the detection limit (Schertz et al. 1991; APHA 2005). Since < 0.3% of all values fell below the minimum detection limit, the data below the limit would not have complicated trend analysis (Hessel and Hirsh 2002).

Results

Water Conservation Area 1

Prior to and after STA1-West became operational in both the wet and dry seasons, slopes were negative for most nutrients. In the dry season Kendall tau analysis showed negative slopes for SRP, K, Ca, Mg, Cl, Fe, and SO_4 prior to STA1-West becoming fully operational in the period from 1986–1999 (Table 2; Fig. 2). After STA1-West became fully operational in both the wet and dry seasons, Kendall tau analysis showed negative slopes for Fe and SO_4 (Table 3; Fig. 3). Additionally, in the post STA dry season, the Tobit NH_4 slope was positive. In the post STA period Tobit slopes were negative for TP in the wet season, whereas results were mixed for dry season. The plots

Table 2. Kendall tau and Tobit trend analysis over both dry and wet seasons for nutrient concentrations in water in Water Conservation Area 1 (WCA1) prior to Stormwater Treatment Area 1-West (STA1-West) full start-up.

Nutrient	STA	Season		Kendall tau			Tobit	
			n	p	% Slope yr^{-1}	p	% Slope yr^{-1}	
TKN	Pre STA	Dry	368	0.967	0.000	**0.014***	**−128.530**	
TKN	Pre STA	Wet	370	**< 0.001***	**0.000**	**< 0.001***	**−246.171**	
NH$_4$	Pre STA	Dry	386	**0.012***	**0.000**	**< 0.001***	**−230.273**	
NH$_4$	Pre STA	Wet	322	**0.012**	**0.000**	**0.012***	**−5.953**	
NO$_3$	Pre STA	Dry	93	0.604	0.000	**0.004***	**−160.092**	
NO$_3$	Pre STA	Wet	107	**0.028***	**0.000**	**< 0.001***	**−380.335**	
TP	Pre STA	Dry	318	0.382	0.000	**< 0.001***	**−4.572**	
TP	Pre STA	Wet	327	0.012	−0.001	**< 0.001***	**−15.906**	
SRP	Pre STA	Dry	362	**< 0.001***	**−0.001**	**< 0.001***	**1.258**	
SRP	Pre STA	Wet	394	**< 0.001***	**0.000**	**< 0.001***	**−14.756**	
K	Pre STA	Dry	233	**< 0.001***	**−0.397**	NS	−10.008	
K	Pre STA	Wet	233	**< 0.001***	**−0.358**	NS	−1.004	
Ca	Pre STA	Dry	233	**0.003***	**−3.796**	**0.005***	**−35.193**	
Ca	Pre STA	Wet	233	**0.010**	**−2.704**	**0.005***	**−42.671**	
Mg	Pre STA	Dry	233	**0.011***	**−0.739**	0.425	−3.248	
Mg	Pre STA	Wet	233	0.119	−0.369	NS	−2.889	
Na	Pre STA	Dry	233	0.076	−1.873	**< 0.001***	**−13.498**	
Na	Pre STA	Wet	233	**0.008***	**−2.641**	**< 0.001***	**−14.182**	
Cl	Pre STA	Dry	232	**< 0.001***	**−30.166**	**< 0.001***	**−78.529**	
Cl	Pre STA	Wet	223	**< 0.001***	**−20.011**	**< 0.001***	**−42.883**	
Fe	Pre STA	Dry	216	**< 0.001***	**−24.096**	**< 0.001***	**−173.172**	
Fe	Pre STA	Wet	220	**< 0.001***	**−48.778**	**< 0.001***	**−8.753**	
SO$_4$	Pre STA	Dry	219	**< 0.001***	**−1.261**	**0.013***	**−6.915**	
SO$_4$	Pre STA	Wet	223	**0.019***	**−0.380**	**< 0.001***	**−42.884**	

[a] WCA1 pre-STA start-up period of record is from 1986 to 1999.
[b] *= significant at ($p \leq 0.05$).
[c] NS not significant at $p \leq 0.05$.

for surface water TP indicate little change in water quality in both the pre and post STA implementation periods (Fig. 4). The post implementation period from 2005 to 2010, however, was a period when mean surface water TP declined (f(x)= −0.002788 ln(x) +0.015319, R^2 = 0.455).

Water Conservation Area 2

Prior to and after STA2 became fully operational in both the wet and dry seasons, slopes were negative for most but not all nutrients. Prior to STA2 becoming fully operational in the dry season, Kendall tau slopes were negative for TKN, SRP, Mg, Fe, and SO$_4$ in the period from 1994–1999 (Table 4; Fig. 5). Alternatively, the Kendall tau NO$_3$ slope and the Tobit NO$_3$, and Cl slopes were positive in the pre-STA2 period. After

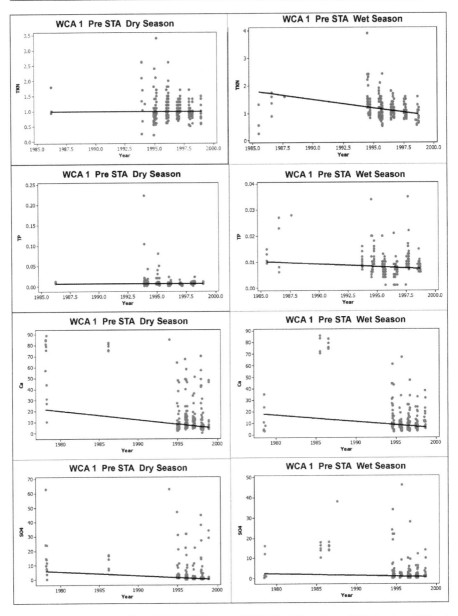

Figure 2. Kendall tau trend Total Kjeldahl Nitrogen (TKN), Total Phosphorus (TP), Calcium (Ca) and sulfate (SO$_4$) during both dry and wet seasons in Water Conservation Area 1 (WCA1) prior to Stormwater Treatment Area 1-West (STA1-West) full start-up.

STA2 became fully operational in the period from 2000–2011, Kendall tau slopes were negative for dry season TP, SRP, Ca, and Cl (Table 5; Fig. 6). In the wet season, Kendall tau slopes were negative for TP, SRP, and Na. In the wet season, Tobit NO$_3$,

Table 3. Kendall tau and Tobit trend analysis over both dry and wet seasons for nutrient concentrations in water in Water Conservation Area 1 (WCA1) post Stormwater Treatment Area 1-West (STA1-West) full start-up.

Nutrient	STA	Season		Kendall tau		Tobit	
			n	p	% Slope yr^{-1}	p	% Slope yr^{-1}
TKN	Post STA	Dry	775	0.301	−0.033	< 0.001*	−185.416
TKN	Post STA	Wet	571	< 0.001*	−0.167	< 0.001*	−250.326
NH$_4$	Post STA	Dry	831	0.002*	0.000	< 0.001*	81.608
NH$_4$	Post STA	Wet	618	0.507	0.000	< 0.001*	−205.815
NO$_3$	Post STA	Dry	263	< 0.001*	0.000	−0.009*	−5.134
NO$_3$	Post STA	Wet	228	0.835	0.000	0.024*	−4.595
TP	Post STA	Dry	426	< 0.001*	0.000	0.001*	−3.788
TP	Post STA	Wet	608	0.007*	0.000	0.009*	−6.669
SRP	Post STA	Dry	548	0.724	0.000	0.612	−0.338
SRP	Post STA	Wet	704	< 0.001*	0.000	0.398	−1.268
K	Post STA	Dry	634	0.625	0.000	0.090	0.096
K	Post STA	Wet	500	< 0.001*	−0.511	0.210	−0.738
Ca	Post STA	Dry	540	0.217	0.686	0.049*	−2.408
Ca	Post STA	Wet	404	0.005*	−2.337	0.001*	−5.886
Mg	Post STA	Dry	529	0.585	4.873	0.078	−4.189
Mg	Post STA	Wet	404	< 0.001*	−6.388	0.023*	−7.436
Na	Post STA	Dry	641	0.603	0.000	0.045*	−1.019
Na	Post STA	Wet	507	< 0.001*	−0.723	0.002*	−2.046
Cl	Post STA	Dry	529	0.695	0.281	0.399	−1.448
Cl	Post STA	Wet	404	< 0.001*	−3.204	0.004	−5.581
Fe	Post STA	Dry	278	0.046*	−8.179	< 0.001*	−9.670
Fe	Post STA	Wet	202	0.185	−11.024	< 0.001*	−7.837
SO$_4$	Post STA	Dry	529	< 0.001*	−0.250	< 0.001*	−3.980
SO$_4$	Post STA	Wet	389	< 0.001*	−0.406	< 0.001*	−7.519

[a] WCA1 post- STA period of record is from 2000 through 2011.
[b] *= significant at ($p \leq 0.05$).
[c] NS not significant at $p \leq 0.05$.

Mg, Na, and SO$_4$ slopes were positive. The surface water TP plot indicates a significant decline in the pre-STA implementation period (Fig.7).

Water Conservation Area 3

Prior to STA3/4 becoming fully operational in the period from 1994–1999, Kendall tau analysis showed neither a negative or positive slope for dry season TKN and TP (Fig. 8). The dry season Kendall tau slope was positive for SO$_4$, and negative for Mg (Table 6). Prior to STA3/4 becoming fully operational in the wet season, Kendall tau NH$_4$, Cl, and SO$_4$ slopes were negative. Additionally, in the wet season, Tobit slopes were negative for NO$_3$, TP, and Fe and positive for Na.

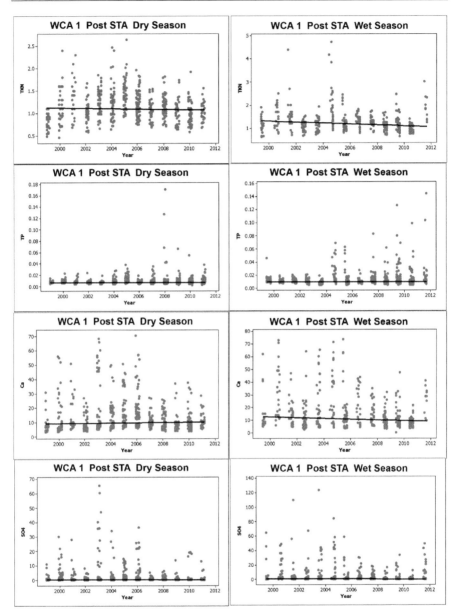

Figure 3. Kendall tau trend Total Kjeldahl Nitrogen (TKN), Total Phosphorus (TP), Calcium (Ca) and sulfate (SO_4) during both dry and wet seasons in Water Conservation Area 1 (WCA1) post Stormwater Treatment Area 1-West (STA1-West) full start-up.

After STA3/4 became fully operational in the period from 2000–2011 dry season Kendall tau TP and SRP were negative (Table 7; Fig. 9). After STA3/4 became fully operational in the dry season, NH_4 and Fe Tobit slopes were negative. In the dry season the Tobit TP slope was positive. In the wet season, the Kendall tau TP slope was negative. After STA3/4 became fully operational in the wet season, Kendall tau

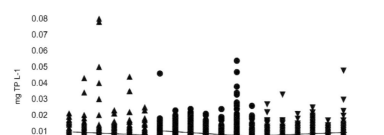

Water Conservation Area 1

Figure 4. Water Conservation Area 1 surface water TP pre (▲) and post (●) STA1-West and post (▼) STA1-East and STA1-West implementation (Pre-STA: y = −0.0001x + 0.2568, R² = 0.0012, post STA1 W: y = −0.0002x + 0.3417, R² = 0.0054 and post STA1 E & W: y = −0.0001x + 0.1226, R² = 0.007) indicating little change in TP concentration.

Table 4. Kendall tau and Tobit trend analysis over both dry and wet seasons for nutrient concentrations in water sampled at marsh stations in Water Conservation Area 2 (WCA2) prior to Stormwater Treatment Area 1 West (STA1-West) full start-up.

Nutrient	STA	Season		Kendall tau			Tobit	
			n	p	% Slope yr⁻¹		p	% Slope yr⁻¹
TKN	Pre STA	Dry	160	**0.046***	**−1.278**		**< 0.000***	**−3168.553**
TKN	Pre STA	Wet	245	**< 0.000***	**−3.040**		**< 0.000***	**−3608.510**
NH₄	Pre STA	Dry	160	0.627	0.000		**< 0.000***	**−167.810**
NH₄	Pre STA	Wet	246	0.449	0.000		**< 0.000***	**−46.002**
NO₃	Pre STA	Dry	122	0.914	0.000		**< 0.000***	**−816.733**
NO₃	Pre STA	Wet	183	**< 0.000***	**0.078**		**< 0.000***	**2383.928**
TP	Pre STA	Dry	148	0.346	−0.007		**< 0.000***	**−83.155**
TP	Pre STA	Wet	230	**0.003***	**−0.026**		**< 0.000***	**−156.755**
SRP	Pre STA	Dry	157	**< 0.000***	**−0.007**		**< 0.000***	**−80.753**
SRP	Pre STA	Wet	243	**< 0.000***	**−0.011**		**< 0.000***	**−20.862**
K	Pre STA	Dry	439	0.826	0.000		NS	0.125
K	Pre STA	Wet	552	**< 0.000***	**−13.136**		NS	−2.005
Ca	Pre STA	Dry	438	0.214	−27.829		**< 0.000***	**−35.113**
Ca	Pre STA	Wet	552	**< 0.000***	**−115.588**		**< 0.000***	**−85.692**
Mg	Pre STA	Dry	437	**0.038***	**−14.493**		**< 0.000***	**−19.133**
Mg	Pre STA	Wet	552	**< 0.000***	**−49.684**		NS	−5.777
Na	Pre STA	Dry	429	0.231	−34.043		**< 0.000***	**−38.810**
Na	Pre STA	Wet	559	**< 0.000***	**−144.712**		**< 0.000***	**−28.343**
Cl	Pre STA	Dry	436	0.059	−85.167		**< 0.000***	**7.964**
Cl	Pre STA	Wet	552	**< 0.000***	**−177.449**		**< 0.000***	**−85.580**
Fe	Pre STA	Dry	316	**0.000***	**−20.829**		NS	−22.778
Fe	Pre STA	Wet	383	0.329	4.663		**< 0.000***	**−826.780**
SO₄	Pre STA	Dry	428	0.915	0.000		**< 0.000***	**−5.667**
SO₄	Pre STA	Wet	550	**< 0.000***	**−120.482**		**< 0.000***	**−17.554**

[a] WCA2 pre-STA start-up period of record is from 1994 through 1999.
[b] * = significant at ($p \leq 0.05$).
[c] NS not significant at $p \leq 0.05$.

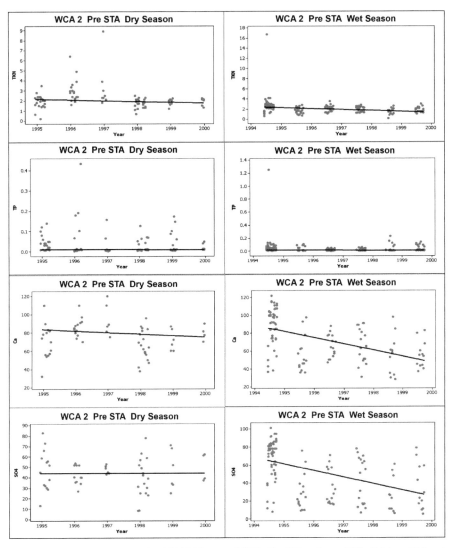

Figure 5. Kendall tau trend Total Kjeldahl Nitrogen (TKN), Total Phosphorus (TP), Calcium (Ca) and sulfate (SO_4) during both dry and wet seasons in Water Conservation Area 2 (WCA2) prior to Stormwater Treatment Area 2 (STA2) full start-up.

Mg and Cl slopes were positive and Tobit slopes for TP, SRP, and Fe were negative. In the wet season NH_4, Mg, Cl, and SO_4 Tobit slopes were positive. The mean plots for surface water TP indicate only a marginally significant decline in the post STA implementation period. Little change in TP is observed in either the pre or post STA implementation period (Fig. 10).

Table 5. Kendall tau and Tobit trend analysis over both dry and wet seasons for nutrient concentrations in water sampled at marsh stations in Water Conservation Area 2 (WCA2) post Stormwater Treatment Area 2 (STA2) full start-up.

Nutrient	STA	Season		Kendall tau		Tobit	
			n	p	% Slope yr^{-1}	p	% Slope yr^{-1}
TKN	Post STA	Dry	177	0.889	−0.639	< 0.000*	**38.668**
TKN	Post STA	Wet	285	0.451	−1.520	< 0.000*	**−738.260**
NH$_4$	Post STA	Dry	194	0.132	0.000	< 0.000*	**55.268**
NH$_4$	Post STA	Wet	332	< 0.000*	**0.000**	< 0.000*	**−543.615**
NO$_3$	Post STA	Dry	140	0.220	0.000	< 0.000*	**18.148**
NO$_3$	Post STA	Wet	248	0.001*	**0.039**	< 0.000*	**51.382**
TP	Post STA	Dry	169	< 0.000*	**−0.004**	0.514	3.295
TP	Post STA	Wet	265	< 0.000*	**−0.013**	< 0.000*	**−14.786**
SRP	Post STA	Dry	176	0.001*	**−0.003**	0.042*	**−4.709**
SRP	Post STA	Wet	269	< 0.000*	**−0.005**	< 0.000*	**−33.911**
K	Post STA	Dry	121	0.390	0.000	NS	−0.870
K	Post STA	Wet	243	0.376	−6.568	NS	6.429
Ca	Post STA	Dry	121	< 0.000*	**−13.915**	< 0.000*	**−28.395**
Ca	Post STA	Wet	242	0.894	−57.794	0.585	3.085
Mg	Post STA	Dry	121	0.222	−7.246	< 0.000*	**−34.054**
Mg	Post STA	Wet	242	0.595	−24.842	< 0.001*	**1.008**
Na	Post STA	Dry	122	0.647	−17.021	0.073	1.263
Na	Post STA	Wet	243	0.009*	**−72.356**	0.011*	**7.562**
Cl	Post STA	Dry	120	0.009*	**−88.725**	< 0.000*	**−19.930**
Cl	Post STA	Wet	241	0.488	−10.415	< 0.000*	**−0.263**
Fe	Post STA	Dry	20	0.107	2.332	0.476	6.137
Fe	Post STA	Wet	65	0.559	0.000	0.085	10.167
SO$_4$	Post STA	Dry	29	0.209	−6.964	< 0.000*	**−14.009**
SO$_4$	Post STA	Wet	241	0.089	8.093	< 0.000*	**15.760**

[a] WCA2 post-STA period of record is from 2000 through 2011.
[b] *= significant at ($p \leq 0.05$).
[c] NS not significant at $p \leq 0.05$.

Everglades National Park

Prior to the first STA (STA1-West) becoming operational in 1999, in both the wet and dry seasons, Kendall tau slopes were negative for most but not all nutrients (Tables 8 and 9; Figs. 11 and 12). In the dry season the Kendall tau NO$_3$ slope was positive. Tobit slopes varied among nutrients in the pre STA period. In the dry season, Tobit NH$_4$ and Ca slopes were positive. In the wet season, Tobit TKN, NH$_4$, and SO$_4$ slopes were positive. After STA1-West became operational, in the dry season, Kendall tau slopes were positive for K, Ca, Mg, and Cl. After STA1-West became operational, in the dry season, Tobit NH$_4$, Ca, Mg, and Cl slopes were positive. In the wet season Tobit slopes were negative for TKN and SO$_4$ but positive for NH$_4$ and Cl. The mean

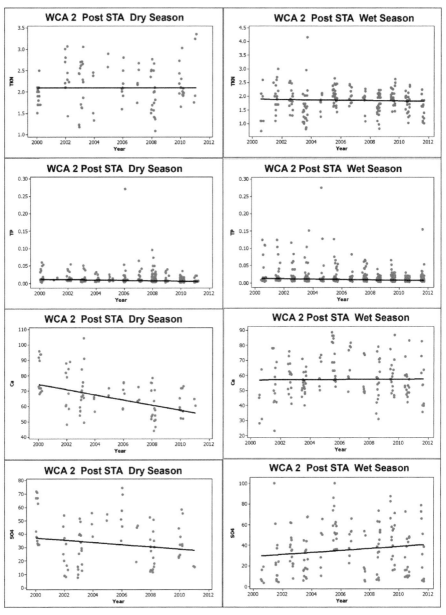

Figure 6. Kendall tau trend Total Kjeldahl Nitrogen (TKN), Total Phosphorus (TP), Calcium (Ca) and sulfate (SO_4) during both dry and wet seasons in Water Conservation Area 2 (WCA2) post Stormwater Treatment Area 2 (STA) full start-up.

plots indicate little change in TP concentration in the pre and post STA implementation periods after plots (Fig. 13).

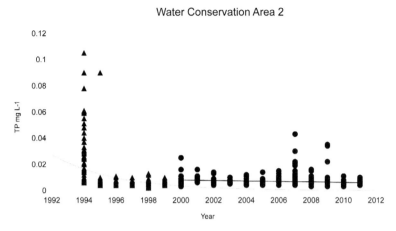

Figure 7. Water Conservation Area 2A surface water TP pre (▲) and post (●) STA 2 implementation (Pre-STA: y = 1.39E + 219 exp (–2.55E-001 x), R^2 = 0.3334, post STA: y = –2.28E-004x + 4.64E-001, R^2 = 0.0333).

Discussion

The analysis trend indicates the concentrations of most nutrients in the WCAs and Park are decreasing, even though Kendall tau and Tobit results did not always agree. Tobit slopes were usually steeper than Kendall tau slopes, whether positive or negative indicating that Tobit analysis may be more sensitive to water quality changes. Nutrients, especially P, are taken up by periphyton and plants within hours to days (Noe et al. 2001; 2003; Childers et al. 2003; Gaiser et al. 2006; Gaiser 2009). Since the Everglades is a nutrient poor ecosystem, nutrients are tightly cycled and then slowly deposited as detritus in the soil leading to low nutrient concentrations in the water column. When nutrients are continually added threshold P concentrations can be reached resulting in changes in periphyton and plant community structure and function (McCormick et al. 2001; 2009; Miao et al. 2001).

The concentration of nutrients in EPA water may fluctuate with season. In the dry season excessive drying may expose large areas of organic soils changing from anaerobic to aerobic conditions allowing rapid nutrient mineralization and addition of those nutrients to wetland water. Depending on STA efficacy, large amounts of nutrients from EAA soils may pass through STAs, and be delivered to the WCAs. The high amount of rainfall in the wet season often dilutes nutrients in EPA water. Season affected several nutrient concentration trends in EPA water, including SO_4 concentration in post STA startup in WCA1, WCA2, and the Park.

Prior to STA implementation, initial improvements in water quality associated with incoming stormwater, resulted in trends of decreasing N and P concentrations in each of the WCAs suggesting that improved BMPs drove initial improvements in water quality in the EPA with STAs playing an important later role. Although seasonal trends for the period were decreasing, the annual analysis for WCA1 indicated little decline in the post-STA implementation period. This is likely due to the conditions in the 2011 dry season, as the period up to 2011 was a period to significant decline in mean surface water TP concentration. WCA2 indicated steeper declines in nutrient

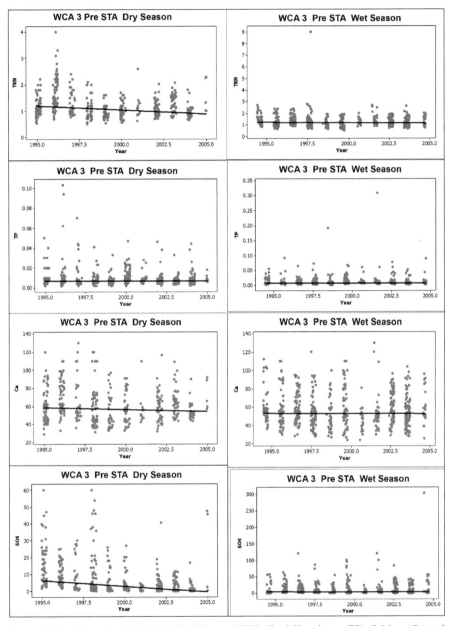

Figure 8. Kendall tau trend Total Kjeldahl Nitrogen (TKN), Total Phosphorus (TP), Calcium (Ca) and sulfate (SO_4) during both dry and wet seasons in Water Conservation Area 3 (WCA3) prior to Stormwater Treatment Area 3/4 (STA3/4) full start-up.

Table 6. Kendall tau and Tobit trend analysis over both dry and wet seasons for nutrient concentrations in water sampled at marsh stations in Water Conservation Area 3 (WCA3) prior to Stormwater Treatment Area (STA) full start-up.

Nutrient	STA	Season		Kendall tau		Tobit	
			n	p	% Slope yr^{-1}	p	% Slope yr^{-1}
TKN	Pre STA	Dry	775	**< 0.001***	**0.0000**	0.7250	−2.542
TKN	Pre STA	Wet	928	**0.029***	**0.0000**	0.0815	0.182
NH$_4$	Pre STA	Dry	906	0.9837	−0.2785	NA	NA
NH$_4$	Pre STA	Wet	1003	**< 0.001***	**−0.0678**	NA	NA
NO$_3$	Pre STA	Dry	253	NA	NA	**< 0.001***	**−364.591**
NO$_3$	Pre STA	Wet	280	NA	NA	**< 0.001***	**−187.378**
TP	Pre STA	Dry	87	**< 0.001***	**0.0000**	**< 0.001***	**−5.162**
TP	Pre STA	Wet	166	0.9412	0.0000	**< 0.001***	**−6.853**
SRP	Pre STA	Dry	669	0.2408	0.0000	**< 0.001***	**−5.351**
SRP	Pre STA	Wet	800	**< 0.001***	**0.0000**	0.1553	−1.960
K	Pre STA	Dry	439	0.1076	0.0000	NS	NA
K	Pre STA	Wet	552	0.3291	0.0000	NS	NA
Ca	Pre STA	Dry	438	0.1203	−0.3417	**< 0.001***	**−12.367**
Ca	Pre STA	Wet	552	0.6574	−0.1750	0.1399	3.092
Mg	Pre STA	Dry	437	**0.017***	**−4.0064**	**0.033***	**−1.580**
Mg	Pre STA	Wet	552	0.4437	−0.4496	0.9651	−0.027
Na	Pre STA	Dry	439	0.6502	−1.4704	0.8770	0.290
Na	Pre STA	Wet	552	0.4887	−0.3021	**0.034***	**4.391**
Cl	Pre STA	Dry	438	0.6914	0.0000	0.5966	−1.678
Cl	Pre STA	Wet	552	0.9541	0.7695	0.2738	2.925
Fe	Pre STA	Dry	316	0.4547	−0.6589	**< 0.001***	**−9.734**
Fe	Pre STA	Wet	383	0.6842	0.0000	**< 0.001***	**−40.087**
SO$_4$	Pre STA	Dry	428	**< 0.001***	**8.4676**	**0.009***	**−6.925**
SO$_4$	Pre STA	Wet	550	**0.011***	**−4.4206**	0.6685	−0.993

[a] WCA3 pre-STA start-up period of record is from 1994 through 1999.
[b] * = significant at ($p \le 0.05$).
[c] NS not significant at $p \le 0.05$.

concentrations in the pre-STA implementation period (indicated by both Kendall tau and Tobit trend analysis, and mean plots) primarily driven by high TP concentrations in early years.

Analyzing different time periods resulted in different trend slopes. Our results are based on the time period prior to and after the STAs became fully operational. Hanlon et al. (2010) found increasing trends for the TP concentration in water flowing from WCA3 through the S12A, S12B, S12C, S12D, and S333 structures into the Park from 1977–1989 and decreasing trends from 1990–1995. They also found decreasing TP, TKN, and TN trends in water flowing into the Park from 1977–2005. Zapata-Rios found that from 2003–2007 Kendall tau the TP concentration trend increased in Park marsh water. In contrast, our Tobit analysis indicated that the TKN and TP concentration in

Table 7. Kendall tau and Tobit trend analysis over both dry and wet seasons for nutrient concentrations in water sampled at marsh stations in Water Conservation Area 3 (WCA3) post Stormwater Treatment Area 3/4 (STA3/4) full start-up.

Nutrient	STA	Season			Kendall tau		Tobit	
			n	p	% Slope yr^{-1}	p	% Slope yr^{-1}	
TKN	Post STA	Dry	207	0.3288	−0.2836	NS	−2.281
TKN	Post STA	Wet	563	0.2285	0.1991	NS	−1.012
NH$_4$	Post STA	Dry	209	NA	NA	**< 0.001***	**−90.583**
NH$_4$	Post STA	Wet	563	NA	NA	**< 0.001***	**23.269**
NO$_3$	Post STA	Dry	120	0.0612	0.0000	NA	NA
NO$_3$	Post STA	Wet	256	0.5725	0.0000	NA	NA
TP	Post STA	Dry	121	**< 0.001***	**−0.0102**	**< 0.001***	**29.154**
TP	Post STA	Wet	360	**< 0.001***	**−0.0078**	**< 0.001***	**−1.324**
SRP	Post STA	Dry	199	**< 0.001***	**−0.0029**	NS	−8.252
SRP	Post STA	Wet	552	**< 0.001***	**0.0000**	**0.027***	**−4.433**
K	Post STA	Dry	122	0.7080	0.2857	0.1098	−3.052
K	Post STA	Wet	243	0.8755	0.0000	NA	NA
Ca	Post STA	Dry	121	0.4843	−6.7139	0.1271	−2.148
Ca	Post STA	Wet	242	0.9196	−0.4983	0.8745	−6.406
Mg	Post STA	Dry	121	0.4267	12.4827	0.3286	−6.026
Mg	Post STA	Wet	242	**0.015***	**19.9371**	**< 0.001***	**15.833**
Na	Post STA	Dry	122	0.1943	2.7502	**0.008***	**−14.723**
Na	Post STA	Wet	243	0.2772	1.7562	0.9881	−0.039
Cl	Post STA	Dry	120	0.1974	13.2917	**0.044***	**−1.062**
Cl	Post STA	Wet	241	**0.008***	**13.8249**	**< 0.001***	**10.325**
Fe	Post STA	Dry	29	0.8852	−7.1528	**< 0.001***	**−30.487**
Fe	Post STA	Wet	65	0.7806	22.2056	**< 0.001***	**−165.433**
SO$_4$	Post STA	Dry	29	0.2500	−0.3601	0.4848	−25.071
SO$_4$	Post STA	Wet	241	0.0540	−1.7780	**< 0.001***	**1.037**

[a] WCA3 post-STA period of record is from 2000 through 2011.
[b] * = significant at ($p \leq 0.05$).
[c] NS not significant at $p \leq 0.05$.

the Park decreased since October 1999, when the first STA became operational. Yet the mean slope plot indicates that changes in un-impacted sites were not significant for this same period (Entry and Gottlieb 2014). Both the time and season analyzed directly led to differences in the patterns observed.

Chloride can be regarded as conservative tracer and is used as an indicator of canal water intrusion into marshes. The Park has lower TKN and TP concentrations, but higher Cl and SO$_4$ concentrations in water after the STAs became fully operational (Tables 8 and 9). An increasing Cl$^-$ trend suggests an increasing volume of canal water containing lower nutrient, but elevated Cl concentrations is being delivered to the Park (Entry and Gottlieb 2014). Kendall tau and Tobit analyses (and the surface water mean plots) showed that TP was neither increasing nor decreasing after STAs became operational. STAs are not directly affecting nutrients and water entering the Park in

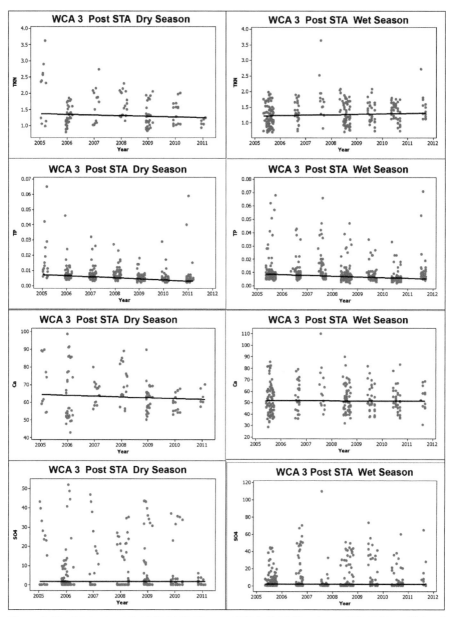

Figure 9. Kendall tau trend Total Kjeldahl Nitrogen (TKN), Total Phosphorus (TP), Calcium (Ca) and sulfate (SO$_4$) during both dry and wet seasons in Water Conservation Area 3 (WCA3) post Stormwater Treatment Area 3/4 (STA3/4) full start-up.

the same way they contribute to the WCAs. Although nutrients derived internally from the WCAs, as well as allocthonous nutrients affect the Park, much of the nutrient load is absorbed by the WCA's periphyton and marsh vegetation prior to water entering the Park. As the TP concentration approaches 10 µg L^{-1} (or some theoretical lower

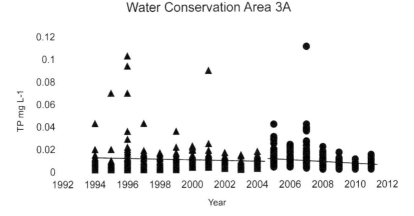

Figure 10. Water Conservation Area 3A surface water TP plot indicating little change in TP concentration during pre (▲) and post (●) STA3/4 implementation periods (Pre-STA: y = –0.0001x + 0.2903 , R^2 = 0.0038, post STA: y = 3.00E + 097 exp(–1.14E-001x), R^2 = 0.1642).

Table 8. Kendall tau and Tobit trend analysis over both dry and wet seasons for nutrient concentrations in water sampled at marsh stations in Everglades National Park prior to Stormwater Treatment Area (STA-1 West) full start-up.

Nutrient	STA	Season		Kendall tau			Tobit	
			n	p	% Slope yr^{-1}		p	% Slope yr^{-1}
TKN	Pre STA	Dry	558	< 0.001*	–0.263		< 0.001*	–265.491
TKN	Pre STA	Wet	642	< 0.001*	–0.177		< 0.001*	18.992
NH$_4$	Pre STA	Dry	587	< 0.001*	–0.003		< 0.001*	–11.737
NH$_4$	Pre STA	Wet	643	0.185	0.000		< 0.001*	5.701
NO$_3$	Pre STA	Dry	552	< 0.001*	–0.012		< 0.001*	–42.662
NO$_3$	Pre STA	Wet	594	0.054	0.000		< 0.001*	–16.289
TP	Pre STA	Dry	552	< 0.001*	–0.001		< 0.001*	–5.254
TP	Pre STA	Wet	607	< 0.001*	–0.002		0.496	–0.679
SRP	Pre STA	Dry	582	0.312	0.000		0.347	–1.984
SRP	Pre STA	Wet	642	0.861	0.000		NS	–0.661
K	Pre STA	Dry	233	0.617	0.072		NS	–0.546
K	Pre STA	Wet	450	0.323	0.110		0.499	0.152
Ca	Pre STA	Dry	233	< 0.001*	–6.166		< 0.001*	–19.185
Ca	Pre STA	Wet	450	0.469	–0.586		< 0.001*	–0.299
Mg	Pre STA	Dry	233	0.255	–0.522		0.425	–1.772
Mg	Pre STA	Wet	439	0.673	–0.139		< 0.001*	–1.384
Na	Pre STA	Dry	233	< 0.001*	–8.440		< 0.001*	–7.362
Na	Pre STA	Wet	450	< 0.001*	–5.798		< 0.001*	–5.632
Cl	Pre STA	Dry	232	< 0.001*	–16.662		< 0.001*	–42.834
Cl	Pre STA	Wet	439	< 0.001*	–12.262		< 0.001*	–11.377
Fe	Pre STA	Dry	216	< 0.001*	–75.940		< 0.001*	–94.457
Fe	Pre STA	Wet	412	< 0.001*	–73.208		< 0.001*	–332.712
SO$_4$	Pre STA	Dry	219	< 0.001*	–1.531		0.013*	–3.772
SO$_4$	Pre STA	Wet	438	< 0.001*	–1.094		< 0.001*	11.980

[a] Everglades National Park pre-STA start-up period of record is from 1986 through 1999.
[b] * = significant at ($p \le 0.05$).
[c] NS not significant at $p \le 0.05$.

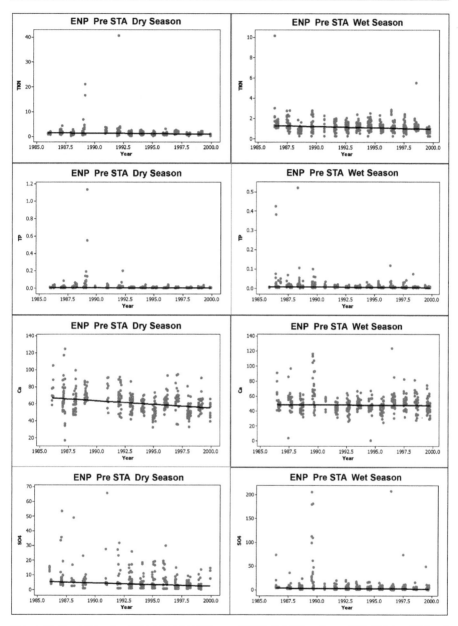

Figure 11. Kendall tau trend Total Kjeldahl Nitrogen (TKN), Total Phosphorus (TP), Calcium (Ca) and sulfate (SO₄) during both dry and wet seasons in Everglades National Park post Stormwater Treatment Area 1-West (STA1-West) full start-up.

limit), as seen in the pristine Everglades, we are less likely to see further declines in marsh water TP concentrations.

Although TKN mean was lower in the post STA implementation period, Kendall tau slopes were positive for TKN, and Cl indicating that the concentrations in water

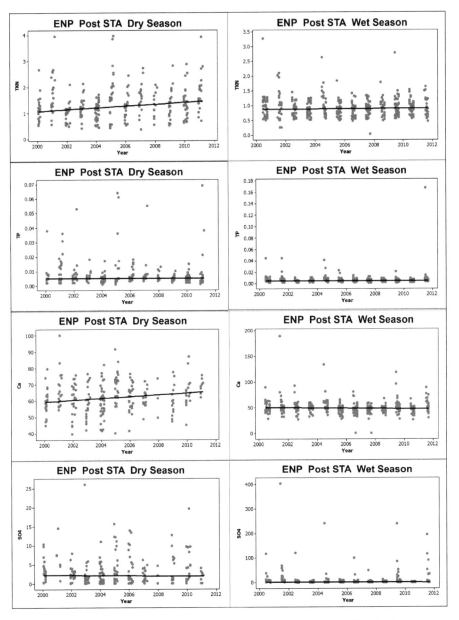

Figure 12. Kendall tau trend Total Kjeldahl Nitrogen (TKN), Total Phosphorus (TP), Calcium (Ca) and sulfate (SO$_4$) during both dry and wet seasons in Everglades National Park post Stormwater Treatment Area 1-West (STA1-West) full start-up.

in the Park may be increasing. From an estuarine and marine standpoint, changes in N may be extremely important given the dominance of P in marine systems.

Acquiring a greater understanding of treatment options, their efficacy, and costs can lead to an integrated strategy to improve Everglades water quality and

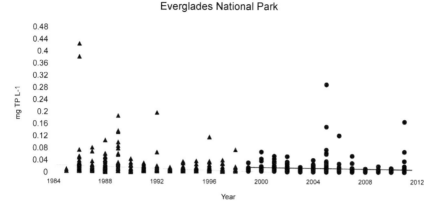

Figure 13. Everglades National Park surface water TP pre (▲) and post (●) STA implementation (Pre-STA= –2.7888 ln(x) + 21.1985 , R^2 = 0.0143, Post-STA: y = –0.0001x + 0.1570, R^2 = 0.0004). Plot does not include data from Taylor Slough sites.

Table 9. Kendall tau and Tobit trend analysis over both dry and wet seasons for nutrient concentrations in water sampled at marsh stations in Everglades National Park post Stormwater Treatment Area (STA -1 West) full start-up.

Nutrient	STA	Season		Kendall tau			Tobit	
			n	p	% Slope yr^{-1}	p	% Slope yr^{-1}	
TKN	Post STA	Dry	362	**< 0.001***	**0.282**	**< 0.001***	**–240.662**	
TKN	Post STA	Wet	495	0.510	0.024	**< 0.001***	**–23.300**	
NH$_4$	Post STA	Dry	415	**0.036***	**0.016**	**< 0.001***	**15.667**	
NH$_4$	Post STA	Wet	506	**< 0.001***	**0.004**	**< 0.001***	**33.588**	
NO$_3$	Post STA	Dry	321	0.331	–0.009	**< 0.001***	**–55.252**	
NO$_3$	Post STA	Wet	475	**< 0.001***	**–0.002**	0.343	1.821	
TP	Post STA	Dry	350	0.347	0.000	**< 0.001***	**–6.187**	
TP	Post STA	Wet	485	0.058	0.000	0.797	0.276	
SRP	Post STA	Dry	362	**< 0.001***	**0.000**	0.790	–0.397	
SRP	Post STA	Wet	495	**< 0.001***	**–0.001**	NS	–0.571	
K	Post STA	Dry	237	**< 0.001***	**0.891**	0.355	–0.628	
K	Post STA	Wet	337	0.097	0.197	NS	0.047	
Ca	Post STA	Dry	236	**0.005***	**4.611**	**< 0.001***	**3.619**	
Ca	Post STA	Wet	373	0.089	–2.030	0.231	–1.211	
Mg	Post STA	Dry	236	**< 0.001***	**18.177**	**< 0.001***	**7.361**	
Mg	Post STA	Wet	372	0.347	2.042	0.087	0.237	
Na	Post STA	Dry	237	**< 0.001**	**2.419**	0.116	0.866	
Na	Post STA	Wet	378	0.266	0.362	0.788	–0.129	
Cl	Post STA	Dry	236	**< 0.001***	**11.058**	**< 0.001***	**6.637**	
Cl	Post STA	Wet	371	0.172	1.858	**< 0.001***	**0.417**	
Fe	Post STA	Dry	19	0.567	54.243	NS	0.000	
Fe	Post STA	Wet	33	0.724	907.963	NS	0.000	
SO$_4$	Post STA	Dry	220	0.401	–0.222	**< 0.001***	**–32.714**	
SO$_4$	Post STA	Wet	344	**< 0.001***	**–0.861**	**< 0.001***	**–2.002**	

[a] Everglades National Park post-STA period of record is from 2000 through 2011.
[b] *= significant at ($p \leq 0.05$).
[c] NS not significant at $p \leq 0.05$.

corresponding biological integrity. Historically BMPs have only been able to reduce nutrient concentration in water delivered to the STAs to 80–150 µg TP L^{-1} (Baker et al. 2012). If BMPs improve nutrient concentrations in stormwater delivered to the STAs, nutrient loading and corresponding surface water nutrient concentrations in the EPA will also likely decrease. The STAs were primarily designed to remove P, but constructed wetlands also reduce most plant macronutrients, especially N in stormwater via denitrification. Benefits other than P reduction should also be accounted for when evaluating and optimizing STA performance.

What is known is that the combination of treatment using BMPs and STAs resulted in improved water quality in the EPA. What is not known is if long-term trends of improved water quality will continue. Long-term slopes suggest that a significant amount of nutrients are removed by both BMPs and STAs but short-term slopes indicate that nutrient concentrations in some areas of the EPA may be increasing. If the Everglades ecosystem is to be restored, managers must continue to improve both the efficacy of STAs and BMPs to remove nutrients. A better understanding of the structure and function of microbiological communities will increase the probability of restoration success as well as protect existing resources. In addition to understanding the biogeochemistry of the system, an analysis of land use and nutrient contribution to canal water in northern basins should also be included to develop a more holistic strategy. Alternative water storage and treatment as well as potential land use changes should be included in the nutrient control strategies. Since microbial communities react to water quality changes quicker than other taxa, we suggest molecular techniques be employed more frequently in routine monitoring. Molecular techniques are now rapid, reliable, and economical. Characterization of algal, bacterial and fungal structure and function along nutrient gradients and associated with restoration projects will provide early metrics of restoration, minimizing risk of undesirable changes in vegetation structure and landscape pattern.

References

[APHA]. American Public Health Association. 2005. Standard Methods for the Examination of Water and Wastewater, American Public Health Association, American Water Works Association, and Water Environment Federation, Washington, D.C.

Baker, W., A. Ramsey and P. Wade. 2012. Nutrient Source Control Programs. *In*: 2012 South Florida Environmental Report. 62 pp. Available at: http://www.sfwmd.gov/portal/page/portal/pg_grp_sfwmdsfer/portlet_prevreport/2012sfer_draft/chapters/v1_ch4.pdf.

Bruland, G.B., S. Grunwald, T.Z. Osborne, K.R. Reddy and S. Newman. 2006. Spatial distribution of soil properties in Water Conservation Area 3 of the Everglades. Soil Sci. Soc. Am. J. 70: 1662–1676.

Childers, D.L., R.F. Doren, R. Jones, G.B. Noe, M. Rugge and L.J. Scinto. 2003. Decadal change in vegetation and soil phosphorus pattern across the Everglades landscape. J. Environ. Qual. 32: 344–362.

Daroub, S.H., T.A. Lang, O.A. Diaz and S. Grunwald. 2009. Long-term water quality trends after implementing best management practices in south Florida. J. Environ. Qual. 38: 1683–1693.

Daroub, S.H., S. Van Horn, T.A. Lang and O.A. Diaz. 2011. Best management practices and long-term water quality trends in the Everglades Agricultural Area. Crit. Rev. Environ. Sci. Technol. 41(S1): 608–632.

David, P.G. 1996. Changes in plant communities relative to hydrologic conditions in the Florida Everglades. Wetlands 16: 15–23.

Davis, J.H., Jr. 1943. The natural features of southern Florida, especially the vegetation, and the Everglades. Florida Department of Conservation Geological Bulletin 25: 1–311.

Davis, S.E., C. Corronado-Molina, D.L. Childers and J.W. Day. 2003. Temporarily dependant C, N and P dynamics associated with the decay of *Rhizophora mangle* L. Leaf litter in oligotrophic mangrove wetlands of the Southern Everglades. Aquatic Bot. 75: 199–215.

DeBusk, W.F., K.R. Reddy, M.S. Koch and Y. Wang. 1994. Spatial distribution of nutrients in a northern Everglades marsh: water conservation area 2A. Soil Sci. Soc. Am. J. 58: 543–552.

DeBusk, W.F., S. Newman and K.R. Reddy. 2001. Spatio–temporal patterns of soil phosphorus enrichment in Everglades Water Conservation Area 2A. J. Environ. Qual. 30: 1438–1446.

Dierberg, F.E., T.A. DeBusk, S.D. Jackson, M.J. Chimney and K. Pietro. 2002. Submerged aquatic vegetation-based treatment wetlands for removing phosphorus from agricultural runoff: response to hydraulic and nutrient loading. Water Research 36: 1409–1422.

Dodds, W.K. 2003. The role of periphyton in phosphorus retention in shallow freshwater aquatic systems. J. Phycol. 39: 840–849.

Doren, R.F., T.V. Armentano, L.D. Whiteaker and R.D. Jones. 1997. Marsh vegetation patterns and soil phosphorus gradients in the Everglades ecosystem. Aquatic Bot. 56: 145–163.

Egler, F.E. 1952. Southeast Saline Everglades Vegetation, Florida and its Management. Vegetation Acta Geobotanica 3: 213–265.

Entry, J.A. 2012a. Water quality characterization in the northern Florida Everglades. Wat. Air Soil Pollut. 223: 3237–3247.

Entry, J.A. 2012b. Water quality gradients in the northern Florida Everglades. Wat. Soil Air Pollut. 223: 6109–6121.

Entry, J.A. and A. Gottlieb. 2014. The impact of stormwater treatment areas on water quality in the Everglades protection area. Environ. Monit. Assess. 186: 1023–1037.

Gaiser, E. 2009. Periphyton as an early indicator of restoration in the Florida Everglades. Ecol. Indicat. 6: S37–S45.

Gaiser, E.E., D.L. Childers, R.D. Jones, J.H. Richards, L.J. Scinto and J.C. Trexler. 2006. Periphyton responses to eutrophication in the Florida Everglades: cross-system patterns of structural and compositional change. Limnol. Ocean. 50: 342–355.

Germain, G. and K. Pietro. 2011. Performance and optimization of the Everglades Stormwater Treatment Areas. *In*: 2011 South Florida Environmental Report. Available at: http://www.sfwmd.gov/portal/page/portal/pg_grp_sfwmd_sfer/portlet_prevreport/2011_sfer/v1/vol1_table_of_contents.html.

Gleason, P.J. and W. Spackman. 1974. Calcareous periphyton and water chemistry in the Everglades. pp. 146–181. *In*: P.J. Gleason (ed.). Environments of South Florida: Present and Past, Memoir 2. Miami Geological Society, Coral Gables, FL.

Gottlieb, A., J.H. Richards and E.E. Gaiser. 2006. Comparative study of periphyton community structure in long and short hydroperiod Everglades marshes. Hydrobiologia 569: 195–207.

Hagerthey, S.E., S. Newman, K. Ruthey, E.P. Smith and J. Godin. 2008. Multiple regime shifts in a subtropical peatland: community-specific thresholds to eutrophication. Ecol. Monograph 78: 547–565.

Hanan, E.J. and M.S. Ross. 2010. Across-scale patterning of plant-soil-water interactions surrounding tree islands in Southern Everglades landscapes. Landscape Ecol. 25: 463–476.

Hanlon, E.A., X.H. Fan, B. Gu, K.W. Migliaccio, Y.C. Li and T.W. Dreschel. 2010. Water quality trends at inflows to Everglades National Park, 1977–2005. J. Environ. Qual. 39: 1724–1733.

Harvey, J.W., J.E. Saiers and J.T. Newlin. 2005. Solute movement and storage mechanisms in wetland of the Everglades, south Florida. Water Resour. Res. 41: W05009. 1–14.

Harvey, J.W., G.B. Noe, L.G. Larsen, D.J. Nowacki and L.E. McPhillips. 2011. Field flume reveals aquatic vegetation's role in sediment and particulate phosphorus transport in a shallow aquatic ecosystem. Geomorphology 126: 297–313.

Helsel, D.R. and R.M. Hirsch. 2002. Statistical methods in water resources. *In*: Techniques of Water-Resources Investigations of the United States Geological Survey Book 4, Hydrologic Analysis and Interpretation. U.S. Geological Resources Chapter 3A. U.S. Geological Survey, Washington D.C. Available at: http://water.usgs.gov/pubs/twri/twri4a3/.

Ivanoff, D. and H. Chen. 2012. Performance and optimization of the Everglades Stormwater Treatment Areas. *In*: 2012 South Florida Environmental Report. Available at: http://www.sfwmd.gov/portal/page/portal/pg_grp_sfwmd_sfer/portlet_prevreport/2012_sfer_draft/chapters/v1_ch5.pdf.

King, R.S., C.J. Richardson, D.L. Urban and E.A. Romanowicz. 2004. Spatial dependency of vegetation-environment linkages in an anthropogenically influenced ecosystem. Ecosystems 7: 75–97.

Lang, T.A., O. Oladeji, M. Josan and S. Daroub. 2010. Environmental and management factors that influence drainage water P loads from Everglades Agricultural Area farms of South Florida. Agric. Ecosyst. Environ. 138: 170–180.

Larsen, L., N. Aumen, C. Bernhardt, V. Engel, T. Givnish, S. Hagerthey, J. Harvey, L. Leonard, P. McCormick, C. McVoy, G. Noe, M. Nungesser, K. Rutchey, F. Sklar, T. Troxler, J. Volin and D. Willard. 2011. Recent and historic drivers of landscape change in the Everglades ridge, slough, and tree island mosaic. Crit. Rev. Environ. Sci. Technol. 41: 344–381.

Larsen, L.G. and J.W. Harvey. 2010. How vegetation and sediment transport feedbacks drive landscape change in the Everglades and wetlands. Worldwide Am. Nat. 173: E66–E79.

Larsen, L.G., J.W. Harvey, J.P. Crimaldi and G.B. Noe. 2009. Predicting organic floc transport dynamics in shallow aquatic ecosystems: insights from the field, laboratory, and numerical modeling. Wat. Resour. Res. 45: W01411, doi:10.1029/2008WR007221.

Liston, S.E. and J.C. Trexler. 2005. Spatial and temporal scaling of macroinvertebrate communities inhabiting floating periphyton mats in the Florida Everglades. J. N. Am. Benthol. Soc. 24: 832–844.

Lorenzen, B., H. Brix, K.L. McKee, I.A. Mendelson and S.L. Miao. 2000. Seed germination of two Everglades species: *Cladium jamaicense* and *Typha domingensis*. Aquatic Bot. 66: 169–180.

Loveless, C.M. 1959. A Study of the vegetation of the Florida Everglades. Ecology 40: 1–9.

McCormick, P.V., M.B. O'Dell, R.B.E. Shuford, J.G. Backus and W.C. Kennedy. 2001. Periphyton response to experimental phosphorus enrichment in a subtropical wetland. Aquatic Bot. 71: 119–139.

McCormick, P.V., S. Newman and L.W. Vilchek. 2009. Landscape responses to wetland eutrophication: Loss of slough habitat in the Florida Everglades, USA. Hydrobiologia 621: 105–114.

McCormick, P.V., J.W. Harvey and E.S. Crawford. 2011. Influence of changing water sources and mineral chemistry on the Everglades ecosystem. Crit. Rev. Environ. Sci. Technol. 41: 28–63.

Miao, S.L., S. Newman and F.H. Sklar. 2000. Effects of habitat nutrients and seed sources on growth and expansion of Typha domingensis. Aquatic Bot. 68: 297–311.

Miao, S.L., P.V. McCormick, S. Newman and S. Rajagopalan. 2001. Interactive effects of seed availability, water depth, and phosphorus enrichment on cattail colonization in an Everglades wetland. Wetlands Ecol. Manage. 9: 39–47.

Morris, D.R., R.A. Gilbert, D.C. Reicosky and R.W. Gesh. 2004. Oxidation potentials of soil organic matter in Histosols under different tillage methods. Soil Sci. Soc. Am. J. 68: 817–826.

Noe, G.B., D.L. Childers and R.D. Jones. 2001. Phosphorus biogeochemistry and the impact of phosphorus enrichment: why is the Everglades so unique. Ecosystems 4: 603–624.

Noe, G.B., L.J. Scinto, J. Taylor, D. Childers and R.D. Jones. 2003. Phosphorus cycling and partitioning in an oligotrophic Everglades wetland ecosystem: a radioisotope tracing study. Freshwater Biol. 48: 1993–2008.

Noe, G.B., J.W. Harvey, R.W. Schaffranek and L.G. Larsen. 2010. Controls of suspended sediment concentration, nutrient content and transport in a subtropical wetland. Wetlands 30: 39–54.

Osborne, T.Z., S. Newman, P. Kalla, D.J. Scheidt, G.L. Bruland, M.J. Cohen, L.J. Scinto and L.R. Ellis. 2011. Landscape patterns of significant soil nutrients and contaminants in the Greater Everglades Ecosystem: past, present and future. Crit. Rev. Environ. Sci. Technol. 41: 121–148.

Osborne, T.Z., L.N. Kobziar and P.W. Inglett. 2013. Investigating the role of fire in shaping and maintaining wetland ecosystems. Fire Ecology 9: 1–5.

Pietro, K., R. Bearzotti, G. Germain and N. Iricanin. 2008. STA Performance and Optimization. *In*: 2008 South Florida Environmental Report. 132 pp. Available at: http://www.sfwmd.gov/portal/page/portal/pg_grp_sfwmd_sfer/portlet_sfer/tab2236041/volume1/chapters/v1_ch_5.pdf.

Qian, Y., S.L. Miao, B. Gu and Y.C. Li. 2009a. Estimation of postfire nutrient loss in the Florida Everglades. J. Environ. Qual. 38: 451–464.

Qian, Y., S.L. Miao, B. Gu and Y.C. Li. 2009b. Effects of burn temperature on ash nutrient forms and availability of cattail (*Typha domingensis*) and sawgrass (*Cladium jamaicense*) growing along a nutrient gradient in the Florida Everglades. J. Environ. Qual. 38: 1812–1820.

Ross, M., J. Meeder, J. Sah, P. Ruiz and G. Telesnicki. 2000. The Southeast Saline Everglades revisited: 50 years of coastal vegetation change. J. Veg. Sci. 11: 101–112.

SAS Institute Inc. 2010. SAS user's guide: statistics—version 9.3 edition. Statistical Analysis System (SAS) Institute Inc., Cary, NC.

Schertz, T.L., R.B. Alexander and D.J. Ohe. 1991. The computer program estimate trend (ESTREND), a system for the detection of trends in water quality data. U.S. Geological Survey Water-Resources Investigations Report 91-4040. 63 pp.

[SFWMD]. South Florida Water Management District. 2010. Field Sampling Quality Manual, SFWMD-FIELD-QM-001-06. South Florida Water Management District, Water Quality Monitoring Division. West Palm Beach, FL.

Sharp, L. and J.H. Solorzano. 1980. Determination of total dissolved phosphorus and particulate phosphorus in natural waters. Limnol. Oceanogr. 25(4): 754–758.

Snedecor, G.W. and W.G. Cochran. 1994. Statistical Methods. 7th ed. Iowa State University Press, Ames, Iowa.

[USFWS]. United States Fish and Wildlife Service. 2007a. Arthur R. Marshall Loxahatchee National Wildlife Refuge—Enhanced Monitoring and Modeling Program Annual Report. LOX06-008, U.S. Fish and Wildlife Service. Boynton Beach, FL.

[USFWS]. United States Fish and Wildlife Service. 2007b. Arthur R. Marshall Loxahatchee National Wildlife Refuge—Enhanced Monitoring and Modeling Program Annual Report. LOX07-005, U.S. Fish and Wildlife Service. Boynton Beach, FL.

Walters, C., L. Gunderson and C.S. Holling. 1992. Experimental policies for water management in the Everglades. Ecol. Appl. 2: 189–202.

Wang, X., L.O. Sternberg, M.S. Ross and V.C. Engel. 2011. Linking water use and nutrient accumulation in tree island upland hammock plant communities in the Everglades National Park, USA. Biogeochemistry 104(1-3): 133–146.

Zapata-Rios, X., R.G. Rivero, G.M. Naja and P. Goovaerts. 2012. Spatial and temporal phosphorus distribution changes in a large wetland ecosystem. Water Resources Research 48: W09512, doi:10.1029/2011WR011421.

Spatial Distribution of Soil Nutrients in the Everglades Protection Area

Todd Z. Osborne,[1,2,*] *Susan Newman,*[3] *K. Ramesh Reddy,*[2]
L. Rex Ellis[4] and *Michael S. Ross*[5]

Role of Soils in Ecosystem Function

Soils are an important, although often underappreciated, component of almost any ecosystem and play both a structural and functional role in ecosystem dynamics. This is especially the case in the Everglades where unique environmental conditions and biogeochemical properties of soils are integral to the maintenance and function of the ecosystem.

[1] Whitney Laboratory for Marine Bioscience, University of Florida, St. Augustine, FL 32080.
  Email: Osbornet@ufl.edu
[2] Wetland Biogeochemistry Laboratory, Soil and Water Science Department, University of Florida, Gainesville, FL 32611.
  Email: Krr@ufl.edu
[3] Everglades Systems Assessment Section, South Florida Water Management District, West Palm Beach, FL 33411.
  Email: Snewman@sfwmd.gov
[4] Environmental Pedology Laboratory, Soil and Water Science Department, University of Florida, Gainesville, FL 32611.
  Email: Rellis@sjrwmd.com
[5] Southeast Environmental Research Center, Florida International University, Miami, Fl 33199.
  Email: Rossm@fiu.edu
* Corresponding author

Factors Governing Formation and Soil Type

South Florida wetland soils reflect a developmental history dating back about 6,000 years, to a time when rapidly rising seas and a moistening climate combined to initiate the formation of the Everglades on a pitted limestone surface. The range of soils we see today can be explained by spatial variation in the five classical soil-forming factors, i.e., parent material, geomorphology, climate, and the biota acting over time (Jenny 1941). The modern day Everglades sits atop a unique geologic framework, the result of thousands of years of shell and coral deposition in a shallow tropical sea. Present day soils were formed over thousands of years of organic matter (OM) and marl deposition in a shallow tropical freshwater wetland basin (Petuch and Roberts 2007). Figure 1 illustrates the major processes affecting soil development within this highly regulated wetland. In the model, hydrology influences soil development in two ways: (1) through *in situ*, biological processes that affect the balance between OM or calcite production, which builds the soils up, and peat oxidation, which break them down; and (2) by redistribution of materials from one place to another in the marsh, i.e., erosion and deposition.

The freshwater Everglades includes several soil environments, differentiated primarily on the basis of hydrologic regime and vegetation. In long hydroperiod (8 to 12 months yr^{-1}) marshes, emergent graminoid or floating-leaved aquatic macrophytes are the dominant primary producers, decomposition of roots and detrital materials is slow, and peats are formed. In shorter hydroperiod (3 to 8 months yr^{-1}) prairies, macrophyte productivity is lower, and periphytic and benthic algal communities predominate (Harvey et al. 2006; Davis et al. 2005b; Ewe et al. 2006). High photosynthesis in the

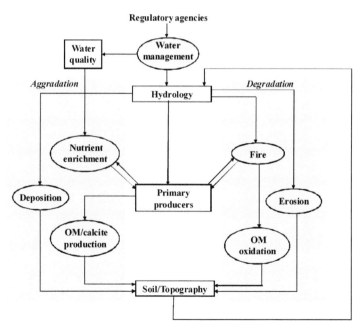

Figure 1. Conceptual model of soil development in a managed wetland. Figure adapted from Ross et al. (2006).

water column or in the benthos increases alkalinity, resulting in the precipitation of calcium carbonate and contributing to the formation of mineral soils (marls) (Gleason 1972; Gleason and Stone 1994). For example, in Everglades National Park, peat soils exist in the Shark Slough drainage feature due to lower bedrock elevation and resultant longer hydroperiods, whereas to the east and west of Shark Slough, higher elevation bedrock results in shorter hydroperiods and thus marl soils dominate (Lodge 2010; Osborne et al. 2011b).

As a general rule, significant ecosystem scale elevational gradients in the Everglades system are rare (McVoy et al. 2011; Holt et al. 2006), however, peat and marl soils reflect very different hydroperiods in close proximity. The exception to this rule occurs in the local scale transition areas from tree islands to marsh or from sawgrass ridge to open water slough (discussed later).

While rainfall is plentiful in the Everglades, very low slopes and paucity of adjacent easily erodible material results in little mineral soil material inputs to Everglades marshes. Hence, the landscape within and adjacent to the Everglades is fairly stable with respect to erosion under normal hydrologic conditions (Larsen and Harvey 2010; McVoy et al. 2011). The absence of topographic relief implies rare erosion-producing flow velocities. However, there is evidence that material transport within Everglades sloughs may help maintain the corrugated landscape in the ridge and slough mosaic (Larsen and Harvey 2010). DeAngelis and White (1994) suggest that erosional processes brought about by sea level rise may be significant in coastal soils (wave action) and climatic changes may contribute to erosion of tree island soils from intensive rainfall (compressed rainy season).

Peatlands

The majority of Everglades soils, such as the Loxahatchee and Everglades peats, are highly organic in nature with loss on ignition (LOI) values ranging from 50–97% (Reddy et al. 2005; Scheidt and Kalla 2007). Peat soils are formed because of the predominantly anaerobic conditions brought about by hydrology. Despite the oligotrophic nature of the ecosystem, the substantial plant productivity (Davis 1991; Craft and Richardson 1993; Clark and Reddy 2003) coupled with slow anaerobic decomposition, results in accretion of OM in the soil pool (Fig. 2; Reddy and DeLaune 2008). Because soil development takes place within a continually changing spatial context; that is, soil formation in one location is not independent of the landscape around it. When the Everglades were first forming, the context was framed by the pitted bedrock surface and the drainage network that connected the interior to the coastal estuaries. As soils formed on this base, they smoothed out much of the fine-scale surface roughness, especially where accretion was rapid. In other places, however, the spatial context was such that a patterned landscape characterized by heterogeneity of an intermediate scale emerged. The chemical composition of OM which makes up the soil determined rates of accretion as different plant types contributed OM of variable quality to the soil (Osborne et al. 2007; Cohen et al. unpubl. data). Higher soil elevations or ridges, which are dominated by sawgrass *Cladium jamaicense* Crantz. These ridges are only 10–20 cm higher than that of the surrounding open water sloughs (Watts et al. 2010; Lodge 2010). This elevation difference delineates the two competing landscape

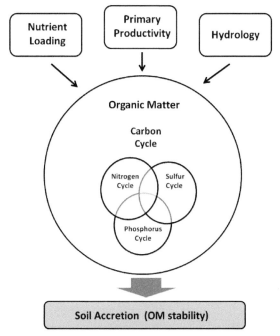

Figure 2. Conceptual model depicting the interactions of plant community, nutrient loading, and hydrology with organic matter production. Biogeochemical cycles of major nutrients such as nitrogen, phosphorus and sulfur all revolve around the cycling of organic carbon. Adapted from Reddy and DeLaune (2008).

ecotypes and is reinforced via a feedback mechanism by the quality of the OM derived from the dominant vegetation types within ridges and sloughs (Clark and Reddy 2003; DeBusk and Reddy 2003). Detrital material derived from sawgrass has a significantly higher portion of the recalcitrant biopolymer lignin as opposed to detritus derived from *Nymphea odorata* (dominant deep water species) and associated submerged aquatic vegetation commonly found in sloughs (Osborne et al. 2007; Clark and Reddy 2003). Thus potential soil accretion rates in sawgrass ridges are nearly an order of magnitude higher than in sloughs because of vegetation-derived differences in OM production rates and litter lignin content. Seasonal hydrology serves as a feedback mechanism to reduce elevation on ridges when drawdown allows aerobic conditions to accelerate lignin degradation (Criquet et al. 2000; Freeman et al. 2001). The hydrologically mediated interplay of soil accretion and oxidation also serves to regulate ecosystem nutrient availability via OM stabilization and mineralization, respectively (Fig. 1). In this "Ridge-and-Slough" landscape, the same processes illustrated in Fig. 4 yielded a repeated series of deeper and shallower organic soils, arranged in elongated landforms that parallel the presumed, predominant long-term flow direction.

Within the context of ridge and slough landscapes of the central and southern Everglades, the influence of sedimentation and deposition (Fig. 2) in peat soil dynamics has been an active topic of recent research (Noe et al. 2001; Harvey et al. 2009; Larsen and Harvey 2010; Leonard et al. 2006; Bazante et al. 2006). While sediment redistribution may be an important process maintaining the local balance between ridges and sloughs, there is no net change at the large scale, i.e., material eroded from one location is likely to be deposited somewhere nearby. Furthermore, when the landscape is in balance, accretion rates in sloughs should equal those in ridges (Cohen et al. 2011), though the respective rates may be arrived at by different combinations of production, sedimentation/erosion, and decomposition.

Marl Prairies

Marls are formed rather than peats when (1) vegetation production is low, (2) production of calcareous periphyton is high, (3) a calcium carbonate source is abundantly available, (4) the water column is well-lit, and (5) a significant dry-down occurs during most years (Gleason et al. 1974; Browder et al. 1994). The vegetation community structure of marl prairies is one that is suited to very shallow soils with seasonal inundation (Osborne et al. 2011b; Davis et al. 2005a; Browder and Ogden 1999; Obeysekera et al. 1999; Olmstead and Loope 1984). Marls do not form in P-enriched waters, because the blue-green algae that dominate calcareous periphyton mats are not tolerant of high P (Gaiser et al. 2004). Once established and stabilized, they are not prone to entrainment by water, as the silt particles that predominate are not easily dislodged. Depending on vegetation density, fire frequency may be high in marl prairies, but their effects on soils may be minimal due to the low soil OM content.

Marl soils are present on either side of Shark Slough in the southern Everglades, as well as in fresh- and brackish-water prairies of the Southeast Saline Everglades, where they formed when sea level was lower and freshwater runoff was higher. Today, those marshes are experiencing mangrove encroachment, and formation of mangrove detritus based peat has ensued on top of the marl base. The thickest south Florida marls are found in these prairies (Perrine series), with depths sometimes approaching 1 m.

Relationships between marl accretion and hydrology, water quality, and other physical drivers have not been quantified, though marl production is thought to be strongly associated with the productivity of calcareous periphyton (Gleason 1972). Marl accretion is typically about 1 mm per year (Meeder et al. 1996; Browder et al. 1994), which is only about one-half to one-third of average rates for marsh peats. Nevertheless, sequestration of inorganic carbon (C) in fresh water marls (as calcite) is substantial (Table 1). In fact, assuming the compiled values in Table 1, annual estimates of organic C sequestered in peats (1290 kg ha^{-1}) and inorganic C sequestered in marls (1118 kg ha^{-1}) in the Everglades are very similar. Marl soils are significantly more stable with respect to change over time, as the mechanism of removal is erosion or dissolution in contrast to the mechanism for peat removal, which is oxidation.

Table 1. Physical data from several peat and marl soils.

	Peat	**Marl**
Mean annual accretion (mm)	2.3[1]	1.0[2]
Total carbon (%)	51.33[3]	13.82[3]
Dry bulk density (g/cm^3)	0.11[3]	0.81[3]

[1] Craft and Richardson 1995 (Cesium dating, WCA-2 and WCA-3).
[2] Meeder et al. 1996 (Cs and Pb dating, C-111 basin, Perrine Series).
[3] Ross et al. unpublished data (southern Shark Slough).

High Tree Islands

The Everglades landscape also includes scattered forest fragments, or tree islands. Those occupied by swamp forest trees are flooded for much of the year, and experience a soil development process that resembles that of the marsh peatlands (Ross and Sah 2011). Others occupy raised, rarely flooded surfaces, and support "hardwood hammocks" with upland tree species. Hardwood hammocks in the southern Everglades grow on two distinct soil types (Ross et al. unpublished manuscript). The first type is common in the seasonally flooded marl prairie landscape, and consists of shallow, organic, relatively low-P soils formed directly on limestone outcrops. In contrast, hammocks on islands embedded in long hydroperiod marsh have deeper, alkaline, mineral soils with extremely high P concentrations. The first group resembles the organic soils found in Florida Keys hammocks, where soil development consists primarily of a reprocessing of dead roots and remains of aboveground plant tissue. Maintenance of these soils requires sustaining high aboveground production, because decomposition rates in such well-aerated settings are rapid. Development of the second type of tree island soil, which is restricted to islands enveloped throughout the year by flooded or saturated marsh peats, is not yet completely understood. Several explanations for their high mineral content have been proffered. One is that the mineral component is the residuum of rapid OM decomposition and weathering of the bedrock. A second is that the source is bone material transported to the islands by aboriginal occupants or visitors, and which suffuse many of these profiles. A third hypothesis is that the minerals are fixed through subsurface precipitation of calcium and other cations, drawn to the islands in the transpiration stream of trees. Some support for this mechanism is provided by Wetzel et al. (2011), who reported high Na and Cl ionic concentrations in groundwater beneath an elevated tree island in WCA-3A. Each of the above mechanisms probably contribute to soil development in high islands in interior peatlands.

The organic soils in the marl prairie tree islands are very vulnerable to fire, whose effects can be exacerbated by reductions in seasonal water levels. In contrast, the mineral soils in the high slough islands probably maintain more stable moisture conditions due to supplementation from adjacent marsh waters, and would not be consumed by the rare fire that reaches them. Prolonged high water could have indirect effects on soils in either setting, via its influence on the forest canopy. Soil dynamics in both types are apparently dependent on a continuous tree layer which contributes organic materials, draws water and nutrients into the islands through the transpiration stream, attracts animals, intercepts aerosols, and controls the belowground microclimate. Flooding has eliminated forest canopies in some parts of the Everglades (Hoffmockel et al. 2008), and changes in soils are now under investigation (Ewe et al. 2010).

Functional Role of Soils

Soil, by way of its physical and biogeochemical properties can regulate two key ecosystem properties in the Everglades: plant community structure and nutrient cycling within the ecosystem. Plant community structure and soil relationships were discussed previously, so here the focus will be on biogeochemical properties. In OM accreting systems such as the Everglades, the soil can serve as a sink or storage pool for ecologically significant nutrients and elements (DeBusk et al. 1994; Newman et al. 1997; Bruland et al. 2006; Scheidt and Kalla 2007). Soils can also serve as a source in the biogeochemical cycling of these nutrients, and thus their ecological role as biogeochemical modulators is significant (Reddy et al. 2005; Reddy and DeLaune 2008). As plants grow and access essential nutrients from the water column or the soil, they are stored in the organic tissues of the plant. After senescence, plant litter (both above and below ground), the most significant contributor to soil accretion, still contains relatively large amounts of these nutrients. The cycle from soil to plant to soil may be repeated indefinitely if not for the potential for some portion of these nutrients to be buried below the root zone and exported to the surrounding water column. The export process can occur several ways, however, mineralization of OM is usually a key step. The nutrients bound up in the soil OM can be released to the water column when soil is oxidized by microbial activity (Wright and Snyder 2009; Wright and Reddy 2003; Stephens 1956) or fire (Brix et al. 2010). Similarly, the reverse is also possible. Elevated levels of nutrients or other constituents can be incorporated into soil OM via microbial uptake and plant growth. In some cases, simple concentration gradient driven mass transfer may enable soil enrichment of nutrients or export to the water column.

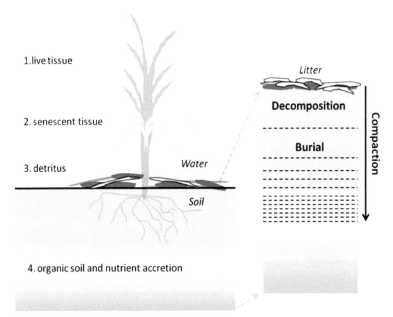

Figure 3. Schematic describing dominant processes of soil accretion from plant detrital material in the Everglades, insert depicts deposition or peat building material. As annual deposition continues, previous years material is further decomposed and compressed under new material. Water levels modulate this process by reducing decomposition (flooded) or accelerating it (drained). Adapted from Reddy and DeLaune (2008).

Anthropogenic Impacts to Everglades Soils

Human Impacts to Soils

The climate of a given year affects the amount of water in the system, but management and regulation have an overarching influence on the distribution of the water and its quality. Nutrient enrichment affects the rates of both aggregation and degradation processes, while fire can cause soil degradation under certain conditions. The feedback loops are critical, as soil accretion or loss alter future hydrologic conditions, and thereby future soil development or loss.

Hydrologic Management Impacts

The interaction of geology, hydrology, and biology dictate local soil accretion/loss and resultant elevation (Fig. 4). A comparison of early land surveys to current elevations indicates that subsidence of 1–3 meters has taken place in extensive areas south of Lake Okeechobee (Ingebritsen et al. 1999). Subsidence in drained and tilled peat soils was first quantified by researchers at the USDA Belle Glade station (Stephens and Johnson 1951). Soil loss in the EAA has slowed considerably from the rapid rates characteristic of the first few decades of water management (Wright and Snyder 2009), but recent surveys indicate that significant soil loss has continued in some nearby portions of the Water Conservation Areas (Scheidt and Kalla 2007). During the last 1200 yrs, the accretion rate of peat in the northern Everglades has been found to be approximately 1.6 mm yr^{-1} (Gleason and Stone 1994; DeAngelis 1994) under inundated (anaerobic) conditions. Under aerobic conditions, rapid decomposition can cause soil loss (i.e., negative accretion) of approximately 3 cm yr^{-1}, highlighting the significantly asymmetrical rates of the two competing processes of production and decomposition (Stephens and Johnson 1951; Snyder and Davidson 1994; Maltby and Dugan 1994) and the integral role of hydrology in soil dynamics.

As in other wetlands worldwide, subsidence of Everglades peats increase with the depth of the water table. Data and modeling presented by Stephens and Stewart (1977) indicated that peat subsidence increased as a linear function of depth when the water table receded from 30 to 80 cm below the surface, with annual subsidence increasing by about 7 mm per 10 cm increase in depth. The model did not predict changes in the peat surface in flooded soils. To date, severe soil oxidation, and thus loss of elevation, has been noted in northern Water Conservation Area (WCA)-3A and WCA-2A and WCA-2B (Scheidt et al. 2000; Scheidt and Kalla 2007). By calculating the distance between peat elevations from the pre-drainage surface, using the Natural System Regional Simulation Model (NSRSM) and the current condition (USACE), mean subsidence of 1.7 m has been estimated across the Everglades Agriculture Area (EAA) and between 0.01 m and 0.9 m in the Everglades (Aich and Dreschel in press). This is on par with the range of 1.2 m to 1.5 m in the EAA presented in Snyder (2005) and extreme localized subsidence > 3 m in some areas under sugarcane (*Saccharum officinarum* L.) and vegetable production (Snyder and Davidson 1994). These observations highlight the devastating effects of prolonged dewatering on organic soil elevation in the Everglades marshes. However, in general, compared to

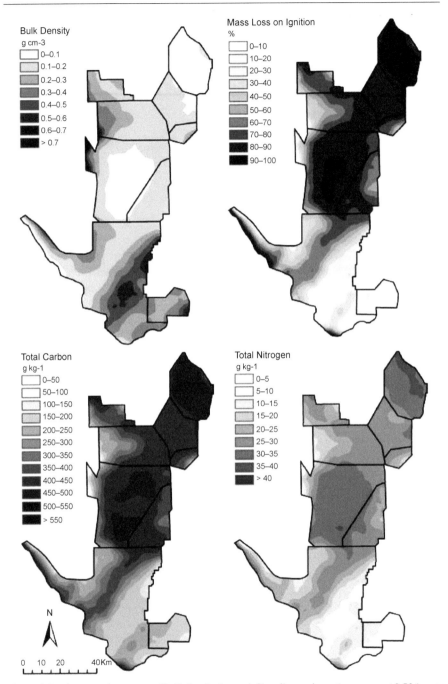

Figure 4. Landscape scale patterns of bulk density (upper left), soil organic matter as percent LOI (upper right), total carbon (lower left) and total nitrogen (lower right) in the Everglades Protection Area. Figure adapted from Reddy et al. 2005.

Color image of this figure appears in the color plate section at the end of the book.

marsh peats, the well-drained soils of the upland tree islands would not be especially sensitive to enhanced decomposition rates associated with reduced water tables.

The most devastating loss of peat occurs under extremely overdrained conditions which result in peat fires. Newspaper accounts from Miami's early years are filled with reports of uncontrolled fires in the wetland during the dry and early wet season (December to June), and there is little doubt that drainage contributed greatly to the destructiveness of those fires. Craighead (1971) reported major conflagrations in 1945, 1947, 1951–52, 1962, and 1965. Davis (1946) described a site in which a 1945 fire burned enough peat to leave six inches of ash behind. Parker (1974) described ash layers two inches thick, the remains of peat fires that burned to the water table. These leading scientists contended that several south Florida soils, i.e., rocklands in the East Everglades and sands in western Broward County, included extensive peat-coverage prior to drainage. Even in the interior of the Everglades, fires burn across the surface during dry years, but peat is consumed only when the water table has been artificially lowered by drainage, or during very persistent or extreme climatic droughts. Unfortunately, no large-scale surveys of soil loss have been initiated to date, leaving information on the interactions of hydrology, fire behavior, and peat loss largely anecdotal at this point.

Nutrient Loading Impacts

Beginning with the first attempts to drain the northern Everglades to bring the rich organic soils south of Lake Okeechobee into agricultural production, humans have altered the unique environmental conditions that supported the ecosystem and the accretion of large expanses of histosols (organic soils) found there (Davis 1994). The draining of the waterlogged soils of the EAA, caused not only the great loss of soil (up to 3 m in some areas) in the upper basin, but has also contributed, by way of OM mineralization, to elevated nutrient runoff to the remaining Everglades Protection Area (EPA) downstream (WCAs, ENP). Further, liberal use of agro-chemicals and fertilizers has contributed to increased presence of anthropogenic organic compounds and nutrients in waters flowing into the Everglades today (SFWMD 2007), requiring the use of extensive stormwater treatment areas or STAs—treatment marshes designed to remove and store nutrients from agricultural waters prior to release into the environment (Pietro et al. 2007). About 50% or more of the nutrient laden water from the EAA is diverted to the St. Lucie and Caloosahatchee Estuaries via canals (Redfield and Efron 2007; Richardson and Huvane 2008). The remaining portion enters the Everglades through an extensive system of highly manipulated drainage canals and water control structures. These inputs of agricultural runoff bring trace metals, organic chemicals, and nutrients that can be detrimental to the system. One of the most detrimental and well documented of which is phosphorus (P) (Davis 1991; Craft and Richardson 1993; Reddy et al. 1993).

Another factor that has significantly impacted the Everglades is the rapid increase in human population and the concomitant urban sprawl. People have lived in the Everglades region for more than a thousand years, however, within the last century their activities profoundly impacted this ecosystem. The Everglades have been reduced to roughly half the original spatial extent by aggressive encroachments from both residential and agricultural land uses (Walker 2001). The ecosystem has been under continuing threat from increasing population density (Walker and Solecki 1997). Currently, the threat of unchecked urban sprawl has been curtailed due to the federally mandated Urban Growth Boundary (UGB), a line establishing the urban development limits to prevent any more encroachment on the Everglades.

With population and metropolitan growth comes urban problems such as overloaded sewage treatment discharge and water shortages. Wilcox et al. (2004) demonstrated over 60% of the water being removed by municipal pumping in Miami-Dade county originated in the Everglades. Meanwhile, over the past three decades, the drainage canals conveying urban and agricultural runoff to the Everglades contained water with high total P concentration ranging from 100–1000 $\mu g\,L^{-1}$ (Sklar et al. 2005).

The 1970's through 1990's witnessed significant ecosystem decline seemingly linked to P enrichment. This resulted in much attention and research concerning P loading to the Water Conservation Areas and subsequent ecosystem changes such as vegetative community shifts and habitat degradation that were attributed in the most part to excess soil P (SFWMD 1992; Davis 1994; Noe et al. 2001; McCormick et al. 2002). Due to overwhelming evidence of P enrichment in the northern Everglades, several studies have been conducted to investigate P enrichment in soils (Koch and Reddy 1992; Amador and Jones 1993; Craft and Richardson 1993; Reddy et al. 1993; DeBusk et al. 1994; Qualls and Richardson 1995; Amador and Jones 1995; Newman et al. 1996; 1997; 1998; Miao and Sklar 1998; Noe et al. 2002; Noe et al. 2003; Daoust and Childers 2004; Chambers and Penderson 2006) and in some cases, changes to soil condition over time (Childers et al. 2003; Reddy et al. 2005; Bruland et al. 2006; Rivero et al. 2009; Newman et al. in press). These studies overwhelmingly contend that inflow waters from agriculture operations upstream have caused P enrichment of soils in many areas across the Everglades landscape (Scheidt and Kalla 2007; Hagerthey et al. 2008; Osborne et al. 2011a).

As mentioned previously, the Everglades was historically an oligotrophic system, and as such, the availability of essential nutrients such as P was very limited. Other than internal recycling of P from soils to plants, sources of P inputs were few, with atmospheric deposition being the most significant source (Scheidt et al. 2000; Noe et al. 2001). The limited availability of P translates to very tight biogeochemical cycling of P forms in the ecosystem, leaving bioavailable forms in very low concentrations and a majority of the P sequestered in plant or microbial tissues and particulate detrital material. This nutrient limitation brought about ecosystem dominance of plant communities adapted to growing in low P conditions (Davis 1991; Miao and DeBusk 1999; Miao and Sklar 1998).

A majority of studies and the scientific community working in the Everglades concur that even small additions of P can have a dramatic effect on ecosystem productivity and functioning (Gaiser et al. 2005; Childers et al. 2003; Chiang et al. 2000). Nowhere are these effects more readily observed than in the changes to vegetation. Autochthonous nutrient inputs have resulted in significant alterations to the indigenous system with large expansions of cattail (*Typha domingensis*; Davis 1994; Newman et al. 1998; Richardson et al. 2008). Cattails are adapted to grow rapidly in the presence of available P, and in doing so, out-compete native vegetation such as sawgrass (*C. jamaicense*; Davis 1991; Davis 1994; Miao and Sklar 1998; Miao and DeBusk 1999).

These changes to vegetation have far reaching consequences to the ecosystem. Beyond the visual changes, changes in vegetation result in alterations to the heterotrophic food web by way of altered OM quality. This also affects soil accretion processes. Accelerated decomposition of cattail detritus over sawgrass has been observed (Craft and Richardson 1997; DeBusk and Reddy 1998) suggesting rapid turnover of soil building materials. Based on ^{137}Cs activities in soil cores, Craft and Richardson (1993) reported a mean accretion rate of 2.3 mm per year in un-enriched portions of WCA-2 and WCA-3A. Fastest accretion was found in persistently flooded areas (lower WCA-3A, central WCA-2A) and the slowest was found in over-drained areas (northern WCA-3A, WCA-2B). In the same study, the most rapid peat accretion (~4 mm per year) occurred at a P-enriched site within the cattail invasion front in WCA-2A.

Phosphorus re-absorption during senescence has also been found to be less efficient in cattails vs. sawgrass, resulting in greater amounts of P being deposited to the detrital pool (Miao and Sklar 1998; Osborne et al. 2007). Further, rapid OM decomposition due to lower lignin content of cattails also increases the turnover rate of P, resulting in a positive feedback mechanism to encourage *Typha* expansion. Other ecological implications of cattail expansion include changes to water quality via dissolved oxygen content, fish and wading bird habitat degradation, accelerated biogeochemical cycling of nutrients and metals of concern, such as mercury (Osborne et al. 2011a), and dramatic changes to the calcareous periphyton communities (Gaiser et al. 2005; Gaiser et al. 2006).

As with sawgrass and other low P adapted plants of the Everglades, the unique calcareous periphyton found almost ubiquitously throughout the system is adapted to low P availability (McCormick and Stevenson 1998). Periphyton is composed of filamentous cyanobacteria and diatoms assembled together in a laminar sheet that can be attached to the soil surface, plant surfaces below the water column, or floating on the water surface. During the process of marl formation (calcium carbonate precipitation), P is often co-precipitated, becoming part of the soil (McCormick et al. 2001; Gleason et al. 1974). Periphyton is indicative of pristine conditions and is an integral part of ecosystem functioning in that it is responsible for significant primary production, P sequestration, C sequestration, and actively regulates other biogeochemical processes in the water column (McCormick and O'Dell 1996; Noe et al. 2001; Gaiser et al. 2006). When water column P concentrations consistently exceed background concentrations, cyanobacteria are replaced by filamentous green algae (McCormick et al. 2001; Gaiser et al. 2005; 2006). The effect on marl formation and accretion by way of the loss of calcium carbonate precipitating cyanobacterial component is detrimental.

In the Everglades surface water quality and vegetative communities are intricately inter-related with soil characteristics (Daoust and Childers 2004; Davis et al. 2005b; Ogden 2005; Hagerthey et al. 2008). Diminished surface water quality has resulted in altered vegetation and periphyton communities, which have in turn altered litter quality, microbial activity (Wright and Reddy 2001), mineral precipitation, and ultimately soil accretion. These altered plant communities may cause further changes in soil type and thickness as these different plant communities eventually decompose and form altered soil (Scheidt and Kalla 2007). Of considerable concern is the reduction of C storage and soil accretion, reduction of P storage, and resulting increase of available P. The later serves as a significant positive feedback for further vegetation and periphyton changes and increased microbial decomposition of litter and soils.

Scheidt and Kalla (2007) report that in 2005, $24.5 \pm 6.4\%$ and $49.3 \pm 7.1\%$ of the Everglades area soil P content exceeded 500 mg kg^{-1} and 400 mg kg^{-1}, respectively. These numbers are to be compared with those obtained in 1995–1996 ($16.3 \pm 4.1\%$ exceeding 500 mg kg^{-1} and $33.7 \pm 5.4\%$ exceeding 400 mg kg^{-1}). The Everglades areas with the highest soil TP concentration are generally the peat soils located in WCA-3A north of Alligator Alley, northern WCA-2A, and the edges of Loxahatchee National Wildlife Refuge close to the rim canal. When expressed on a volume basis ($\mu g\ cm^{-3}$) to differentiate among the soil types, the peat soils with higher TP content are located in WCA-2A and at the edges of Loxahatchee NWR. The areas in the Everglades National Park with higher bulk density have a higher P content. The locations in southern WCA 3A that contained above 500 mg kg^{-1} (mass basis) have a low TP content on a volume basis. The marked increase in overall area of P enrichment from 1995–2005 is cause for concern. While water quality has improved markedly since 1992 (Entry and Gotlieb 2014), P enrichment of soils continues to spread spatially. Reddy et al. (2011) described how this legacy phosphorus explains the continued advancement of enrichment fronts across the Everglades landscape. Legacy P cycling suggests that P enriched soils can be a continual source of P until the enrichment gradient is eased and soil P is equilibrated throughout the system. This has significant implications for the continued shifts in vegetation currently observed in enriched areas.

Biogeochemical Characteristics of Everglades Soils

The following representation of spatial distributions of soil nutrients and is a compilation of information derived from the Everglades Soil Mapping (ESM) program (Reddy et al. 2005; Osborne et al. 2011). Landscape scale patterns in soil nutrients and contaminants were modeled from 1352 individual soil sampling sites (Fig. 5) sampled in 2003–2004 (Reddy et al. 2005; Corstanje et al. 2006; Bruland et al. 2007; Rivero et al. 2007; Osborne et al. 2008; Grunwald 2008; Osborne et al. 2011a; Osborne et al. 2011b; Reddy et al. 2011). We discuss only a few of the most ecologically relevant soil parameters here that were measured during that investigation, and we begin with the most microbially relevant element, C (in the form of OM), as it is the energetic source for most of the microbially mediated biogeochemical reactions. Following the discussion of C will be nitrogen (N), P and sulfur (S).

Carbon

One of the most recognizable and ecologically significant attributes of the Everglades ecosystem is the vast expanse of organic soils it contains (Davis 1946; Stephens 1956; Davis and Ogden 1994; Bruland and Richardson 2006). The large peat deposits of the northern and central Everglades represent more than 5000 years of soil accretion, and as such, are a repository for nutrients accumulated in this organic matrix over that time period. The extensive Everglades Histosols are, first and foremost, a

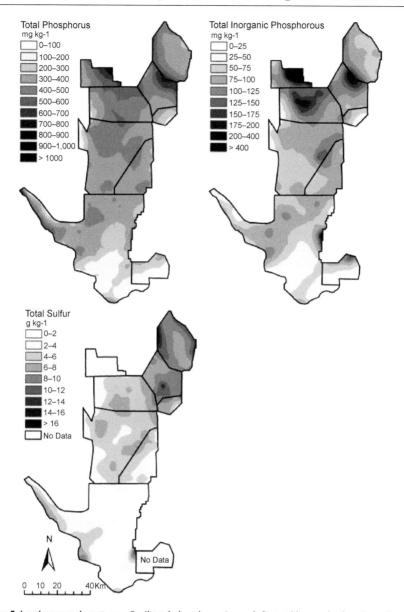

Figure 5. Landscape scale patterns of soil total phosphorus (upper left), total inorganic phosphorus (upper right), and total sulfur (lower left) in the Everglades Protection Area. Figure adapted from Reddy et al. 2005.

Color image of this figure appears in the color plate section at the end of the book.

considerable storage of C. The mean total C content of peat soils in the northern and central Everglades is 47%, translating to approximately half of the loss on ignition (LOI) value (Fig. 5). However, TC for Everglades National Park (ENP) soils can be much greater due to high inorganic C of marl soils (Reddy et al. 2005; Osborne et al. 2011). Arguably, this alone makes the soils of the EPA a worthy of conservation. Of more importance perhaps, is the organic nature of these soils, which in conjunction with the hydrology of the Everglades provides a unique biogeochemical environment for storage of environmentally significant nutrients. To facilitate understanding of the environmental implications of such a massive storage of nutrients in soil across the Everglades landscape, a discussion of current soil patterns is necessary.

The spatial distribution of soil OM indicates that the northern and central Everglades contain a majority of the organic soils in the EPA (Fig. 4) (Reddy et al. 2005). The highest OM soils are found in the Arthur R. Marshall Loxahatchee National Wildlife Refuge (WCA-1) and trend down in organic content to the southern portion of WCA-3. The ENP, due to the predominance of more shallow wet prairies and marl prairies, contains lesser peat deposits. Peat soils of the ENP are located in the Shark River Slough, a predominant drainage feature on the landscape. Similar spatial trends of soil OM were reported by other researchers monitoring soil attributes in the EPA (Scheidt et al. 2000; Scheidt and Kalla 2007). Models of both data sets suggest a trend toward lower OM soils as one moves from the north to the south. The major exception to that trend is an area of interest in the northwestern corner of WCA-3A (Fig. 4). This area is also noted to be an area of significant soil subsidence due to chronic over drainage (Scheidt et al. 2000) reported significant soil depth decreases in this area from 1946 to 2005. Comparison of historic soil depth documentation in the Everglades (Davis 1946) to soil depth measurements conducted under Scheidt and Kalla (2007) suggests extensive subsidence in northern WCA-3A. Scheidt et al. (2000) estimated that up to 28% of peat soils have been lost from public lands in the EPA between 1946 and 1996. As of 2005, about $25.1 \pm 2.0\%$ of the Everglades had a soil thickness of less than 1.0 foot (Scheidt and Kalla 2007). The differential between the time involved in accreting organic soils and the relatively short time required to oxidize them is reason for great concern in Everglades restoration. This hysteresis in soil creation and loss has been noted as a driving force in shaping the present day ecosystem (DeAngelis 1994) and one that makes it highly unstable (Maltby and Dugan 1994).

Because Everglades soils contain large storages of limiting nutrients and contaminants, the oxidation of these soils only exacerbates current eutrophication and contamination problems. Soil subsidence via oxidation releases those nutrients and contaminants that were bound in the organic substrates of soils and protected by the anaerobic conditions prevalent under flooded conditions. This suggests that chronic over drainage due to water diversion or management may be a significant factor responsible for the spread of nutrient and contaminant enrichment in areas less directly affected by agricultural runoff, such as northern WCA-3A.

In the EPA, bulk density is a very accurate indicator of soil OM, which in turn is, as expected, closely related to TC and total N (TN) (Fig. 4) by positive linear relationships ($R^2 = 0.88$ and 0.93, $p < 0.0001$ respectively).

Nitrogen

Much of the attention concerning nutrient enrichment of Everglades soils is focused on P due to the overwhelming evidence of ecosystem change associated with elevated P concentration in water and soil. While not of immediate concern to ecosystem managers, N can be a limiting nutrient in areas where excessive P enrichment occurs.

Patterns of soil total N (TN) are very similar to LOI and TC (Fig. 4). Therefore, areas of concern for loss of OM from oxidation and areas of significant storage of OM are similar for N. Perhaps the most important issue surrounding soil N in the Everglades is that of oxidation of OM and export of N downstream. Florida Bay, an estuarine and marine habitat of great economic and environmental significance in south Florida is the water body directly below the freshwater Everglades. It receives water via Shark Slough from ENP and the WCAs to the north. Any mineralized N from upstream can have tremendous negative impacts on the N sensitive waters of Florida Bay.

Phosphorus

Excessive amounts of the limiting nutrient P have been the focal point of much research concerning Everglades restoration (Davis and Ogden 1994). Eutrophication of extensive areas of the northern Everglades via P laden runoff from the EAA was one of the significant catalysts in the movement to restore the Everglades ecosystem and is included in several conceptual ecological models and performance measures used in framing restoration planning (Ogden 2005; RECOVER 2006). The degradation of the northern Everglades marshes due to P enrichment is extremely well documented (LOTAC II 1990; Scheidt et al. 2000; Noe et al. 2001; McCormick et al. 2002; Childers et al. 2003; Hagerthey et al. 2008), and as such, significant scientific evidence exists to aid restoration efforts with respect to P. A P control program was initiated in the 1990s in order to prevent further nutrient-induced degradation of the Everglades. To help attain these goals, best management practices (BMPs) were instituted in the EAA and extensive treatment marshes, Stormwater Treatment Areas (STAs), were constructed on the boundary between the EAA and the WCAs. Because soils are known to be long-term integrators of water quality and significant storage pools for P, Florida defines P impact as soils exceeding 500 mg kg^{-1} within the top 0–10 cm soil layer (Recover 2007; Reddy et al. 1998). In addition, CERP instituted a restoration goal of maintaining or reducing long-term average soil TP concentrations at 400 mg kg^{-1}. These goals were based upon several studies identifying enriched or impacted soils and correlation with soil TP and resulting expansion of *Typha* sp. (DeBusk et al. 1994; Doren et al. 1996; Newman et al. 1998; DeBusk et al. 2001; Payne et al. 2003).

Results of the ESM sampling effort in the EPA indicate several areas of concern with respect to soil TP (Fig. 5). Although the sampling designs were somewhat different, as discussed previously, similar spatial patterns in TP were found for the 0–10 cm surface soils. Both studies suggest that TP enrichment of soils in WCA-1 is contained in the peripheral areas as P laden agricultural waters often do not penetrate to the interior of the marsh (Suratt et al. 2012; Corstanje et al. 2006). It is important to note that several more sensitive ecosystem components respond to P enrichment before increasing TP is manifested in the soils. These initial changes include loss of water column dissolved oxygen and changes to periphyton and macrophyte communities (McCormick et al. 2002). Water Conservation Area 2A, the site of much historical eutrophication work, maintains a distinct nutrient gradient in the northern portion extending south from the S-10 series outfall structures. However, the ESM data suggest that the area of impact may be smaller than that of other reports (Scheidt and Kalla 2007). This could be due to variability of sampling locations or nuances of the geostatistical modeling. Interestingly, there appears to be a new area of enrichment, present in both data sets, on the western corner of WCA-2A that was not prevalent in previous sampling efforts in 1990 and 1998 by Reddy et al. (1998) and DeBusk et al. (2001) respectively.

Many studies also indicate significant enrichment in northern WCA-3A, proximate to the Miami Canal outfalls. A smaller area of enrichment is also noted proximate to the L-28 extension canal outfall in west central WCA-3A, which is in within the Miccosukee Tribe of Indians Federal Reservation. Both models indicate that northern WCA-3A has enriched soils, however, no clear gradients exist. This suggests P remobilization from upstream sources is likely the source. Scheidt and Kalla (2007) point out that when soil TP is reported across the EPA on a volumetric basis, normalized for bulk density (a highly variable measure of soil density) many of the enriched areas seen in central WCA-3A are no longer categorized as enriched. Similarly, for marl soils in the ENP, which appear to be highly enriched, when TP is expressed volumetrically, these soils are not considered enriched. Scheidt et al. (2000) point out that > 500 mg kg^{-1} is not necessarily indicative of enrichment in mineral/marl soils of the ENP. The ENP continues to be the least impacted unit of the EPA with respect to TP (Osborne et al. 2011) (Fig. 5).

Although the ESM program was a single phase sampling effort, historical data sets have been used to investigate changes to soil TP. Initial spatial mapping of ESM data by hydrologic unit such as WCA-1 (Corstanje et al. 2006), WCA-2 (Rivero et al. 2007) and WCA-3 (Bruland et al. 2006) have been compared to reconstructed spatial data sets to infer changes to soil properties over time. Marchant et al. (2009) compared ESM data from WCA-1 (Corstanje et al. 2006) to the first spatial sampling in 1992 (Newman et al. 1997) revealing significant changes in TP. Likewise, Grunwald et al. (2008) compared 2003 ESM spatial trends (Rivero et al. 2007) with two prior spatial samplings of this unit in 1990 (DeBusk et al. 1994) and 1998 (DeBusk et al. 2001). This analysis revealed extensive areas of both increase and decrease in soil TP suggesting both a reduction in P loading to WCA-2A and evidence for internal cycling of P within the unit. This finding has significant implication to future restoration efforts as while P loading was decreased, significant enrichment continues to occur due to internal cycling of P. It has been suggested that WCAs have acted much like the STAs capturing P in soils (Entry and Gotlieb 2014) over time and this P is now cycling internally within the WCAs (Reddy et al. 2011).

Bruland et al. (2007) compared 1992 spatial data from WCA-3 (Reddy et al. 1994) to the 2003 ESM spatial data (Bruland et al. 2006) to find significant areas of increase and decrease in TP across the unit. The WCA-3 study of Bruland et al. (2007) was the first comparison study to come from the ESM program and is illustrative of the value of these landscape scale investigations to restoration efforts. For example, comparison of the 2003 ESM data (Bruland et al. 2006) to the 1992 work by Reddy et al. (1994) indicated TP increase in 53% of the 0–10 cm soils in WCA-3 (Fig. 5). Also, 30% of the surface soils measured in 2003 were considered enriched (> 500 mg kg^{-1}) in contrast to 21% of soils measured in 1992. This equates to roughly 1% per year increase in spatial area of soil enrichment. Calculated changes in spatial distributions of soil TP suggest that significant enrichment occurred in northern WCA-3A, possibly associated with the Miami canal outfalls. It has been suggested that soil oxidation in that area due to subsidence and possibly fire contributed internal loading of P (Scheidt et al. 2000; Bruland et al. 2007; Scheidt and Kalla 2007). Significant soil subsidence in northern WCA-3A supports this assertion (Scheidt and Kalla 2007). Of equal interest is the noted decrease in surface soil TP in western WCA-3A at the outfall of the L-28I canal, which in 1992 was a significant area of P enrichment. In 2003, this area indicates significant decrease in surface soil TP, suggesting that changes in water quality and deliver via this canal has significant positive impacts to the surrounding area with respect to P enrichment of soils. Finally, Bruland et al. (2007) echoes the discussion of Scheidt and Kalla (2007) concerning the values of assessing soils with respect to TP on a volumetric basis versus mass basis, as this method can significantly change the outcome of the assessments for a given investigation.

Sulfur

Sulfur is a plant essential macronutrient; however, its presence in the Everglades in excess can be a stressor on the system. Sulfur naturally exists in the environment in several forms, for example, as a constituent in OM. Mineralization of organic S in an aerobic environment results in the environmentally ubiquitous sulfate ion (SO_4^-) see Chapters 4, 5, 15, 16 and 17. Sulfate, an oxidized form of S, can be utilized as an alternate electron acceptor by $SO4^-$ reducing bacteria (SRB), via a process known as $SO4^-$ reduction, that results in reduction of $SO4^-$ to sulfide (S^{2-}) (Reddy and DeLaune 2008). Sulfate reduction is a dominant biogeochemical pathway in brackish and salt marshes worldwide, including the mangrove swamps of the southern Everglades, however, in freshwaters, excessive $SO4^-$ reduction can have deleterious effects.

Sulfur is an element of concern in the Everglades as the process of sulfate reduction, mediated by SRB has been linked to mercury methylation in several studies (Gilmour et al. 1992; Fink and Rawlik 2000; Jeremiason et al. 2006; Axelrad et al. 2008). Perhaps more significant, the introduction of an alternative electron acceptor en mass to a soil environment characterized by high OM and very low oxygen availability can result in accelerated C mineralization and concomitant nutrient remobilization. As discussed previously, soil oxidation is greatly reduced when soils are flooded, due to the effective reduction of available oxygen for respiration. In the event that considerable $SO4^-$ is available for respiration, nutrient regeneration and soil oxidation can continue undeterred by redox condition. A lesser negative effect of the introduction of S to the Everglades ecosystem in large quantities is the byproduct of SO_4^- reduction, S^{2-}. Sulfide in very low doses is not harmful, but as reduced S species (S^{2-}, H_2S) build up, it can be toxic to both plants and other aquatic organisms. Interestingly, there is a positive caveat to S^{2-} production which is the affinity it has for mercury. Mercuric-sulfide (HgS) complex effectively binds mercury and thus reduces Hg available for methylation (Gilmour et al. 1998; Benoit et al. 1999). This reaction is most likely to take place in anaerobic sites in floc and soils versus the water column as available oxygen rapidly catalyzes S^{2-} oxidation to SO_4^-.

There is a strong argument that stormwater discharged from the EAA is the main source contributing SO_4^- to the EPA (Scheidt et al. 2000; Stober et al. 2001; Bates et al. 2002; Gabriel et al. 2008). Bates et al. (2002) identified large soil pool of S in EAA soils which were linked to elemental S application for fertilizer enhancement.

However, Shueneman (2001) reported that S mineralized from soil subsidence, not agricultural use of fertilizers, is the primary source in the EAA. Axelrad et al. (2008) reported that groundwater was not a significant source of SO_4^- to the EPA and Gabriel et al. (2008) reports that while atmospheric deposition of SO_4^- from multiple sources such as marine aerosols does contribute to the SO_4^- loading in the EPA (0.5–5 mg L^{-1}), it does not constitute a major source in relation to EAA canal sources (5–200 + mg L^{-1}) (Gilmour et al. 2007). Conveyance of SO_4^- laden agriculture waters into the northern Everglades WCAs is evidenced by surface water SO_4^- and S^{2-} distributions reported by Scheidt and Kalla (2007). The highest concentrations occur in WCA-2A, which receives up to 100 mg L^{-1} SO_4^-, as compared to marsh background of less than 0.2 mg L^{-1}. In 2005 about 57% of the Everglades marsh exceeded the CERP water quality target of 1 mg L^{-1} for SO_4^- (Scheidt and Kalla 2007). Spatial patterns of SO_4^- enrichment in waters of the northern Everglades are indicative of canal water inputs to the WCAs (Entry 2012).

Osborne et al. (2008) reported the spatial distribution of total S (TS) in soils of the EPA and surrounding areas (Fig. 5). Spatial patterns of S enrichment in floc were closely related to agricultural water inputs in the northern Everglades (data not shown). In WCA-1, patterns of soil and floc TS enrichment are very similar and suggest inflow waters, and thus enrichment, is typically limited to the periphery (Fig. 5). However, Wang et al. (2009) noted that SO_4^- from canal discharge is impacting even the interior portions of WCA-1. However, edaphic S enrichment of WCA-2A indicates enrichment zones below the known P enrichment zone in northern portion of the unit. This pattern suggests that much of the S expected in the soils along the eutrophication gradient has been reduced and therefore not observed in the TS analysis. This conclusion is echoed by the patterns in porewater S^{2-} presented by Scheidt and Kalla (2007). These patterns suggest that a significant portion of S in soils has been reduced in soils impacted by P enrichment. Further, these patterns of high porewater S^{2-} are also generally associated with higher concentrations of surface water MeHg, suggesting that at least some portion of Hg methylation may be occurring within the soil profile. There appears to be a general spatial association of TS and total mercury (THg) in WCA-1 and WCA-2, however, this association breaks down in WCA-3 and is not present in the ENP. This may be explained in part due to differences in soil OM. Of special interest is the marked enrichment of TS in the northern portion of WCA-3 where surface water S species have been reported to be low (Scheidt and Kalla 2007). This enrichment front is due north of the largest enrichment area of THg in soils, and with current soils subsidence activity in that area, the potential for continued S remobilization, in concert with loading of agricultural waters, suggests that future migration of S enrichment south may provide for significant increases in Hg methylation in the future.

Finally, TS enrichment of the ENP, the receiving body of waters from the northern Everglades appears to be relatively low as previously reported by Chambers and Pederson (2006). One area of interest is the headwaters of Taylor Slough, a secondary drainage feature on the landscape and a significant focal point of conservation. The area of enrichment is also spatially similar to that of recent P enrichment (Osborne et al. 2013) and requires further investigation. Other enriched areas in the ENP are associated with zones of marine influence in the mangrove interface, and are not of immediate concern, however, predicted sea level rise may subject those areas to significant SO_4^- loading and ensuing effects in the future. Recent work by Chambers et al. (2013a; Chambers et al. 2013b) suggests that increased SO_4^- with saltwater intrusion can have a variety of effects on soil microbial communities and alter, at least in the short term, nutrient cycling and diagenesis of OM.

Geographic Focus Areas

Several issues pertaining to soils at a landscape scale identified in the literature reviewed here give reason for concern. Extensive areas in WCA-1 continue to be impacted by P and S enrichment even in light of recent improvements to water quality from the EAA. Similarly, WCA-2A appears to have some stabilization in soil P enrichment along the eutrophic gradient in the northern portion of the unit; however, recent spatial and temporal comparisons indicate translocation of TP and suggest possible internal cycling as mechanisms for increased spatial extent of P impact to soils in light of overall reduced soil P concentrations (Marchant et al. 2009; Reddy et al. 2011), as well as, continued water quality inputs in excess of CERP goals. Sulfur enrichment in WCA-2A, combined with current sulfide distribution in surface soils suggests that extensive sulfate reduction is occurring in the eutrophic areas, further complicating P mobilization problems (Osborne et al. 2011a).

Of great concern is the extensive soil subsidence in WCA-3A, as described by Scheidt and Kalla (2007). Landscape trends suggest that if this subsidence is not mitigated hydrologically, soil losses will continue. Soil oxidation in the Everglades has great impact on the surrounding landscape as nutrients and contaminants stored in these peat deposits will be mobilized causing further exacerbation of eutrophication and contamination downstream. As P enrichment is significant in this area, subsidence here has great potential to increase local enrichment, along with the resultant expansion of cattail, and increase P export/translocation as well. Spatial patterns of S and mercury (Hg) suggest extensive accumulation of both of these potentially detrimental elements in soils south of the most severe soil subsidence (Scheidt and Kalla 2007; Reddy et al. 2005). This could be an effect of previous subsidence and mobilization; however, the most relevant threat is remobilization via soil oxidation in these areas of concentration.

Positive findings of minimal P enrichment in southern WCA-3 and ENP give credence to the prioritizing of these areas to protect from future encroachment by excessive nutrients and contaminants. However, while soil concentrations of P are low in the ENP, recent evidence suggests accelerated P enrichment in Taylor Slough, a key habitat area of ENP (Reddy et al. 2008). This finding suggests that vigilance in soil resource protection is required to maintain the relatively unimpacted areas in their present condition while concomitantly working to restore the enriched areas.

Lastly, the hydrologic impacts on tree island soils, particularly in WCA-3A and 3B, and the potential impacts of sea-level rise/saltwater intrusion on soils at the oligohaline interface cannot be overlooked. However, these issues associated with soil reserves have not been sufficiently documented or vary over relatively small spatial scales.

Management Consideration for Everglades Soils

Restoration Coordination and Verification (RECOVER) has developed performance measures for evaluating modeled system-wide performance and assessing actual system wide performance of CERP in meeting its goals and objectives. These performance measures are tools based on a set of indicators developed through conceptual ecological models (CEM) that identified key stressors and attributes of the natural system. These CEM can be found in the Monitoring and Assessment Plan (MAP), the primary tool by which the RECOVER program assesses the performance of the Comprehensive Everglades Restoration Plan (CERP or Plan). The 2009 RECOVER Monitoring and Assessment Plan (MAP) (RECOVER 2009) is an updated version of MAP 2004 describing 1) the monitoring components and supporting research of the MAP 2) summarizing the assessment process and 3) developing the conceptual ecological models necessary to establish the system wide performance measures. The scientific and technical information in the MAP allow RECOVER to assess CERP status and performance.

Soil is recognized as an important component in the functioning of the Everglades ecosystem in the CERP system-wide MAP. This program includes performance measures that address soil condition directly (P in soil, soil loss) or indirectly (water inundation pattern, drought) as soil is impacted by other stressors, however, MAP does not presently provide for soil condition monitoring.

Phosphorus

Total P concentration in soil is an effective means to evaluate long-term ecosystem impacts and is crucial to assessment of CERP activities. The performance target set by CERP is to decrease the areal extent of TP concentrations exceeding 500 mg kg^{-1} and maintain and reduce long-term average concentrations to 400 mg kg^{-1} or less in the upper 10 cm of soil (RECOVER 2007). Based on analysis of available data and discussions among South Florida Water Management District, the United States Environmental Protection Agency, and the United States Fish and Wildlife Service scientists, a numeric sediment TP concentration target in the range of 200 to 400 mg kg^{-1} in the top 10 cm of soil column, including the overlying layer of flocculent or unconsolidated sediment, will be used to delineate areas minimally impacted (biologically) from those in which the structure and function of the native biological communities have been significantly altered by P enrichment. This target is variable in space and only applicable to peat soils (RECOVER 2007). Scheidt and Kalla (2007) reported that during 2005 soil P exceeded 500 mg kg^{-1}, Florida's definition of "impacted", in 24% of the Everglades, and it exceeded 400 mg kg^{-1}, CERP's restoration goal, in 49% of the Everglades. These proportions are higher than the 16% and 34%, respectively, observed in 1995–1996.

Carbon

Drainage of the ecosystem, compartmentalization and reduction of the total water quantity stored in the ecosystem has exaggerated the dry seasons and dry years that can follow (Stober et al. 2001). The CERP establishes a performance measure that addresses extreme low and high water levels and ultimately benefits Everglades soils. The performance measure examines the frequency, duration, and percent period of record of extreme events and the peat exposure due to droughts (RECOVER 2007). The intent of the drought intensity component of the performance measure is to use a quantitative graphical display of cumulative desiccation intensity (magnitude and duration of drying event) to determine whether alternative project designs are likely to increase or decrease the potential for further unnatural loss of organic soils. The goal is to reduce the risk of further loss of soil elevations due to excessive drying.

A second performance measure related to soil preservation is the restoration of the sheet flow in the Everglades ridge and slough landscape to re-establish the natural patterns of distribution, timing, continuity and volume of sheet flow (RECOVER 2007). This will significantly help to sustain the microtopography in relation to organic soil accretion and loss (sheet flow interacts with hydroperiod, water depth, fire, and nutrient dynamics to maintain organic soil accretion and loss in a state of dynamic equilibrium).

Other performance measures like dry events in Shark River Slough or the inundation patterns in the Everglades wetlands will also help restore the soil (formation and maintenance) in the Everglades since the soil peat accretion typical of the ridge and slough landscape requires prolonged flooding, characterized by 10 to 12 month annual hydroperiods, and groundwater that rarely drops more than one foot below ground surface.

Approaches to Assessment of Everglades Soils

To assess the effectiveness of restoration projects, extensive monitoring of soil along with other parameters is required. Regular sampling to determine the spatial distribution of soil physicochemical characteristics is important to determine the long-term impacts and changes. Several monitoring design options are currently available. These options should be carefully assessed as to their advantages and disadvantages prior to selecting a design.

Geospatial

The Everglades Protection Area is a spatially large and diverse ecosystem. It is important to monitor the system as a whole versus select areas to determine and follow the changes in soil components (Scheidt and Kalla 2007). A randomized design, similar to the one developed in the R-EMAP study is capable of spatially integrating broad-scale changes for the entire landscape area. The R-EMAP design is utilizing a probability based sampling design (reviewed and recommended by the National Research Council) with a total of 1145 sites. Each iteration of system wide sampling included hundreds of sample locations across the EPA, enabling temporal as well as, spatial interpretation of patterns and trends in soil attributes. The preeminent strength of the probability-based design used by R-EMAP is the ability to make quantitative statements across space about the status of indicators of ecological health with known confidence limits across space. This design is desirable if a broad spatial integration is the highest priority (NRC 2003). Another stratified-random sampling method was selected by Reddy et al. (2005) in conjunction with the South Florida Water Management District and the RECOVER Assessment Team to conduct a system wide soil sampling effort, the Everglades Soil Mapping project (ESM) at 1358 sites in a single phase. This sampling design was based on one implemented in Loxahatchee National Wildlife Refuge in the 1990s (Newman et al. 1997) to ensure that large zones with low variability in physicochemical soil properties were not over-sampled and small zones with high variability were not ignored (DeBusk et al. 2000; Grunwald et al. 2008). This results in higher sampling density in areas previously characterized by high spatial variability of soil P concentration (Richardson et al. 1990).

Traditional and Hybrid

Transects are an excellent approach toward assessing changes along gradients. These gradients are the most likely locations where ecological change will occur; transects maximize the ability to detect this change. If a transect design is chosen, sampling along gradients can either be completely randomized or stratified random sampling (depending on steepness of gradient or on habitat type); sample sites can be fixed for some parameters (with randomized initial selection) such as for ground water wells or individual trees (Reddy et al. 1998).

Another stratified random sampling design is based on primary sampling units (PSU). In this design, the greater Everglades is divided into a series of landscape subunits (LSU). Within each subunit, the design recommends selection of primary sampling units (PSUs) that be selected based upon availability and location of sparse emergent freshwater marsh within each landscape subunit. After the PSUs are selected, sampling locations will be randomly chosen within the appropriate habitat of each PSU. The PSU design is currently used by several researchers in the MAP and RECOVER monitoring of tree islands and ridge slough status. A recent study of short range spatial variability with respect to soil P also utilized this study design (Cohen et al. 2009).

Another design was also suggested for sentinel site sampling. Sentinel sites would be located to be representative of regionally significant ecotypes and intended to provide information for regions of interest either in impact areas or in reference, un-impacted regions. These sentinel sites could be existing fixed station monitoring sites with historical data.

While there are several alternative spatial sampling design options, the utilization of any one in particular is objective driven as there are substantial arguments for the use of each one. The most pressing management question with respect to soil monitoring currently, is how often this type of sampling should occur. R-EMAP was projected to occur in phases that spanned 3–5 years. The last phase was completed in 2005 and at this time continuation of this effort is unknown. The intent of ESM, from the standpoint of RECOVER, was to complete this system wide sampling on a 7–10 year cycle. Smaller scale sampling efforts are definitely more financially feasible while landscape scale monitoring is very resource demanding. However, several studies contend that this level of sampling, at some set temporal interval, is critical for monitoring ecosystem health and assessing restoration success (Scheidt and Kalla 2007; Osborne et al. 2011a).

References

Aich, S. and T.W. Dreschel. 2014. Evaluating Everglades peat carbon loss using geospatial techniques. Florida Scientist (in press).

Amador, J.A. and R.D. Jones. 1993. Nutrient limitations on microbial respiration in peat soils with different total phosphorus content. Soil Biol. Biochem. 25: 793–801.

Amador, J.A. and R.D. Jones. 1995. Carbon mineralization in pristine and phosphorus-enriched peat soils of the Florida Everglades. Soil Sci. 159: 129–141.

Axelrad, D.M., T. Lange, M. Gabriel, T.D. Atkenson, C.D. Pollman, W.H. Orem, D.J. Scheidt, P.I. Kalla, P.C. Frederick and C.C. Gilmour. 2008. Mercury and sulfur monitoring, research and environmental assessment in south Florida. *In*: 2008 South Florida Environmental Report. South Florida Water Management District, West Palm Beach, FL.

Bates, A.L., W.H. Orem, J.W. Harvey and E.C. Spiker. 2002. Tracing sources of sulfur in the Florida Everglades. J. Environ. Qual. 31: 287–299.

Bazante, J., G. Jacobi, H.M. Solo-Gabriele, D. Reed, S. Mitchell-Bruker, D.L. Childers, L. Leonard and M. Ross. 2006. Hydrologic measurements and implications for tree island formation within Everglades National Park. J. Hydrol. 329: 606–619.

Benoit, J.M., C.C. Gilmour, R.P. Mason and A. Heyes. 1999. Sulfide controls on mercury speciation and bioavialbility to methylating bacteria in sediment pore waters. Environ. Sci. Technol. 33: 951–957.

Brady, N.C. and R.R. Weil. 2003. Elements of the Nature and Properties of Soils. 12th Ed. Prentice Hall Inc., Upper Saddle River, NJ.

Browder, J. and J.C. Ogden. 1999. The natural South Florida system II: Pre-drainage ecology. Urban Ecosyst. 3: 125–158.

Browder, J.A., P.J. Gleason and D.R. Swift. 1994. Periphyton in the Everglades: spatial variation, environmental correlates, and implications. pp. 379–418. *In*: S.M. Davis and J.C. Ogden (eds.). Everglades, the Ecosystem and its Restoration. St. Lucie Press, Delray Beach, FL.

Bruland, G.L. and C.J. Richardson. 2006. Comparison of soil OM in created, restored, and paired natural wetlands in North Carolina. Wetlands Ecol. Manag. 14: 245–251.

Bruland, G.L., S. Grunwald, T.Z. Osborne, K.R. Reddy and S. Newman. 2006. Spatial distribution of soil properties in Water Conservation Area 3 of the Everglades. Soil Sci. Soc. Am. J. 70: 1662–1676.

Bruland, G.L., T.Z. Osborne, K.R. Reddy, S. Grunwald, S. Newman and W.F. DeBusk. 2007. Recent changes in soil total phosphorus in the Everglades: Water Conservation Area 3. Environ. Monit. Assess. 129: 379–395.

Chambers, R.M. and K.A. Pederson. 2006. Variation in soil phosphorus, sulfur, and iron pools among south Florida. Hydrobiologi. 569: 63–70.

Chen, M., L.Q. Ma and Y.C. Li. 2000. Concentrations of P, K, Al, Fe, Mn, Cu, Zn, and As in marl soils from south Florida. Soil and Crop Sciences Society of Florida Proceedings 59: 124–129.

Chiang, C., C.B. Craft, D.W. Rogers and C.J. Richardson. 2000. Effects of 4 years of nitrogen and phosphorus additions on Everglades plant communities. Aquatic Bot. 68: 61–78.

Childers, D.L., R.F. Doren, R. Jones, G.B. Noe, M. Rugge and L.J. Scinto. 2003. Decadal change in vegetation and soil phosphorus pattern across the Everglades landscape. J. Environ. Qual. 32: 344–362.

Clark, M.W. and K.R. Reddy. 2003. Decomposition dynamics of dominant vegetation in ridge and slough habitats of the Everglades. Final Report. South Florida Water Management District, West Palm Beach, FL.

Cohen, A.D. and W. Spackman. 1984. The petrology of peats from the Everglades and coastal swamps of southern Florida. pp. 353–374. *In*: P.J. Gleason (ed.). Environments of South Florida: Present and Past II. Miami Geological Society, Miami, FL.

Cohen, M.J., S. Lamsal, T.Z. Osborne, J.C. Bonzongo, S. Newman and K.R. Reddy. 2009. Mapping mercury concentrations in the Greater Everglades. Soil Sci. Soc. Am. J. 73: 675–685.

Cohen, M.J., J. Heffernan, D. Watts and T.Z. Osborne. 2011. Reciprocal biotic control on hydrology, nutrient gradients, and landform in the Greater Everglades. Crit. Rev. Environ. Sci. Technol. 41: 395–429.

Corstanje, R., S. Grunwald, K.R. Reddy, T.Z. Osborne and S. Newman. 2006. Assessment of the spatial distribution of soil properties in a northern everglades marsh. J. Environ. Qual. 35: 938–949.

Craft, C.B. and C.J. Richardson. 1993. Peat accretion and phosphorus accumulation along a eutrophic gradient in northern Everglades. Biogeochem. 22: 133–156.

Craft, C.B. and C.J. Richardson. 1997. Relationships between soil nutrients and plant species composition in Everglades peatlands. J. Environ. Qual. 26: 224–232.

Craighead, F.C. 1971. The Trees of South Florida. University of Miami Press, Coral Gables, Florida.

Criquet, S., A.M. Farnet, S. Tagger and J. Le Petit. 2000. Annual variations of phenoloxidase activities in evergreen oak litter: influence of certain biotic and abiotic factors. Soil Biol. Biochem. 32: 1505–1513.

Daoust, R.J. and D.L. Childers. 2004. Ecological effects of low-level phosphorus additions on two plant communities in a neotropical freshwater wetland ecosystem. Oecologia 141: 672–686.

Davis, J.H., Jr. 1943. The natural features of southern Florida. Geol. Bull. 25. FL Geol. Survey, Tallahassee, FL.

Davis, J.H., Jr. 1946. The peat deposits of Florida. Geol. Bull. 30. FL Geol. Survey, Tallahassee, FL.

Davis, S.M. 1991. Growth, decomposition, and nutrient retention of *Cladium jamaicense* Crantz. and *Typha domingensis* Pers. in the Florida Everglades. Aquatic Bot. 40: 203–224.

Davis, S.M. 1994. Phosphorus inputs and vegetation sensitivity in the Everglades. pp. 357–378. *In*: S.M. Davis and J.C. Odgen (eds.). Everglades: The Ecosystem and its Restoration. St. Lucie Press, Delray Beach, FL.

Davis, S.M. and J.C. Ogden (eds.). 1994. Everglades: The Ecosystem and its Restoration. St. Lucie Press, Delray Beach, FL.

Davis, S.M., E.E. Gaiser, W.F. Loftus and A.E. Huffman. 2005a. Southern marl prairies conceptual ecological model. Wetlands 25: 821–831.

Davis, S.M., D.L. Childers, J.L. Lorenz, H.R. Wanless and T.E. Hopkins. 2005b. A conceptual ecological model of ecological interactions in the mangrove estuaries of the Florida Everglades. Wetlands 25: 832–842.

DeAngelis, D.L. 1994. Synthesis: Spatial and temporal characteristics of the environment. pp. 307–322. *In*: S.M. Davis and J.C. Odgen (eds.). Everglades: The Ecosystem and its Restoration. St. Lucie Press, Delray Beach, FL.

DeAngelis, D.L. and P.S. White. 1994. Ecosystems as products of spatially and temporally driving forces, ecological processes and landscapes: a theoretical perspective. pp. 9–27. *In*: S.M. Davis and J.C. Odgen (eds.). Everglades: The Ecosystem and its Restoration. St. Lucie Press, Delray Beach, FL.

DeBusk, W.F. and K.R. Reddy. 1998. Turnover of detrital organic carbon in a nutrient-impacted Everglades marsh. Soil Sci. Soc. Am. J. 62: 1460–1468.

DeBusk, W.F. and K.R. Reddy. 2003. Nutrient and hydrology effects on soil respiration in a northern Everglades marsh. J. Environ. Qual. 32: 702–710.

DeBusk, W.F., K.R. Reddy, M.S. Koch and Y. Wang. 1994. Spatial patterns of soil phosphorus in Everglades Water Conservation Area 2A. Soil Sci. Soc. Am. J. 58: 543–552.

DeBusk, W.F., S. Newman and K.R. Reddy. 2001. Spatio-temporal patterns of soil phosphorus enrichment in Everglades Water Conservation Area 2A. J. Environ. Qual. 30: 1438–1446.

Doren, R.F., T.V. Armentano, L.D. Whiteaker and R.D. Jones. 1996. Marsh vegetation patterns and soil phosphorus gradients in the Everglades Ecosystem. Aquatic Bot. 56: 145–163.

Entry, J.A. 2012a. Water quality trends in the Loxahatchee National Wildlife Refuge. Wat. Air Soil Pollut. 223: 4515–4215.

Entry, J.A. 2012b. Water quality gradients in the Northern Florida Everglades. Wat. Air Soil Pollut. 223: 6109–6121.

Entry, J.A. and A. Gotlieb. 2014. The impact of stormwater treatment areas and agricultural best managment practices on water quality in the Everglades Protection Area. Environ. Monit. Assess. 186: 1023–1037.

Ewe, S.M.L., E.E. Gaiser, D.L. Childers, V.H. Rivera-Monroy, D. Iwaniec, J. Fourquerean and R.R. Twilley. 2006. Spatial and temporal patterns of aboveground net primary productivity (ANPP) in the Florida Coastal Everglades LTER (2001–2004). Hydrobiologia 569: 459–474.

Ewe, S.M.L., B. Gu, J. Vega, K. Vaughan, S. Aich and F. Sklar. 2010. Landscape-scale trends and patterns of ghost tree islands in the Everglades. Abstract, Greater Everglades Ecosystem Restoration, Naples, FL, July 2010.

Fink, L. and P. Rawlik. 2000. The Everglades mercury problem. Everglades Consolidated Report. South Florida Water Management District, West Palm Beach, FL.

Freeman, C., N. Ostle and H. Kang. 2001. An enzymatic 'latch' on a global carbon store. Nature 409: 149.

Gabriel, M., G. Redfield and D. Rumbold. 2008. Appendix 3B-2: Sulfur as a regional water quality concern in south Florida. *In*: 2008 South Florida Environmental Report. South Florida Water Management District, West Palm Beach, FL.

Gaiser, E.E., L.J. Scinto, J.H. Richards, K. Jayachandran, D.L. Childers, J.C. Trexler and R.D. Jones. 2004. Phosphorus in periphyton mats provides the best metric for detecting low-level P enrichment in an oligotrophic wetland. Water Res. 38: 507–516.

Gaiser, E.E., J.C. Trexler, J.H. Richards, D.L. Childers, D. Lee, A.L. Edwards, L.J. Scinto, K. Jayachandran, G.B. Noe and R.D. Jones. 2005. Cascading ecological effects of low-level phosphorus enrichment in the Florida Everglades. J. Environ. Qual. 34: 717–723.

Gaiser, E.E., D.L. Childers, R.D. Jones, J.H. Richards, L.J. Scinto and J.C. Trexler. 2006. Periphyton responses to eutrophication in the Florida Everglades. Cross-system patterns of structural and compositional change. Limnol. Oceanogr. 51: 617–630.

Gilmour, C.C., E.A. Henry and R. Mitchell. 1992. Sulfate stimulation of mercury methylation in freshwater sediments. Environ. Sci. Technol. 26: 2287–2294.

Gilmour, C.C., G.S. Riedel, M.C. Ederington, J.T. Bell, J.M. Benoit, G.A. Gill and M.C. Stordal. 1998. Methylmercury concentrations and production rates across a trophic gradient in the northern Everglades. Biogeochemistry 40: 327–345.

Gilmour, C.C., W. Orem, D. Krabbenhoft and I.A. Mendelssohn. 2007. Preliminary assessment of sulfur sources, trends, and effects in the Everglades. *In*: 2007 South Florida Environmental Report. South Florida Water Management District, West Palm Beach, FL.

Gleason, P.J. 1972. The origin, sedimentation and stratigraphy of a calcitic mud located in the southern freshwater Everglades. PhD dissertation, Pennsylvania State University, University Park, PA.

Gleason, P.J. and P.A. Stone. 1994. Age, origin, and landscape evolution of the Everglades peatland. pp. 149–198. *In*: S.M. Davis and J.C. Ogden (eds.). Everglades: The Ecosystem and its Restoration. St. Lucie Press, Delray Beach, FL.

Gleason, P.J., A.D. Cohen, P. Stone, W.G. Smith, H.K. Brooks, R. Goodrick and W. Spackman, Jr. 1974. The environmental significance of Holocene sediments from the Everglades and saline tidal plains. pp. 297–351. *In*: P.J. Gleason (ed.). Environments of South Florida, Present and Past. Miami Geological Society, Coral Gables, FL.

Grunwald, S., T.Z. Osborne and K.R. Reddy. 2008. Temporal trajectories of phosphorus and pedo-patterns mapped in Water Conservation Area 2, Everglades, Florida, USA. Geoderma 146: 1–13.

Hagerthey, S.E., S. Newman, K. Rutchey, E.P. Smith and J. Godin. 2008. Multiple regime shifts in a subtropical peatland community: Specific thresholds to eutrophication. Ecol. Monogr. 78: 547–565.

Harvey, J.W., R.W. Schaffranek, G.B. Noe, L.G. Larsen, D.J. Nowacki and B.L. O'Connor. 2009. Hydroecological factors governing surface water flow on a low-gradient floodplain. Water Resources Research 45: W03421, doi:10.1029/2008WR007129.

Hofmockel, K., C.J. Richardson and P.N. Halpin. 2008. Effects of hydrologic management decisions on Everglades tree islands. pp. 191–214. *In*: C.J. Richardson (ed.). The Everglades Experiments: Lessons for Ecosystem Restoration. Springer, New York.

Holt, P.R., R.J. Sutton and D. Vogler. 2006. South Florida Digital Elevation Model, Version 1.1 U.S. Army Corps of Engineers, Jacksonville District, Jacksonville, FL.

Ingebritsen, S.E., C. McVoy, B. Glaz and W. Park. 1999. Subsidence threatens agriculture and complicates ecosystem restoration. pp. 95–106. *In*: D. Galloway, D.R. Jones and S.E. Ingebritsen (eds.). Land Subsidence in the United States. USGS Circular 1182.

Jenny, H. 1941. Factors of Soil Formation. Dover Publications, New York, New York.

Jeremiason, J., D. Engstrom, E. Swain, E. Nater, B. Johnson, J. Almendinger, B. Monson and R. Kolka. 2006. Sulfate addition increases methylmercury production in an experimental wetland. Environ. Sci. Technol. 40: 3800–3806.

Jones, L.A. 1948. Soils, geology, and water control in the Everglades region. Bulletin 442, University of Florida Agricultural Experiment Station and Soil Conservation Service, Gainesville, FL.

Keddy, P.A. 2000. Wetland Ecology: Principles and Conservation. Cambridge University Press, Cambridge, UK.

Koch, M.S. and K.R. Reddy. 1992. Distribution of soil and plant nutrients along a trophic gradient in the Florida Everglades. Soil Sci. Soc. Am. J. 56: 1492–1499.

Lake Okeechobee Technical Advisory Council (LOTAC) II. 1990. Final Report to the Governor, State of Florida, Secretary, Department of Environmental Regulation, Governing Board, South Florida Water Management District, West Palm Beach, Florida.

Larsen, L. and J.W. Harvey. 2010. How vegetation and sediment transport feedbacks drive landscape change in the Everglades and wetlands worldwide. American Naturalist. (in press).

Leonard, L., A. Croft, D. Childers, S. Mitchell-Bruker, H. Solo-Gabriele and M.S. Ross. 2006. Characteristics of surface flows in the Ridge and Slough landscape of Everglades National Park: implications for particulate transport. Hydrobiologia 569: 5–22.

Lodge, T.E. 2010. The Everglades Handbook: Understanding the Ecosystem, Third Edition. CRC Press, Boca Raton, FL.

Maltby, E. and P.J. Dugan. 1994. Wetland ecosystem protection, management, and restoration. pp. 29–46. *In*: S.D. Davis and J.C. Ogden (eds.). Everglades: The Ecosystem and its Restoration. St. Lucie Press, Delray Beach, FL.

Marchant, B.P., S. Newman, R. Corstanje, K.R. Reddy, T.Z. Osborne and R.M. Lark. 2009. Spatial monitoring of a non-stationary soil property: phosphorus in a Florida water conservation area. Eur. J. Soil Sci. 60: 757–769.

McCormick, P.V. and M.B. O'Dell. 1996. Quantifying periphyton responses to phosphorus in the Florida Everglades: a synoptic-experimental approach. J. North Am. Benthol. Soc. 15: 450–468.

McCormick, P.V. and R.J. Stevenson. 1998. Periphyton as a tool for ecological assessment and management in the Florida Everglades. J. Phycol. 34: 726–733.

McCormick, P.V., M.B. O'Dell, R.B.E. Shuford III, J.G. Backus and W.C. Kennedy. 2001. Periphyton responses to experimental phosphorus enrichment in a subtropical wetland. Aquatic Bot. 71: 119–139.

McCormick, P.V., S. Newman, S.L. Miao, D.E. Gawlik, D. Marley, K.R. Reddy and T.D. Fontaine. 2002. Effects of anthropogenic phosphorus inputs on the Everglades. pp. 83–126. *In*: J. Porter and K. Porter (eds.). The Everglades, Florida Bay, and Coral Reefs of the Florida Keys: An Ecosystem Sourcebook. CRC Press, Boca Raton, FL.

McVoy, C.W., W.P. Said, J. Obeysakera, J. Vanarman and T.W. Dreschel. 2011. Pre-drainage Everglades Landscapes and Hydrology. University Press of Florida, Gainesville, FL.

Meeder, J.F., M.S. Ross, G. Telesnicki, P.L. Ruiz and J.P. Sah. 1996. Vegetation analysis in the C-111/ Taylor Slough basin. SERC Research Reports, Florida International University. Available from: http:// digitalcommons.fiu.edu/sercrp/6.

Miao, S.L. and F.H. Sklar. 1998. Biomass and nutrient allocation of sawgrass and cattail along a nutrient gradient in the Florida Everglades. Wet. Ecol. Manage. 5: 245–263.

Miao, S.L. and W.F. DeBusk. 1999. Effects of phosphorus enrichment on structure and function of sawgrass and cattail communities in Florida wetlands. pp. 275–295. *In*: K.R. Reddy, G.A. O'Connor and C.L. Schelske (eds.). Phosphorus Biogeochemistry of Subtropical Ecosystems. Lewis Publishers, Boca Raton, FL.

Mitsch, W.J. and J.G. Gosselink. 2000. Wetlands, 3rd Ed. John Wiley and Sons, New York, NY.

Newman, S., J.B. Grace and J.W. Koebel. 1996. Effects of nutrients and hydroperiod on *Typha*, *Cladium*, and *Eleocharis*: implications for Everglades restoration. Ecol. Appl. 6: 774–783.

Newman, S., K.R. Reddy, W.F. DeBusk, Y. Wang, G. Shih and M.M. Fisher. 1997. Spatial distribution of soil nutrients in a northern Everglades marsh: Water conservation area 1. Soil Sci. Soc. Am. J. 61: 1275–1283.

Newman, S., J. Schuette, J.B. Grace, K. Rutchey, T. Fontaine, K.R. Reddy and M. Pietrucha. 1998. Factors influencing cattail abundance in the northern Everglades. Aquatic Bot. 60: 265–280.

Noe, G.B., D.L. Childers and R.D. Jones. 2001. Phosphorus biogeochemistry and the impact of phosphorus enrichment: Why is the everglades so unique? Ecosystems 4: 603–624.

Noe, G.B., D.L. Childers, A.L. Edwards, E.E. Gaiser, K. Jayachandran, D. Lee, J. Meeder, J.H. Richards, L.J. Scinto, J.C. Trexler and R.D. Jones. 2002. Short-term changes in phosphorus storage in an oligotrophic Everglades wetland ecosystem receiving experimental nutrient enrichment. Biogeochemistry 59: 239–267.

Noe, G.B., L.J. Scinto, J. Taylor, D.L. Childers and R.D. Jones. 2003. Phosphorus cycling and partitioning in an oligotrophic Everglades wetland ecosystem: a radioisotope tracing study. Freshwater Biol. 48: 1993–2008.

Obeysekera, J., J.A. Browder, L. Hornung and M.A. Harwell. 1999. The natural South Florida system I: climate, geology, and hydrology. Urban Ecosyst. 3: 223–244.

Ogden, J.C. 2005. Everglades ridge and slough conceptual ecological model. Wetlands 25: 810–831.

Olmsted, I.C. and L.L. Loope. 1984. Plant communities of Everglades National Park. pp. 167–184. *In*: P.J. Gleason (ed.). Environments of South Florida: Past and Present II. Miami Geological Society, Coral Gables, FL.

Osborne, T.Z., P.W. Inglett and K.R. Reddy. 2007. Linkages between potential dissolved and particulate organic matter from vegetation in a subtropical wetland. Aquatic Bot. 86: 53–61.

Osborne, T.Z., S. Newman and K.R. Reddy. 2008. Spatial distribution of Total sulfur in the soils of the Northern and Southern Everglades. Final Report # 45000-12699/14856, South Florida Water Management District, West Palm Beach, FL.

Osborne, T.Z., S. Newman, P. Kalla, D.J. Scheidt, G.L. Bruland, M.J. Cohen, L.J. Scinto and L.R. Ellis. 2011a. Landscape patterns of significant soil nutrients and contaminants in the Greater Everglades Ecosystem: Past, Present, and Future. Crit. Rev. Environ. Sci. Technol. 41: 121–148.

Osborne, T.Z., G.L. Bruland, S. Newman, K.R. Reddy and S. Grunwald. 2011b. Spatial distributions and eco-partitioning of soil biogeochemical properties in Everglades National Park. Environmental Monitoring and Assessment. (in press).

Osborne, T.Z., L.N. Kobziar and P.W. Inglett. 2013. Investigating the role of fire in shaping and maintaining wetland ecosystems. Fire Ecol. 9: 1–5.

Parker, G.G. 1974. Hydrology of the pre-drainage system of the Everglades in south Florida. pp. 18–27. *In*: P.J. Gleason (ed.). Environments of South Florida: Present and Past. Memoir II. Miami Geol. Soc., Coral Gables, FL.

Payne, G., K. Weaver and T. Bennett. 2003. Chapter 5: Development of a numeric phosphorus Criterion for the Everglades Protection Area. *In*: SFWMD (eds.). 2003 Everglades Consolidated Report, South Florida Water Management District, West Palm Beach, FL.

Petuch, E.J. and C.E. Roberts. 2007. The Geology of the Everglades and Adjacent Areas. CRC Press, Boca Raton, FL.

Pietro, K., R. Bearzotti, M. Chimney, G. Germain, N. Iricanin and T. Piccone. 2007. STA performance, compliance and optimization. pp. 5-1-5-128. *In*: Vol. 1, South Florida Environmental Report, South Florida Water Management District, West Palm Beach, FL.

Qualls, R.G. and C.J. Richardson. 1995. Forms of soil phosphorus along a nutrient enrichment gradient in the northern Everglades. Soil Sci. 160: 183–198.

[RECOVER]. 2006. Method for *in situ* surveys and physical measurements. Available from: http://www.evergladesplan.org/pm/recover/recover_docs/qasr/qasr_app_f_insitu_surveys.pdf .

[RECOVER]. 2007. Development and Application of Comprehensive Everglades Restoration Plan System-wide Performance Measures. Available from: http://www.evergladesplan.org/pm/recover/perf_systemwide.aspx. United States Army Corps of Engineers (Jacksonville District) and South Florida Water Management District, Jacksonville and West Palm Beach, FL.

[RECOVER]. 2009. CERP Monitoring and Assessment Plan. Available from: http://www.evergladesplan.org/pm/recover/recover_map_2009.aspx. United States Army Corps of Engineers (Jacksonville District) and South Florida Water Management District, Jacksonville and West Palm Beach, FL.

Reddy, K.R. and R.D. DeLaune. 2008. Biogeochemistry of Wetlands: Science and Applications. CRC Press Taylor and Francis Group, Boca Raton, FL.

Reddy, K.R., W.F. DeBusk, Y. Wang, R. DeLaune and M. Koch. 1991. Physico-chemical properties of soils in the Water Conservation Area 2 of the Everglades. Soil Science Department, Institute of Food and Agricultural Sciences, Gainesville, FL.

Reddy, K.R., R. DeLaune, W.F. DeBusk and M.S. Koch. 1993. Long term nutrient accumulation rates in the Everglades. Soil Sci. Soc. Am. J. 57: 1147–1155.

Reddy, K.R., Y. Wang, W.F. DeBusk and S. Newman. 1994. Physico-chemical properties of soils in Water Conservation Area 3 (WCA-3) of the Everglades. Final Report. South Florida Water Management District, West Palm Beach, FL.

Reddy, K.R., Y. Wang, W.F. DeBusk, M.M. Fisher and S. Newman. 1998. Forms of soil phosphorus in selected hydrologic units of the Florida Everglades. Soil Sci. Soc. Am. J. 62: 1134–1147.

Reddy, K.R., S. Newman, S. Grunwald, T.Z. Osborne, R. Corstanje, G.L. Bruland and R.G. Rivero. 2005. Everglades soil mapping final report. South Florida Water Management District, West Palm Beach, FL.

Reddy, K.R., T.Z. Osborne, N. Aumen and M.S. Zimmerman. 2008. Longterm changes in soil phosphorus in selected hydrologic units of the Everglades. Final Report USDOI NPS.

Reddy, K.R., S. Newman, T.Z. Osborne, J.R. White and H.C. Fitz. 2011. Phosphorus cycling in The Everglades Ecosystem: Legacy phosphorus implications for management and restoration. Critical Reviews in Environmental Science and Technology Special Issue No. 1. (in press).

Redfield, G. and S. Efron. 2007. Chapter 1B: Cross-cutting issues in the 2006 South Florida Environmental Report—Impacts of the 2004 hurricanes on South Florida Environment. South Florida Environmental Report 2006. South Florida Water Management District, West Palm Beach, FL.

Richardson, C.J. and J.K. Huvane. 2008. Ecological status of the Everglades: environmental and human factors that control the peatland complex on the landscape. pp. 13–58. *In*: C.J. Richardson (ed.). The Everglades Experiments: Lessons for Ecosystem Restoration. Springer, New York.

Richardson, J.R., W.L. Bryant, W.M. Kitchens, J.E. Mattson and K.R. Pope. 1990. An evaluation of refuge habitats and relationships to water quality, quantity, and hydroperiod. A Synthesis Report. A.R.M. Loxahatchee National Wildlife Refuge, Boynton Beach, FL.

Rivero, R.G., S. Grunwald, T.Z. Osborne, K.R. Reddy and S. Newman. 2007. Characterization of the spatial distribution of soil properties in water conservation area 2A, Everglades, Florida. Soil Sci. 172: 149–166.

Rivero, R.G., S. Grunwald, M.W. Binford and T.Z. Osborne. 2009. Integrating spectral indices into prediction models of soil phosphorus in a subtropical wetland. Remote Sen. Environ. 113: 2389–2402.

Ross, M.S. and J.P. Sah. 2011. Forest resource islands in a sub-tropical marsh: soil-site relationships in Everglades hardwood hammocks. Ecosystems 14: 632–645.

Ross, M.S., S. Mitchell-Bruker, J.P. Sah, S. Stothoff, P.L. Ruiz, D.L. Reed, K. Jayachandran and C.L. Coultas. 2006. Interaction of hydrology and nutrient limitation in the Ridge and Slough landscape of the southern Everglades. Hydrobiologia 569: 37–59.

Scheidt, D.J. and P.I. Kalla. 2007. Everglades ecosystem assessment: Water management and quality, eutrophication, mercury contamination, soils and habitat: monitoring for adaptive management: A R-EMAP status report. USEPA Region 4, Athens, GA.

Scheidt, D.J., Stober, R. Jones and K. Thornton. 2000. South Florida ecosystem assessment: Everglades water management, soil loss, eutrophication and habitat. Report No. 904-R-00-003. US Environmental Protection Agency, Athens, GA.

[SFWMD]. (South Florida Water Management District. 1992. Surface water improvement and management plan for the Everglades. Supporting information Document. South Florida Water Management District, West Palm Beach, Florida, USA.

[SFWMD]. 2007. South Florida Environmental Report. South Florida Water Management District. West Palm Beach, FL. www.sfwmd.gov/sfer.

Shueneman, T.J. 2001. Characterization of sulfur sources in the EAA. Soil and Crop Science Society of Florida Proceedings 60: 49–52.

Sklar, F., L. Brandt, D. DeAngelis, C. Fitz, D. Gawlik, S. Krupa, C. Madden, Mazzotti, C. McVoy, S. Miao, D. Rudnick, K. Rutchney, K. Tarboton, L. Vilchek and W.Y. 2000. Hydrological needs-effects of hydrology on the Everglades. South Florida Water Management District, West Palm Beach, FL.

Snyder, G.H. 2005. Everglades Agricultural Area soil subsidence and land use projections. Proceedings of the Soil and Crop Science Society of Florida 64: 44–51.

Snyder, G.H. and J.M. Davidson. 1994. Everglades agriculture: past, present and future. pp. xx–xx. *In*: S.D. Davis and J.C. Ogden (eds.). Everglades: The Ecosystem and its Restoration. St. Lucie Press, Delray Beach, FL, pp. 826.

Stephens, J.C. 1956. Subsidence of organic soils in the Florida Everglades. Soil Sci. Soc. Am. Proc. 20: 77–80.

Stephens, J.C. and L. Johnson. 1951. Subsidence of organic soils in the upper Everglades region of Florida. Soil Science Society of Florida Proceedings 57: 20–29.

Stephens, J.C. and E.H. Stewart. 1977. Effect of climate on organic soil subsidence. Inter. Assoc. Hydrol. Sci. 121: 647–655.

Stober, Q.J., K. Thornton, R. Jones, J. Richards, C. Ivey, R. Welch, M. Madden, J. Trexler, E. Gaiser, D. Scheidt and S. Rathbun. 2001. South Florida ecosystem assessment: Phase I/II—Everglades stressor interactions: hydropatterns, eutrophication, habitat alteration, and mercury contamination. USEPA Region 4, Athens, GA. EPA 904-R-01-003.

Surratt, D., D. Shinde and N. Aumen. 2012. Recent cattail expansion and possible relationships to water management: changes in upper Taylor Slough (Everglades National Park, Florida, USA). Environ. Manage. 49: 720–733.

Wang, H., M.G. Waldon, E.A. Meselhe, J.C. Arceneaux, C. Chen and M.C. Harwell. 2009. Surface water Sulfate Dynamics in the Northern Everglades. J. Environ. Qual. 38: 734–741.

Watts, D.L., M.J. Cohen, J.B. Heffernan and T.Z. Osborne. 2010. Hydrologic modification and the loss of self-organized patterning in the ridge slough mosaic of the Everglades. Ecosystems (in press).

Wetzel, P.R., F.H. Sklar, C.A. Coronado-Molina, T. Troxler, S.L. Krupa, P.L. Sullivan, S. Ewe, R.M. Price, S. Newman and W.H. Orem. 2011. Biogeochemical processes on tree islands in the Greater Everglades: Initiating a new paradigm. Crit. Rev. Environ. Sci. Technol. 41(S1): 670–701.

Wright, A.L. and K.R. Reddy. 2001. Heterotrophic microbial activity in Northern Everglades wetland soils. Soil Sci. Soc. Am. J. 65(6): 1856–1864.

Wright, A.L. and G.H. Snyder. 2009. Soil subsidence in the Everglades Agricultural Area. Soil and Water Science Department, Florida Cooperative Extension Service, Institute of Food and Agricultural Sciences, University of Florida. Publication #SL 311.

Wu, Y., F.H. Sklar and K. Rutchey. 1997. Analysis and simulations of fragmentation patterns in the Everglades. Ecol. App. 7: 268–276.

Mercury Mass Budget Estimates and Cycling in the Florida Everglades

Guangliang Liu,[1,a,*] *Yong Cai,*[1,b] *Georgio Tachiev*[2,c] *and*
Leonel Lagos[2,d]

Introduction

The accumulation of elevated levels of mercury (Hg) in fishes and wildlife is of concern for the Florida Everglades, a subtropical freshwater wetland ecosystem (Ware et al. 1990). Since the detection of elevated Hg concentrations in Everglades fish and wildlife, continuous efforts have been made to investigate the source (Guentzel et al. 2001), transport (Zhang and Lindberg 2000; Drexel et al. 2002; Liu et al. 2008b; Liu et al. 2008c; Liu et al. 2009), transformation (e.g., reduction/oxidation and in particular methylation/demethylation) (Gilmour et al. 1998; Hurley et al. 1998; Marvin-DiPasquale and Oremland 1998; Cleckner et al. 1999; Li et al. 2010), and bioaccumulation of Hg (Cleckner et al. 1998; Loftus 2000) in the Everglades. These studies have significantly advanced our understanding of Hg biogeochemical cycling in this wetland ecosystem.

---

[1] Department of Chemistry & Biochemistry and Southeast Environmental Research Center, Florida International University, Miami, FL 33199, USA.
[a] Email: liug@fiu.edu
[b] Email: cai@fiu.edu
[2] Applied Research Center, Florida International University, Miami, FL 33174, USA.
[c] Email: georgio.tachiev@fiu.edu
[d] Email: lagosl@fiu.edu
* Corresponding author

As a large ecosystem spreading over 7,000 km^2, the Everglades includes four management units, located from north to south Water Conservation Area 1 (WCA 1, also known as the Arthur R. Marshall Loxahatchee National Wildlife Refuge, LNWR), WCA 2, WCA 3, and the Everglades National Park (ENP). These four management units are interconnected by an extensive system of canals, levees, and water control structures. The water flows have not only interrupted slow overland sheetflow of water in the Everglades marsh, but also provided a conduit for pollutant (e.g., Hg) transport across management units (Scheidt and Kalla 2007). In such a large ecosystem with subareas connected by water flows, it is important to investigate the distribution and cycling of Hg at the whole ecosystem level, in addition to studying individual processes of Hg transport and transformation.

Poised to understand mercury cycling in the Everglades at a large scale, a few studies have surveyed the scope of Hg contamination in the Everglades (Stober et al. 2001; Axelrad et al. 2006; Scheidt and Kalla 2007; Cohen et al. 2009). Examples of these studies include the U.S. Environmental Protection Agency (EPA), Everglades Regional Environmental Monitoring and Assessment Program (R-EMAP) and the US Geological Survey (USGS) Aquatic Cycling of Mercury in the Everglades (ACME) project (Stober et al. 2001; Gilmour et al. 2006). These projects provided valuable data for exploring large scale mercury cycling in the Everglades, as they often collected and analyzed samples in a large area or even throughout the ecosystem.

As an important means of investigating the large scale mercury cycling in an ecosystem, constructing an ecosystem-scale Hg mass budget, including inputs, outputs, and mass storage of total Hg (THg) and methylmercury (MeHg), is particularly helpful for obtaining a complete picture of the magnitude and scope of Hg contamination in this system. This mass budget information will provide important implications for adaptive management of Hg contamination in the Everglades, and on the ongoing restoration of this ecosystem. For instance, comparisons of the Hg mass budgets among the four management units would help assess the influence of restoration activities (e.g., modified water deliveries) on Hg cycling in this system.

In this study, we constructed mass budget for THg and MeHg in each management unit of the Everglades, including input, output, and mass storage of Hg in each ecosystem component, by taking advantage of the probability-based sampling design and ecosystem-wide sampling of R-EMAP and combining R-EMAP with other datasets. The information obtained here will help evaluate the status of Hg contamination in the Everglades.

Methodology for Constructing Mercury Mass Budget

A mass inventory method was used to construct mercury mass budget in the Everglades. The mercury inventory was defined here to include two parts. The first part represented mercury inputs (e.g., from atmospheric wet and dry deposition and water inflow) and outputs (e.g., through water outflow and evasion into the atmosphere). The second part of the mercury inventory was the mercury storage in each ecosystem component. The ecosystem components considered here include surface water, soil (top 10 cm layer only as this layer of soil is more relevant to Hg biogeochemical cycling, in comparison to deeper soil), flocculent detrital material (floc, which is a layer of suspended organic materials on top of the Everglades soil and contains mostly detritus from plants and algal inputs from periphyton), periphyton, macrophyte, and mosquitofish. Corresponding to the four management units of the Everglades, WCAs 1, 2, 3, and ENP, the mercury mass budget was constructed separately for each management unit.

The calculations here were based primarily on the R-EMAP datasets, especially the R-EMAP III wet season dataset which was generated in the November 2005. R-EMAP adopted a probability-based, ecosystem-wide sampling design to collect samples throughout the Everglades (shown in Fig. 1 are sampling sites distributed throughout the entire Everglades system during the R-EMAP), which was particularly useful for calculating ecosystem-wide mass budget for mercury. The R-EMAP III sampling events occurred in May 2005 for the dry season and November for the wet season. Consistent with this R-EMAP sampling schedule, we defined here the 2005 dry season from mid November 2004 through mid May 2005 and the wet season from mid May 2005 through mid November 2005. During our calculations, the inputs and outputs were calculated to reflect the total amounts of mercury entering into or leaving out of the system during the whole 2005 wet season, while the storage was used to

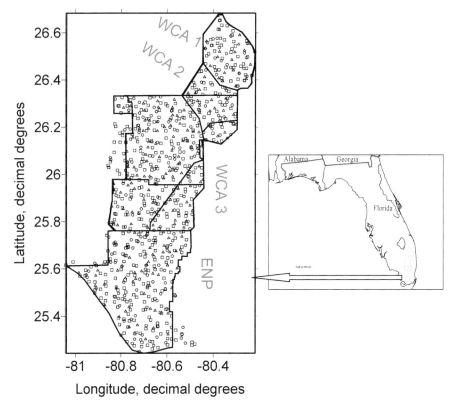

Figure 1. A map showing the distribution of sampling sites in the four management units of the Florida Everglades during the Everglades Regional Environmental Monitoring and Assessment Program (R-EMAP) (Δ) Phase I (1995-96), (○) Phase II (1999), and (□) Phase III (2005) sampling events. The four management units are Water Conservation Area 1 (WCA 1, also known as Arthur R. Marshall Loxahatchee National Wildlife Refuge, LNWR), WCA 2, WCA 3, and the Everglades National Park, ENP.

reflect a relative perspective of instantaneous mass among ecosystem components at the time of sampling. The mercury storage was calculated based on the data of mercury concentrations and masses/volumes of ecosystem components obtained in R-EMAP 2005 November sampling event.

Calculating mass budget to characterize ecosystem-wide Hg cycling requires the combination and comprehensive analysis of multiple datasets. In addition to R-EMAP, other datasets have been generated for Hg in the Everglades, originating from projects focusing on different aspects of Hg biogeochemical cycling in this ecosystem. These datasets include the USGS Aquatic Cycling of Mercury in the Everglades (ACME) project, the Mercury Deposition Network (MDN, http://nadp.sws.uiuc.edu/mdn/), and the DBHYDRO database maintained by South Florida Water Management District (SFWMD) (Mercury Deposition Network; SFWMD DBHYDRO accessed on 02/06/2010). These datasets were used as supplementary data sources during our calculations, if necessary data were not available in the R-EMAP database. Detailed information about sampling design, sampling protocols, analytical procedures, and original data for all the datasets used in the calculations can be found elsewhere (Stober et al. 2001; Gilmour et al. 2006; Scheidt and Kalla 2007; Liu et al. 2008a).

Calculation of Mass Storage of Mercury in the Ecosystem Components of the Everglades

THg mass storage

For the following ecosystem components, namely surface water, soil (top 10 cm layer only), floc, periphyton, and mosquitofish, total mercury mass stored in each ecosystem component in each management unit of the Everglades was calculated by utilizing the probability sampling design adopted in R-EMAP. One of the useful features in this probability sampling design is that the probability of a sample being included in the sampling event is known and can be expressed as inclusion probability density function. As a result, the area that each sample point represents is predetermined and the area is expressed as the reciprocal of the inclusion probability density function. According to the Horvitz-Thompson Theorem (Horvitz and Thompson 1952; Cordy 1993; Stober et al. 2001), the following equations can be used to calculate THg mass in each ecosystem component for each management unit of the Everglades.

$$M_{SW}^{THg} = \frac{\sum_{i=1}^{n} \dfrac{C_{SW\,i}^{THg} \times d_{SW\,i}}{\pi_i}}{\sum_{i=1}^{n} \left(\dfrac{1}{\pi_i}\right)} \times A \times 10^3 \tag{1}$$

$$M_{SD}^{THg} = \frac{\sum_{i=1}^{n} \dfrac{C_{SD\,i}^{THg} \times d_{SDi} \times BD_{SDi}}{\pi_i}}{\sum_{i=1}^{n} \left(\dfrac{1}{\pi_i}\right)} \times A \times 10^6 \tag{2}$$

$$M_{FC}^{THg} = \frac{\sum\limits_{i=1}^{n} \dfrac{C_{FC\ i}^{THg} \times d_{FCi} \times BD_{FCi}}{\pi_i}}{\sum\limits_{i=1}^{n} (\dfrac{1}{\pi_i})} \times A \times 10^6 \tag{3}$$

$$M_{PE}^{THg} = \frac{\sum\limits_{i=1}^{n} \dfrac{C_{PE\ i}^{THg} \times BM_{PEi}}{\pi_i}}{\sum\limits_{i=1}^{n} (\dfrac{1}{\pi_i})} \times A \tag{4}$$

$$M_{FS}^{THg} = \frac{\sum\limits_{i=1}^{n} \dfrac{C_{FS\ i}^{THg} \times W_{FSi} \times BM_{FSi}}{\pi_i}}{\sum\limits_{i=1}^{n} (\dfrac{1}{\pi_i})} \times A \tag{5}$$

where M_{SW}^{THg}, M_{SD}^{THg}, M_{FC}^{THg}, M_{PE}^{THg}, M_{FS}^{THg} are THg mass (ng) stored in surface water, soil (top 10 cm layer only), floc, periphyton, and mosquitofish, respectively. $C_{SW\ i}^{THg}$, $C_{SD\ i}^{THg}$, $C_{FC\ i}^{THg}$, $C_{PE\ i}^{THg}$, $C_{FS\ i}^{THg}$ are THg concentration (ng/L for water and ng/g for the other ecosystem components) in surface water, soil, floc, periphyton, and mosquitofish at station i, respectively. d_{SWi}, d_{SDi}, d_{FCi} are thickness (m) of surface water, soil (0.1 m for all stations as only the top 10 cm layer was considered in the calculations), and floc at station i. BD_{SDi} and BD_{FCi} are bulk density (g/ml) of soil and floc at station i. BM_{PEi} (g/m^2) and BM_{FSi} (fish/m^2) are areal biomass of periphyton and mosquitofish at station i. W_{FSi} (g) is average weight of mosquitofish analyzed for station i. π_i is inclusion probability density function of station i being included in the sampling design. A is (m^2) of a management unit. An average value was estimated for BM_{PEi} or BM_{FSi} for each management unit, since these parameters were not extensively determined for all sampling stations during the 2005 R-EMAP sampling. Table 1 listed the parameters used for calculating Hg mass storage and the determination of the values for these parameters are discussed in the Results and Discussion.

For macrophyte ecosystem component, neither the Hg data in macrophyte nor the biomass of macrophyte is available in the 2005 R-EMAP database. As a result, THg mass in macrophyte was estimated based on the average THg concentration in macrophyte and the biomass of macrophyte previously reported in the literature (Browder 1982; Stober et al. 2001; Fink 2003). As macrophyte has never been extensively sampled and analyzed for Hg throughout the Everglades, there is lacking information in terms of macrophyte THg data for different management units. In the calculations, we used the same average THg concentration in macrophyte for all of the four management units, not taking into account the spatial variations in macrophyte THg concentrations, which would result in a certain degree of uncertainty for the calculated THg mass in macrophyte. Table 1 lists the average macrophyte Hg

Table 1. Input parameters for calculating mass storage of THg and MeHg in different ecosystem components in the management units of the Everglades (WCA 1, WCA 2, WCA 3, and ENP) at the time of R-EMAP III wet season sampling.

Parameter	Definition	Value				Reference
		WCA 1	WCA 2	WCA 3	ENP	
A	Area (m²) of the region	5.72×10^8	5.44×10^8	2.44×10^9	3.37×10^9	(FDEP 2000)
$d_{SW i}$	Thickness (m) of surface water at station i	R-EMAP Phase III data	R-EMAP Phase III data	R-EMAP Phase III data	R-EMAP Phase III data	This study (Scheidt and Kalla 2007)
$d_{SD i}$	Thickness (m) of soil at station i					
$d_{FC i}$	Thickness (m) of floc at station i					
$BD_{SD i}$	Bulk density (g/ml) of soil at station i					
$BD_{FC i}$	Bulk density (g/ml) of floc at station i					
$BM_{PE i}$	Areal biomass (g/m²) of periphyton at station i	5.0	82	45	739	(Gaiser et al. 2006; Gaiser 2008)
BM_{PK}	Areal biomass (g/m²) of macrophyte	2380	1507	2656	1438	(Browder 1982; Childers et al. 2003; Fink 2003)
$BM_{FS i}$	Areal biomass (fish/m²) of mosquitofish at station i	4.80	17.4	26.6	14.5	(Gaff et al. 2000; Gaff et al. 2004)
$W_{FS i}$	Average weight (g) of mosquitofish analyzed for station i	R-EMAP Phase III data	R-EMAP Phase III data	R-EMAP Phase III data	R-EMAP Phase III data	This study (Scheidt and Kalla 2007)
π_i	Inclusion probability of station i					
$C^{THg}_{SW i}$	THg concentration (ng/L) in water at station i					
$C^{THg}_{SD i}$	THg concentration (ng/g) in soil at station i					
$C^{THg}_{FC i}$	THg concentration (ng/g) in floc at station i					
$C^{THg}_{PE i}$	THg concentration (ng/g) in periphyton at station i					
$C^{THg}_{FS i}$	THg concentration (ng/g) in mosquitofish at station i					
$C^{MeHg}_{SW i}$	MeHg concentration (ng/L) in water at station i					
$C^{MeHg}_{SD i}$	MeHg concentration (ng/g) in soil at station i					
$C^{MeHg}_{FC i}$	MeHg concentration (ng/g) in floc at station i					
$C^{MeHg}_{PE i}$	MeHg concentration (ng/g) in periphyton at station i					
C^{THg}_{PK}	Average THg concentration (ng/g) in macrophyte	7.30 (ACME and REMAP data)	7.30 (ACME and REMAP data)	7.30 (ACME and REMAP data)	7.30 (ACME and REMAP data)	(Stober et al. 2001; Fink 2003)
C^{MeHg}_{PK}	Average MeHg concentration (ng/g) in macrophyte	0.51 (ACME data)	0.51 (ACME data)	0.51 (ACME data)	0.51 (ACME data)	(Fink 2003)

concentrations for the Everglades and macrophyte biomass for each management unit and the literature or data sources used to estimate these parameters.

MeHg mass storage

The calculations of MeHg mass storage in surface water, soil (top 10 cm layer only), floc, and periphyton were performed in a similar way with THg by using Equations 1–4. For Equations 1–4, substitute $C_{SW\ i}^{MeHg}$, $C_{SD\ i}^{MeHg}$, $C_{FC\ i}^{MeHg}$, and $C_{PE\ i}^{MeHg}$ for $C_{SW\ i}^{THg}$, $C_{SD\ i}^{THg}$, $C_{FC\ i}^{THg}$, and $C_{PE\ i}^{THg}$, respectively to calculate MeHg mass storage, where $C_{SW\ i}^{MeHg}$, $C_{SD\ i}^{MeHg}$, $C_{FC\ i}^{MeHg}$, and $C_{PE\ i}^{MeHg}$ are MeHg concentrations (ng/L for water and ng/g for the other ecosystem components) in surface water, soil, floc, and periphyton at station i, respectively. For macrophyte ecosystem component, MeHg data are not available in the R-EMAP III database and limited data are reported in the literature. We estimated that in macrophyte MeHg accounts for 7% of THg based on the results of USGS analyses for cattail and sawgrass samples from the WCAs (Fink 2003). For mosquitofish, calculations of MeHg mass storage were not performed and we assume that THg mass storage could be used as MeHg data to evaluate MeHg mass storage in mosquitofish, as previous studies suggest that over 95% of Hg present in mosquitofish is in the form of MeHg (Stober et al. 2001).

Quantification of Mercury Input and Output for Each Management Unit of the Everglades

THg input and output

THg input to a management unit of the Everglades was estimated for three processes, namely wet deposition, dry deposition, and water inflow. Mass of THg input through wet deposition was estimated based on the monitoring data at MDN site FL11 (Everglades National Park Research Center), while dry deposition of Hg was based on the estimation in the literature (Keeler et al. 2001). For each management unit, the THg input through water inflows was estimated by multiplying the cumulative flows into that management during the 2005 wet season by the average THg concentrations in the flows. The water flow and mercury concentration data were obtained from the SFWMD DBHYDRO database and related reports. THg output from a management unit was estimated for two processes, evasion and water outflow. The mass of THg evasion was estimated by assuming an evasion rate of 2 ng/m^2/h (same for all management units due to lacking spatial data) and 10 h per day during which evasion occurs (Krabbenhoft et al. 1998; Lindberg and Zhang 2000; Zhang and Lindberg 2000; Lindberg and Meyers 2001; Lindberg et al. 2002; Marsik et al. 2005). The THg output through water outflows was estimated by multiplying the cumulative flows out of a management unit during the 2005 wet season by the average THg concentrations in the flows. Table 2 illustrates the parameters used for calculating inputs and outputs of THg and MeHg in the four management units of the Everglades during the 2005 wet season. The values of parameters and the procedures used for the calculations of deposition, inputs via water inflows, outputs via water outflows, and evasion of THg can be found in the table and Results and Discussion.

Table 2. Input parameters for calculating inputs and outputs of THg and MeHg in different ecosystem components in the management units of the Everglades (WCA 1, WCA 2, WCA 3, and ENP) during the 2005 wet season.

Parameter	Definition	Value				Reference
		WCA 1	WCA 2	WCA 3	ENP	
M_{BD}^{THg}	THg deposition (ng) including wet and dry deposition[a]	1.14×10^{13}	1.09×10^{13}	4.73×10^{13}	6.72×10^{13}	MDN (Mercury Deposition Network. accessed on 02/06/2010); (Keeler et al. 2001)
M_{EV}^{THg}	THg evasion (ng)[b]	2.06×10^{12}	1.96×10^{12}	8.53×10^{12}	1.21×10^{13}	This study
V_{IF}	Volume (L) of water inflow[c]	2.28×10^{11}	9.44×10^{11}	1.65×10^{12}	1.68×10^{12}	DBHYDRO (SFWMD DBHYDRO. accessed on 02/06/2010); (Abtew et al. 2006; Alleman et al. 2006; Abtew et al. 2007)
V_{OF}	Volume (L) of water outflow[c]	1.96×10^{11}	9.09×10^{11}	1.58×10^{12}	2.40×10^{11}	
C_{IF}^{THg}	Average THg concentration (ng/L) in inflow water[d]	1.20	0.96	1.80	1.60	
C_{OF}^{THg}	Average THg concentration (ng/L) in outflow water[d]	0.96	1.00	1.60	1.40	
C_{IF}^{MeHg}	Average MeHg concentration (ng/L) in inflow water[d]	0.12	0.28	0.22	0.12	
C_{OF}^{MeHg}	Average MeHg concentration (ng/L) in outflow water[d]	0.28	0.10	0.12	0.10	

MeHg input and output

For calculations of MeHg input, atmospheric deposition was not considered a direct source of MeHg input to the Everglades. There are two reasons for not considering atmospheric deposition as a major source of MeHg input. First, although measurements of MeHg in precipitation are sparse, previous limited studies have shown that the concentration of MeHg in rainwater is typically low, with inorganic Hg being the dominant form (Holz et al. 1999; Lawson and Mason 2001). In South Florida, MeHg concentrations in rainfall are generally considered environmentally insignificant (Guentzel et al. 1995; Guentzel 1997; Fink et al. 1998). Second, previous studies have shown that the MeHg in the Everglades is primarily produced *in situ*, in particular in soil, floc, and periphyton (Gilmour et al. 1998; Cleckner et al. 1999; Mauro et al. 2002). Although Hg methylation has been observed in some waters (Mason et al. 1993; Ullrich et al. 2001), MeHg production in the Everglades water column is negligible (Mauro et al. 2002). Therefore, *in situ* production in soil, floc, and periphyton was considered the primary source of MeHg load in each management unit. The inside-system production of MeHg has been reflected in the calculated mass storage of MeHg in the respective ecosystem component, and thus was not considered as input or output pathway. The MeHg input to each management unit was calculated for water inflow to that management unit, by multiplying the volumes of water inflows by the MeHg concentrations in the inflows. For calculating MeHg output, loss of MeHg through evasion was not considered in this study, since it is considered to be a minor pathway of MeHg input and output (Fink 2003). The MeHg output from a management unit was calculated based on the volumes of water outflows and the MeHg concentrations in the outflows.

Uncertainty Analysis for Hg Mass Budget Calculations

Uncertainty analysis was performed for the calculations of mass storage of Hg in the ecosystem components, but not for Hg input and output calculations. The main reason for this treatment is that we have statistically sufficient data to do uncertainty analysis for mass storage calculations, as the R-EMAP III database used here were generated through extensive sampling throughout the entire Everglades. In contrary, the calculations of the inputs (e.g., atmospheric deposition and water inflows) and outputs (e.g., outflows and evasion) of THg and MeHg were conducted mainly based on the values reported in the literature that are limited in terms of the number of data points and the spatial coverage of the data. For example, concentrations of THg and MeHg in macrophyte and areal biomass of macrophyte were taken from the previous reports which only dealt with a localized area in the Everglades. For these data, it is difficult to conduct statistically sound uncertainty analysis due to insufficient data. Our arbitrary estimate is that the uncertainties associated with the calculations using these data should be comparable to (or higher than) the magnitude of the uncertainties estimated below for mass storage calculations.

For uncertainty analysis associated with calculations of Hg mass storage, we followed the principles described in the U.S. Guide to the Expression of Uncertainty in Measurement (GUM) to estimate the uncertainty for calculated THg and MeHg mass storage (NCSL 1997). We used the relative standard error (*RSE*) to estimate the relative uncertainty (U_r) based on the GUM and the Horvitz-Thompson Theorem (Horvitz and Thompson 1952; Cordy 1993; Stober et al. 2001), which provides general estimating formula for the accompanying variance expressions as below, in addition to the means.

$$U_r = 100 \times \frac{\sqrt{n} \times \sqrt{\dfrac{\sum\limits_{i=1}^{n}(\dfrac{Z_i - \hat{Z}}{\pi_i})^2}{n-1}}}{\sum\limits_{i=1}^{n}\dfrac{Z_i}{\pi_i}} \tag{6}$$

where Z_i, expressed as a function of measured parameters that are used to calculate Hg storage in the ecosystem components,

$$
\begin{aligned}
&= C_{SW\,i}^{Hg} \times d_{SW\,i} && \text{for water}\\
&= C_{SD\,i}^{Hg} \times d_{SD\,i} \times BD_{SD\,i} && \text{for soil}\\
&= C_{FC\,i}^{Hg} \times d_{FC\,i} \times BD_{FC\,i} && \text{for floc}\\
&= C_{PE\,i}^{Hg} \times BM_{PE\,i} && \text{for periphyton}\\
&= C_{FS\,i}^{Hg} \times W_{FS\,i} \times BM_{FS\,i} && \text{for fish}
\end{aligned}
$$

$\hat{Z}$ is the mean of Z_i and calculated as follows

$$\hat{Z} = \frac{\sum\limits_{i=1}^{n}\dfrac{Z_i}{\pi_i}}{\sum\limits_{i=1}^{n}(\dfrac{1}{\pi_i})} \tag{7}$$

Mass Inventories of Mercury Cycling in the Everglades: Results of Mass Budget and Discussion

Tables 1 and 2 illustrate the parameters used for calculating Hg mass storage and Hg transport (input and output) in the Everglades, respectively. For the parameters the values of which were determined based on the R-EMAP database, usually the median values determined in the R-EMAP Phase III study were used for calculations. This is because, as a massive dataset spatially covering the entire Everglades, the R-EMAP III database showed the heterogeneity of Hg distribution throughout the Everglades, as evidenced by the wide range of Hg concentrations and the high values of skewness of normal distribution. Table 3 listed the descriptive statistics on Hg concentrations in different ecosystem components, including minimum, maximum, quartiles, median, mean, range of mean at 95% confidence level (lower confidence level, LCL, and upper confidence level, UCL), and skewness of normal distribution, which shows the heterogeneity of Hg distribution throughout the Everglades.

For the parameters the values of which could not determined based on the R-EMAP database, other data sources are used. In this case, these parameters are often reported for a localized area in the Everglades and not representative of the ecosystem-scale

Table 3. Descriptive characteristics of Hg concentrations in different ecosystem components of the Everglades, determined in the R-EMAP 2005 wet season sampling, showing the wide ranges and skewness of Hg concentrations.

Matrix	Hg species	Minimum	1st Quartile	Mean	Median	3rd Quartile	Maximum	Std Deviation	LCL Mean	UCL Mean	Skewness
Water	THg	1.1	1.7	2.6	2.2	3	8.3	1.3	2.3	2.8	2
	MeHg	0.038	0.12	0.29	0.21	0.34	2.8	0.32	0.23	0.34	4.8
Soil	THg	17	97	140	140	190	350	72	130	160	0.32
	MeHg	0.04	0.2	0.87	0.49	0.95	12	1.4	0.61	1.1	5.1
Floc	THg	34	98	130	130	170	300	54	120	150	0.58
	MeHg	0.24	1.6	3.6	3	5.6	13	2.8	3.1	4.2	1.2
	MeHg	0.04	0.81	2.1	1.7	2.7	9.4	1.8	1.7	2.5	1.8
Periphyton	THg	3.5	11	23	21	32	92	16	20	27	1.6
	MeHg	0.04	0.64	2	1.5	2.6	9.4	1.8	1.6	2.3	1.9
Mosquitofish	THg	4.8	54	100	87	130	320	69	90	120	1.1

status. For instance, macrophyte THg concentration was assumed to be 7.3 ng/g for all the four management units for the 2005 wet season, based on previous R-EMAP and USGS ACME results (Stober et al. 2001; Fink 2003). This universal value apparently cannot reflect the distribution of macrophyte Hg (as spatial variations are expected for macrophyte Hg in the Everglades), but was used in the calculations due to limited data reported. Similarly, periphyton areal biomass was arbitrarily set to 5.0, 82, 45, and 739 g/m² for the 2005 wet season for WCA 1, 2, 3, and ENP, respectively, based on the previously reported data (McCormick et al. 1998; Stober et al. 2001; Gaiser et al. 2006; Gaiser 2008). Macrophyte areal biomass was assigned to 2380, 1507, 2656, and 1438 g/m² in the 2005 wet season for WCA 1, 2, 3, and ENP, respectively (Browder 1982; Childers et al. 2003; Fink 2003). The areal biomass of mosquitofish (density) was estimated by using the ALFISH model outputs to be 4.80, 17.4, 26.6, and 14.5 fish/m² for WCA 1, 2, 3, and ENP, respectively (Gaff et al. 2000; Stober et al. 2001; Gaff et al. 2004). As the ALFISH is a model developed to assess the spatial pattern of fish densities through the greater Everglades freshwater marshes, the outputs of the ALFISH model should be able to capture the spatial variability in fish density and thus reflect the variations in mosquitofish biomass among the four management units. The median weight of mosquitofish determined in the R-EMAP was then used to calculate the areal biomass of mosquitofish at each sampling station. For calculation of Hg mass storage in soil, the soil thickness was set to 10 cm, in consideration that only the surface layer of soil is likely involved in Hg distribution during a period of a wet season (about 6 months). For all calculation, the areas of WCA 1, 2, 3, and ENP were set to 572, 544, 2370, and 3368 km², respectively (FDEP 2000).

Mass Budget of THg

Our calculations of THg mass storage in the ecosystem components (Fig. 2) indicate that, probably as a result of long-term accumulation of THg inputs, significant amount of THg has been stored in the system, in particular in soil and floc. As the largest Hg sink, the Everglades soil stored from 784 kg of THg for WCA 1 to 7381 kg of THg for ENP, greatly exceeding THg mass in any other component and any THg input or output. The THg masses in floc can be up to 10–15% of soil THg amounts, suggesting floc could be an important ecosystem component for Hg storage. Compared to soil and floc, the mass storage of THg in periphyton, macrophyte, and fish was small, usually amounting to 1% or less for each ecosystem component. The mass storage of THg in the ecosystem components we estimated agrees well with previous studies (Stober et al. 2001; Cohen et al. 2009), except for fish for which greater THg mass was obtained in this study. This is probably because the fish biomass we estimated here was higher. As we only took the top 10 cm layer of soil into account when calculating Hg mass storage in soil, the overall THg mass in the Everglades soil will be much larger than the value calculated here if deeper soil is considered (although Hg in deeper soil might be of less relevance in terms of Hg biogeochemical cycling).

Our calculations on the inputs and outputs of Hg suggest that atmospheric deposition is the predominant source of THg inputs to the Everglades, accounting for more than 90% of Hg entering the Everglades. Compared to atmospheric deposition, THg input through water inflows accounts for only about 2–8% of total THg inputs.

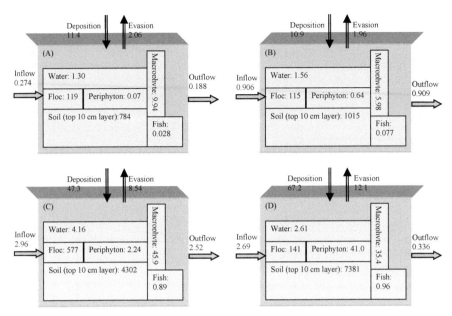

Figure 2. Mass budget of THg in (A) WCA 1, (B) WCA 2, (C) WCA 3, and (D) ENP of the Everglades. THg instantaneous mass (kg) stored in each ecosystem component at the time of sampling (November 2005) is indicated by the number inside the box. Numbers accompanying arrows depict THg inputs and outputs (kg) during the 2005 wet season (May through November). Mosquitofish was abbreviated to fish in the figure.

These results agree with previous studies which indicate that atmospheric deposition is the primary source of Hg input to the Everglades (Arfstrom et al. 2000; Stober et al. 2001; Atkeson and Axelrad 2004; Axelrad et al. 2006). Among the four management units, even for the WCAs, which are adjacent to the Everglades Agricultural Area (EAA) and receive runoff from EAA, THg inputs through water inflows are still relatively of minor importance compared to atmospheric deposition. For instance, THg inputs through water inflows account for 7.6% and 5.9% of THg input for WCA 2 and for WCA 3, respectively.

The calculation results on THg mass budget indicate that THg output from the Everglades ecosystem is less than THg input, suggesting that the Everglades is a sink for Hg and Hg will be accumulated in the system with time. The two major pathways of Hg removal from the Everglades system, through water outflows and evasion, account for a relative small fraction of Hg inputs. For instance, THg output through water outflows accounts for 1.6, 7.7, 5.0, and 0.5% of THg inputs for WCA 1, WCA 2, WCA 3, and ENP, respectively. The amount of Hg output through evasion is much larger than through water outflows, being about 17.5% of THg inputs for all four management units, but still a small fraction of THg inputs. These results indicate that the Everglades ecosystem is a sink for THg, where most of the THg inputs (either from atmospheric deposition or from water inflows) will be accumulated over time. If we compare THg masses stored in the Everglades (all ecosystem components combined, 914 kg for WCA 1, 1138 kg for WCA 2, 4931 kg for WCA 3, and 7602 kg for ENP) to the amount of annual THg input (including deposition and water inflow) to the respective management unit, it is clear that the THg input to the system is significantly less, amounting only to 1–2% of the THg mass storage. This disparity suggests the accumulation of Hg in this ecosystem and underscores the importance of understanding the internal cycling of Hg that has been stored in the system when investigating Hg contamination in the Everglades.

As the Everglades is a large ecosystem with many water structures (e.g., canals and levees) distributed throughout the system, we calculated Hg input to and output of each management unit through water flow, aiming to examine the role of water flow in transferring Hg from one management unit to another. We did so by calculating 1) the total volumes of water inflows and outflows entering and leaving each management unit based on the monthly inflows and outflows of major water structures and 2) the average Hg concentrations in water inflows and outflows. By using the monthly inflows and outflows recorded in the SFWMD DBHYDRO database and reported in the South Florida Environmental Report (2006 and 2007) (Abtew et al. 2006; Abtew et al. 2007), we summed up the inflows and outflows occurred during the period from May 2005 to November 2005 to calculate the flows for the 2005 wet season. By averaging Hg concentrations monitored for each structure during the period of the 2005 wet season, which were also recorded in the DBHYDRO database, we calculated the average Hg concentrations in water inflows and outflows. The outflow for the ENP was estimated based on the results of water flows to Florida Bay and the Keys from the Everglades. Flows from Taylor River Slough, C111 structure, Shark River Slough, and three major creeks that flow into the bay, Trout Creek and Taylor Creek (both flowing into the eastern bay) and McCormick Creek (flowing into the central bay), were used for calculation (Alleman et al. 2006).

The input and output calculation results for each management unit suggest that THg transport between different management units through water flow indeed occurs, but the influence of water flow Hg transport among management units on the overall Hg cycling is not clearly defined. This is because that the amount of Hg across-management unit transport is generally rather small, but could be in a considerable amount depending on the climate and hydrogeology conditions. For instance, during the 2005 wet season about 3 kg of THg entered into WCA 3 and ENP through water inflows. This amount, however, could increase in an extremely wet year with substantially increased water flows entering these management units, and in this case the across-unit Hg transport through water flow may become more important in determining the overall Hg cycling. During the calculations, it was also found that the mercury amount in the water outflows of WCA1 and WCA2 did not match the mercury amount in the water inflows to WCA2 and WCA3, respectively. This is probably because the water inputs to WCA 2 are not coming solely from WCA 1, and water to WCA 3 is not only from WCA 2. An overall pattern for across management unit Hg transport is the increasing THg amounts in water inflows and outflows from north to south, indicating that water flows from one water management unit to another play a role in the Hg cycling.

It can be seen from the calculations that atmospheric deposition is the main source of Hg input to the Everglades. It should be pointed out that the dry deposition were estimated based on the modeling results in the literature due to the lacking data of realistic measurements, which could bear a high degree of uncertainty (Keeler et al. 2001). Even for the wet deposition, the data should be viewed with caution, although the data were estimated from the monitoring data from the Mercury Deposition Network

(MDN, http://nadp.sws.uiuc.edu/mdn/). We used the MDN site FL11, which is located in the Everglades National Park Research Center (Mercury Deposition Network. accessed on 02/06/2010), to estimate wet Hg deposition. There are other MDN stations, e.g., FL34 and FL04, in a north-south transect throughout the four management units in the Everglades, and Hg wet deposition varies among these stations due to differences in rainfall volume and Hg concentration in rainfall. Considering that there are only three stations in the large area of the entire Everglades system, it would be difficult to accurately estimate wet Hg deposition for each management unit, even if we used the data from these three sites. Therefore, we used only FL11 for Hg wet deposition input, since this site has the longest period of record monitoring mercury in wet deposition. Additional considerations include the extreme weather events during the sampling period, which sometimes could affect the measurement of Hg deposition. For instance, during the R-EMAP 2005 wet season, the landfall of several hurricanes, in particular Katrina in August and Rita in September, significantly affected the functioning of Hg monitoring at station FL11. It was observed that, during these severe storm events, the collectors were non-functioning or recorded significant precipitation with little THg (Rumbold et al. 2007). Therefore, the monitoring results of Hg wet deposition in 2005 downloaded from the MDN network should be given special attention when they are used for estimation of wet Hg deposition.

Mass Budget of MeHg

The mass budget calculations of MeHg suggest that soil and floc are the major ecosystem components where MeHg is stored (Fig. 3), which is the similar pattern for THg. Different than THg, the relative importance of floc in MeHg cycling became increasingly greater than in THg cycling. Except for ENP, a significant amount of MeHg is stored in floc, with 4.53, 2.33, and 9.65 kg calculated for WCAs 1, 2, and 3, respectively. These values amount to about 50% of the MeHg mass storage in soil for each corresponding management unit, indicating the importance of floc in MeHg cycling. For ENP, the situation is quite different. Although the absolute amount of MeHg storage in floc in ENP is of a comparable order of magnitude in comparison to the MeHg amounts in WCAs, the fraction of floc MeHg against total MeHg storage (or MeHg storage in soil as soil is the largest sink) is much lower for ENP than for WCAs. The calculations indicate that 3.91 kg of MeHg is entrapped in the floc layer in ENP, well comparable to 2.33–9.65 kg for WCAs, but this 3.91 kg of floc MeHg accounts for only less than 10% of MeHg mass in soil in ENP (significantly lower than 50% for WCAs). One of possible reasons for this distinction between ENP and WCAs could be the difference in floc thickness and therefore the total mass (or volume) of floc component. It was observed that the median floc thickness in ENP was 0.83 cm for the 2005 wet season, which was significantly thinner than in the WCAs where the median floc thickness ranged from 2.83 to 4.17 cm during the same sampling period. Compared to soil and floc, periphyton generally stored a minor amount of MeHg (0.005 to 0.15 kg or less than 2% of soil MeHg for WCAs), except for ENP where 4.32 kg (or more than 10% of soil MeHg) was stored. Macrophyte stored about 5% of total MeHg in each management unit. The mass storage of MeHg estimated here was comparable with a previous study (Stober et al. 2001), although the floc MeHg

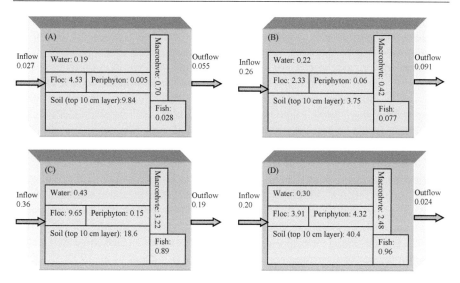

Figure 3. Mass budget of MeHg in (A) WCA 1, (B) WCA 2, (C) WCA 3, and (D) ENP of the Everglades. MeHg instantaneous mass (kg) stored in each ecosystem component at the time of sampling (November 2005) is indicated by the number inside the box. Numbers accompanying arrows depict MeHg inputs and outputs (kg) during the 2005 wet season (May through November). Mosquitofish was abbreviated to fish in the figure.

mass calculated here appeared higher than before. The comparisons between different studies across different years should be performed with cautions, as there is a great deal of variability in MeHg mass estimation for different years (with varying water depth, floc thickness, periphyton biomass, and macrophyte biomass).

Similar to THg, we calculated the amounts of MeHg input to and output of each management unit based on the monthly inflows and outflows of major structures entering and leaving each unit and the MeHg concentrations in these water inflows and outflows. The results suggest that both inputs through water inflow and outputs through water outflow of MeHg are fairly small (usually less than 1%), compared to MeHg masses that are being stored in the ecosystem. However, there is one exception to this pattern. The MeHg inputs to WCA 2 through water inflows amounted to about 4% of total MeHg storage in WCA 2, indicating that MeHg input through water inflows could not be neglected in some cases. This increased MeHg input through water inflows could be related to the fact that WCA 2 receives runoff from EAA which could carry MeHg from EAA to WCA 2. Nonetheless, *in situ* production, presumably in soil, floc, and periphyton, is generally the predominant source of MeHg that is cycled in the Everglades. After being produced in soil, floc, and periphyton, MeHg can be transported into the water column, adsorbed and deposited in soil and floc, uptaken by periphyton and macrophyte, and transferred into the food web. The cycling of MeHg (from production to bioaccumulation) inside the ecosystem determines the magnitude of MeHg bioaccumulation in fish.

For MeHg, we did not consider evasion a pathway of MeHg output and only considered water outflow as MeHg output pathway. The reason is that MeHg is not expected to be lost from water directly through evasion, and MeHg loss from water likely involves degradation of MeHg into inorganic Hg and subsequent evasion of inorganic Hg (presumably gaseous Hg). Even for estimation of THg evasion, the uncertainty could be large, as a wide range of evasion rates were reported in the literature. In our calculations, we estimated the mass of THg evasion by assuming an evasion rate of 2 $ng/m^2/h$, 10 h per day during which evasion occurs, and 180 days for the 2005 wet season (Krabbenhoft et al. 1998; Lindberg and Zhang 2000; Zhang and Lindberg 2000; Lindberg and Meyers 2001; Lindberg et al. 2002; Marsik et al. 2005). Based on these assumptions, the calculated Hg evasion rate for the 2005 wet season was 3.6 $\mu g/m^2$/season. As the processes of Hg evasion, which includes production of dissolved gaseous Hg, Hg^0 oxidation, and Hg^0 emission, are affected by a variety of factors such as temperature, solar radiation, and water depth (Gill et al. 1995; Krabbenhoft et al. 1998; Lindberg and Zhang 2000; Zhang and Lindberg 2000), the assumed evasion rate of 2 $ng/m^2/h$ here should be treated as a reference value. The assumed evasion rate was indeed backed by previous studies. A series of studies have directly measured gaseous Hg fluxes over open water surface and estimated the evasion rates of Hg from water in the Everglades (Lindberg et al. 1999; Lindberg and Zhang 2000; Zhang and Lindberg 2000; Lindberg et al. 2005). The results of these studies suggest that the Hg evasion rates are around 1–3 $ng/m^2/h$ (daytime), with Hg evasion at night being negligible. For instance, a direct evasion flux of Hg from Everglades water was measured to be 2.7 $ng/m^2/h$ (Lindberg and Zhang 2000). An annual evasion rate of Hg from the Everglades of 2.2 $\mu g/m^2$/year was estimated based on the measured dissolved gaseous mercury concentrations (Krabbenhoft et al. 1998). It should be noted that Hg transpiration through macrophyte was not included in the estimation of Hg evasion. This is because it is unknown about the types and coverage of macrophyte in each management unit of the Everglades and the capability of each type of macrophyte transferring Hg and therefore it is not feasible to estimate Hg transpiration through macrophyte, as Hg transpiration involves uptake of Hg by macrophyte roots and transport of Hg from macrophyte to the atmosphere (Zhang and Lindberg 2000; Lindberg and Meyers 2001; Lindberg et al. 2002; Lindberg et al. 2005; Marsik et al. 2005).

As Hg bioaccumulation is of particular concern for its ecological and health risks, the Hg mass storage in fish is an important index for evaluating Hg bioaccumulation and the impact and risk of Hg contamination in this important ecosystem. As the predominant form of Hg in mosquitofish is MeHg, accounting for more than 95% of THg, we did not measure MeHg concentration in mosquitofish and used THg as a surrogate for mosquitofish MeHg instead. Our results suggest that the mass storage of THg in mosquitofish accounts for only a small fraction of THg in the system. It should be pointed out that the Hg storage in mosquitofish estimated here can only be viewed as the amount of Hg stored in mosquitofish (this specific fish species), rather than Hg storage in fish in general, as we only estimated the biomass of mosquitofish and used the mosquitofish biomass and Hg concentrations in mosquitofish we determined to calculate Hg storage. It is extremely difficult, if not impossible, to obtain sufficient data on the species, density, and biomass of all fish and wildlife species (particularly the larger predatory species) and the ecosystem-wide concentration profile of Hg in these species. As a consequence, it is nearly impossible to estimate the overall Hg mass storage in all fish and wildlife species. Therefore our purpose of estimating Hg mass storage in mosquitofish was to select mosquitofish as an example to demonstrate how to calculate Hg storage in higher-trophic levels of biological components (e.g., fish and wildlife). We used mosquitofish as a representative species, because we had data on ecosystem-wide mosquitofish weights and Hg concentrations in mosquitofish, which made it possible to estimate mosquitofish Hg storage. Nonetheless, our calculations here indicated that Hg mass storage in mosquitofish is significantly lower by several orders of magnitude, compared to that in soil and floc. Even if all fish and wildlife species are taken into account, the overall Hg mass storage in all higher-trophic levels of fish and wildlife, which could be possibly several-fold higher than the mosquitofish Hg storage estimated here, may still be significantly less that the Hg mass storage in the other ecosystem components (e.g., soil and floc).

Uncertainty of Hg Mass Storage Calculations

Table 4 illustrates the results of uncertainty analysis associated with the calculations of THg and MeHg mass storage in the ecosystem components. The uncertainties associated with the calculated mass storage of THg and MeHg are usually less than 30%. But for floc, the uncertainties associated with THg and MeHg mass storage can range from 30 to about 60%, probably due to the large variability in floc thickness throughout the Everglades system. Again, uncertainty analysis was not performed for the calculations of Hg input and output due to insufficient data points and limited spatial coverage of the data extracted from the literature.

Table 4. Uncertainty (relative standard error, %) associated with mass storage calculations of THg and MeHg in the ecosystem components in WCA 1, WCA 2, WCA 3, and ENP of the Everglades. The calculations are based on the R-EMAP 2005 wet season data.

	Uncertainty for THg mass (%)				Uncertainty for MeHg mass (%)			
	WCA 1	WCA 2	WCA 3	ENP	WCA 1	WCA 2	WCA 3	ENP
Water	17	17	7.3	11	33	19	12	14
Soil	10	15	4.1	6.7	37	29	12	18
Floc	42	37	30	31	57	57	30	30
Periphyton	25	25	10	14	28	28	11	19
Mosquitofish	17	25	11	11				

Concluding Remarks

Constructing an ecosystem-scale Hg mass budget, including inputs, outputs, and mass storage, is an important means of investigating the large scale mercury cycling in an ecosystem. By taking advantage of the probability-based sampling design and ecosystem-wide sampling of R-EMAP and combining R-EMAP with other datasets, we constructed mass budget for THg and MeHg in each management unit of the Everglades. The calculation results of Hg mass storage indicate that the Everglades is a sink for Hg, where most of Hg input to the system, primarily from atmospheric deposition, will be accumulated, in particular in soil and floc. The calculation results of Hg input and output suggest that Hg transport across different management units of the

Everglades through water flow indeed occurs, but the influence of Hg transport through water flow on the overall Hg cycling is dependent on the climate and hydrogeology conditions. This mass budget information will provide important implications for adaptive management of Hg contamination in the Everglades, and on the ongoing restoration of this ecosystem.

Acknowledgements

This work was sponsored by R-EMAP, EPA's South Florida Geographic Initiative, the Monitoring and Assessment Plan of the Army Corps of Engineers Comprehensive Everglades Restoration Plan, the Critical Ecosystem Studies Initiative of Everglades National Park, U.S. Department of the Interior, the Mercury Science Program of the Florida Department of Environmental Protection, and the US Department of Energy.

References

Abtew, W., R.S. Huebner and V. Ciuca. 2006. Appendix 5-3: Water Year 2005 Monthly Inflows and Outflows, 2006 South Florida Environmental Report. South Florida Water Management District and Florida Department of Environmental Protection, West Palm Beach, FL.

Abtew, W., R.S. Huebner, C. Pathak and V. Ciuca. 2007. Appendix 2-8: Water Year 2006 Monthly Inflows and Outflows, 2007 South Florida Environmental Report. South Florida Water Management District and Florida Department of Environmental Protection, West Palm Beach, FL.

Alleman, R., T. Barnes, R. Bennett, R. Chamberlain, T. Coley, T. Conboy, P. Doering, D. Drum, A. Goldstein, M. Gostel, M. Hedgepeth, G. Hu, M. Hunt, S. Kelly, C. Madden, A. McDonald, R. Robbins, D. Rudnick, E. Skornick, T. Stone, P. Walker and Y. Wan. 2006. Management and Restoration of Coastal Ecosystems, 2006 South Florida Environmental Report. South Florida Water Management District and Florida Department of Environmental Protection, West Palm Beach, FL.

Arfstrom, C., A.W. Macfarlane and R.D. Jones. 2000. Distributions of mercury and phosphorous in Everglades soils from Water Conservation Area 3A, Florida, U.S.A. Water Air Soil Pollut. 121: 133–159.

Atkeson, T. and D. Axelrad. 2004. Mercury Monitoring, Research and Environmental Assessment, 2004 Everglades Consolidated Report. South Florida Water Management District and Florida Department of Environmental Protection, West Palm Beach, FL.

Axelrad, D.M., T.D. Atkeson, C.D. Pollman and T. Lange. 2006. Mercury Monitoring, Research and Environmental Assessment in South Florida, 2006 South Florida Environmental Report. South Florida Water Management District and Florida Department of Environmental Protection, West Palm Beach, FL.

Browder, J.A. 1982. Biomass and primary production of microphytes and macrophytes in periphyton habitats of the southern Everglades. T-662. South Florida Research Center, Homestead, FL, p. 49.

Childers, D.L., R.F. Doren, R. Jones, G.B. Noe, M. Rugge and L.J. Scinto. 2003. Decadal change in vegetation and soil phosphorus pattern across the Everglades landscape. J. Environ. Qual. 32: 344–362.

Cleckner, L., P. Garrison, J. Hurley, M. Olson and D. Krabbenhoft. 1998. Trophic transfer of methyl mercury in the northern Florida Everglades. Biogeochemistry 40: 347–361.

Cleckner, L., C. Gilmour, J. Hurley and D. Krabbenhoft. 1999. Mercury methylation in periphyton of the Florida Everglades. Limnol. Oceanogr. 44: 1815–1825.

Cohen, M.J., S. Lamsal, T.Z. Osborne, J.C.J. Bonzongo, S. Newman and K.R. Reddy. 2009. Soil total mercury concentrations across the Greater Everglades. Soil Sci. Soc. Am. J. 73: 675–685.

Cordy, C. 1993. An extension of the Horvitz-Thompson theorem to point sampling from a continuous universe. Probability and Statistics Letters 18: 353–362.

Drexel, R.T., M. Haitzer, J.N. Ryan, G.R. Aiken and K.L. Nagy. 2002. Mercury(II) sorption to two Florida Everglades peats: evidence for strong and weak binding and competition by dissolved organic matter released from the peat. Environ. Sci. Technol. 36: 4058–4064.

[FDEP]. 2000. South Florida Water Quality Protection Program (WQPP). Florida Department of Environmental Protection Southeast District.

Fink, L. 2003. Appendix 2B-5: Evaluation of the Effect of Surface Water, Pore Water and Sediment Quality on the Everglades Mercury Cycle, 2003 Everglades Consolidated Report. South Florida Water Management District and Florida Department of Environmental Protection, West Palm Beach, FL.

Fink, L., D. Rumbold and P. Rawlik. 1998. The Everglades Mercury Problem, Everglades Interim Report. South Florida Water Management District, West Palm Beach, FL, p. 70.

Gaff, H., D.L. DeAngelis, L.J. Gross, R. Salinas and M. Shorrosh. 2000. A dynamic landscape model for fish in the Everglades and its application to restoration. Ecol. Model. 127: 33–52.

Gaff, H., J. Chick, J. Trexler, D. DeAngelis, L. Gross and R. Salinas. 2004. Evaluation of and insights from ALFISH: a spatially explicit, landscape-level simulation of fish populations in the Everglades. Hydrobiologia 520: 73–86.

Gaiser, E. 2008. Aquatic Fauna and Periphyton Production Data Collection, Science Coordination Group Science Workshop. South Florida Ecosystem Restoration Task Force.

Gaiser, E.E., D.L. Childers, R.D. Jones, J.H. Richards, L.J. Scinto and J.C. Trexler. 2006. Periphyton responses to eutrophication in the Florida Everglades: cross-system patterns of structural and compositional change. Limnol. Oceanogr. 51: 617–630.

Gill, G.A., J.L. Guentzel, W.M. Landing and C.D. Pollman. 1995. Total gaseous mercury measurements in Florida: The FAMS project (1992–1994). Water Air Soil Pollut. 80: 235–244.

Gilmour, C.C., G.S. Riedel, M.C. Edrington, J.T. Bell, J.M. Benoit, G.A. Gill and M.C. Stordal. 1998. Methylmercury concentrations and production rates across a trophic gradient in the northern Everglades. Biogeochemistry 40: 327–345.

Gilmour, C., D. Krabbenhoft, W. Orem, G. Aiken, E. Roden and I. Mendelssohn. 2006. Appendix 2B-2: Status Report on ACME Studies on the Control of Hg Methylation and Bioaccumulation in the Everglades, 2006 South Florida Environmental Report.

Guentzel, J.L. 1997. The atmospheric sources, transport, and deposition of mercury in Florida. PhD dissertation, Florida State University, Tallahassee, FL.

Guentzel, J.L., W.M. Landing, G.A. Gill and C.D. Pollman. 1995. Atmospheric deposition of mercury in Florida: The FAMS project (1992–1994). Water Air Soil Pollut. 80: 393–402.

Guentzel, J.L., W.M. Landing, G.A. Gill and C.D. Pollman. 2001. Processes influencing rainfall deposition of mercury in Florida. Environ. Sci. Technol. 35: 863–873.

Holz, J., J. Kreutzmann, R.D. Wilken and R. Falter. 1999. Methylmercury monitoring in rainwater samples using *in situ* ethylation in combination with GC-AFS and GC-ICP-MS techniques. App. Organomet. Chem. 13: 789–794.

Horvitz, D.G. and D.J. Thompson. 1952. A generalization of sampling without replacement from a finite universe. J. Am. Stat. Assoc. 663–685.

Hurley, J., D. Krabbenhoft, L. Cleckner, M. Olson, G. Aiken and P. Rawlik. 1998. System controls on the aqueous distribution of mercury in the northern Florida Everglades. Biogeochemistry 40: 293–311.

Keeler, G.J., F.J. Marsik, K.I. Al-Wali and J.T. Dvonch. 2001. Appendix 7-6: Status of the Atmospheric Dispersion and Deposition Model. 2001 Everglades Consolidated Report. South Florida Water Management District and Florida Department of Environmental Protection, West Palm Beach, FL.

Krabbenhoft, D., J. Hurley, M. Olson and L. Cleckner. 1998. Diel variability of mercury phase and species distributions in the Florida Everglades. Biogeochemistry 40: 311–325.

Lawson, N.M. and R.P. Mason. 2001. Concentration of mercury, methylmercury, cadmium, lead, arsenic, and selenium in the rain and stream water of two contrasting watersheds in western Maryland. Water Res. 35: 4039–4052.

Li, Y.B., Y.X. Mao, G.L. Liu, G. Tachiev, D. Roelant, X.B. Feng and Y. Cai. 2010. Degradation of methylmercury and its effects on mercury distribution and cycling in the Florida Everglades. Environ. Sci. Technol. 44: 6661–6666.

Lindberg, S.E. and H. Zhang. 2000. Air water exchange of mercury in the Everglades II: measuring and modeling evasion of mercury from surface waters in the Everglades Nutrient Removal Project. Sci. Total Environ. 259: 135–143.

Lindberg, S.E. and T.P. Meyers. 2001. Development of an automated micrometeorological method for measuring the emission of mercury vapor from wetland vegetation. Wetl. Ecol. Manag. 9: 333–347.

Lindberg, S.E., H. Zhang and T.P. Meyers. 1999. Everglades Mercury Air/Surface Exchange Study (E-MASE). Prepared for the South Florida Water Management District, West Palm Beach, FL.

Lindberg, S.E., W. Dong and T. Meyers. 2002. Transpiration of gaseous elemental mercury through vegetation in a subtropical wetland in Florida. Atmos. Environ. 36: 5207–5219.

Lindberg, S.E., W. Dong, J. Chanton, R.G. Qualls and T. Meyers. 2005. A mechanism for bimodal emission of gaseous mercury from aquatic macrophytes. Atmos. Environ. 39: 1289–1301.

Liu, G., Y. Cai, T. Philippi, P. Kalla, D. Scheidt, J. Richards, L. Scinto and C. Appleby. 2008a. Distribution of total and methylmercury in different ecosystem compartments in the Everglades: implications for mercury bioaccumulation. Environ. Pollut. 153: 257–265.

Liu, G.L., Y. Cai, P. Kalla, D. Scheidt, J. Richards, L.J. Scinto, E. Gaiser and C. Appleby. 2008b. Mercury mass budget estimates and cycling seasonality in the Florida everglades. Environ. Sci. Technol. 42: 1954–1960.

Liu, G.L., Y. Cai, T. Philippi, P. Kalla, D. Scheidt, J. Richards, L. Scinto and C. Appleby. 2008c. Distribution of total and methylmercury in different ecosystem compartments in the Everglades: implications for mercury bioaccumulation. Environ. Pollut. 153: 257–265.

Liu, G.L., Y. Cai, Y.X. Mao, D. Scheidt, P. Kalla, J. Richards, L.J. Scinto, G. Tachiev, D. Roelant and C. Appleby. 2009. Spatial variability in mercury cycling and relevant biogeochemical controls in the Florida Everglades. Environ. Sci. Technol. 43: 4361–4366.

Loftus, W.F. 2000. Accumulation and fate of mercury in an Everglade aquatic food web. PhD Dissertation, Florida International University, Miami, FL.

Marsik, F.J., G.J. Keeler, S.E. Lindberg and H. Zhang. 2005. Air-surface exchange of gaseous mercury over a mixed sawgrass-cattail stand within the Florida Everglades. Environ. Sci. Technol. 39: 4739–4746.

Marvin-DiPasquale, M.C. and R.S. Oremland. 1998. Bacterial methylmercury degradation in Florida Everglades peat sediment. Environ. Sci. Technol. 32: 2556–2563.

Mason, R.P., W.F. Fitzgerald, J. Hurley, A.K. Hanson, Jr., P.L. Donaghay and J.M. Sieburth. 1993. Mercury biogeochemical cycling in a stratified estuary. Limnol. Oceanogr. 38: 1227–1241.

Mauro, J., J. Guimarães, H. Hintelmann, C. Watras, E. Haack and S. Coelho-Souza. 2002. Mercury methylation in macrophytes, periphyton, and water—comparative studies with stable and radio-mercury additions. Anal. Bioanal. Chem. 374: 983–989.

McCormick, P.V., R.B.E. Shuford, III, J.G. Backus and W.C. Kennedy. 1998. Spatial and seasonal patterns of periphyton biomass and productivity in the northern Everglades, Florida, U.S.A. Hydrobiologia 362: 185–208.

Mercury Deposition Network. accessed on 02/06/2010. http://nadp.sws.uiuc.edu/mdn/.

[NCSL]. 1997. American National Standard for Expressing Uncertainty-U.S. Guide to the Expression of Uncertainty in Measurement. National Conference of Standards Laboratories, ANSI/NCSL Z540-2-1997.

Rumbold, D., N. Howard, F. Matson, S. Atkins, J. Jean-Jacques, K. Nicholas, C. Owens, K. Strayer and B. Warner. 2007. Appendix 3B-1: Annual Permit Compliance Monitoring Report for Mercury in Downstream Receiving Waters of the Everglades Protection Area, 2007 South Florida Environmental Report, West Palm Beach, FL.

Scheidt, D. and P. Kalla. 2007. Everglades ecosystem assessment: water management and quality, eutrophication, mercury contamination, soils and habitat: monitoring for adaptive management: A R-EMAP status report. EPA 904-R-07-001. USEPA Region 4, Athens, GA, p. 98.

[SFWMD DBHYDRO]. accessed on 02/06/2010. http://my.sfwmd.gov/dbhydroplsql/show_dbkey_info.main_menu.

Stober, Q.J., K. Thornton, R. Jones, J. Richards, C. Ivey, R. Welch, M. Madden, J. Trexler, E. Gaiser, D. Scheidt and S. Rathbun. 2001. South Florida Ecosystem Assessment: Phase I/II (Technical Report)—Everglades Stressor Interactions: Hydropatterns, Eutrophication, Habitat Alteration, and Mercury Contamination. EPA 904-R-01-003. USEPA Region 4, Athens, GA.

Ullrich, S.M., T.W. Tanton and S.A. Abdrashitova. 2001. Mercury in the aquatic environment: a review of factors affecting methylation. Crit. Rev. Environ. Sci. Technol. 31: 241–293.

Ware, F., H. Royals and F. Lange. 1990. Mercury contamination in Florida largemouth bass. Proceedings of Annual Conference of Southeastern Association of Fish and Wildlife Agencies 44: 5–12.

Zhang, H. and S.E. Lindberg. 2000. Air water exchange of mercury in the Everglades I: the behavior of dissolved gaseous mercury in the Everglades Nutrient Removal Project. Sci. Total Environ. 259: 123–133.

Sea Level Rise in the Everglades: Plant-Soil-Microbial Feedbacks in Response to Changing Physical Conditions

Lisa G. Chambers[1,*] *Stephen E. Davis*[2] *and Tiffany G. Troxler*[3]

Introduction

Coastal wetlands occupy the intertidal zone between the freshwater Everglades and the Gulf of Mexico. At the interface with the marine environment, these wetlands are dominated by mangrove forests, tidal creeks, and mudflats. Behind the coastal fringe zone is an ecotone that transitions into a freshwater/upland community characterized by an increasing diversity of halophytic to glycophytic trees, shrubs, and herbaceous vegetation. These productive coastal wetlands provide numerous, important ecosystem services. Directly benefiting humans, coastal wetlands function as habitat and nursery ground for shellfish and other commercially important fisheries, stabilize the coastline, and lessen the impact of storm surges (Aburto-Oropeza et al. 2008; Costanza et al. 2008; Gedan et al. 2011). Moreover, some of the most important functions of coastal

[1] Department of Earth and Atmospheric Science, Saint Louis University, 3642 Lindell Blvd., St. Louis, MO 63108.
   Email: chamberslg@slu.edu
[2] Everglades Foundation, 18001 Old Cutler Rd. suite 625, Palmetto Bay, FL 33157.
   Email: sdavis@evergladesfoundation.org.
[3] Southeast Environmental Research Center, Florida International University, 11200 SW 8th Street, Miami, FL 33199.
   Email: troxlert@fiu.edu
* Corresponding author

wetlands are less tangible processes, including nutrient cycling, pollution removal, and carbon (C) storage (Craft et al. 2009; Gedan et al. 2009; Barbier 2013). Because coastal wetlands are a transitional ecosystem between the land and the ocean, sea level rise represents a significant threat to their global distribution and extent. In order to persist, the soil platform of a coastal wetland must maintain a specific elevation niche relative to the ocean, normally between mean sea level and mean high tide (Morris et al. 2002).

This chapter addresses the connections between sea level rise and Everglades soil microbial processes, with a focus on soil C dynamics and the relationship between vegetation, soil, and microbial ecology. Specifically, salinity and inundation are expected to increase under projected sea level rise, hydrologic drivers that are known to alter the structure and productivity of coastal wetland plant communities (e.g., Williams et al. 1999; Donnelly and Bertness 2001; Teh et al. 2008; Smith 2009). Plant community change can affect the quality and quantity of organic C inputs to the system, which in turn, can influence soil microorganisms and biogeochemical processes (Neubauer et al. 2013; Morrissey et al. 2014a). The feedback between plant and soil microbial dynamics, vegetation change, and evidence of such shifts in the Everglades will be discussed. Because soil microbes play a key role in regulating how much organic matter is buried or stored in coastal wetlands, processes which promote vertical accretion and resilience to rising sea levels, we will also describe the current understanding of how sea level drivers may directly impact soil microbial activity and diversity. Finally, the chapter will conclude with a discussion of how sea level rise impacts are assessed and the current evidence of sea level rise in the Everglades.

South Florida and Sea Level Rise Throughout History

Changes in global sea level are a natural part of the geologic history of the earth and have always been a key abiotic driver of the ecology of Florida. South Florida, in particular, is considered as being "land from sea" because this region has undergone a repeated history of submergence and emergence from the ocean that is recorded in a stratigraphy of alternating freshwater and marine sediments (Willard and Bernhardt 2011). For instance, during the Eemian interglacial period, which began about 120,000 years ago, sea levels were at least 6.6 meters higher than present and south Florida was a shallow bay (Muhs et al. 2011). During this period, the coral reefs formed that would later become the Florida Keys and sediments began to collect on the Florida platform. Approximately 18,000 years ago, at the height of the last glacial period, sea levels dropped to approximately 120 m lower than today, exposing the carbonate platform that includes present-day south Florida (Wanless et al. 1994). This was followed by a period of step-wise rises in sea level that deposited and recycled sediments, allowed for organic matter accumulation, and eventually led to the evolution of the Everglades during a period of relatively stable sea levels, approximately 3,200 years before present (Locker 1996). Sea level began to rise again in approximately 1850 at a rate of ~1.7 mm y^{-1}, and has accelerated since 1993 to a rate of between 2.8 and 3.1 ($\pm$ 0.7) mm y^{-1}. Half of the current rate of sea level rise is attributed to thermal expansion of the ocean and half to melting land ice (IPCC 2007).

The flat, emergent wetland-dominated landscape of the Everglades has earned it the nickname, The River of Grass. Over its recent geologic history, the size and shape of the Everglades have changed in concert with sea levels. Currently, approximately 1/3 of the greater Everglades are within 1.5 m of sea level, and half of Everglades National Park lies below 0.6 m of sea level (Titus and Richman 2001). Further, the slope of the Everglades averages just 5–8 cm (2–3 inches) per mile (Lodge 2010; McVoy et al. 2011). These geomorphologic conditions make south Florida's natural environments highly vulnerable to increased sea levels. When considering the population that lives immediately adjacent to the Everglades (approximately 6 million people, U.S. Census Bureau), the flat, gently sloping landscape renders the Everglades a virtual "canary in a coal mine" for how both society and the natural environment will respond to sea level rise.

As sea level rises, the coastal zone along the Everglades will be increasingly exposed to salinity and inundation. The general thinking is coastal wetlands may respond by: 1) keeping pace with sea level through vertical accretion, 2) migrating landward to maintain an optimal elevation relative to sea level, or 3) submerging. It has been predicted that if sea level rises between 0.18–0.59 m before the end of this century, 30% of the world's coastal wetlands will be lost, either to submergence or the inability to migrate landward (IPCC 2007). An increase in mean sea level of 1 m will inundate approximately 4,050 km^3 (4 million acres) of wetlands in coastal areas (Neumann et al. 2000). Therefore, understanding the controls over the fate of C stored in coastal soils will be critical in predicting the fate of our coastlines—not to mention the vast reservoir of nutrients (particularly nitrogen (N) and phosphorus (P)) stored in those soils.

In south Florida, sea levels are conservatively predicted to rise 0.60 m by 2060 (Zhang et al. 2011). This will cause salinity and inundation to increase in fresh and brackish water areas of the Florida coastal zone and will increase the risk of storm surge-induced flooding and saltwater exposure in oligohaline areas of the Everglades at the top of the estuarine ecotone (Teh et al. 2008; Pearlstine et al. 2010). The Everglades also has the added risk factor of being a highly modified and managed system with a large number of canals, ditches, and dams used to divert water for flood control, agricultural water supply, and human consumption. For example, the construction of canals on the east and west sides of Lake Okeechobee in the northern Everglades caused a 10-fold increase in the portion of freshwater flowing directly into the Atlantic Ocean and Gulf of Mexico, bypassing the Everglades. Much of this freshwater had previously flowed south through the Everglades and into Florida Bay, effectively countering saltwater along the coastal interface (Nuttle et al. 2000). The change in the quantity, timing, and distribution of freshwater delivery to the coastal zone by this and many other hydrologic alterations throughout the Everglades are believed to be amplifying saltwater intrusion and the rate of the landward migration of coastal habitats, especially in the eastern Everglades (Ross et al. 2000).

Soil Elevation and the Carbon Cycle

Carbon storage and accumulation is vital to the health of the coastal Everglades because it is a major component of the soils and sediments that serve as the wetland platform.

Everglades soils are predominately histosols and entisols. Histosols, or peat soils, found in the Everglades coastal zone can be up to 5.5 m deep with total C contents of 15–42% and organic matter contents of 32–89% (Whelan et al. 2005; Castaneda 2010; Chambers et al. 2014). These areas often support riverine and fringe mangroves. Everglades enitsols are typically either of a marl-type, derived from dry-down and deposition of calcitic periphyton mats, or derived from marine sediments resulting from storm and tidal influences in the coastal zone. These inorganic C-dominated soils are typically shallower and support "dwarf" mangroves, sawgrass, or may be unvegetated.

Many coastal wetlands have persisted through centuries of sea level fluctuations due to the natural feedback mechanism of vertical marsh accretion: the accumulation of soil C that leads to an increase in the elevation of the wetland platform (Morris et al. 2002; McKee et al. 2007). This process is a dynamic interplay between sea level and primary production governed by both autochthonous contributions to soil elevation (i.e., belowground production and litter accumulation) and deposition of allochthonous mineral sediments on the soil surface (Morris et al. 2002). The vegetation directly adds organic C to the soil, while also enhancing further sediment deposition (containing both organic and inorganic C) through the slowing of water velocities by aboveground biomass. Studies indicate both organic and inorganic C are important to vertical marsh accretion (Day et al. 2000), with the latter generally comprising a greater percentage of the soil as the proximity to the ocean or rivers increases. Soil C accumulation in coastal wetlands is a balance between the C inputs (i.e., imports and CO_2 fixation) and outputs (i.e., exports, CO_2, and CH_4 flux) from the system. Major C reservoirs include plant biomass, detritus, peat, microbial biomass, and dissolved C (Fig. 1). In order for coastal wetlands to "keep pace" with sea level rise, production and sediment input/

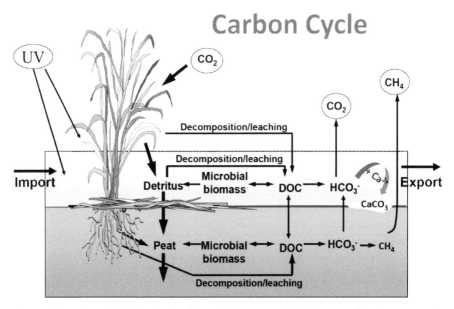

Figure 1. A generalized depiction of the carbon cycle in wetlands, including the major storage reservoirs and transformations. Modified from: Reddy and R. Delaune 2008.

retention must exceed losses on a magnitude that mimics or exceeds the current rate of sea level rise (Morris et al. 2002; Day et al. 2011).

Several lines of evidence suggest that conditions are generally met for coastal wetland elevations to be sustainable in the context of current rates of sea level rise. For example, a meta-analysis of the accretion rates from 15 mangrove forests throughout the world indicates approximately 80% of the systems studied are currently accreting at, or faster than, the rate of global sea level rise (Alongi 2008). At least one area of the Everglades is responding in a similar manner. In the coastal fringe mangroves of Shark River Slough, average accretion rates between 1924 and 2009 were 2.5 to 3.6 mm y^{-1}, while sea level was rising at a rate of 2.2 mm y^{-1} (Smoak et al. 2013). However, rates of sea level rise are not uniform across our coasts, coastal vegetation is not restricted to mangrove forests, and human impacts in the coastal zone, such as freshwater diversion, alter dynamics that may otherwise lead to accretion rates that exceeds rates of sea level rise.

The relative importance of organic versus inorganic sediment accumulation necessary to combat coastal submergence has been studied extensively elsewhere (e.g., Nyman et al. 1990; Day et al. 2011), but little information is available for the Everglades. Often, allochthonous deposition is the critical factor for wetland accretion in meso-tidal systems characterized by high sediment loads; these types of coastal wetlands are thought to be the most stable during periods of sea level rise (Morris et al. 2002). Where sediment supplies are low, as is the case in many areas of the coastal Everglades, root biomass and productivity likely become the major contributor to soil accretion. A recent study of mangroves in Belize and Florida showed approximately 50% of production occurs belowground and mangrove roots can account for 1–12 mm of surface elevation increase per year—equivalent to as much as 55% of the annual vertical change in soil elevation (McKee 2011). Furthermore, mangrove roots are fairly refractory and have slow rates of decomposition in the field, especially relative to mangrove leaves (Middleton and McKee 2001). Root decomposition studies by Huxham et al. (2010) and McKee et al. (2007) found about 24–60% of root material remained after 1-year of decomposition in the field, depending on species and tidal position. However, root accumulation rates differ by mangrove forest types (e.g., fringe, dwarf, etc.) and are sensitive to variations in the supply of N or P (McKee et al. 2007). A study conducted in Taylor Slough mangroves found that both resource limitations and greater inundation were correlated with increased fine root production, which was a primary driver of soil formation and accretion (Castaneda-Moya et al. 2011). This suggests that differential plant growth responses to nutrient availability or sea level rise stressors, such as inundation, may be key determinants of coastal wetland resilience. The response of mangrove root productivity to nutrient addition is especially important in the coastal Everglades because Florida Bay tends to be the primary source of nutrients (Childers et al. 2006), making increased nutrient supply a possible indirect effect of sea level rise.

While root dynamics are thought to be important drivers of soil elevation in the coastal Everglades, storm events that deliver large quantities of inorganic sediment can be significant in localized areas (Castañeda-Moya et al. 2010), as well as seasonal groundwater dynamics that may affect the shrink-swell dynamics of the peat soils (Whelan et al. 2005). For example, a single hurricane (Wilma 2005) deposited between

0.5 and 4.5 cm of sediment across Shark River Slough and exceeded the annual vertical accretion rate by 8–17 times (Castañeda-Moya et al. 2010). Benthic mats (dominated by algae and cyanobacteria) can also contribute as much as 0.5–5 mm in soil elevation each year, accounting for 9–35% of the total increase in soil elevation (McKee 2011). In addition to the direct contributions benthic mats and inorganic sediments make to soil accretion, their presence may also increase nutrient availability in the wetland (Castañeda-Moya et al. 2010), which will feed back into primary production.

Heterotrophic soil microbes, on the other hand, continually act in opposition to the accumulation of organic matter by utilizing C in senescing leaves and roots, as well as organic exudates from roots, as energy sources. These organisms assist in the break-down of organic polymers into monomers, which can then be assimilated by bacteria and function as electron donors during respiration. The process of organic matter mineralization results in the release of organic C in a particulate (POC) or dissolved (DOC) form, or as carbon dioxide (CO_2) or methane (CH_4) gas (Schlesinger 1997). This C may be exported via aquatic transport during the ebb tide, or lost to the atmosphere via soil efflux or diffusion and ebullition through the water column (Dittmar et al. 2006; Bouillon et al. 2008). Laboratory intact soil core studies (excluding vegetation) indicate organic C loss from coastal wetland soils predominately occurs through microbial respiration and the subsequent release of CO_2 back to the atmosphere. Across a salinity gradient of tidal freshwater, brackish, and salt marsh soils, an average of > 96% of C lost was emitted as CO_2, ≤ 3% as CH_4, and < 1% as DOC (Chambers et al. 2013). At the ecosystem scale, mangrove research has found a significant portion of C (up to 50% of litter production) is exported as leaves, detrital material, and dissolved C, with the contribution of C to the coastal zone dependent upon the tidal amplitude, season, geomorphology, and productivity (Jennerjahn and Ittekkot 2002; and references therein). Dissolved inorganic C (DIC) export appears to be important in mangrove systems and may exceed DOC export by a factor of 3 to 10 (Bouillon et al. 2008). Finally, a large portion of mangrove organic C remains buried in the soil, with a global average burial rate of 26.1 Tg OC (Breithaupt et al. 2012).

Depending on the influence of tides and the presence of litter consumers (e.g., crabs), other internal pathways contributing to C loss may prevail. Early work by Robertson et al. (1992) showed that sesarmid crabs can account for the consumption or burial of up to 28% of the total annual litterfall in a mangrove forest, making it easier for by-products to be mineralized by microorganisms or to be exported tidally. Other studies tracking the fate of leaf litter have shown that > 30% of the dry mass of leaf litter is leachable, contributing to C loss but also representing a source of nutrients that can facilitate bacterial colonization and increased palatability of the leaf detritus (Davis et al. 2003; Davis et al. 2006). Increased exposure of coastal wetlands to tidal flushing due to sea level rise could affect the relative importance of this type of internal cycling, thus affecting net soil accretion.

Considering all of this, it remains unclear if the necessary soil accretion rates can be maintained under the accelerated rate of sea level rise scientists are predicting (e.g., Church and White 2006; Haigh et al. 2014), and what the consequences will be of the water management activities and human development within the greater Everglades watershed. Despite pole-ward expansion of mangroves with global warming and the ability to migrate landward, Alongi (2008) predicts a 10–15% global loss of mangroves under accelerated rates of sea level rise by 2100. Some research forecasts accretion in many coastal wetlands world-wide will be out-paced by rising sea levels, leading to submergence and land loss. This fate can be driven by physical constraints, such as a small tidal range or low sediment supply in the coastal zone (Nicholls et al. 1999), or a biological feedback. For example, submergence is often observed when areas of salt-sensitive coastal vegetation are not colonized by more salinity tolerant species quickly enough. The existing vegetation becomes stunted or dies as a result of water-logging, osmotic stress, the accumulation of toxic hydrogen sulfide (HS^-) from intruding seawater, or salinity induced nutrient inhibition (Koch et al. 1990; Bradley and Morris 1991; Batzer and Shartiz 2006). Because live plants play a vital function in maintaining soil structure, slowing water velocities to allow for sediment deposition, and providing a source of soil organic matter, a decrease in vegetation health can rapidly cascade into peat collapse, ponding, and accelerated submergence (Nyman et al. 1990; DeLaune et al. 1994). There is also speculation that rising sea levels could exacerbate erosion along the coastline of the Everglades, which also contributes to submergence (Wanless et al. 1994).

Peat Collapse in the Everglades?

Much of the coastal Everglades include peat soils (Craft and Richardson 2008), which have a very low bulk density (often exceeding 85% pore space by volume; Nyman et al. 1990), making them susceptible to collapse. Peat collapse can occur in a variety of systems under varying circumstances (Day et al. 2011), but is typically initiated by a loss of soil structure, such as by de-watering, accelerated decomposition, or root death. This reduces the strength of the soil matrix and causes the surface soil material to cave-in upon the subsurface soil, resulting in a rapid loss of elevation (DeLaune et al. 1994). Recent geological research in the Everglades has shed light on the potential for collapse of freshwater organic peat soils as they are exposed to increasingly saline conditions (Fig. 2). In a report for Everglades National Park, Wanless and Vlaswinkel (2005) describe a series of "collapse" events that occurred in Cape Sable and Whitewater Bay that they attributed to disturbance caused by channelization, storm surge, sea level rise, and freshwater diversion. These collapse events have led to the exposure of previously freshwater wetlands to more inundated and saline conditions, often converting emergent freshwater marsh areas to open water before mangroves have time to become established and stabilize the soil. Although similar to general "submergence," in which the depth of inundation increases without a comparable response in vertical accretion and the system slowly converts to open water, peat collapse results in a more rapid transition to open water conditions and can occur away from the aquatic edge, such as near the inland ecotonal boundary of salt-tolerant vegetation, as observed in the Everglades.

Freshwater peat collapse could limit the landward colonization of mangrove propagules in response to sea level rise by reducing the soil elevation to a point that it is too deep for new vegetation to become established; this occurrence may be spatially heterogeneous, based on the location of freshwater peat deposits throughout the Everglades. Shallow open water habitat or embayments may develop instead, potentially interrupting the dispersal mechanisms for more salt-tolerant mangrove

Figure 2. Photographs showing evidence of peat collapse in a sawgrass (*Cladium jamaicense*) marsh surrounded by an expanding mangrove forest in lower Shark River Slough, Everglades National Park. [Photo credit: South Florida Water Management District.]

vegetation to migrate landward. Alternatively, successfully colonization of peat collapse areas would hinge upon the ability of vegetation to rapidly accrete new peat soil or mineral sediment, as described earlier, to keep pace with increasing sea level. Areas of south Florida where freshwater diversions have been extensive may be especially vulnerable to wetland loss via peat collapse as sea level rises.

Wanless and Vlaswinkel (2005) describe a few plausible explanations for the observed peat collapse in the Everglades. First, over-drainage of some areas may lead to soil drying, organic matter oxidation, and compaction. Next, hurricanes, which occur frequently in south Florida on geological timescales, physically damage vegetation, leading to soil exposure. These soils are then vulnerable to oxidation and collapse until vegetation is reestablished and organic soils are re-stabilized. Finally, Wanless and Vlaswinkel (2005) describe a second scenario where increasing salinity (from sea-level rise or storm surge) stresses, and then kills obligate freshwater vegetation at the top of the estuarine ecotone. Until more salt-tolerant vegetation is reestablished, the organic peats are also vulnerable to oxidation and collapse.

The scenarios posed by Wanless and Vlaswinkel (2005) rely on vegetation reestablishment in order to protect peat soils from oxidizing and "collapsing", as it is the establishment of plants and their investment of belowground biomass (i.e., roots) that serve to stabilize the soils. Studies of peat collapse events in coastal Louisiana also indicate death of belowground biomass is a key instigator and warning sign of impending collapse (Turner et al. 2004). However, accelerated soil mineralization may already be underway while the vegetation is still in place. Additionally, some areas of the coastal Everglades are composed of mineral sediments and soils that developed under the influence of marl-forming periphyton. The response of these types of soils to sea level rise, de-watering, and accelerated mineralization are not understood. In this chapter, we propose a slightly different alternative that considers changes in both the plant-mediated control over peat soil formation and maintenance, as well as the underlying biogeochemical mechanisms behind peat degradation, compaction, and collapse. Recent experimental data combined with concepts in wetland soil biogeochemistry (described later in this chapter) may shed more light on the validity of the peat collapse concept and provide us with better predictive capability as to how Everglades peat soils will respond to sea-level rise.

Vegetation Change

The distribution of vegetation communities across the Everglades is a function of salinity, depth, hydroperiod, and nutrient availability (Daoust and Childers 2004; Barr et al. 2010; Castaneda-Moya et al. 2013; Troxler et al. 2013). In the coastal zone, vegetation species typically orient in identifiable zones parallel to the shoreline or tidal creeks; these zones are dictated mainly by each species' tolerance to salt and water-logging (Fig. 3). In general, the Everglades coastal fringe is dominated by red mangrove (*Rhizophora mangle*), followed by black mangrove (*Avicennia germinans*) and white mangrove (*Laguncularia racemosa*) as you move landward. Mangroves are typically able to tolerate high salinities (~30 + ppt) and are adapted to wide fluctuations in both water level and salinity. A sharp ecotone often separates mangroves from the less salinity tolerant communities, such as hardwood hammocks, that are located further inland. This boundary is maintained by a combination of the salinity gradient and a positive feedback mechanism by which each species modifies its' own environment to promote its permanence (Jiang et al. 2011). In some areas of the coastal Everglades, particularly the southeastern region near Taylor Slough, the mangroves have a less abrupt inland ecotone, first becoming interspersed with herbaceous and succulent

Figure 3. Typical vegetation zonation patterns observed in the coastal Everglades, with mangroves occupying the land fringe, followed by one or more ecotonal boundaries into less salt tolerant vegetation [Photo credit: Lisa G. Chambers].

halophytes (e.g., *Batis maritima, Borrichia frutescens* and *Juncus romoerianus*), then giving way to Gulf Coast spikerush (*Elecocharis cellulosa*) and less salt tolerant sawgrass (*Cladium jamaicense*) communities (Ross et al. 2000). There are also areas of unvegetated mud flats and limestone pinnacles (Smith et al. 2013). If the coastal Everglades follow the typical model for landward migration, the seaward edge of the mangrove forests will submerge and give-way to subtidal habitats such as tidal flats and seagrass beds. Upstream, oligohaline and freshwater marshes will give way to mangroves. However, the risk of soil collapse in predominantly peat-based oligohaline or freshwater areas needs to be considered; this could lead to the establishment of open water conditions prior to mangrove establishment (Wanless and Vlaswinkel 2005).

Salinity as a Driver of Vegetation Change

Across the coastal Everglades landscape, salinity is a key driver of primary productivity and species composition (Castañeda-Moya et al. 2011; 2013; Troxler et al. 2013). However, the salinity gradient is highly dynamic, varying daily (with tides), seasonally (wet and dry season), and over the longer term (e.g., drought periods). During the wet season (May–November), freshwater flow from the northern Everglades penetrates well into the mangrove zone, while reduced flows during the dry season (December–April) allow tidal water to penetrate further upstream. For example, under dry season conditions in lower Taylor Slough, surface water salinity can exceed 40 ppt, and porewater salinity levels in some sawgrass (*C. jamaicense*) marshes can exceed 30 ppt (McIvor et al. 1994; Troxler et al. 2012). During the wet season, the surface water of these same areas is typically fresh (< 0.5 ppt). Sawgrass is only weakly salt-tolerant and can be negatively affected by even small increases in salinity, showing signs of physiological stress at salinities as low as 5 ppt (Rejmankova and Macek 2008). During periods of drought in Taylor Slough, the replacement of sawgrass by more salt-loving species such as mangroves has been observed (Ross et al. 2000). Mangrove expansion in other areas of the Everglades has also been noted and thought to be correlated with increased water levels associated with sea level rise (Smith et al. 2013). Extended droughts, managed flow reductions, and the construction of canals and ditches can all initiate inland mangrove expansion by allowing the tidal prism to migrate upstream and carry with it mangrove propagules to facilitate mangrove expansion (Fig. 4; Ross et al. 2000).

The observed expansion of mangroves can have numerous implications for soil microbiology, although, to our knowledge, this area of research has received little attention in the literature. For example, while both mangroves and sawgrass conservatively store nutrients in tissues of leaves, resulting in high C to nutrient ratios, the quantity and quality of organic matter provided by mangrove wood and leaf litter may be less labile than sawgrass litter. This consideration led Bianchi et al. (2013) to hypothesize that an increased C storage capacity of wetlands would ensue with continued pole-ward expansion of mangroves into coastal zones previously dominated by graminoids (e.g., *Spartina alterniflora*). The landward migration of mangroves in the Everglades could alter the impact of disturbance events such as fire, lightning strikes, and hurricanes on light penetration and soil temperature, and could also alter the rate of vegetation-induced sediment deposition. Finally, soil redox potential may

Figure 4. Mangroves expanding landward along tidal creeks in the coastal Everglades, likely facilitated by saltwater intrusion and propagule dispersal [Photo credit: Lisa G. Chambers].

be differ under cover of these different plant communities, which affects the utilization of various electron acceptors by soil microbes and influences the overall rate of C mineralization (Reddy and DeLaune 2008). Within mangrove forests themselves, there can be species zonation and spatial heterogeneity that produce differences in rhizosphere oxidation, which can affect C mineralization pathways, the availability of nutrients, rates of soil respiration, and other physio-chemical conditions in the soil (Alongi et al. 2000). In contrast to mangroves, soil redox potential in monotypic sawgrass communities seem to vary with water level, but otherwise remain fairly consistent throughout the surficial (0–25 cm) soil and are not significantly affected by nutrient gradients (Qualls et al. 2001).

In addition to shifting species composition, salinity can also have a direct physiological effect on vegetation and conditions in the soil. In the Everglades, studies have found a linear decrease in the light-use efficiency of mangroves as salinity increases, suggesting decreased productivity with saltwater intrusion (Barr et al. 2010). Sawgrass (*C. jamaicense*) aboveground net primary production has also been reported to be negatively correlated with surface water salinity, especially periods of high maximum salinities, which primary production rates seem slow to recover from (Childers et al. 2006; Troxler et al. 2013). Increasing root phosphatase activity has been detected for sawgrass plants associated with relatively low salinity levels (0.5–5 ppt), indicating a strong demand for P (Rejmankova and Macek 2008). Field and laboratory studies conducted elsewhere on marsh vegetation have documented plant mortality and reduced growth in several other common Gulf of Mexico species (e.g., *Sagittaria lancifolia, Panicum hemitomon, Leersia oryzoides*) as salinity increases, with each species having slightly different levels of salt-sensitivity (McKee and Mendelssohn 1989). Wetland vegetation that is not adapted to saltwater often suffers from osmotic stress (Batzer and Shartiz 2006) and an accumulation of hydrogen sulfide (HS$^-$) in the soil porewater. Hydrogen sulfide, a by-product of sulfate reduction, acts as a phytotoxin and suppresses plant metabolism, reduces growth, and inhibits nutrient uptake (Koch et al. 1990).

In some cases, vegetation shifts are thought to be initiated by extreme salinity events, such as storm surges. Even if the intensity and frequency of storm surges is not affected by climate change, sea level rise will still result in an increase in the height of the surge, and thus increase the area of land inundated by a high water event. However, studies suggest episodes of coastal flooding will increase in the coming decades (Najjar et al. 2000). Based on historic data and predictive models, the return period of storm surges throughout Florida is expected to be condensed, such that a 1-in-50 year surge will be experienced roughly every 5 years (Park et al. 2011). This coincides with evidence that the number of maximum water level events (meteorological and storm related) has increased in frequency in south Florida since 1961 (Obeysekera et al. 2011).

It is thought that much of the current distribution of mangroves is a product of past hurricanes (Doyle et al. 2003). Storm surges, like droughts, can accelerate the landward migration of the mangrove ecotone by carrying and depositing propagules further inland. The sharp vegetation boundary between the coastal mangroves and hardwood hammocks seems especially vulnerable to saltwater intrusion events. Models predict that just a 1-day saltwater intrusion event of salinities > 15 ppt could initiate a transition to a mangrove dominated system within a previously hammock community as a result of salinity stress (Teh et al. 2008). Meanwhile, other research in the Everglades has documented catastrophic damage to mangroves as a result of hurricanes (e.g., wind throws, defoliation, smothering by sediments), leading to a transition to mudflats near the coastline (Smith et al. 2009).

Phosphorus as a Driver of Vegetation Change

Because of the Everglades' legacy as a nutrient-limited system, differentiating between the impacts of sea level rise (salinity and inundation), and the associated increase in nutrient supply (especially P) accompanying saltwater intrusion, can be challenging. Most estuarine wetlands tend to be N-limited, meaning plant productivity is constrained by the availability of N needed for biomass synthesis (Vitousek and Howarth 1991). While N is still an important element for growth and production in the Everglades, P typically regulates the species composition and trophic state of the greater ecosystem (Noe et al. 2001). In its pristine, pre-drainage state, P in the Everglades was naturally low (< 10 ug L^{-1}), with much of it being derived from the atmosphere (Belanger et al. 1989). The P limitation was a key driver in evolution of Everglades ecology, favoring the establishment of a unique assemblage of species with low P requirements (e.g., periphyton, sawgrass). Today, the northern Everglades are subject to P loading, mostly from agricultural sources, which has led to a shift in plant communities. This is especially evident in the Water Conservation Areas where periphyton biomass has declined and areas previously occupied almost exclusively by *C. jamaicense* are now monotypic stands of *Typha domingesis* (Davis 1991; McCormick et al. 1998).

Understanding the implications of P loading in the northern Everglades is relevant to a discussion of sea level rise because P concentrations in Florida Bay are naturally higher than the un-impacted areas of the Everglades, making the ocean the primary source of P in the coastal zone (Childers et al. 2006; Rivera-Monroy et al. 2007). There is increasing evidence of the importance of marine-derived P in shaping mangrove forest structure and productivity in the Everglades. For example, the reestablishment of a mangrove community damaged by Hurricane Donna (1960) was evaluated to determine the driving forces of structural development. The study found that neither soil salinity nor sulfide concentrations reached levels known to influence species composition, but concentrations of N and P mirrored productivity rates. Both basal area and wood production were highest at the coast (where N and P availability was greatest) and decreased further inland (Chen and Twilley 1999). Research also indicates the importance of marine-derived P in the partitioning of C within mangroves. Trees growing in upstream portions of the estuary and regions with low tidal exchange allocated more biomass belowground, rather than aboveground, in response to the P gradient (Castañeda-Moya et al. 2013). Belowground productivity contributes significantly to soil accretion and preserves soil structure to combat peat collapse (DeLaune et al. 1994; Turner et al. 2004). How an influx of P with saltwater intrusion might affect accretion rates, belowground productivity, and soil stability in the coastal Everglades has not been investigated.

Vegetation-Microbial Interactions

Vegetation change directly impacts soil microbiology by altering the amount and timing of C available to microbes, the lability of the C substrate, and the structural habitat for microbes. This is in addition to the direct impact of the physical changes caused by sea level rise (e.g., increasing salinity, altered nutrient availability, and changing depth and duration of flooding). In general, plant species richness and diversity tend to decrease with increasing salinity (e.g., Odum 1988; Wieski et al. 2010; Sharpe and Baldwin 2012). In the Everglades, sea level rise threatens the future of as many as 21 species of rare, low-lying coastal plants that lack adjacent suitable habitat for species migration (Saha et al. 2011). The decrease in wetland structural complexity caused by salinity also reduces the variety of terrestrial organic matter to serve as a microbial substrate, and could prompt a similar decline in the diversity of the microorganisms that rely on them. Indeed, a variety of studies have shown different plant species and functional guilds support unique microbial assemblages (e.g., Grayston et al. 1998; Troxler et al. 2012), suggesting changes in coastal plant community distribution driven by sea level rise will likely cascade into an alteration of soil microbial ecology. However, no studies to date have directly studied the effect of reduced plant species diversity due to sea level rise on soil microbial diversity.

Shifts in vegetation composition can also impact microbes due to differences in the bioavailability of their litter material. For example, polyphenolic compounds such as condensed tannins and lignins, reduce the ability of microbes to degrade those tissues, creating the need for expensive exoenzyme production to break-down these refractory materials (Field and Lettinga 1992; Berg et al. 1996). Leached, plant-derived polyphenolic compounds may also have an inhibitory effect on microbial activity (Field and Lettinga 1992). In general, species with high lignin content, high leaf dry matter content, and greater specific leaf area (common characteristics of woody species) have slower decay rates (Prescott 2011). Inputs of organic matter from plants occur as leaf and root litter, and dissolved organic matter (DOM) from litter leaching. While components of freshly leached DOM can be easily degraded and stimulate microbial respiration, litter DOM can also contain less labile components that reflect those chemically-complex compounds found in litter tissue (see Cornwell et al. 2008 for overview). While both the complexity and nutritional composition (structure of C-C bonds and C:N:P) of plants either in leaves or roots are important for soil organic matter (SOM) dynamics, microbial community composition also has a proximate control on SOM (Melillo et al. 1982). For instance, fungal and actinomycete bacterial species are among the most efficient in degrading more complex C compounds, but are associated with degradation in aerobic, low nutrient environments (Goodfellow and Williams 1983; McCarthy et al. 1987; Güsewell and Gessner 2009; Peltoniemi et al. 2009). With adequate nutrient supply, microbes may also synthesize metabolically expensive enzymes to acquire complex C molecules of recalcitrant tissues (Moorehead and Sinsabaugh 2006). Recent research suggests SOM quantity and quality (as indicated by C:N) is inversely related to salinity in oligohaline wetlands (Morrissey et al. 2014b). In the Everglades, the expansion of mangroves into areas previously dominated by hardwood hammocks or graminoid is expected to result in significant changes in the quality of liter material and alter SOM dynamics in the coastal zone.

Microbial Ecology and Biogeochemistry

The connection between sea level rise and microbiology is an emerging area of research with broad implications given the strong connection between hydrology, plant production, soil microbes, and soil C storage. Soil microbes are often the first organisms in a wetland to respond to environmental changes due to their large surface-to-volume ratio and rapid turnover rate. Although microbial changes are less visible than vegetation shifts, they often occur within a matter of hours or days following an event and could involve alterations in community function, composition, and diversity. For example, periphyton mats in the Everglades (a mixture of cyanobacteria, algae, and microinvertbrates), are highly sensitive to changes in P concentrations and begin to show changes in species composition within a few weeks—long before changes can be observed in the soil or vegetation (McCormick et al. 2001). In coastal wetlands, saltwater intrusion, increased inundation, and storm surge events linked to sea level rise could all directly alter soil microbial community structure, activity, and subsequently the balance of soil C that controls how resilient a wetland is to sea level rise (e.g., Chambers et al. 2011; 2013; Neubauer et al. 2013).

From the most fundamental perspective, salinity increases the ionic strength and conductivity of the microbial environment in coastal wetlands. In order to survive in saline conditions, microorganisms must be capable of osmoregulation, which can involve either the accumulation of potassium chloride in the cytoplasm, or the biosynthesis and/or accumulation of compatible solutes (Oren 2008). Salt intolerant species without these capabilities will experience osmotic stress, disruptions in metabolic function, or even cell lysis upon exposure to salinity (e.g., Reitz and Haynes 2003; Wichern et al. 2006). There is a long history of research pertaining to the effects of salt accumulation in upland soils that demonstrates salt can reduce the size of the soil microbial community and microbial activity, as indicated by lower rates of CO_2 and CH_4 production (Pattnaik et al. 2000; Muhammad et al. 2006; Gennari et al. 2007). It is believed the higher conductivity in the soil-water environment causes osmotic/ionic stress to the organisms and leads to an overall decrease in the rate of C cycling in these systems (Frankenberger and Bingham 1982; Gennari et al. 2007). Few studies have investigated the direct impacts of ionic strength on soil microbiology in wetlands or aquatic systems. A laboratory study where a freshwater wetland soil was exposed to an increase in conductivity from 0 to ~28 mS cm^{-1} through the addition of NaCl demonstrated a 30% decline in microbial respiration over a period of 3 weeks (Chambers et al. 2011). Whether this decline resulted from a general reduction in microbial activity or a shift in community structure remains unclear.

When addressing the impact of sea level rise on wetland soil microbiology, another environmental change may be even more critical than the increase in ionic strength—an increase in the sulfate, SO_4^{2-}, concentration. Unlike inland salinity discussed above, seawater contains a consistent ratio of ions, of which SO_4^{2-} is the third most abundant. Sulfate functions as a terminal electron acceptor (TEA) that soil microbes can utilize for anaerobic respiration. In most terrestrial ecosystems, the presence of SO_4^{2-} is of little consequence to the microbial biota because TEAs are plentiful. The most commonly used TEA by heterotrophic bacteria is oxygen, which utilizes C as an energy source (electron donor) and produces energy in a relatively efficient manner. The presence of water reduces the diffusion of oxygen into the soil by 10,000 times, requiring microbes in wetlands and coastal ecosystems to rely on alternative TEAs, which they utilize in a specific sequence based on their availability and potential energy yield (Patrick and DeLaune 1977; Table 1). Sulfate is near the bottom of the energy cascade and is used only when the environment becomes sufficiently reduced (as indicated by an oxidation reduction potential (Eh) of < -100 mV) and produces only -0.7×10^{-3} kJ mol^{-1} of energy.

Sulfate reduction is typically the dominant pathway of microbial respiration in brackish and saline marshes and mangroves (Howarth 1984; Weston et al. 2006; Kristensen et al. 2008). This suggests that although an increase in ionic strength from seawater intrusion may slightly suppress microbial activity by causing osmotic stress, there will also be a stimulatory effect from the influx of SO_4^{2-} providing an abundance of new TEAs for anaerobic respiration. This was demonstrated under laboratory conditions where CO_2 production rates showed a short-term increase (20–32%) in proportion to the concentration of sulfate added to a freshwater wetland

Table 1. Theoretical energy yields (the more negative the value, the greater the net energy gain) calculated as reactions coupled with glucose oxidation ($C_6H_{12}O_6 \rightarrow CO_2$) and H_2 oxidation ($H_2 \rightarrow H^+$).

Alternative Electron Acceptor	Dominant C End-Product	Eh Range (@ pH 7)	Theoretical Energy Yield (ΔG°_R)
NO_3^-	CO_2	250 to 350 mV	-14.5×10^{-3} kJ mol^{-1}
Mn^{4+}	CO_2	220 to 300 mV	-3.0×10^{-3} kJ mol^{-1}
Fe^{3+}	CO_2	120 to 180 mV	-1.7×10^{-3} kJ mol^{-1}
SO_4^{2-}	CO_2	-100 to -250 mV	-0.7×10^{-3} kJ mol^{-1}
CO_2	CH_4	< -250 mV	-0.1×10^{-3} kJ mol^{-1}

soil (Chambers et al. 2011). Additions of 10 ppt seawater to freshwater sediment cores can cause SO_4^{2-} reduction to become the dominant pathway for microbial respiration after only 12 days, and account for 95% of all organic C oxidation after 35 days of exposure (Weston et al. 2006). However, the stimulation in the overall rate of C loss through respiration may be short-lived as other factors, such as the availability of labile C substrates or nutrients, become limiting (Chambers et al. 2011). A longer-term manipulative field study in a freshwater tidal marsh found CO_2 flux actually declined in treatments exposed to increased salinity for 3.5 years; this decline was correlated with a reduction in the quality of the SOM, further demonstrating the importance of plant-microbial interactions (Neubauer et al. 2013). Interestingly, while the stimulatory effect of SO_4^{2-} on respiration appears to be somewhat transient, the suppression of methanogensis through competitive inhibition by sulfate reducers appears to persist over time (Chambers et al. 2011; Neubauer et al. 2013). In the laboratory, a pulse of brackish water (13 ppt) in a freshwater wetland soil reduced CH_4 flux by 97% in just 5 days (Chambers et al. 2013). In the field, oligohaline water additions to a freshwater wetland soil caused a 2 to 3-fold decrease in CH_4 production that persisted for 3.5 y (Neubauer et al. 2013).

Hydroperiod is another environmental driver of soil microbial processes. In the coastal Everglades, hydroperiod fluctuates seasonally based on rainfall, and daily, based on semi-diurnal tides. During low water (low tide) conditions, more oxygen can diffuse into the soil to promote aerobic respiration. For this reason, low tide CO_2 production rates can be between 50–300% higher in coastal wetlands than high tide CO_2 production rates, with the variability attributed to differences in the hydraulic conductivity of the soil (Chambers et al. 2013). However, as sea level rises, we can expect deeper, more prolonged periods of inundation. Even wetlands that are accreting vertically at a pace comparable to sea level rise tend to do so in a step-wise manner, creating a lag phase in which inundation is greater than under static sea level conditions (Kirwan and Temmerman 2009). Longer periods of water-logging generally slow down microbial activity because organisms must rely exclusively on anaerobic pathways, which tend to be slower and less efficient than aerobic respiration. In a mesocosm study that simulated sea level rise in an Everglades mangrove peat soil, soil organic C loss was 90% higher under control water levels, as compared increased inundation, when combined with elevated salinities (Chambers et al. 2014). This may reduce the amount of organic C lost through the microbial pathway as hydroperiod increases, promoting soil C storage and accretion. However, the same study also demonstrated that prolonged inundation may actually result in a loss of soil material, as seen by a decrease in surface (0–5 cm) soil bulk density (Chambers et al. 2014). While the mechanism for this reduction in bulk density is not known, it was correlated with an increase in porewater dissolved organic C, causing speculation it may be a product of excessive leaching during water-logging or increased shear stress due to the deeper water column above the soil (Chambers et al. 2014).

It is generally thought microbial density and diversity is comparable in freshwater and saltwater systems, but the identity of the individual organisms themselves differs with salinity (Capone and Kiene 1988). However, new evidence contradicts this axiom, finding higher microbial biomass-associated C in salt marsh soils, compared to freshwater and brackish marsh soils (Chambers et al. 2013). Other research has found a direct correlation between bacterial abundance and salinity in freshwater tidal marshes (Morrissey et al. 2014b). Only one study of microbial community composition has been performed in the Everglades along a salinity transect from 0 to 49 ppt. Here, the diversity of the microbial community remained similar, but the identity of the microbes diverged significantly, based primarily on salinity, and secondarily on P availability (Ikenaga et al. 2010).

Ultimately, while soil microbial communities may be rapid indicators of wetland ecosystem response, the feedback between plant and microbial communities will modulate this response. For example, the presence or absence of plants will have significant influences on soil redox potential (i.e., the size of the oxidized rhizosphere), quantity and quality of organic C, and will interact with enzyme synthesis. Under conditions in which salinity drives a decline of plant productivity, vegetation death, or vegetation community shifts, diversity and function of soil microbial communities will be fundamentally altered. Unfortunately, plant-microbial interactions are not well understood in wetlands, especially in coastal peatlands.

The Future of the Coastal Everglades

As salinity and inundation patterns in the Everglades change in response to sea level rise and human-driven alterations in hydrology, coastal zone ecology is changing as well. The movement of ecotones (regions bridging two distinct community types) is often a reliable way of monitoring environmental change because ecotones develop and migrate in response to specific environmental gradients. Several studies have used historic aerial photography and various bio-indicators to document shifts in the location of coastal ecotones in the Everglades over time. This provides a glimpse of how the ecosystem has responded to past sea level changes and a basis for predicting future ecosystem responses.

The land boundary of the Everglades, and all of Florida, has changed significantly throughout geologic time in response to sea level. Soil cores indicate the current seaward edge of the Everglades formed from red mangrove (*R. mangle*) derived peats that began accumulating approximately 3,500 years B.P. (Parkinson et al. 1989). As sea level rose, the mangrove soil platform accreted vertically and expanded landward. Meanwhile, the establishment and growth of oyster reefs off the coast also allowed mangroves to expand in the seaward direction, creating mangrove islands (Parkinson et al. 1989). The distribution of mangroves, and peat accumulation from relic mangrove forests, are considered good indicators of historic sea level because they always occupy the upper portion of the tidal range (Scholl 1964).

Current research shows that mangroves within the Everglades are continuing to respond to changing sea levels, mainly through the expansion of their coverage at the expanse of inland marsh habitat (Doyle et al. 2003). A look at aerial photographs of the Ten Thousand Islands National Wildlife Refuge on the western edge of the Everglades has shown a 35% increase in mangrove coverage in the past 78 years. The construction of canals near the coast is a driving force in the expansion of mangroves within previously low salinity marshes because they provide a conduit for saltwater intrusion and propagule dispersal (Krauss et al. 2011). In a region near Taylor Slough, known as the Southeast Saline Everglades, researchers have documented the movement of inland ecotones since the 1940s. Here the boundary between the mangrove-graminoid community and the interior sawgrass marsh has shifted inland 3.3 km, which is believed to be in response to a combination of reduced freshwater flows and encroaching seas (Ross et al. 2000; Troxler 2012). In the southeast Everglades, extensive water diversions and flow alterations are accelerating saltwater intrusion and may be a useful model for predicting sea level rise effects elsewhere. According to mollusk records in the coastal soils near Biscayne Bay, prior to local drainage efforts that began ~70 years ago, the marsh/mangrove ecotone was migrating landward at a rate of 0.14 m y^{-1}, but since drainage, the rate has increased to 3.1 m y^{-1} (Gaiser et al. 2006). This migration is correlated with an increase in salinity from 2 ppt to 13.2 ppt. In addition to general mangrove expansion, models also predict the mangroves themselves will have reduced height and contain a greater proportion of red mangroves as sea level rises (Doyle et al. 2003).

While there has been significant research demonstrating that the inland ecotones of the coastal Everglades are migrating landward with rising sea levels, there are large uncertainties about how the balance of coastal erosion and soil accretion will determine the position of the land boundary as sea level rises. Some studies suggest that coastal erosion directly resulting from sea level rise is low, with most documented erosional events accompanying hurricanes (Doyle et al. 2003; Wanless et al. 1994). At present, mangrove soil accretion rates are exceeding sea level rise rates in at least one area of the Everglades (Smoak et al. 2012), but more research is needed to understand how accretion may vary spatially. The fact that seawater serves as the primary source of nutrients to the coastal zone (Childers et al. 2006) suggests that saltwater intrusion could increase productivity (Chen and Twilley 1999), and subsequently soil accretion. However, increasing salinity may have the opposite effect due to the complicated relationship with concomitant physical forcings such as inundation depth, sediment supply, and disturbances (e.g., fire, wind, storm surge). The interaction of salinity-induced collapse of freshwater peats and mangrove expansion is also worth considering as a driver shaping the coastal Everglades of the future.

As discussed earlier, Wanless and Vlaswinkel (2005) suggest that collapsed areas of freshwater peat—as observed in the Cape Sable area of Everglades National Park—may coalesce through time, resulting in larger open water areas. In fact, it is believed that this phenomenon contributed to the formation of Whitewater Bay in Everglades National Park (Wanless and Vlaswinkel 2005). One could hypothesize that unless collapsed areas receive new sediment or are colonized by mangroves that can re-stabilize the soil, they may continue to grow larger and transition directly into subtidal habitat as sea levels continue to rise. With 8,744 km^2 of south and southwest Florida being located below the 1.5 m elevation contour (Titus and Richman 2001) and rates of sea level rise thought to be accelerating (Church and White 2006), the fate of the seaward boundary of the Everglades remains highly uncertain.

Conclusion

Overall, there is ample evidence to support the idea that coastal ecology in the Everglades is changing in concert with rising sea levels. In the southeast Everglades, the signature of sea level rise has been blurred with significant hydrologic modifications that are decreasing freshwater flows and accelerating saltwater intrusion (Ross et al. 2000). However, ecotone shifts and elevated salinities are being documented across the entire ecosystem. Changes in vegetation communities are often the most apparent manifestation of migrating environmental gradients and directly impact soil microbiology by altering the physical and chemical environment. However, the structure and function of microbial communities will likely respond far in advance of vegetation shifts. Few studies have addressed the direct impacts of sea level rise on soil micro-biota, but evidence such as changes in soil respiration rates (CO_2 production) suggest seawater intrusion can accelerate heterotrophic microbial activity for the short-term and suppress methanogenesis for the long-term. The important question is how changes in vegetation type, productivity, and microbial activity will affect the overall balance of C in coastal wetlands. An increase in C inputs and net decrease in microbial respiration would create a positive feedback to promote vertical accretion and increase wetland resilience. In the most likely scenario, different regions of the Everglades will have unique responses to sea level based on the current health of the ecosystem, the supply of inorganic sediments, nutrient availability, topography, and occurrence of extreme events. Future research should focus on quantifying the impact to microbial populations and vital microbial processes such as nutrient cycling, C storage, and the plant-soil interactions that modulate the stability of coastal peatlands vulnerable to sea-level rise, such as the Everglades.

Acknowledgements

We are grateful to all FCE LTER scientists for their contributions to discussions of this topic and acknowledge financial support provided by National Science Foundation grants DEB-1237517 and DBI-0620409. This is contribution number 668 from the Southeast Environmental Research Center at Florida International University.

References

Aburto-Oropeza, O., E. Ezcurra, G. Danemann, V. Valdez, J. Murray and E. Sala. 2008. Mangroves in the Gulf of California increase fishery yields. Proc. Natl. Acad. Sci. USA 105: 10456–10459.

Alongi, D.M. 2008. Mangrove forests: Resilience, protection from tsunamis, and responses to global climate change. Est. Coast. Shelf Sci. 76: 1–13.

Alongi, D.M., F. Tirendi and B.F. Clough. 2000. Below-ground decomposition of organic matter in forests of the mangroves *Rhizophora stylosa* and *Avicennia marina* along the arid coast of Western Australia. Aquatic Bot. 69: 97–122.

Barbier, E.B. 2013. Valuing ecosystem services for coastal wetland protection and restoration: progress and challenges. Resources 2: 213–230.

Barr, J.G., V. Engel, J.D. Fuentes, J.C. Zieman, T.L. O'Halloran, T.J. Smith and G.H. Anderson. 2010. Controls on mangrove forest-atmosphere carbon dioxide exchanges in western Everglades National Park. J. Geophys. Res. Biogeosci. 115: G02020.

Batzer, D.P. and R.R. Sharitz (eds.). 2006. Ecology of Freshwater and Estuarine Wetlands. University of California Press, Berkeley.

Belanger, T.V., D.J. Scheidt and J.R. Platko, II. 1989. Effects of nutrient enrichment on the Florida Everglades USA. Lake Reservoir Manage. 5: 101–112.

Berg, B., G. Ekbohm, M. Johansson, C. McClaughtery, F. Rutigliano and A.V. DeSanto. 1996. Maximum decomposition limits of forest litter types: a synthesis. Can. J. Bot. 74: 659–672.

Bianchi, T.S., M.A. Allison, J. Zhao, X. Li, R.S. Comeaux, R.A. Feagin and R.W. Kulawardhana. 2013. Historical reconstruction of mangrove expansion in the Gulf of Mexico: linking climate change with carbon sequestration in coastal wetlands. Est. Coast. Shelf Sci. 119: 7–16.

Bouillon, S., A.V. Borges, E. Castaneda-Moya, K. Diele, T. Dittmar, N.C. Duke, E. Kristensen, S.Y. Lee, C. Marchand, J.J. Middelburg, V.H. Rivera-Monroy, T.J. Smith and R.R. Twilley. 2008. Mangrove production and carbon sinks: A revision of global budget estimates. Global Biogeochem. Cyc. 22: GB003052.

Bradley, P.M. and J.T. Morris. 1991. The influence of salinity on the kinetics of NH4$^+$ uptake in Spartina-alterniflora. Oecologia 85: 375–380.

Breithaupt, J.L., J.M. Smoak, T.J. Smith, C.J. Sanders and A. Hoare. 2012. Organic carbon burial rates in mangrove sediments: Strengthening the global budget. Global Biogeochem. Cyc. 26: GB3011.

Capone, D.G. and R.P. Kiene. 1988. Comparison of microbial dynamics in marine and fresh-water sediments: contrasts in anaerobic carbon catabolism. Limnol. Oceanogr. 33: 725–749.

Castañeda, E. 2010. Landscape patterns of community structure, biomass, and net primary productivity of mangrove forests in the Florida coastal Everglades as a function of resources, regulators, hydroperiod, and hurricane disturbance. Ph.D. Dissertation, Louisiana State University, Baton Rouge, LA.

Castañeda-Moya, E., R.R. Twilley, V.H. Rivera-Monroy, K. Zhang, S.E. Davis and M. Ross. 2010. Spatial patterns of sediment deposition in mangrove forests of the Florida Coastal Everglades after the passage of Hurricane Wilma. Est. Coast. 33: 45–58.

Castañeda-Moya, E., R.R. Twilley, V.H. Rivera-Monroy, B. Marx, C. Coronado-Molina and S.E. Ewe. 2011. Patterns of root dynamics in mangrove forests along environmental gradients in the Florida Coastal Everglades, USA. Ecosystems 14: 1178–1195.

Castañeda-Moya, E., R.R. Twilley and V.H. Rivera-Monroy. 2013. Allocation of biomass and net primary productivity of mangrove forests along environmental gradients in the Florida Coastal Everglades, USA. For. Ecol. Manage. 307: 226–241.

Chambers, L.G., K.R. Reddy and T.Z. Osborne. 2011. Short-term response of carbon cycling to salinity pulses in a freshwater wetland. Soil Sci. Soc. Am. J. 75: 2000–2007.

Chambers, L.G., T.Z. Osborne and K.R. Reddy. 2013. Effect of salinity pulsing events on soil organic carbon loss across an intertidal wetland gradient: a laboratory experiment. Biogeochemistry 115: 363–383.

Chambers, L.G., S.E. Davis, T. Troxler, J. Boyer, A. Downey-Wall and L. Scinto. 2014. Biogeochemical effects of saltwater intrusion and increased inundation on Everglades peat soil. Hydrobiologia 726: 195–211.

Chen, R.H. and R.R. Twilley. 1999. Patterns of mangrove forest structure and soil nutrient dynamics along the Shark River Estuary, Florida. Estuaries 22: 955–970.

Childers, D.L., J.N. Boyer, S.E. Davis, C.J. Madden, D.T. Rudnick and F.H. Sklar. 2006. Relating precipitation and water management to nutrient concentrations in the oligotrophic "upside-down" estuaries of the Florida Everglades. Limnol. Oceanogr. 51: 602–616.

Church, J.A. and N.J. White. 2006. A 20th century acceleration in global sea-level rise. Geophys. Res. Let. 33: 4.

Cornwell, W.K., J.H.C. Cornelissen and K. Amatangelo, E. Dorrepaal, V.T. Eviner, O. Godoy, S.E. Hobbie, B. Hoorens, H. Kurokawa, N. Perez-Harguindeguy, H.M. Quested, L.S. Santiago, D.A. Wardle, I.J. Wright, R. Aerts, S.D. Allison, P. van Bodegom, V. Brovkin, A. Chatain, T.V. Callaghan, S. Diaz, E. Garnier, D.E. Gurvich, E. Kazakou, J.A. Klein, J. Read, P.B. Reich, N.A. Soudzilovskaia, M.V. Vaieretti and M. Westoby. 2008. Plant species traits are the predominant control on litter decomposition rates within biomes worldwide. Ecol. Let. 11: 1065–1071.

Costanza, R., O. Perez-Maqueo, M.L. Martinez, P. Sutton, S.J. Anderson and K. Mulder. 2008. The value of coastal wetlands for hurricane protection. Ambio 37: 241–248.

Craft, C., J. Clough, J. Ehman, S. Joye, R. Park, S. Pennings, H.Y. Guo and M. Machmuller. 2009. Forecasting the effects of accelerated sea-level rise on tidal marsh ecosystem services. Front. Ecol. Environ. 7: 73–78.

Craft, C.B. and C.J. Richardson. 2008. Soil characteristics of the everglades peatland. pp. 59–72. *In*: C.J. Richardson (ed.). The Everglades Experiments: Lessons for Ecosystem Restoration. Springer, New York.

Daoust, R.J. and D.L. Childers. 2004. Ecological effects of low-level phosphorus additions on two plant communities in a neotropical freshwater wetland ecosystem. Oecologia 141: 672–686.

Davis, S.E., C. Coronado-Molina, D.L. Childers and J.W. Day, Jr. 2003. Temporal variability in C, N, and P dynamics associated with red mangrove (*Rhizophora mangle* L.) leaf decomposition. Aquatic Bot. 75: 199–215.

Davis, S.E., D.L. Childers and G.B. Noe. 2006. The contribution of leaching to the rapid release of nutrients and carbon in the early decay of oligotrophic wetland vegetation. Hydrobiologia 569: 87–97.

Davis, S.M. 1991. Growth, decomposition, and nutrient retention of Cladium jamaicense crantz and Typha-domingensis pres in the Florida Everglades. Aquatic Bot. 40: 203–224.

Day, J.W., G.P. Shaffer, L.D. Britsch, D.J. Reed, S.R. Hawes and D. Cahoon. 2000. Pattern and process of land loss in the Mississippi Delta: a spatial and temporal analysis of wetland habitat change. Estuaries 23: 425–438.

Day, J.W., G.P. Kemp, D.J. Reed, D.R. Cahoon, R.M. Boumans, J.M. Suhayda and R. Gambrell. 2011. Vegetation death and rapid loss of surface elevation in two contrasting Mississippi delta salt marshes: The role of sedimentation, autocompaction and sea-level rise. Ecol. Eng. 37: 229–240.

DeLaune, R.D., J.A. Nyman and W.H. Patrick. 1994. Peat collapse, pending and wetland loss in a rapidly submerging coastal marsh. J. Coast. Res. 10: 1021–1030.

Dittmar, T., N. Hertkorn, G. Kattner and R.J. Lara. 2006. Mangroves, a major source of dissolved organic carbon to the oceans. Global Biogeochem. Cyc. 20: GB1012.

Donnelly, J.P. and M.D. Bertness. 2001. Rapid shoreward encroachment of salt marsh cordgrass in response to accelerated sea-level rise. Proc. Natl. Acad. Sci. USA 98: 14218–14223.

Doyle, T.W., G.F. Girod and M.A. Books. 2003. Modeling mangrove forest migration along the southwest Coast of Florida under climate change. pp. 211–221. *In*: Z.H. Ning, R.E. Turner, T.W. Doyle and K. Abdollahi (eds.). Integrated Assessment of the Climate Change Impacts on the Gulf Coast Region. GRCCC and LSU Graphic Services, Baton Rouge, LA.

Field, J.A. and G. Lettinga. 1992. Toxicity of tannic compounds to microorganisms. pp. 673–692. *In*: R.W. Hemingway and E. Laks (eds.). Plant Polyphenols: Synthesis, Properties, Significance. Plenum Press, New York.

Frankenberger, W.T. and F.T. Bingham. 1982. Influence of salinity on soil enzyme-activities. Soil Sc. Soc. Am. J. 46: 1173–1177.

Gaiser, E.E., A. Zafiris, P.L. Ruiz, F.A.C. Tobias and M.S. Ross. 2006. Tracking rates of ecotone migration due to salt-water encroachment using fossil mollusks in coastal South Florida. Hydrobiologia 569: 237–257.

Gedan, K.B., B.R. Silliman and M.D. Bertness. 2009. Centuries of human-driven change in salt marsh ecosystems. Ann. Rev. Mar. Sci. 1: 117–141.

Gedan, K.B., M.L. Kirwan, E. Wolanski, E.B. Barbier and B.R. Silliman. 2011. The present and future role of coastal wetland vegetation in protecting shorelines: answering recent challenges to the paradigm. Clim. Chan. 106: 7–29.

Gennari, M., C. Abbate, V. La Porta, A. Baglieri and A. Cignetti. 2007. Microbial response to Na_2SO_4 additions in a volcanic soil. Arid Land Res. Manage. 21: 211–227.

Goodfellow, M. and S.T. Williams. 1983. Ecology of actinomycetes. Ann. Rev. Microbiol. 37: 189–216.

Grayston, S.J., S.Q. Wang, C.D. Campbell and A.C. Edwards. 1998. Selective influence of plant species on microbial diversity in the rhizosphere. Soil Biol. Biochem. 30: 369–378.

Güsewell, S. and M.O. Gessner. 2009. N:P ratios influence litter decomposition and colonization by fungi and bacteria in microcosms. Funct. Ecol. 23: 211–219.

Haigh, I.D., T. Wahl, E.J. Rohling, R.M. Price, C.B. Pattiaratchi, F.M. Calafat and S. Dangendorf. 2014. Timescales for detecting a significant acceleration in sea level rise. Nature Comm. 5: 3635.

Howarth, R.W. 1984. The ecological significance of sulfur in the energy dynamics of salt-marsh and coastal marine-sediments. Biogeochem. 1: 5–27.

Huxham, M., J. Langat, F. Tamooh, H. Kennedy, M. Mencuccini, M.W. Skov and J. Kairo. 2010. Decomposition of mangrove roots: Effects of location, nutrients, species identity and mix in a Kenyan forest. Est. Coast. Shelf Sci. 88: 135–142.

Ikenaga, M., R. Guevara, A.L. Dean, C. Pisani and J.N. Boyer. 2010. Changes in community structure of sediment bacteria along the Florida coastal Everglades marsh-mangrove-seagrass salinity gradient. Microb. Ecol. 59: 284–295.

[IPCC]. 2007. Climate Change 2007: A Synthesis Report. Valencia, Spain.

Jennerjahn, T.C. and V. Ittekkot. 2002. Relevance of mangroves for the production and deposition of organic matter along tropical continental margins. Naturwissenschaften 89: 23–30.

Jiang, J., D.L. DeAngelis, T.J. Smith, S.Y. Teh and H.L. Koh. 2012. Spatial pattern formation of coastal vegetation in response to external gradients and positive feedbacks affecting soil porewater salinity: a model study. Land. Ecol. 27: 109–119.

Kirwan, M. and S. Temmerman. 2009. Coastal marsh response to historical and future sea-level acceleration. Quat. Sci. Rev. 28: 1801–1808.

Koch, M.S., I.A. Mendelssohn and K.L. McKee. 1990. Mechanism for the hydrogen sulfide-induced growth limitation in wetland macrophytes. Limnol. Oceanogr. 35: 399–408.

Krauss, K.W., A.S. From, T.W. Doyle, T.J. Doyle and M.J. Barry. 2011. Sea-level rise and landscape change influence mangrove encroachment onto marsh in the Ten Thousand Islands region of Florida, USA. J. Coast. Conserv. 15: 629–638.

Kristensen, E., S. Bouillon, T. Dittmar and C. Marchand. 2008. Organic carbon dynamics in mangrove ecosystems: A review. Aquatic Bot. 89: 201–219.

Locker, S.D., A.C. Hine, L.P. Tedesco and E.A. Shinn. 1996. Magnitude and timing of episodic sea-level rise during the last deglaciation. Geology 24: 827–830.

Lodge, T.E. 2010. The Everglades Handbook: Understanding the Ecosystem. 3rd Edition. CRC Press, Boca Raton, FL.

McCarthy, A.J. 1987. Lignocellulose-degrading actinomycetes. FEMS Microbiology Reviews 46: 145–163.

McCormick, P.V., R.B.E. Shuford, J.G. Backus and W.C. Kennedy. 1998. Spatial and seasonal patterns of periphyton biomass and productivity in the northern Everglades, Florida, USA. Hydrobiologia 362: 185–208.

McCormick, P.V., M.B. O'Dell, R.B.E. Shuford, III, J.G. Backus and W.C. Kennedy. 2001. Periphyton responses to experimental phosphorus enrichment in a subtropical wetland. Aquatic Bot. 71: 119–139.

McIvor, C.C., J.A. Ley and R.D. Bjork. 1994. Changes in freshwater inflow from the Everglades to Florida Bay including effects on biota and biotic processes: a review. pp. 117–146. *In*: S.M. Davis and J.C. Ogden (eds.). Everglades: The Ecosystem and its Restoration. St. Lucie Press, Delray Beach, FL.

McKee, K.L. 2011. Biophysical controls on accretion and elevation change in Caribbean mangrove ecosystems. Est. Coast. Shelf. Sci. 91: 475–483.

McKee, K.L. and I.A. Mendelssohn. 1989. Response of a fresh-water marsh plant community to increased salinity and water level. Aquatic Bot. 34: 301–316.

McKee, K.L., D.R. Cahoon and I.C. Feller. 2007. Caribbean mangroves adjust to rising sea level through biotic controls on change in soil elevation. Global Ecol. Biogeograph. 16: 545–556.

McVoy, C.W., W.P. Said, J. Obeysekera, J.A. Van Arman and T.W. Dreschel. 2011. Landscapes and Hydrology of the Predrainage Everglades. University Press of Florida, Gainesville, FL.

Melillo, J.M., J.D. Aber and J.F. Muratore. 1982. Nitrogen and lignin control of hardwood leaf litter decomposition dynamics. Ecology 63: 621–626.

Middleton, B.A. and K.L. McKee. 2001. Degradation of mangrove tissues and implications for peat formation in Belizean island forests. J. Ecol. 89: 818–828.

Moorehead, D.L. and R.L. Sinsabaugh. 2006. A theoretical model of litter decay and microbial interaction. Ecological Monogr. 76: 151–174.

Morris, J.T., P.V. Sundareshwar, C.T. Nietch, B. Kjerfve and D.R. Cahoon. 2002. Responses of coastal wetlands to rising sea level. Ecology 83: 2869–2877.

Morrissey, E.M., D.J. Berrier, S.C. Neubauer and R.B. Franklin. 2014a. Using microbial communities and extracellular enzymes to link soil organic matter characteristics to greenhouse gas production in a tidal freshwater wetland. Biogeochem. 117: 473–490.

Morrissey, E.M., J.L. Gillespie, J.C. Morina and R.B. Franklin. 2014b. Salinity affects microbial activity and soil organic matter content in tidal wetlands. Global Change Biol. 20: 1351–1361.

Muhammad, S., T. Muller and R.G. Joergensen. 2006. Decomposition of pea and maize straw in Pakistani soils along a gradient in salinity. Biol. Fertil. Soil. 43: 93–101.

Muhs, D.R., K.R. Simmons, R.R. Schumann and R.B. Halley. 2011. Sea-level history of the past two interglacial periods: new evidence from U-series dating of reef corals from south Florida. Quat. Sci. Rev. 30: 570–590.

Najjar, R.G., H.A. Walker, P.J. Anderson, E.J. Barron, R.J. Bord, J.R. Gibson, V.S. Kennedy, C.G. Knight, J.P. Megonigal, R.E. O'Connor, C.D. Polsky, N.P. Psuty, B.A. Richards, L.G. Sorenson, E.M. Steele and R.S. Swanson. 2000. The potential impacts of climate change on the mid-Atlantic coastal region. Clim. Res. 14: 219–233.

Neubauer, S.C. 2013. Ecosystem responses of a tidal freshwater marsh experiencing saltwater intrusion and altered hydrology. Est. Coast. 36: 491–507.

Neubauer, S.C., R.B. Franklin and D.J. Berrier. 2013. Saltwater intrusion into tidal freshwater marshes alters the biogeochemical processing of organic carbon. Biogeosciences 10: 8171–8183.

Neumann, J.E., G. Yohe, R. Nicholls and M. Manion. 2000. Sea-level rise and global climate change: A review of impacts to U.S. coasts. Pew Center on Global Climate Change.

Nicholls, R.J., F.M.J. Hoozemans and M. Marchand. 1999. Increasing flood risk and wetland losses due to global sea-level rise: regional and global analyses. Global Environ. Chang 9: S69–S87.

Noe, G., D.L. Childers and R.D. Jones. 2001. Phosphorus biogeochemistry and the impacts of phosphorus enrichment: Why are the Everglades so unique? Ecosystems 4: 603–624.

Nuttle, W.K., J.W. Fourqurean, B.J. Cosby, J.C. Zieman and M.B. Robblee. 2000. Influence of net freshwater supply on salinity in Florida Bay. Wat. Resour. Res. 36: 1805–1822.

Nyman, J.A., R.D. Delaune and W.H. Patrick. 1990. Wetland soil formation in the rapidly subsiding Mississippi River delatic plain-mineral and organic-matter relationships. Est. Coast. Shelf Sci. 31: 57–69.

Obeysekera, J., M. Irizarry, J. Park, J. Barnes and T. Dessalegne. 2011. Climate change and its implications for water resources management in south Florida. Stochastic Environm. Res. Risk Assess. 25: 495–516.

Odum, W.E. 1988. Comparative ecology of tidal fresh-water and salt marshes. Ann. Rev. Ecol. Syst. 19: 147–176.

Oren, A. 2008. Microbial life at high salt concentrations: phylogenetic and metabolic diversity. Saline Syst. 4: 2.

Park, J., J. Obeysekera, M. Irizarry, J. Barnes, P. Trimble and W. Park-Said. 2011. Storm surge projections and implications for water management in South Florida. Clim. Chan. 107: 109–128.

Parkinson, R.W. 1989. Decelerating Holocene sea-level rise and its influence on southwest Florida coastal evolution—a transgressive regressive stratigraphy. J. Sed. Petrol. 59: 960–972.

Patrick, W.H.J. and R.D. DeLaune. 1977. Chemical and biological redox systems affecting nutrient availability in the coastal wetlands. Geosci. Mar. 18: 131–137.

Pattnaik, P., S.R. Mishra, K. Bharati, S.R. Mohanty, N. Sethunathan and T.K. Adhya. 2000. Influence of salinity on methanogenesis and associated microflora in tropical rice soils. Microbiol. Res. 155: 215–220.

Pearlstine, L.G., E.V. Pearlstine and N.G. Aumen. 2010. A review of the ecological consequences and management implications of climate change for the Everglades. J. N. Am. Benthol. Soc. 29: 1510–1526.

Peltoniemi, K., H. Fritze and R. Laiho. 2009. Response of fungal and actinobacterial communities to water-level drawdown in boreal peatland sites. Soil Biol. Biochem. 41: 1902–1914.

Prescott, C. 2010. Litter decomposition: what controls it and how can we alter it to sequester more carbon in forest soils? Biogeochemistry 101: 133–149.

Qualls, R.G., C.J. Richardson and L.J. Sherwood. 2001. Soil reduction-oxidation potential along a nutrient enrichment gradient in the Everglades. Wetlands 21: 403–411.

Reddy, K.R. and R.D. DeLaune. 2008. Biogeochemistry of Wetlands. CRC Press, Boca Raton, FL.

Rejmankova, E. and P. Macek. 2008. Response of root and sediment phosphatase activity to increased nutrients and salinity. Biogeochemistry 90: 159–169.

Rietz, D.N. and R.J. Haynes. 2003. Effects of irrigation-induced salinity and sodicity on soil microbial activity. Soil Biol. Biochem. 35: 845–854.

Rivera-Monroy, V.H., K. de Mutsert, R.R. Twilley, E. Castaneda-Moya, M.M. Romigh and S.E. Davis. 2007. Patterns of nutrient exchange in a riverine mangrove forest in the Shark River Estuary, Florida, USA. Hydrobiologia 17: 169–178.

Robertson, A.I., D.M. Alongi and K.G. Boto. 1992. Food chains and carbon fluxes. pp. 293–326. *In*: A.I. Robertson and D.M. Alongi (eds.). Tropical Mangrove Ecosystems. American Geophysical Union, Washington, D.C.

Ross, M.S., J.F. Meeder, J.P. Sah, P.L. Ruiz and G.J. Telesnicki. 2000. The Southeast Saline Everglades revisited: 50 years of coastal vegetation change. J. Veg. Sci. 11: 101–112.

Saha, A.K., S. Saha, J. Sadle, J. Jiang, M.S. Ross, R.M. Price, L. Sternberg and K.S. Wendelberger. 2011. Sea level rise and South Florida coastal forests. Clim. Chan. 107: 81–108.

Schlesinger, W.H. 1997. Biogeochemistry: An Analysis of Global Change. Academic Press, San Diego, CA.

Scholl, D. 1964. Recent sedimentary record in mangrove swamps and rise in sea level over part of the southwestern coast of Florida: part 1. Mar. Geol. 1: 344–366.

Sharpe, P.J. and A.H. Baldwin. 2012. Tidal marsh plant community response to sea-level rise: a mesocosm study. Aquatic Bot. 101: 34–40.

Smith, S.M. 2009. Multi-decadal changes in Salt Marshes of Cape Cod, MA: photographic analyses of vegetation loss, species shifts, and geomorphic change. Northeastern Naturalist 16: 183–208.

Smith, T.J., G.H. Anderson, K. Balentine, G. Tiling, G.A. Ward and K.R.T. Whelan. 2009. Cumulative impacts of hurricanes on Florida mangrove ecosystems: sediment deposition, storm surges and vegetation. Wetlands 29: 24–34.

Smith, T.J., A.M. Foster, G. Tiling-Range and J.W. Jones. 2013. Dynamics of mangrove-marsh ecotones in subtropical coastal wetlands: fire, sea-level rise, and water levels. Fire Ecol. 9: 66–77.

Smoak, J.M., J.L. Breithaupt, T.J. Smith and C.J. Sanders. 2013. Sediment accretion and organic carbon burial relative to sea-level rise and storm events in two mangrove forests in Everglades National Park. Catena 104: 58–66.

Teh, S.Y., D.L. DeAngelis, L.D.L. Sternberg, F.R. Miralles-Wilhelm, T.J. Smith and H.L. Koh. 2008. A simulation model for projecting changes in salinity concentrations and species dominance in the coastal margin habitats of the Everglades. Ecol. Model. 213: 245–256.

Titus, J.G. and C. Richman. 2001. Maps of lands vulnerable to sea level rise: modeled elevations along the US Atlantic and Gulf coasts. Clim. Res. 18: 205–228.

Troxler, T.G. 2012. Ecological monitoring of southern Everglades wetlands, mangrove transition zone and "white zone" interactions with Florida Bay. Annual Report to the South Florida Water Management District, West Palm Beach, FL, 65 pp.

Troxler, T.G., M. Ikenaga, L. Scinto, J. Boyer, R. Condit, R. Perez, G. Gann and D. Childers. 2012. Patterns of soil bacteria and canopy community structure related to tropical peatland development. Wetlands 32: 769–782.

Troxler, T.G., D.L. Childers and C.J. Madden. 2013. Drivers of decadal-scale change in southern Everglades wetland macrophyte communities of the coastal ecotone. Wetlands 10.1007/s13157-013-0446-5.

Turner, R.E., E.M. Swenson, C.S. Milan, J.M. Lee and T.A. Oswald. 2004. Below-ground biomass in healthy and impaired salt marshes. Ecol. Res. 19: 29–35.

U.S. Census Bureau. 2013. http://quickfacts.census.gov/qfd/states.

Vitousek, P.M. and R.W. Howarth. 1991. Nitrogen limitations on land and in the sea—how can it occur. Biogeochemistry 13: 87–115.

Wanless, H.R. and B.M. Vlaswinkel. 2005. Coastal landscape and channel evolution affecting critical habitats at Cape Sable, Everglades National Park, Florida. Final Report to Everglades National Park Service, U.S. Department of Interior. 197 pp.

Wanless, H., R. Parkinson and L. Tedesco. 1994. Sea level control on stability of Everglades wetlands. pp. 198–224. *In*: S. Davis and J. Ogden (eds.). Everglades: The Ecosystem and its Restoration. St. Lucie, Boca Raton, FL.

Weston, N.B., R.E. Dixon and S.B. Joye. 2006. Ramifications of increased salinity in tidal freshwater sediments: Geochemistry and microbial pathways of organic matter mineralization. J. Geophys. Res. Biogeosci. 111: G01009.

Whelan, K.R.T., T.J. Smith, III, D.R. Cahoon, J.C. Lynch and G.H. Anderson. 2005. Groundwater control of mangrove surface elevation: shrink and swell varies with soil depth. Estuaries 28: 833–843.

Wichern, J., F. Wichern and R.G. Joergensen. 2006. Impact of salinity on soil microbial communities and the decomposition of maize in acidic soils. Geoderma 137: 100–108.

Wieski, K., H.Y. Guo, C.B. Craft and S.C. Pennings. 2010. Ecosystem functions of tidal fresh, brackish, and salt marshes on the Georgia Coast. Est. Coast. 33: 161–169.

Willard, D.A. and C.E. Bernhardt. 2011. Impacts of past climate and sea level change on Everglades wetlands: placing a century of anthropogenic change into a late-Holocene context. Clim. Chan. 107: 59–80.

Williams, K., K.C. Ewel, R.P. Stumpf, F.E. Putz and T.W. Workman. 1999. Sea-level rise and coastal forest retreat on the west coast of Florida, USA. Ecology 80: 2045–2063.

Zhang, K.Q. 2011. Analysis of non-linear inundation from sea-level rise using LIDAR data: a case study for South Florida. Clim. Chan. 106: 537–565.

SECTION II

Periphyton

CHAPTER

6

The Importance of Species-Based Microbial Assessment of Water Quality in Freshwater Everglades Wetlands

Evelyn E. Gaiser,[1,*,a] *Andrew D. Gottlieb,*[2,c] *Sylvia S. Lee*[1,d] *and Joel C. Trexler*[1,b]

Introduction

Benthic microbial communities (commonly, periphyton) are important components of shallow aquatic ecosystems, occurring as a mixed consortium of algae, bacteria, fungi and detritus associated with benthic substrata. Due to relatively high carbon turnover rates, periphyton can assume a regulatory role in ecosystems, controlling gas and nutrient concentrations, organic and mineral soil accrual, and the quality and abundance of food for aquatic consumers (Stevenson et al. 1996). Periphyton communities also respond very rapidly to environmental changes, driven by rapid cellular responses of competing species that differ in resource use efficiency, stress resistance, or dispersal capability (Wetzel 2005; Thomas et al. 2006). For this reason, periphyton community attributes are widely incorporated into water quality assessment protocols (Stevenson

[1] Department of Biological Sciences, Southeast Environmental Research Center, Florida International University, Miami, FL 33199.
[a] Email: gaisere@fiu.edu
[b] Email: trexlerj@fiu.edu
[2] Harvard Drive, Lake Worth, FL 33460, USA.
[c] Email: adgottlieb71@gmail.com
[d] Email: slee017@fiu.edu
* Corresponding author

and Bahls 1999), and are a core component of regulatory monitoring of water quality in the United States (Charles et al. 2002; Hill et al. 2003) and other nations (Smith and McBride 1990; Kelly et al. 1998).

In Everglades wetlands, periphyton is a prominent feature, forming thick mats on most submersed surfaces (Browder et al. 1994; Gaiser et al. 2011). These mats are dominated by a matrix of filamentous cyanobacteria that excrete copious extrapolymeric substances (EPS), trap detritus and accrete calcium carbonate in this limestone-based ecosystem. Cyanobacteria are joined by diatoms, green algae, bacteria and fungi capable of surviving inherent stresses of high surface light intensities, occasional desiccation, and very low intrinsic concentrations of phosphorus (P), the limiting nutrient in Everglades wetlands (Gaiser et al. 2011). Periphyton has been found to accrue at rates of 10–20 g C m^{-2} yr^{-1}, attaining an average standing stock of 150 g C m^{-2} (but up to 10,000 g C m^{-2}) across Everglades marshes and sloughs (Troxler et al. 2013). These accumulation rates and standing stocks are considerably higher than most wetlands, although similar to other carbonate-based ecosystems of the Caribbean (La Hée and Gaiser 2012; Gaiser et al. 2011). The high rates of production in P-limited settings has been attributed to highly efficient exchange among dissolved, adsorbed and organic forms by autotrophic and heterotrophic microbes in the mats (Hagerthey et al. 2011).

The longstanding need for reliable indicators of water quality and hydrological change in the Everglades has motivated innovation in the development of biological criteria for wetland assessment and regulation (McCormick and Stevenson 1998). A wealth of studies has provided support for periphyton as a key indicator of water quality change, especially P, in the Everglades (McCormick and O'Dell 1996; McCormick et al. 1996; Gaiser et al. 2004; 2005). In particular, periphyton total P (TP) content, abundance, enzyme activity, and species composition have shown rapid and reliable responses to P load and concentration, so these variables have been widely employed in water quality assessment (Penton and Newman 2007; Gaiser 2009). Due to low ambient P concentrations and high adsorptive capacity of carbonate bedrock, marl and living periphyton, P enrichment stimulates ecosystem changes that precede detection in the water column (Gaiser et al. 2004; 2005). As a result, microbial criteria (i.e., periphyton TP, rather than water column TP) are invaluable in the oligotrophic Everglades because they provide indication of problems long before irreversible changes emerge, such as the invasion of monospecific stands of cattail that suppress carbon turnover (Hagerthey et al. 2010; Gaiser et al. 2012; Surratt et al. 2012). In accordance with national protocols in the United States (Stevenson and Bahls 1999), Everglades water quality assessment incorporates evaluation of periphyton abundance, nutrient content and species composition, and a stoplight-reporting system is utilized to express trends relative to restorative projects in a straightforward manner to regulatory agencies (Gaiser 2009).

The remarkable progress in biological, and specifically periphyton-based, approaches to aquatic ecosystem assessment, and the accumulation of datasets both locally and nationally with which to evaluate incorporated regulatory tools, has provided an opportunity to gauge the effectiveness of employed approaches in detecting water quality trends (i.e., Charles et al. 2002; Porter 2008). Such evaluations of algal water quality criteria overwhelmingly point to the utility of biological (species-based) criteria for detection of water quality degradation (or improvement) (Birk et al. 2012).

This chapter follows the guidance of these cross-continental studies to evaluate the efficacy of approaches to periphyton-based water quality assessment employed in the Everglades. We specifically compare the accuracy of chemically-based approaches such as periphyton total P concentration to biological ones based on diatom species composition. We further explore how species-based assessment of benthic algae contributes to an improved understanding of food web function, reducing uncertainty about the future of key Everglades biota. Rather than reviewing the copious literature on Everglades periphyton and periphyton-based assessment, we point the reader to several such summaries (Browder et al. 1994; Gaiser et al. 2011; Hagerthey et al. 2011), and instead begin by introducing a new conceptual model for Everglades periphyton that is being used to guide current assessment methodology. We then provide an evaluation of this methodology and implications for assessment of water quality and food web function both in the Everglades and elsewhere.

Conceptual Models

Several conceptual ecological models describing the role of periphyton in the Everglades have been developed and used as a road map for scientific hypothesis testing and to guide restoration assessment (Davis et al. 2004; Gaiser 2009). Most models begin by describing the primary environmental factors regulating periphyton characteristics. In the Everglades, these include pH, conductivity, inflowing nutrient concentrations and hydrologic attributes such as water depth, hydroperiod and flow (collectively, "hydropattern") (Gaiser et al. 2011). In the southern Everglades wetlands dominated by calcareous periphyton, nutrient inputs (primarily P) and hydropattern have an overriding influence on periphyton structure and function (Gottlieb et al. 2006; Gaiser et al. 2006; Lee et al. 2013). In the simplified conceptual model provided here (Fig. 1), we focus on these two controlling variables to align with Everglades restoration goals to re-establish more natural hydropatterns without compromising water quality. These connections between hydrology, water quality and biota (periphyton) vary across a range of spatial and temporal scales, supporting geographically distributed approaches to assessment. Periphyton can also act as an ecosystem engineer, modifying environmental features. Examples include the formation of calcareous (marl) soils, which can, over time, influence hydropattern through topographic modification (Gleason and Spackman 1974), and the sequestration of water column nutrients by periphyton microbes, that modifies downstream nutrient loads (Gaiser et al. 2004; Thomas et al. 2006).

As periphyton is influenced by hydrology and water quality, so, too, is it interacting with upper trophic levels, providing both food and habitat structure for consumers that can then also alter periphyton abundance and composition. Variation in the spatial distribution of small aquatic invertebrates and fishes on the landscape has been directly and indirectly linked to periphyton abundance, nutrient content and species composition (Liston et al. 2008; Sargeant et al. 2010; 2011). While the thick, calcareous mats common to low-nutrient, short-hydroperiod wetlands are a poor food source for these consumers, they do offer a hiding place from predators and a hydrated refuge in times of drying (Liston et al. 2008). In addition, while the filamentous cyanobacterial matrix of calcareous mats is both mechanically difficult to consume, encased in non-

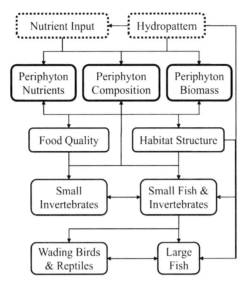

Figure 1. Conceptual model modified from that developed by the CERP RECOVER team that links the two primary drivers, manipulated through restorative projects, to periphyton. Cascading effects of changes in periphyton biomass, nutrient content and composition upon the Everglades food web are also shown.

nutritive mineral structures, and can be toxic, the imbedded diatoms and green algae are preferred food sources, and their detrital products feed bacteria that together provide a microbial shunt moving photosynthetically-derived energy from periphyton into upper trophic levels (Geddes and Trexler 2003; Williams et al. 2006; Belicka et al. 2012). Small aquatic consumers that feed on these edible components and by-products of mats, in turn, become food for large fish, which are themselves the main prey items in the diets of waterfowl, wading birds and crocodilians. Like all food webs, particularly those in structurally complex benthic habitats, linear and unidirectional linkages are unusual, so the framework shown here is a simplification of pathways unraveled through more comprehensive approaches such as structural equation modeling and path analysis (Sargeant et al. 2010; 2011; Abbey-Lee et al. 2012).

A variety of periphyton characteristics are sensitive and reliable indicators of hydropattern and nutrient input, and many of these are consequential to other ecosystem features including periphyton consumers and upper trophic levels. These attributes have been grouped into broad categories of chemical composition, taxonomic composition, biomass and function (Gaiser 2009) that correspond to those incorporated into national Ecological Risk Assessment (ERA) protocols (Stevenson and Smol 2003). In the development of ERA protocols, stressor-response models are used to identify the relationship between environmental stressors of interest and ecological responses to detect and understand the cause for departure from baseline conditions (Stevenson and Smol 2003; Stevenson et al. 2006). A similar approach has been taken to identify key periphyton indicators in the Everglades. Chemical composition, including aggregate periphyton total macronutrient concentrations and ratios, mercury and other metal content, and toxic substance levels have been widely related to known contamination sources (Liu et al. 2011), and of these, periphyton TP concentration has been repeatedly shown in experimental and descriptive studies to provide a highly reliable indication of prior P exposure (Gaiser et al. 2004). Periphyton composition can be described in terms of taxonomic composition, diversity, the proportion of living cells to detritus, indicators of the proportion of autotrophic to heterotrophic components (such as chlorophyll *a* content, microbial abundance), and the ratio of organic to mineral matter in the mat (McCormick and O'Dell 1996; Gaiser et al. 2006; Hagerthey et al. 2011). Like other habitat assessment studies (Potopova and Charles 2007; Hill et al. 2003), taxonomic composition, particularly of the diatom assemblage contained in periphyton, has been found to be one of the accurate indicators of changes in nutrient availability and hydropattern (McCormick and O'Dell et al.; Gaiser et al. 2006; Lee et al. 2013). Because Everglades algal assemblages contain many taxa that appear to be endemic to karst wetlands in the Caribbean and elsewhere (Slate and Stevenson 2007; La Hée and Gaiser 2012), the ratio of these distinctive taxa to more cosmopolitan (or "weedy") ones works well to identify impairment. In addition, the proportion of edible algae (diatoms and green algae) to inedible cyanobacterial filaments, is not only related to nutrient availability but is positively correlated with the abundance and biomass of primary consumers (Sargeant et al. 2011).

Finally, the abundance of periphyton can be measured by the proportion of benthic surfaces coated by mat (percent cover), the volume of water column occupied, wet and dry mass and the biomass, or organic mass (often reported as ash-free dry mass). The latter is frequently used as a measure of periphyton production, and is generally negatively related to nutrient availability, water depth and hydroperiod (Gaiser et al. 2005; Gaiser 2009; Lee et al. 2013). Functional attributes such as specific enzymatic activity and metabolic rates are also strongly tied to nutrient availability (Penton and Newman 2007). Together these variables have provided not only meaningful assessment of hydrologic and water quality changes in the Everglades, but also the implications of these changes to the Everglades ecosystem, including upper trophic levels.

Stressor-Response Models

Here we evaluate the effectiveness of metrics of P enrichment based on periphyton chemistry and species threshold responses. We follow the ERA approach that (1) uses stressor-response relationships in benthic algal communities to identify thresholds and baselines, and (2) employs species-based indicators that have a high sensitivity and low variance in response to stressors (Stevenson 1998). This approach has a long history of scientific support and application, and benthic algal indicators are now broadly incorporated into water quality risk assessment in North America (McCormick and Stevenson 1998; Stevenson et al. 2010) and in wetlands elsewhere (Gaiser and Rühland 2010). There are interesting biological and mathematical reasons why biological (species-based) approaches can improve the accuracy of water quality assessment compared to physical and chemical approaches, many of which are related to the wealth of information provided by potentially hundreds of species that integrate the conditions of their environment over biologically meaningful timescales versus the handful of point measurements of physicochemical variables. There are also financial considerations to including biological criteria in assessment, since species-based approaches typically require more technical expertise and time than other methodologies, so evaluations of their efficacy is also fiscally justifiable.

Past applications of periphyton metrics in Everglades assessment have relied on periphyton TP content, species composition, and biomass to determine water quality (P) impairment, based on comparison to baseline conditions represented in the interior un-impacted areas of marsh (Gaiser et al. 2006; Gaiser 2009). However, this methodology does not support a stressor-response approach to assessment since exposure histories at sites downstream from control structures cannot be measured directly. Instead, we examined the stressor (P)—periphyton response relationship in an experimental P-enrichment study to determine the detection probabilities of a metric based simply on periphyton TP content and one based on species composition. Briefly, four 100-m long x 3 m wide sections of marsh were separated by plastic curtains to create treatment channels oriented parallel to the direction of water flow in three areas of central SRS that were thought to be un-impacted by P enrichment. Channels were randomly assigned to control, +5, +15 and +30 ug L^{-1} treatments and P was delivered continuously in organic form at rates depending on water flow velocities so that incoming water column TP concentrations remained constant. Periphyton was sampled at fixed locations downstream from delivery at bimonthly intervals using the same methods as for assessment, described above. Experimental treatments continued for five years, and results showed that any additions of P above ambient cause cascading ecosystem changes that converge on a similar endpoint regardless of concentration and load (Gaiser et al. 2005). Therefore, to capture dose-dependency, we used data from just the first two years of experimental P enrichment (1998–1999) to determine the frequency distributions of the two variables of interest, periphyton TP and the "weedy" to "endemic" diatom ratio, across dose levels (Gaiser et al. 2004). For the latter, we used a "weedy" to "endemic" diatom ratio based on prior knowledge of species responses to periphyton abundance and nutrient content. The carbonate-rich periphyton mats of the Florida Everglades harbor an unusual diatom community (Slate and Stevenson 2007) that has so far been detected only in similar karst wetlands of the (sub)tropics (La Hée and Gaiser 2012). This community is considered "endemic" to such carbonate wetlands. Conversely, species herein termed "weedy" are most abundant in P-enriched or otherwise disturbed locations, and are widely distributed across ecosystems. We further clarified the "endemic" and "weedy" diatom taxon lists by dropping each species consecutively from the calculation of the ratio, and determining its effect on the detection probability of the correct dose level (explained below). The final list of taxa is provided in Table 1.

The resultant stressor-response model for periphyton TP and the weedy:endemic diatom ratio show increasing mean values with exposure level. Notably, the weedy:endemic ratio in the 10 μg L^{-1} channel is significantly different from the control, while the TP concentration is not (importantly, these values combine a 100-m spatial and 2-year temporal response) (Fig. 2).

The mean and standard deviation for TP in control channels was 130 μg g^{-1} ± 52 and the weedy:endemic diatom ratio was 0.6 ± 0.5. As expected, the frequency distributions of periphyton TP concentration and the weedy:endemic ratio differed significantly across all treatments (Fig. 3).

Table 1. Everglades diatoms categorized as "endemic" or "weedy" based on their contribution to improved detection probability of the weedy:endemic ratio.

Endemic Diatom Taxa	Weedy Diatom Taxa
Achnanthidium minutissimum (Kütz.) Czarn.	*Encyonema silesiacum* (Bleisch) D.G. Mann
Brachysira neoexilis Lange-Bert.	*Eunotia flexuosa* Bréb. ex. Kütz.
Cyclotella meneghiniana Kütz.	*Eunotia naegelii* Migula
Diploneis oblongella (Nägeli) Cleve-Euler	*Gomphonema affine* Kütz.
Encyonema evergladianum Krammer	*Gomphonema auritum* A. Braun ex. Kütz.
Fragilaria synegrotesca Lange-Bert.	*Gomphonema intricatum* var. *vibrio* (Ehrenb.) Cleve
Mastogloia calcarea Lee, Gaiser, Van de Vijver, Edlund & Spaulding	*Navicula angusta* Grunow
Navicula subtilissima Cleve	*Nitzschia amphibia* Grunow
Nitzschia palea var. *debilis* (Kütz.) Grunow	*Navicula cryptocephala* Kütz.
Nitzschia serpentiraphe Lange-Bert.	*Nitzschia palea* (Kütz.) W. Sm.
Sellaphora laevissima (Kütz.) Krammer	*Nitzschia semirobusta* Lange-Bert.
	Rhopalodia gibba (Ehr.) O. Müll.

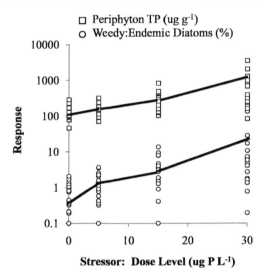

Figure 2. An example of a stressor-response model showing how periphyton TP concentration and the ratio of weedy to endemic diatom species responds to elevated phosphorus inputs. Data are from the first two years of continuous enrichment of phosphorus in three experimental flumes in central Shark River Slough in Everglades National Park. Lines connecting mean values for each enrichment level (+5, 15, 30 µg P L^{-1} above ambient concentrations) are shown, and note that the response (Y) axis is on a log scale.

Using the variance across treatments, we determined the probability of detecting enrichment with distance from source based on a simple one-metric test using periphyton TP concentration and a dual criterion requiring either periphyton TP or the weedy:endemic diatom ratio exceed one standard deviation of the mean. The single criterion method detected enrichment 30, 60 and 80% of cases where marsh was exposed to 5, 15 and 30 µg L^{-1} over ambient levels for 180 continuous days, with decreasing probability of detection with distance from source. We found that the

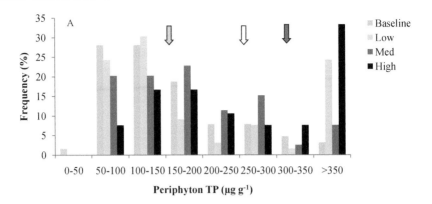

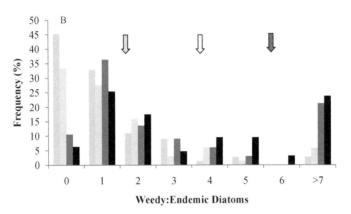

Figure 3. Frequency histogram of (A) periphyton TP values, and (B) ratio of weedy to endemic diatoms, in baseline (control) conditions, and in low (+5 µg P L⁻¹), medium (+15 µg P L⁻¹) and high (+30 µg P L⁻¹) enrichment channels over the first two years of continuous P dosing in three experimental flumes in central Shark River Slough in Everglades National Park. Green, yellow and red arrows represent the mean, and one and two standard deviations from the mean, respectively.

Color image of this figure appears in the color plate section at the end of the book.

additional compositional metric improved detection by an average of 15% across all treatments, times and distances, and was most useful in early detection of low dose settings (Fig. 4).

In summary, periphyton TP and diatom-based assessment used alone yield an 75 and 85% detection probability at highest exposure levels, respectively, and improve detection of this exposure more than 20% when used together.

Assessment Application

We applied the resultant single and dual-criterion metrics to assess P enrichment throughout the Everglades using data from a spatially- and temporally-extensive survey. Briefly, systematic monitoring and assessment of ecological change in the Everglades using periphyton began in 2005, when a generalized random-tessellation stratification

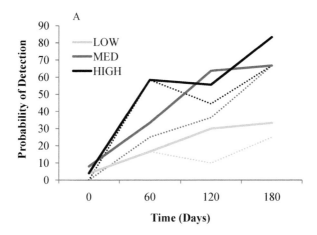

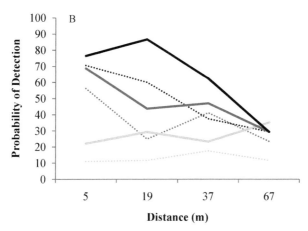

Figure 4. Probability that P-enrichment will be detected by periphyton TP indicator (dotted lines) and by the combined TP and weedy:endemic diatom indicator (solid lines) in low (+5 µg P L⁻¹), medium (+15 µg P L⁻¹) and high (+30 µg P L⁻¹) enrichment channels in the second year of the dosing study, using the one standard deviation criterion. Plot also shows how probability increases with (A) increasing duration of exposure, and (B) decreasing distance, to inflows of a constant concentration.

Color image of this figure appears in the color plate section at the end of the book.

approach was established to document ecosystem response to restoration projects in a spatially balanced manner (Stevens and Olsen 2004). The landscape was divided into 800 m × 800 m grids and a representative sample of these was drawn as primary sampling units (PSU); three sampling sites were randomly selected from the samplable habitat in each PSU. Periphyton was collected within 1 m² quadrants and processed using methods identical to those employed in metric development, and described in detail in Lee et al. (2013). Data reported here are from 62 sites sampled annually during the wet season from 2006–2011; notably diatom species were excluded from analysis after 2011 and therefore cannot be included in this assessment. To assess P enrichment we assigned "baseline" (green), "caution" (yellow) and "altered" (red) condition to each site if it was within one, two and greater than two standard deviations from the mean, respectively, for both periphyton TP and the weedy:endemic ratio. To apply the dual criterion, we denoted "caution" status if one or more metric denoted "caution" or "altered", and "altered" if both metrics indicated an "altered" state. We applied each metric to each year separately, and to the mean values across years.

Both the periphyton TP metric and dual criterion approach denoted significant patterns of "caution" and "altered" conditions throughout the system across years (Fig. 5).

The periphyton TP metric alone showed 39% of sites with values at least one standard deviation greater than background concentrations (17% and 22% "caution" and "altered" sites, respectively). By including the species metric, we find that 56% of sites are impaired (17% and 39% "caution" and "altered", respectively). Among years, the dual metric resulted in an 17% improvement in the ability to indicate above-ambient P exposure.

While the purpose of this study was not to determine the drivers of interannual changes in water quality impairment, some notable patterns appear when examining patterns in the dual metric assessment across years relative to inflowing water quality concentrations. The dual metric suggests impairment in central Water Conservation Area-3 (WCA-3), and impacts along the eastern boundary of Everglades National Park (ENP) likely associated with changes in operations of the L-31 W and S-332 structures (see Gaiser et al. 2013 and Bramburger et al. 2013). An important caveat

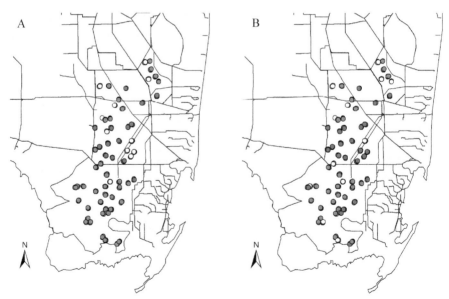

Figure 5. Assessment of phosphorus enrichment across the Everglades watershed, based on (A) periphyton TP concentrations, and (B) a combination of periphyton TP and weedy:endemic diatoms averaged over 2006–2011. Sites are coded green, yellow and red if they are less than one, greater than one and greater than two standard deviations, respectively, from the mean TP concentration (A) or either mean TP or weedy:endemic diatom ratio (B).

Color image of this figure appears in the color plate section at the end of the book.

is that the diatom metric also shows impairment in southwestern Shark River Slough (SRS) and downstream areas of Taylor Slough (TS), but this pattern may be due to the natural increase in P availability in the oligohaline areas of ENP via exposure to marine sources of P (Gaiser et al. 2011; 2012).

The inter-annual pattern in cautionary and altered sites was examined relative to water flows, and TP loads and weighted-mean concentrations at inflow structures obtained from the South Florida Water Management District (Fig. 6). Specifically, we downloaded monthly data on all three variables from DBHYDRO for the period of record for all stations allowing any inflow to SRS (S12-A-D, S333), TS (S174, S332DX, S18C, S332, S175), WCA-2A (S7, S11A, S11B, S11C, S38, S34) and WCA-3 (S140, S190, G406, S8, S150, S11A-C, S9, S9A, S8, G123, G344E-L, G352B, G354C, G393B, G606) and calculated annual means or totals. We found no relationship between the condition metrics and flow or load across basins, but there was a strong relationship between the flow-weighted mean inflow TP concentration over the three months prior to sampling and the TP, Diatom and TP-Diatom metrics (using a second order polynomial model to account for threshold response; $R^2 = 0.29, 0.75, 0.67$, respectively). Notably the relationship between inflowing water TP and periphyton TP was stronger than the diatom metric, even though experimental research shows the diatom metric is more sensitive to exposure (Gaiser et al. 2005). Conclusions based on other experimental work suggest that diatoms retain the signal (i.e., presence of enrichment indicator taxa) of historical TP exposure longer than

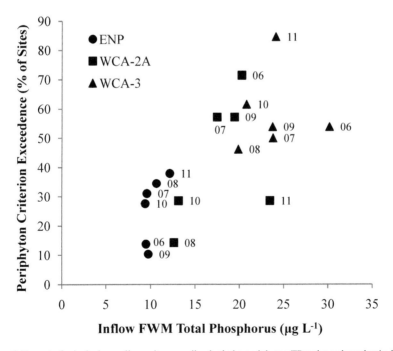

Figure 6. Percent of principal sampling units exceeding both the periphyton TP and weedy:endemic diatom criterion (one standard deviation from mean) by year in Everglades National Park (ENP), Water Conservation Area-2A (WCA-2A) and Water Conservation Area-3 (WCA-3) for 2006–2011, showing positive relationship to the flow-weighted mean (September-August prior to sampling) TP concentration averaged among inflows.

the concentrations within the periphyton mat (Gaiser et al. 2006), suggesting further analysis of lags in response to historical loading are necessary. The high correlation between inflow concentration (at structures) and condition status across each wetland is surprising given the distance from input sources and since it includes sites well to the interior, explaining the shallow slope of the regression.

Developments in Periphyton-based Hydrological Assessment

Much research has been directed at understanding periphyton response to hydrological variability, both through experimental studies and using long-term assessment data that include wet and dry years. Experimental studies have been useful to control for water depth and/or inundation timing while holding nutrient concentrations constant (Thomas et al. 2006; Gottlieb et al. 2006), which help resolve stressor-response relationships in the natural environment where the two drivers are often correlated. These studies show that while biomass often declines with water depth and inundation timing, this relationship has high variance that is often also influenced by a commensurate translocation of nutrients. The strongest independent relationships appear to be in the algal species composition, where filamentous cyanobacteria increase in abundance in dry conditions and diatoms and green algae flourish in longer-hydroperiod settings (Gottlieb et al. 2006). Within the diatom community, Lee et al. (2013) were able to identify a suite of taxa that appear to be reliable indicators of dry conditions and others that are absent from sites with a history of drying. About 63% of variance in diatom composition is driven by hydroperiod, and although this value is high for any group of organisms, it is 20–30% lower than that observed for TP across Everglades wetland regions (Gaiser et al. 2006). This difference suggests that resource (P) limitation may be the primary determinant of composition, rather than the stress conferred by drying. Desiccation resistance may be a common attribute of diatom species because the stress is binary—if a species is not desiccation resistant it will not survive any duration of drying, and even the longest hydroperiod Everglades marshes dry occasionally. In addition, diatoms may benefit by living within cyanobacteria-dominated periphyton mats that provide physical and biochemical buffers against the effects of short-term desiccation. Yet diatoms are responding to hydrologically-mediated changes, providing inferences of hydroperiod within about a 55-day window of accuracy. Maps have been generated that compare diatom-based hydroperiod estimates to those derived from nearby water level gauges and elevation maps, providing a guide to an ecological impression of regional hydrology (see Gottlieb et al., this book). The next step in this research is to develop a hydrologic change assessment tool that first accounts for community variance due to P, and then predicts hydroperiod to the extent possible given the measured range of species preferences. The road block to diatom-based indication of hydrologic restoration continues to be the failure to quantify natural system community targets. Diatoms are poorly preserved in sediment records, offering little evidence for paleohydrologic interpretations of natural system conditions (Sanchez et al. 2013). A solution may be to use targets projected by natural system hydrologic models to show the deviation in diatom-inferred hydroperiod from what would be predicted given the rainfall (and existing climate conditions) of a particular assessment year(s). Assessment of hydrologic restoration must consider both water quality and hydropattern, which is possible using species-based microbial assessment.

Linkages to Upper Trophic Levels

The conceptual ecological models linking periphyton attributes to ecosystem structure and function focus on periphyton as the base of the food web, supplying energy that ultimately reaches megafaunal indicators such as wading birds and alligators. Research on primary and secondary consumers (which eat periphyton and are prey for higher consumers) has shown that periphyton structure, abundance and composition are important regulators of consumer density and diversity, and explain variance in consumer response to hydrological disturbance (Sargeant et al. 2010; 2011). Notably, the periphyton metrics making the largest contribution to interpreting hydrologic controls on consumer densities are biomass and the ratio of edible (green algae and diatoms) to inedible (filamentous blue-green algae) algae (see Trexler et al., chapter 8). Low ratios are found where calcareous periphyton mats are most abundant, in the central and highly P-limited Everglades interior. These low ratios of edible:inedible algae are associated with a very low biomass of primary consumers (Sargeant et al. 2011). On the other hand, higher ratios are found in areas of disturbance, P-enrichment and unusually high water depths, and appear to drive an increase in secondary productivity. There has been great utility in this metric, therefore, for providing support for the unusual biomass pyramid characterizing the native Everglades, where an abundance of primary producers does not translate to high consumer densities (Gaiser et al. 2012). For this reason, inclusion of species-based metrics provides not only an accurate indication of P load history, but also a diagnostic tool for interpreting the distribution of aquatic consumers via prey food quality.

Conclusions

The development of strong conceptual models provides an adaptive and testable framework for the study of periphyton for use in monitoring and assessing environmental change. A combination of approaches including chemical, biological, and physical factors provides insight into changes in water quality and habitat structure and function. Ongoing monitoring and species based approaches provide early warning signs of environmental change that are only later identified using traditional water quality approaches (and that have the potential to be propagated into higher trophic levels). This combination of physical, chemical and biological approaches provides a more thorough understanding of community change at a variety of spatial and temporal scales.

Acknowledgments

Funding was provided to Florida International University for the nutrient dosing study from the U.S. Department of the Interior and the South Florida Water Management District, and for the CERP MAP program from Task Agreement CP040130 from South Florida Water Management District and Cooperative Agreement W912HZ-11-2-0048 from the U.S. Army Corps of Engineers. We thank Jana Newman for coordination of the CERP MAP research program, and Franco Tobias for coordinating field and laboratory research. This work was enhanced through collaborations with the Florida Coastal Everglades Long Term Ecological Research program (#DEB-1237517). This is contribution #671 of the Southeast Environmental Research Center.

References

Abbey-Lee, R.N., E.E. Gaiser and J.C. Trexler. 2013. Relative roles of dispersal dynamics and competition in determining the isotopic niche breadth of a wetland fish. Freshwater Biol. 58: 780–792.

Belicka, L.L., E.R. Sokol, J.M. Hoch, R. Jaffé and J.C. Trexler. 2012. A molecular and stable isotopic approach to investigate algal and detrital energy pathways in a freshwater marsh. Wetlands 32: 531–542.

Birk, S., W. Bonne, A. Borja, S. Brucet, A. Courrat, S. Poikane, A. Solimini, W. van de Bund, N. Zampoukas and D. Herin. 2012. Three hundred ways to assess Europe's surface waters: An almost complete overview of biological methods to implement the Water Framework Directive. Ecol. Indic. 18: 31–41.

Bramburger, A., J. Munyon and E. Gaiser. 2013. Water quality and wet season diatom assemblage characteristics from the Tamiami Trail pilot swales sites. Phytotaxa 127: 163–182.

Browder, J.A., P.J. Gleason and D.R. Swift. 1994. Periphyton in the Everglades: spatial variation, environmental correlates, and ecological implications. pp. 379–418. *In*: S.M. Davis and J.C. Ogden (eds.). The Everglades: The Ecosystem and its Restoration. St. Lucie Press, Delray Beach, FL.

Charles, D., F.C. Knowles and R.S. Davis. 2002. Protocols for the analysis of algal samples collected as part of the US Geological Survey National Water-Quality Assessment Program. Patrick Center for Environmental Research. Philadelphia: Academy of Natural Sciences.

Davis, S., E. Gaiser, W. Loftus and A. Huffman. 2005. Southern marl prairies conceptual ecological model. Wetlands 25: 821–831.

Gaiser, E. 2009. Periphyton as an indicator of restoration in the Everglades. Ecol. Indic. 9: S37–S45.

Gaiser, E. and K. Rühland. 2010. Diatoms as indicators of environmental change in wetlands and peatlands. pp. 473–496. *In*: J. Smol and E. Stoermer (eds.). The Diatoms: Applications in Environmental and Earth Sciences. Camridge University Press, New York, USA.

Gaiser, E., L. Scinto, J. Richards, K. Jayachandran, D. Childers, J. Trexler and R. Jones. 2004. Phosphorus in periphyton mats provides best metric for detecting low-level P enrichment in an oligotrophic wetland. Water Res. 38: 507–516.

Gaiser, E., J. Trexler, J. Richards, D. Childers, D. Lee, A. Edwards, L. Scinto, K. Jayachandran, G. Noe and R. Jones. 2005. Cascading ecological effects of low-level phosphorus enrichment in the Florida Everglades. J. Environ. Qual. 34: 717–723.

Gaiser, E., J. Richards, J. Trexler, R. Jones and D. Childers. 2006. Periphyton responses to eutrophication in the Florida Everglades: cross-system patterns of structural and compositional change. Limnol. Oceanogr. 51: 617–630.

Gaiser, E., P. McCormick and S. Hagerthey. 2011. Landscape patterns of periphyton in the Florida Everglades. Crit. Rev. Environ. Sci. Technol. 41(S1): 92–120.

Gaiser, E., J. Trexler and P. Wetzel. 2012. The Everglades. pp. 231–252. *In*: D. Batzer and A. Baldwin (eds.). Wetland Habitats of North America: Ecology and Conservation Concerns. University of California Press, Berkeley.

Gaiser, E., P. Sullivan, F.A.C. Tobias, A.J. Bramburger and J.C. Trexler. 2013. Boundary effects on benthic microbial phosphorus concentrations and diatom beta diversity in a hydrologically-modified, nutrient-limited wetland. Wetlands. DOI 10.1007/s13157-013-0379-z.

Geddes, P. and J.C. Trexler. 2003. Uncoupling of omnivore-mediated positive and negative effects on periphyton mats. Oecologia 136: 585–595.

Gleason, P.J. and W. Spackman, Jr. 1974. Calcareous periphyton and water chemistry in the Everglades. pp. 146–181. *In*: P.J. Gleason (ed.). Environments of South Florida: Present and Past, Memoir No. 2. Miami Geological Society, Coral Gables, FL.

Gottlieb, A., J. Richards and E. Gaiser. 2006. Comparative study of periphyton community structure in long and short hydroperiod Everglades marshes. Hydrobiologia 569: 195–207.

Hagerthey, S., J. Cole and D. Kilbane. 2010. Aquatic metabolism in the Everglades: dominance of water column heterotrophy. Limnol. Oceanogr. 55: 653–666.

Hagerthey, S., B. Bellinger, K. Wheeler, M. Gantar and E. Gaiser. 2011. Everglades periphyton: a biogeochemical perspective. Crit. Rev. Environ. Sci. Technol. 41(S1): 309–343.

Hill, B.H., A.T. Herlihy, P.R. Kaufmann, S.J. DeCelles and M.A. Van der Borgh. 2003. Assessment of streams of the eastern United States using a periphyton index of biotic integrity. Ecol. Indic. 2: 325–338.

Kelly, M.G., A. Cazaubon, E. Coring, A. Dell'Uomo, L. Ector, B. Goldsmith, H. Guasch, J. Hürlimann, A. Jarlman, B. Kawecka, J. Kwandrans, R. Laugaste, E.-A. Lindstrøm, M. Leitao, P. Marvan, J. Padisák, E. Pipp, J. Prygiel, E. Rott, S. Sabater, H. van Dam and J. Vizinet. 1998. Recommendations for the routine sampling of diatoms for water quality assessments in Europe. J. Appl. Phycol. 10: 215–224.

La Hée, J. and E. Gaiser. 2012. Benthic diatom assemblages as indicators of water quality in the Everglades and three tropical karstic wetlands. Freshwater Sci. 31: 205–221.

Lee, S., E. Gaiser and J. Trexler. 2013. Diatom-based models for inferring hydrology and periphyton abundance in a subtropical karstic wetland: implications for ecosystem-scale bioassessment. Wetlands 33: 157–173.

Liston, S.E., S. Newman and J.C. Trexler. 2008. Macroinvertebrate community response to eutrophication in an oligotrophic wetland: an *in situ* mesocosm experiment. Wetlands 28: 686–694.

Liu, G., M. Naja, P. Kalla, D. Scheidt, E. Gaiser and Y. Cai. 2011. Legacy and fate of mercury and methylmercury in the Florida Everglades. Environ. Sci. Technol. 45: 496–501.

McCormick, P.V. and M.B. O'Dell. 1996. Quantifying periphyton responses to phosphorus in the Florida Everglades: a synoptic-experimental approach. J. N. Am. Benthol. Soc. 15: 450–468.

McCormick, P.V. and R.J. Stevenson. 1998. Periphyton as a tool for ecological assessment and management in the Florida Everglades. J. Phycol. 34: 726–733.

McCormick, P.V., R.S. Rawlick, K. Lurding, E.P. Smith and F.H. Sklar. 1996. Periphyton–water quality relationships along a nutrient gradient in the northern Florida Everglades. J. North Am. Benthol. Soc. 15: 433–449.

Penton, C.R. and S. Newman. 2007. Enzyme activity responses to nutrient loading in subtropical wetlands. Biogeochemistry 84: 83–98.

Porter, S.D. 2008. Algal attributes: an autecological classification of algal taxa collected by the National Water-Quality Assessment Program. US Geological Survey.

Potopova, M. and D.F. Charles. 2007. Diatom metrics for monitoring eutrophication in rivers of the United States. Ecol. Indic. 7: 48–70.

Sanchez, C., E. Gaiser, C. Saunders, A. Wachnicka and N. Oehm. 2013. Exploring siliceous subfossils as a tool for inferring past water level and hydroperiod in Everglades marshes. J. Paleolim. 49: 45–66.

Sargeant, B., J. Trexler and E. Gaiser. 2010. Biotic and abiotic determinants of intermediate-consumer trophic diversity in the Florida Everglades. Mar. Freshwater Res. 61: 11–22.

Sargeant, B., E. Gaiser and J. Trexler. 2011. Indirect and direct controls of macroinvertebrates and small fish by abiotic factors and trophic interactions in the Florida Everglades. Freshwater Biol. 56: 2334–2346.

Slate, J.E. and R.J. Stevenson. 2007. The diatom flora of phosphorus enriched and unenriched sites in an Everglades marsh. Diat. Res. 22: 355–386.

Smith, D.G. and G.B. McBride. 1990. New Zealand's national water quality monitoring network—Design and first year's operation. J. Am. Water Resour. Assocs. 26: 767–775.

Stevens, D.L. and A.R. Olsen. 2004. Spatial balanced sampling of natural resources. J. Am. Stat. As. 99: 262–278.

Stevenson, R.J. 1998. Diatom indicators of stream and wetland stressors in a risk management framework. Environ. Monit. Assess. 51: 107–18.

Stevenson, R.J. and L.L. Bahls. 1999. Periphyton protocols. pp. 6-/1–6/22M. *In*: T. Barbour, J. Gerritsen, B.D. Snyder and J.B. Stribling (eds.). Rapid Bioassessment Protocols for Use in Streams and Wadeable Rivers: Periphyton, Benthic Macro Invertebrates, and Fish. 2nd ed. EPA 841-B-99-002.

Stevenson, R.J. and J.P. Smol. 2003. Use of algae in environmental assessments. pp. 775–804. *In*: J.D. Wehrand and R.G. Sheath (eds.). Freshwater Algae in North America: Classification and Ecology. Academic Press. Massachusetts, USA.

Stevenson, R.J., M. Bothwell and R. Lowe. 1996. Algal Ecology: Benthic Algae in Freshwater Ecosystems. Academic Press, Massachusetts, New York, USA.

Stevenson, R.J., S.T. Rier, C.M. Riseng, R.E. Schultz and M.J. Wiley. 2006. Comparing effects of nutrients on algal biomass in streams in 2 regions with different disturbance regimes and with applications for developing nutrient criteria. Hydrobiologia 561: 149–65.

Stevenson, R.J., Y. Pan and H. Van Dam. 2010. Assessing environmental condition in rivers and streams with diatoms. pp. 57–85. *In:* J.P. Smol and E.F. Stoermer (eds.). The Diatoms: Applications for the Environmental and Earth Sciences, 2nd Edition. Cambridge University Press, New York, USA.

Surratt, D., D. Shinde and N. Aumen. 2012. Recent cattail expansion and possible relationships to water management: changes in Upper Taylor Slough (Everglades National Park, Florida, USA). Environ. Manage. 49: 720–733.

Thomas, S., E. Gaiser, M. Gantar and L. Scinto. 2006. Quantifying the responses of calcareous periphyton crusts to rehydration: A microcosm study (Florida Everglades). Aquat. Bot. 84: 317–323.

Troxler, T., E. Gaiser, J. Barr, J. Fuentes, R. Jaffe, D. Childers, L. Collado-Vides, V. Rivera-Monroy, E. Castañeda-Moya, W. Anderson, R. Chambers, M. Chen, C. Coronado-Molina, S. Davis, V. Engel, C. Fitz, J. Fourqurean, T. Frankovich, J. Kominoski, C. Madden, S. Malone, S. Oberbauer, P. Olivas, J. Richards, C. Saunders, J. Schedlbauer, F. Sklar, T. Smith, J. Smoak, G. Starr, R. Twilley and K. Whelan. 2013. Integrated carbon budget models for the Everglades terrestrial-coastal-oceanic gradient: current status and needs for inter-site comparisons. Oceanography 26: 98–107.

Wetzel, R.G. 2005. Periphyton in the aquatic ecosystem and food webs. pp. 51–69. *In:* A.E.M. Vergdegem, A. van Dam and M. Beveridge (eds.). Periphyton: Ecology, Exploitation, and Management. CABI, London.

Williams, A.J. and J.C. Trexler. 2006. A preliminary analysis of the correlation of food-web characteristics with hydrology and nutrient gradients in the southern Everglades. Hydrobiologia 569: 493–504.

7

Changes in Hydrology, Nutrient Loading and Conductivity in the Florida Everglades, and Concurrent Effects on Periphyton Community Structure and Function

Andrew D. Gottlieb,[1,*] *Evelyn E. Gaiser*[2,a] *and Sylvia S. Lee*[2,b]

Introduction

The Florida Everglades is a vast subtropical wetland that now extends from the headwaters north of Lake Okeechobee, through a network of canals and storm water treatment areas (STAs), eventually into Everglades National Park (ENP) and the estuaries of Florida Bay and the Gulf of Mexico (Fig. 1). Although the Everglades system was historically nearly twice as large, connecting directly to Lake Okeechobee to the north and flowing freely to the Gulf of Mexico, Florida Bay, and the Atlantic Ocean, now the Everglades is a managed system surrounded by control structures

[1] Harvard Drive, Lake Worth, FL 33460, USA.
  Email: adgottlieb71@gmail.com
[2] Department of Biological Sciences, Southeast Environmental Research Center, Florida International University, Miami, FL 33199, USA.
[a] Email: gaisere@fiu.edu
[b] Email: slee017@fiu.edu
* Corresponding author

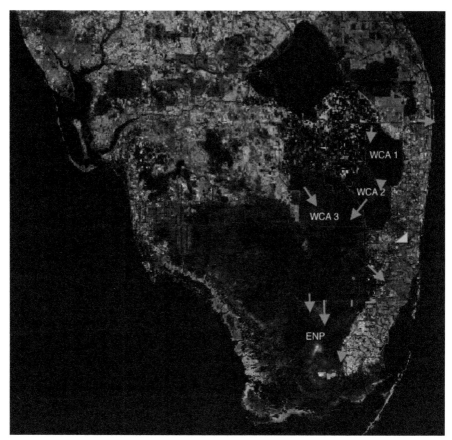

Figure 1. Map of the Florida Everglades including Water Conservation Areas (WCAs) and Everglades National Park (ENP). Primary canals and structure flows noted in red and blue, respectively. (Landsat 2004; SFWMD canals 1997). Arrows indicating Lake flow to the Caloosahatchee and St. Lucie not indicated.

Color image of this figure appears in the color plate section at the end of the book.

of dikes, levees and pumps. Agricultural and urban development and corresponding change in historic spatial extent and connectivity of the Everglades, combined with infrastructure and water management operations, shape the drivers of structure and function of periphyton communities throughout the Everglades landscape (Ogden et al. 2005; Firbank et al. 2003). This chapter describes observed responses of periphyton structure and function to water management-driven changes in hydrology, phosphorus (P) cycling, and conductivity in the Florida Everglades.

Periphyton, a complex microbiological community of algae, bacteria, fungi, and detritus, is nearly ubiquitous in Everglades marshes. In short hydroperiod marshes, periphyton is found as benthic mats covering marl substrates. In intermediate hydroperiod settings, mats form epiphytic aggregations ('sweaters') on the outside of living and dead stems of *Eleocharis* spp. and other rooted macrophytes. In deeper, longer hydroperiod habitats, floating mats are commonly associated with the bladderwort *Utricularia purpurea*. In addition to being common in the landscape, periphyton is also an important primary producer (Gottlieb et al. 2006; McCormick et al. 1997; Vymazal and Richardson 1995) and provides structure for fish and invertebrates (Liston and Trexler 2011; Sargeant et al. 2010; Lamberti 1996). Because periphyton is an important primary producer and an excellent indicator of water quality (Gaiser 2009; McCormick et al. 2001) it is often used as a barometer of ecosystem health (McCormick and Stevenson 2008; Doren et al. 2009). Although some algal species appear to be endemic to south Florida and the Everglades (Gaiser et al. 2006), periphyton is found throughout the world and comparable carbonate mat communities thrive in similar marshes in the Caribbean and Latin America (Rejmankova and Komarkova 2000; La Hee and Gaiser 2012). Understanding the ways in which periphyton responds to operational changes in the Everglades, therefore, may serve as a guide for ecosystem management decisions in other similar karst regions.

In addition to the direct loss of spatial extent, hydrology, or the timing, depth, and distribution of water, and water quality are also affected by land use, water management structures and related operations. Changes in this managed system that increase or decrease hydroperiod or alter nutrient availability also impact the structure of periphyton communities. Structural and functional changes include shifts in algal species composition, the physical integrity of the community, and stoichiometry, each of which can have cascading effects on higher trophic levels, including fish, invertebrates, and wading birds (see Gaiser et al.; Trexler et al., this book).

South Florida's seasonal climate complicates the operations and management of water movement through the Everglades. The rainy season generally lasts from June through October and accounts for approximately 70 percent of annual rainfall (Ali et al. 2000). During the wet season, requirements for flood control to protect large cities like Miami and Ft. Lauderdale compete with the need to recharge the aquifer underlying the Everglades, the dominant water supply for South Florida. Loss of spatial extent and water storage capacity, as well as flood control requirements, force operational discharge of large volumes of fresh water during the rainy season to the ocean (particularly from Lake Okeechobee to the Northern Everglades where approximately 891000 acre feet per year are discharged to the Caloosahatchee River and 215000 acre feet per year are discharged to the St. Lucie River (SFWMM ECBEAA 2010), http://www.evergladesplan.org/pm/recover/recover_docs/band_1_report/012810_band1_app_2.pdf). Large discharges are also made in anticipation of hurricanes or tropical storms, often late in the rainy season when the volume of water necessary for sufficient recharge of the Everglades for the upcoming dry season is uncertain. Sloughs, which historically remained wet, now dry frequently in the current Everglades landscape.

The following chapter discusses the historical lessons learned from the Everglades about how large-scale wetland management and operations influence wetland ecology. This important history is being presented in order to understand modern day conditions relevant to Everglades microbiota. In particular, the effects of structure operations on hydroperiod, nutrient loading and surface water conductivity are reviewed through their impact on periphyton mat structure and function.

The Comprehensive Everglades Restoration Plan (CERP) and related state and federal efforts are aimed at improving hydrology and reducing nutrients entering the Everglades. This chapter provides guidance for structural and operational changes that may be needed to support ecosystem restoration goals. Although examples are specific to the Everglades, the structural features and operational decisions are not unique to the Everglades but instead are applicable to large systems under current or potentially future management. Given increase in global population, future water resource demands, and exposure to climate change, it is likely that similar pressures will be placed on wetlands worldwide.

Effects of Structural and Operational Changes on Hydrology, Nutrient Loading and Conductivity Distributions

Hydrology

The Everglades originally flowed nearly uninterrupted from north to south as a vast, shallow sheet of water. Although some constrictions associated with tree islands and narrowing sloughs may have locally concentrated flows, sheet flow (slow moving, shallow water with a large spatial extent) was the dominant characteristic of the Everglades landscape (Douglas 1947). Given the generally unconstrained historic nature of the system, the depth of the Everglades was nearly the same throughout the flow path (McVoy et al. 2011) (with few local exceptions—Miami rock ridge/ northeast Shark River Slough), but as drainage for agriculture and urban development occurred, the Everglades became compartmentalized. The loss of the essential link to Lake Okeechobee to the north entailed a drastic reduction in water storage capacity. At times, Lake Okeechobee (and wetlands within what is currently the Everglades Agricultural Area or EAA) had the potential to discharge large pulses of water, helping to maintain healthy landscape structure significantly downstream in the Everglades. Storms, such as the one in August and September of 2012, which caused Lake Okeechobee stage to increase from 12.38 feet August 25 to 14.63 feet September 7 (Station LZ40, DBHYDRO) and delivered over one million acre feet to the Lake, provided the volumes necessary for such pulses (Fig. 2, reproduced with permission, SFER 2007, V1 app 2-2).

Historically Lake Okeechobee was not a diked system but rather connected to downstream wetlands. Given similar deliveries at a time of higher stage, large volumes of Lake water would have likely flowed to the south under historical conditions.

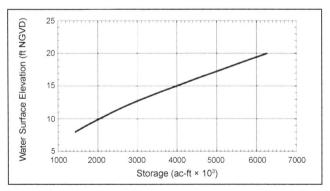

Figure 2. Relationship between Lake Okeechobee water storage and water surface elevation (SFWMD 2007).

These pulses of water helped to maintain landscape connectivity and maintained the distinctive ridge and slough patterning (Harvey et al. 2009; Larsen et al. 2009; Dong 2006; Odum et al. 1996). Now, due to the loss in connectivity to the Lake and marsh spatial extent, less water moves south than did historically, water is conveyed via structure flow, and the timing and distribution of flows is greatly altered (Fig. 3).

In addition to the changes at the headwaters of the system are a series of roadways and levees, which further compartmentalize the Everglades. Levees have caused many parts of the Everglades to depart from historical hydrological conditions. First, at the southern boundary of the water conservation areas (WCAs), the southward flow of water is blocked resulting in longer hydroperiod conditions (and generally deeper water) compared to pre-development conditions (RECOVER 2010). On the other end of the spectrum, because inflows to the WCAs are limited to structural inflows and rainfall, the north end of the WCAs and ENP tend to be too dry, particularly in WCA 3 where inflows are limited to the NW and the S-11 structures (and NE Shark River Slough). Impoundments not only create artificially wet or dry conditions, they also increase dependence on culvert and pump structures. Rather than water flowing as a distributed sheet, flows are directed toward these structures (Ho et al. 2009). This shift in the timing and distribution of flow has direct consequences to biota, and also changes the distribution and concentrations of nutrient input into marshes.

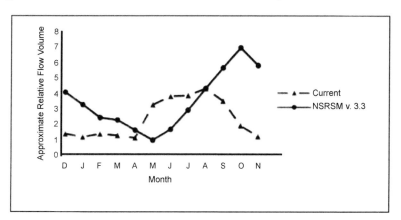

Figure 3. Current condition (triangles) showing reduced flow volume (1431 kaf/yr) and shifts in timing of flows relative to NSRSMv3.3 (2125 kaf/yr) across an east to west transect at the Northern boundary of the Everglades (modified from River of Grass Scenarios, SFWMD 2010 and Band 1 report, 2010).

Nutrient Loading

Because the Everglades is underlain by karst bedrock and is primarily a precipitation-driven system, it is naturally P - limited (Noe et al. 2001). Historical sheet flow over carbonate bedrock and deep peat soils promoted the maintenance of oligotrophy. Shallow, slowly flowing waters moved across the landscape interacting with plants and microbial communities along the way. The spatially distributed flow across a previously larger landscape with deeper water interacting with greater plant surface area allowed rapid assimilation of nutrients by nutrient-limited biota (Lodge 2010; Jarvie et al. 2002). Although increased flows at times may have provided increased loads, water column nutrient concentrations generally remained low, thereby maintaining the unique assemblage of flora and fauna found in the Everglades (Dong 2006).

Changes in land use and water delivery affect the total inflowing load and the internal cycling and spatio-temporal distribution of nutrients. Initially, as drainage and agricultural expansion along the northern boundary and urban and industrial development of the eastern boundary occurred, nutrient loading to the Everglades increased, resulting in marsh surface water nutrient concentrations well above the ambient concentrations of 10 ppb total phosphorus (TP) or less (Lang et al. 2010; Noe et al. 2001). With eventual implementation of best management practices (BMPs) by farmers, and the creation of STAs, nutrient enrichment has decreased considerably but still continues (Baker 2012; Lang et al. 2010; Entry and Gottlieb 2013). Water management operations and natural hydrologic cycles also affect nutrient cycling and hence impact Everglades microbial communities.

Hydrological changes have also resulted in the mobilization of nutrients, particularly the limiting nutrient P on the landscape. As ecosystem susceptibility to desiccation increases, sediment organic matter is oxidized, releasing P into the water column at levels higher than in reduced, flooded settings (Osborne et al. 2011; Watts et al. 2010). Exposure to P may also be enhanced via water diversion through culverts or flow constraining features that can locally increase load and influence microbial activity (Bramburger et al. 2013; Vymazal 2007; RECOVER 2010; 2005). The CERP's proposed construction of bridges to replace flow impediments along the southern boundaries of the WCAs is aimed at increasing flows but also attaining sheet flow that would minimize local loading hot spots associated with structures (RECOVER 2010; 2005). Nutrient enrichment leads to significant shifts in mat community structure, as well as dissolution of the physical mat structure (Gaiser et al. 2005; McCormick et al. 2001).

Conductivity

Changes in conductivity, a measure of the ability of water to pass an electrical current (Wetzel 1975), effect algal community structure and function (Gaiser et al. 2011; McCormick et al. 2011; Ryves et al. 2002). Conductance is strongly correlated with major ions, such as calcium and chloride, as well as alkalinity. Conductivity is primarily a function of water source, evapotranspiration, substrate and bedrock type, and residence time. Dominance of the hydrological cycle by precipitation causes low conductivity throughout much of the freshwater Everglades. In parts of the system, a thin or non-existent peat layer over the limestone bedrock increases interaction between the shallow surface water and mineral-rich groundwater resulting in higher alkalinity (McCormick et al. 2011). In the Big Cypress National Refuge and the Loxahatchee National Wildlife Refuge (LNWR), however, deeper water and deeper layers of peat on top of the bedrock results in softwater environments. In addition, Florida Bay, Biscayne Bay and the Gulf Coast, the southern end members of the Everglades, are estuarine and have high conductivity. Landscape scale imagery and vegetation mapping indicate the gradual intrusion of the estuarine/marine end-member into the freshwater marsh in a visible pattern recognized as the "white zone" (Ross et al. 2000).

Previous studies indicate that changes in conductivity often occur prior to changes in surface water nutrient concentration or community structure (McCormick et al. 2011; Dow and Zampella 2000; Morgan 1987). These changes are not confined to the marine estuarine gradient but are also observed in the freshwater system. The LNWR was historically dominated by low conductivity marsh water. Under current conditions, water no longer enters the LNWR as sheetflow, but passes from a series of STAs into a perimeter canal. If water levels in the marsh are lower than the canal, water enters the marsh. Newman et al. (2011) discovered significant gradients in conductivity and nutrients between the perimeter canal and the interior of the LNWR. Although current operations and management are aimed at tackling the nutrient problem, primarily P, the STAs are not designed to reduce the conductivity of inflows. Reduced sheetflow, agricultural and urban runoff, and excavation of canals all contribute to the observed increases in conductivity and pH, which decrease the solubility of calcium carbonate, Slate and Stevenson (2000) documented an abrupt shift in the algal community associated with a shift in pH at the LNWR–WCA 2 boundary. Conductivity effects on algal community structure have been evaluated for many systems including evaluation of taxa-specific optima and tolerance (Potopova and Charles 2003; Wachnicka et al. 2010).

Effects of Changes in Hydrology, Nutrient Loading, and Conductivity on Periphyton Mat Structure and Function

Hydroperiod Effects on Mat Structure and Function

The timing, duration, and depth of flooding influence periphyton mat structure and function. Because the present Everglades is a compartmentalized system, some regions of the Everglades have experienced a decrease in hydroperiod, while other regions have experienced greater inundation periods compared to predrainage conditions (McVoy et al. 2011). Hydroperiod affects the physical form and community structure of the mat. Epipelic, benthic mats generally dominate short hydroperiod landscapes. Metaphyton and sweaters associated with rooted macrophytes are abundant in areas with intermediate hydroperiods. Epiphytic, floating periphyton mats (often associated with *Utricularea purpurea* in the Everglades) are abundant in long hydroperiod marshes. Gottlieb et al. (2005) found that although there were significant differences in mat community structure (Figs. 4 and 5), benthic short hydroperiod mats and floating long hydroperiod mats generally had similar species composition. Differences were primarily associated with the relative abundance of given taxa rather than the presence or absence of some. Short hydroperiod sites were dominated by cyanobacteria and a few diatom species, whereas long hydroperiod sites had a greater relative abundance of diatom species. The effects of artificial impoundment and compartmentalization on hydroperiod have direct consequences on microbial community structure and function.

Using periphyton community composition data from the SFWMD sampling program, we evaluated community structure (of diatoms and soft algae combined) across the landscape in ENP in Shark River Slough and Taylor Slough. Results were comparable to those found by Gottlieb et al. (2006). Short hydroperiod (group 4) and long hydroperiod communities differed (Fig. 6).

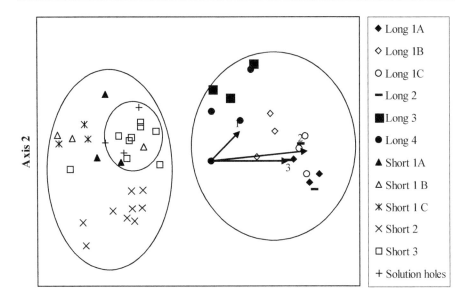

Axis 1

Figure 4. Two dimensional non-metric multidimensional scaling ordination of soft algae communities. Long hydroperiod, short hydroperiod and solution hole communities are plotted along with biplot vector overlays where vector 1 = Soil TP, 2 = water depth, 3 = periphyton TP, water TP and SRP.

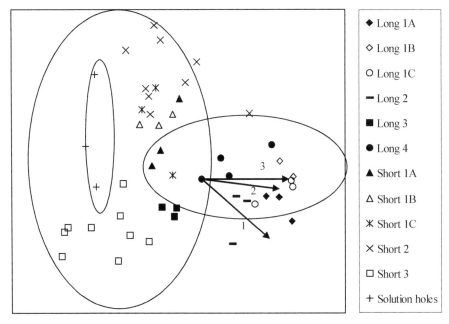

Figure 5. Two dimensional non-metric multidimensional scaling ordination of diatom communities. Long hydroperiod, short hydroperiod and solution hole communities are plotted along with biplot vector overlays of physiolchemical variables where vector 1 = water depth, 2 = water column SRP and TP, 3 = periphyton TP. The direction of the vector indicates increases in the measured parameter(s).

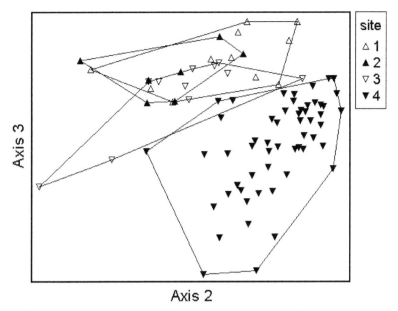

Figure 6. Three dimensional non-metric multidimensional scaling ordination of periphyton mat communities from Shark River Slough and Taylor Slough. A clear separation in mat structure is noted between SRS samples (1, 2, and 3) and shorter hydroperiod Taylor Slough samples (4), along axis 3.

Given that periphyton mat community structure from short and long hydroperiod wetlands differs primarily in the relative abundances of taxa, one would expect the greater the duration of desiccation stress, the more similar long hydroperiod periphyton mat communities become to short hydroperiod periphyton mat communities. Gottlieb et al. (2005) tested the effects of drying on periphyton mat structure. Mats were subjected to varying duration of drying (0, 1, 3, and 8 months) and then rehydrated for one month. NMDS indicated a clear separation between short and long-hydroperiod community structure at the start of the experiment. Community structure continued to differ between short and long-hydroperiod periphyton mats after each of the three desiccation treatments (Fig. 7). Although the communities differed, the directional effect of desiccation duration on both short and long-hydroperiod mat communities was similar. This helps to explain the results why mat communities from long hydroperiod sites dried for 8 months and rehydrated appeared more similar to short hydroperiod mats than long hydroperiod mats, indicating length of drying is a driver of mat community structure (Fig. 7).

Long-hydroperiod periphyton community structure was particularly influenced by the abundance of the filamentous cyanobacteria *Schizothrix calcicola* (Agardh) Gomont 1892 and the diatom *Fragilaria synegrotesca* Lange-Bertalot 1993. Relative abundance of *S. calcicola* increased significantly from 7+/– 4% (standard deviation) to 14+/–3% (Fig. 3). *Fragilaria synegrotesca* responded more rapidly to the desiccation treatment than *S. calcicola*. After 1 month of desiccation and 1 month of rewetting, *F. synegrotesca* relative abundance declined from 18+/–11 to 6+/–5% in long-hydroperiod periphyton mats.

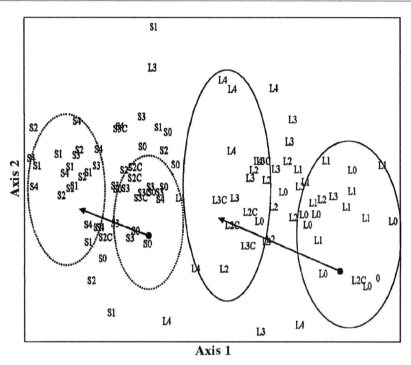

Figure 7. Non-metric multidimensional scaling ordination of Chlorophyceae, Cyanophyceae, and Bacillariophyceae taxa community structure (mean stress = 33.331 and p = 0.0323 where p = proportion of Monte Carlo randomized runs with stress < or = observed stress). L and S indicate samples from long and short hydroperiods and 0–4 indicate the five-desiccation treatments. Ellipses indicate communities at the start and end of the experiment for both short (dotted lines) and long (solid lines) hydroperiod communities. Vector direction indicates increased desiccation duration.

Treatment effects in short-hydroperiod mats were not as strong as those seen for long-hydroperiod mats, yet increasing desiccation led to a significant decline in *Scytonema hofmanni* Agardh abundance. The small decrease in *S. hofmanni* occurred together with a large increase in an unknown cocoid cyanobacteria (Gottlieb et al. 2003). No significant change in *S. hofmanni* abundance occurred in long-hydroperiod samples, but combined relative abundance of *S. calcicola* and *S. hofmanni* increased from an average of 10% to an average of 19% (Fig. 8).

Desiccation duration had a significant effect on diatom abundance in both short- and long-hydroperiod periphyton mat communities (Fig. 8). Diatom cell counts declined significantly from an average of 47% of the long-hydroperiod community at the start of the experiment to 24% after 1 month and decreased further to a low of 12% after 8 months of desiccation. Although diatom abundance was lower initially in short-hydroperiod mats, a significant decline occurred from an average of 3% diatoms at the start of the experiment to less than 1% diatoms after 8 months of desiccation.

In addition to changes in abundance, taxon-specific responses of diatoms to hydroperiod can inform assessment of hydrologic change on ecosystem-scale periphyton community structure. Diatoms are useful indicators of environmental change because they have quick generation times, relatively easily identifiable species

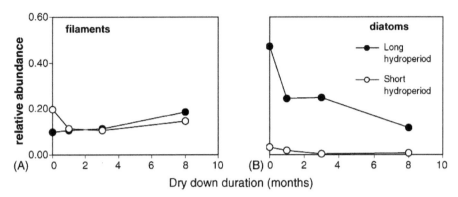

Figure 8. Desiccation effects on cyanobacterial filament (A) and diatom abundance (B). Analysis indicates an increase in mean relative abundance of *Schizothrix calcicola* and *S. hofmanni* filaments combined (A) and a decrease in all diatom taxa combined (B) for both short- and long-hydroperiod periphyton mats.

(compared to cyanobacteria and green algae), and diverse communities that change composition in response to their environment in a reliable manner. Further, similar species identities across long and short hydroperiod regions allow ecosystem-scale application of diatom-based inferences. A diatom transfer function of hydroperiod was developed to obtain taxon-specific hydroperiod optima and to make diatom-based inferences across compartments of the Everglades (Lee et al. 2013). We updated the transfer function using the most recent diatom assemblage data from the CERP Monitoring and Assessment Program (RECOVER 2004). Assemblage data were collected in 2011 from WCA 2, WCA 3, and ENP and included 29 taxa representing 16 genera. Hydroperiod values from the Everglades Depth Estimation Network were adjusted using *in situ* water depth measurements. Hydroperiod from the surveyed basins ranged from 27 to 300 and averaged 174 days flooded (number of sites = 115). The transfer function was tested by simulating prediction errors using the bootstrapping resampling method repeated 1000 times. Based on the updated transfer function, diatoms can be used to infer hydroperiod with a root mean square error of prediction = 69 days, $R^2 = 0.32$, and bootstrapped $R^2 = 0.13$ (Fig. 9). Hydroperiod optima ranged from 134 days for *Diploneis parma* to 236 days for *Navicula radiosa*. Although P has the strongest influence on Everglades diatoms ($R^2 = 0.56$, La Heé and Gaiser 2012), the relationship between hydroperiod and diatom community composition deserves further attention in advance of major hydrologic changes proposed for Everglades restoration. A more complete understanding of diatom responses to hydrologic change, especially over longer time periods that incorporate interannual variability (i.e., wet and dry spells), may improve the predictive ability of diatom-based hydrologic models.

Structure Loading Effects on Periphyton Mat Communities

Nutrient enrichment (primarily P) effects the structure and function of Everglades plant and algal communities. The work of Miao (2000, 2001), Hagerthey (2008), and Newman (1996) defined threshold soil nutrient concentrations (500–650 mg TP/Kg) where sawgrass generally converted to cattail. Similarly, Newman et al. (2004) defined

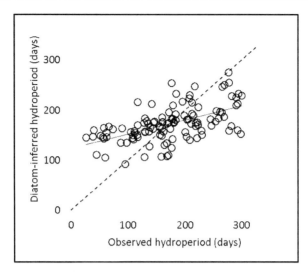

Figure 9. Preliminary relationship between observed and diatom inferred hydroperiod ($R^2 = 0.32$).

thresholds in surface water nutrient concentrations where undesirable biotic changes were likely occur (10 ppb TP threshold). McCormick et al. (2002) and Gaiser et al. (2005) defined P concentrations where periphyton mat community structure shifted and in some cases where mats disintegrated all together. Most studies of nutrient effects on periphyton to date focus on concentrations rather than load. Additional research and analysis about the varying effects of flow vs. load on periphyton structure is needed to better understand the impacts to food web dynamics and soil building processes. Biggs and Close (2006) noted hydrologic factors explained 63.3% of the variance in biomass while nutrients explained 57.6% of the variance (in cobble streams). They recommend that "periphyton data ... always be viewed within the context of the flow history of the site, and not just as a function of nutrient concentrations."

Similar to concentration, nutrient load also plays an important role in defining mat structure and function (Gaiser et al., this book). When flow is constrained (i.e., culverts and bridges), additional volume, accompanied by increased load, passes through the remaining unconstrained flow-way. This increases the load experienced in one location while decreasing it in others, thereby changing the historical distribution of nutrients in the landscape. Understanding the varying effects of concentration versus load can help guide water delivery distribution to the Everglades, thereby minimizing nutrient redistribution and related effects.

Constraints through culverts, canals, and other structures concentrate flow and increase local loading. The varying effects of operations and related loading can be observed at the boundary between the WCAs and ENP. Rather than being delivered as sheet flow across many miles of marsh, water passes into ENP along Tamiami Trail predominantly through the S12 (A-D) and S333 structures. Using information collected by the SFWMD, we evaluated the varying effects (relationships between) of structure flow volume, load, and water column nutrient concentration on mat community structure downstream of Tamiami Trail. Initially all samples below the S12 and S355 structures (on the NW and NE sides of the ENP) were evaluated for differences in community structure. Periphyton samples were collected approximately 2.5, 6.5, and 10 Km below corresponding structures. Nonmetric multidimensional scaling using Bray-Curtis dissimilarity metric was used to visualize differences in community structure. Although there were some differences in community structure by transect location (S12 and 355), the predominant difference was a separation in mat community structure between samples collected from 2002 and before and those collected from 2003–2007 (Fig. 10).

Using the average structure flow and load data, from the three month period prior to periphyton sampling, helps provides understanding of the factors driving variation in mat structure across the landscape. Differences in nutrient and flow conditions are drivers of community structure. The post-2002 period suggests that use of the 343 and 344 structures helped to drive changes in flow and load leading to shifts in mat structure Fig. 10).

When examining one of the two operational periods (2002 and before) and just at the western transect (S12), there was a much stronger relationship between mat structure and flow and load (Fig. 11). The variation in community structure (on axis 3) in 1999 is related to increased P loading, whereas the structural differences in 1997 are related to increased flow. Understanding differential effects of flow and load provided insight into potentially useful indicator communities.

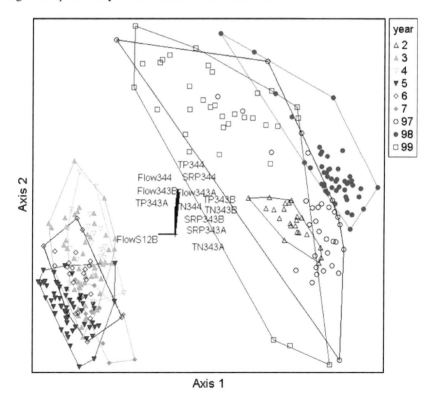

Figure 10. Non-metric multidimensional scaling ordination showing differences in algal community structure among samples collected before 2002 and after 2002. Vectors indicate the magnitude and direction of influence of flows from structures S12B, S343 and S344.

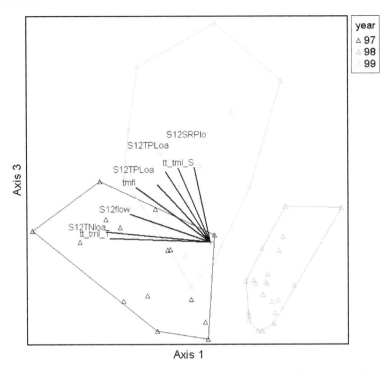

Figure 11. Non-metric multidimensional scaling ordination showing relationship between algal community structure and surface water flow and load (at the S12 structures). Community structure in 1997 is predominantly related to S12 TN load and S12 flows, whereas in 1999 community structure is related to increased S12 TP loads and increased overall structure total monthly flow at the Tamiami Trail.

Relationship Between Load and Mat Nutrient Concentrations

Differences in mat TP by site along each of the transects were significant ($F_{5,171} = 16.1$, $P < .001$). Mat TP concentration at site S12C6 varied widely (range from 46–524 ug/g TP) along with a corresponding high measure of kurtosis (10.1). Unexpectedly, the downstream mat concentrations in the S12 transects were more enriched than those mats closer to the S12 flow structures (Table 1a; Fig. 12). Periphyton mats along the eastern S355 transect followed expected results with mats closer to the structure being more enriched than those downstream ($F_{5,267} = 13.6$, $P < .001$) (Table 1b). The results along the S355 transects are consistent with the transect and enrichment studies conducted by McCormick et al. (2001, 1996). Gaiser et al. (2006) and others demonstrate variation in periphtyon mat TP and species composition with increased nutrient concentrations and loads. To better understand the nutrient patterns identified below, it is important to understand how the system is compartmentalized and how water management structures are operated (in support of both urban and natural system needs).

Tables 1a and 1b. Periphyton mean mat TP downstream of S12 and S355 structures.

1a)

Site	Mean Mat TP	Std Dev
S12A2	78.9	26.6
S12A6	181.6	81.1
S12A10	239.2	104.3
S12C2	94.1	20.1
S12C6	120.0	86.5
S12C10	200.9	106.6

1b)

Site	Mean Mat TP	Std Dev
S355A2	121.4	59.4
S355A6	202.3	111.9
S355A10	175.0	92.5
S355B2	158.0	84.3
S355B6	117.9	56.8
S355B10	81.1	34.0

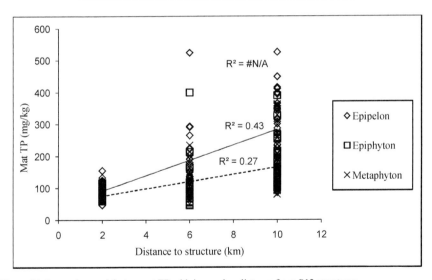

Figure 12. Increasing periphyton mat TP with increasing distance from S12 structures.

Historically the Everglades flowed southeast along topographic gradients paralleling what is now the Miami canal. Water flowed eastward toward Biscayne Bay through the Miami River, as well as through smaller transverse glades during high water periods, but most water hit the prairies and rock ridge and flowed back to the south and west through Shark River Slough (McVoy et al. 2011) (Under extreme high conditions it is likely that both Shark River Slough and the smaller Taylor Slough drainages were connected). Under current compartmentalization and management, surface water flow to the southeast is limited via the L67 canals. This helps to provide flood control for Miami-Dade but also limits the amount of water entering NE Shark River Slough and completely isolates WCA3B. Furthermore, stages in this area were historically kept artificially low to maintain discharge capacity due to concerns for flooding in the 8.5 square mile area to the east of Shark Slough (S3273 stage constraint) (RECOVER 2005). The L-67 levies, Tamiami Trail and the limited structure flow through the S355 structures resulted in enhanced drying and corresponding oxidation and enrichment of NE Shark River Slough (Osborne et al. 2011), thereby establishing the noted P gradient.

Periphyton mats below the S12 structures experience very different hydrologic regimes. Those mats closest to the structures at the 2 km sample points experience much of the flow moving through the corresponding S12A and S12C structures. The downstream sample locations (S12A10 and S12C10) not only receive flow from the S12 structures but also receive surface flow from the S333, and from E (rainfall into NE Shark River Slough) and western flows into central Shark River Slough (through the S343 and S344). Western flows are indicated by increases in abundance of specific desmids more common in softer waters, nearer to Big Cypress.

Conductivity Effects on Periphyton Communities

The ion concentration of surface waters is known to affect the structure and function of periphyton communities (Potopova and Charles 2003). From early studies on conductance effects on stream algae by Johansson (1982) to more recent works in the Florida Everglades by McCormick et al. (2011), Hagerthey et al. (2011), and Gaiser et al. (2011), changes in specific conductance are linked to changes in community structure and function.

Slate and Stevenson (2000) observed changes in algal communities indicating increased pH at the LNWR-WCA2 boundary after wetland impoundment. Similarly in LNWR, changes in mat structure are observed along the canal to interior conductivity gradient (Gaiser et al. 2011; McCormick et al. 2011). Experiments by Gottlieb et al. (2004) showed large shift in periphyton community structure when mixing LNWR low conductivity mats with high conductivity waters from neighboring WCA2A. Seven conductivity treatments were developed with conditions comparable to those observed in LNWR by mixing interior waters from LNWR (80~uS) with WCA2A waters (840~uS). Mats were grown in core tubes within an environmental chamber receiving approximately 1000 u einsteins/s/m^2. Desmids are generally found in greater abundance in softwater systems. The experimental mixing study showed a decline in low conductivity diatom taxa, as well as a decrease in the abundance of desmids with increases in water column conductance (Figs. 13 and 14) (SFER 2005). Results were comparable to those found by others (Table 2).

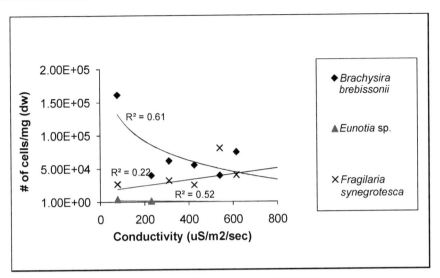

Figure 13. Conductivity treatment effects on diatom abundance (SWS 2004).

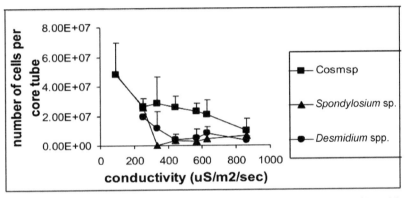

Figure 14. Desmid abundance decreasing with increasing conductivity (where *Cosmsp* and *Spondylosium* sp. denote single species of *Cosmarium* and *Spondylosium* respectively, and *Desmidium* spp. denotes the sum of more than one species of *Desmidium*).

Table 2. Conductivity optima for select algae.

Taxa	Optima (uS)	Literature Optima (uS)	Reference
Brachysira brebissonii	347	40	Potopova and Charles 2003; Ryves et al. 2002
Eunotia flexuosa	126	94	Potopova and Charles 2003; Ryves et al. 2002
Navicula cryptocephala	307	125, 222	Leland et al. 2001; Potopova and Charles 2003
Encyonopsis microcephala	452	380	Potopova and Charles 2003
Nitzschia palea v. debilis	517	460	Potopova and Charles 2003
Euglena spp.	394	398	Leland et al. 2001
Anabaena spp.	493	521	Leland et al. 2001
Chlamydomonas spp.	451	531	Leland et al. 2001

Multivariate Response of Periphyton Communities

Previous sections of this chapter mentioned interactions among nutrient loading, conductivity, and flooding duration across the Everglades landscape. Periphyton community responses reflect the multivariate effects of these habitat characteristics (Fig. 15 and Fig. 16). Periphyton biovolume (mL m^{-2}) is an integrative metric of the multivariate response of periphyton communities: 1) thick, cohesive, high biovolume periphyton is abundant in low nutrient, alkaline surface water conditions in regions with seasonal drying, while 2) loose, low biovolume periphyton is present in high nutrient, less alkaline waters that may have altered flooding patterns (Lee et al. 2013; Lee 2014).

High and low biovolume periphyton provide very different habitats for diatom communities, including structure for colonization, protection from desiccation, UV, and grazers, and coexistence with other microorganisms in the periphyton. In high biovolume periphyton, diatom communities coexist within filamentous cyanobacteria that provide the main cohesive structure of calcareous periphyton. In low biovolume periphyton, however, diatoms usually coexist with non-cohesive groups of green algae (desmids and filamentous taxa). High biovolume periphyton mats are dominated by two endemic taxa, *Mastogloia calcarea* and *Encyonema evergladianum* (refer to Table 1 in Gaiser et al., this volume; Lee et al. 2014). Low biovolume periphyton, however, is inhabited by nutrient-tolerant "weedy" taxa within the genus *Nitzschia* (Fig. 16) (refer to Gaiser et al., this volume). *Gomphonema intricatum* var. *vibrio*, *Encyonema silesiacum* var. *elegans*, *Pinnularia acrosphaeria*, and *Eunotia naegelii* are abundant in the LNWR, where the combination of high P inflows from the EAA and deeper and less alkaline surface waters interactively influence periphyton communities (Swift and Nicholas 1987). *Brachysira microcephala* (*B. neoexilis* in Gaiser et al., this volume) is an endemic taxon that is most abundant in WCA3A, where P levels are lower and conductivity is higher relative to LNWR, but average days since dry is greater than that of Shark River Slough by more than twofold (Lee 2014). Periphyton community

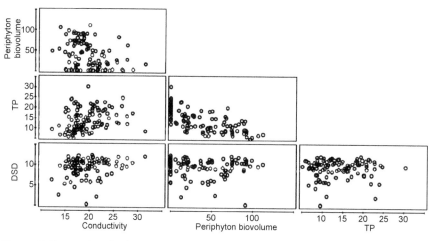

Figure 15. Scatter plots showing relationships between mat TP (μg g^{-1}), Conductivity (μS cm^{-3}), Days Since Dry and an integrative response metric, Periphyton Biovolume (mL m^{-2}). All variables have been square root transformed.

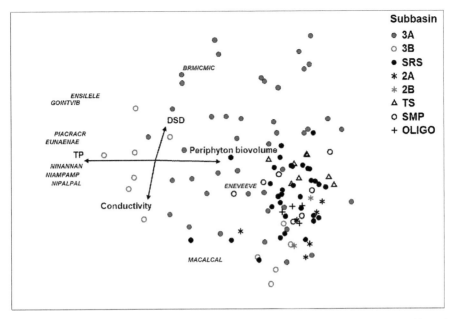

Figure 16. Non-metric multidimensional scaling ordination showing distribution of diatom communities and correlations with mat TP, days since dry, conductivity, and periphyton biovolume across Everglades subbasins. Taxon codes: *Mastogloia calcarea* (MACALCAL), *Encyonema evergladianum* (ENEVEEVE), *Brachysira microcephala* (BRMICMIC), *Encyonema silesiacum* var. *elegans* (ENSILELE), *Gomphonema intricatum* var. *vibrio* (GOINTVIB), *Pinnularia acrosphaeria* (PIACRACR), *Eunotia naegelii* (EUNAENAE), *Nitschia nanana* (NINANNAN), *Nitzschia amphibia* (NIAMPAMP), *Nitzschia palea* (NIPALPAL).

responses reflect the complex interplay of multiple environmental variables influencing natural biogeochemical processes often imposed by anthropogenic modification of nutrient, conductivity, and hydrologic patterns across the spatially heterogeneous Everglades landscape.

Conclusions and Discussion

Utility, Future Research and Monitoring Recommendations

Periphyton monitoring and research in the Florida Everglades helps provide a better understanding of the structural and functional impacts associated with land development and water management on the environment. Periphyton provides utility for ecosystem restoration programs, by helping establish hydrologic and water quality targets. Additionally, monitoring of periphyton community structure is used as a water quality proxy/surrogate both in modern and paleoecological studies.

Periphyton research has shown that mat community structure and function are sensitive to changes in hydroperiod (timing, depth and distribution of water) and water quality. Water quality changes occur not only due to changes in nutrients but also due to changes in salinity/conductivity. Future structural and operational changes in Florida infrastructure should take this into consideration.

The lessons learned regarding locally enhanced loading due to constraining flows and regarding effects of conductivity on both interior and ecotone marshes are applicable to wetlands globally.

Ongoing demand for available freshwater (and land for development) will likely encourage similar pressures in wetland systems globally. The effects of constrained flow, drying/burning and flooding are all evident in the Everglades history.

Future periphyton research should focus on providing additional understanding of periphyton functionality in the ecosystem. Flow versus loading effects should be differentiated to help set optimal flow targets and to better measure sheetflow benefits. Periphyton species and nutrient responses can also be used to identify "hot spots" thereby providing adaptive feedback to managers regarding system operations (water delivery) and project sequencing needs.

Although substantial research has been conducted on P in the Florida Everglades, less nutrient research has been conducted on nitrogen (N) dynamics. The freshwater Everglades is P-limited; therefore changes in N deliveries will likely have greater effects on the (P-rich) marine end member. Hence, future research may examine the effects of changes in hydroperiod and salinity on N dynamics, helping to better understand potential concerns for the bay and reef systems in S Florida.

Further research to understand the feedbacks between algal community structure and function, and other microbial communities is warranted because of the tight coupling which exists between bacteria and algae in periphyton mats. Chemical changes in the periphyton mat likely effect varying taxa differently. For instance, high photosynthesis can drive pH upward thereby negatively influencing components of the bacterial community (Vymazal 2007). Most of the traditional periphyton research focused on the photosynthetic fraction as discussed above but not only can shifts in blue green algal abundance affect N fixation but other components of the microbial community discussed in section three of this book also play a large role in nutrient cycling and carbon dynamics.

Acknowledgments

The authors would like to acknowledge the reviewers, as well as the SFWMD for originally collecting the periphyton samples, and for providing the associated algal community structure and water quality data. Particular thanks to Darlene Marley and Shi Xui for helping to format the extensive periphyton and water quality data sets, respectively. Funding was provided to Florida International University for the CERP MAP program from Task Agreement CP040130 from South Florida Water Management District and Cooperative Agreement W912HZ-11-2-0048 from the U.S. Army Corps of Engineers. We thank Jana Newman for coordination of the CERP MAP research program, and Franco Tobias for coordinating field and laboratory research. This work was enhanced through collaborations with the Florida Coastal Everglades Long Term Ecological Research program (#DEB-1237517). This is contribution #672 of the Southeast Environmental Research Center.

References

Abtew, A.A., W. Van Horn and S.N. Khanal. 2000. Temporal and spatial characterization of rainfall over central and south Florida. J. Am. Water Res. Assoc. 36: 833–848.

Belanger, T.V., D.J. Scheidt and J.R. Platko II. 1989. Effects of nutrient enrichment on the Florida Everglades. Lake Reserv. Manage. 5: 101–111.

Biggs, B., J.F. Close and E. Murray. 1989. Periphyton biomass dynamics in gravel bed rivers: the relative effects of flows and nutrients. Freshwater Biol. 22: 209–231.

Bramburger, A., J. Munyon and E. Gaiser. 2013. Water quality and wet season diatom assemblage characteristics from the Tamiami Trail pilot swales sites. Phytotaxa 127: 163–182.

Browder, J.A., A.S. Black, M. Brown, S. Newman, D. Cottrell, D. Black, R. Pope and P. Pope. 1981. Perspectives on the ecological causes and effects of the variable algal composition of Southern Everglades periphyton. South Florida Research Center, Homestead, Florida, Report T-643.

Childers, D., R.D. Jones, J.C. Trexler, C. Buzelli, S. Dailey, A. Edwards, E. Gaiser, K. Jaychandaran, A. Kenne, D. Lee, J.F. Meeder, J.H.K.A. Pechmann, A. Renshaw, J. Richards, M. Rugge, L.J. Scinto, P. Sterling and W. Van Gelder. 2002. Quantifying the effects of low level phosphorus additions on unenriched Everglades wetlands with *in situ* flumes and phosphorus dosing. pp. 127–152. *In*: J.W. Porter and K.G. Porter (eds.). The Everglades, Florida Bay, and Coral Reefs of the Florida Keys an Ecosystem Sourcebook. CRC Press, Boca Raton, Florida, USA.

Clarke, K.R. 1993. Non-parametric multivariate analyses of changes in community structure. Aust. J. Ecol. 18: 117–143.

Cox, Trevor, F. and M.A. Cox. 2001. Multidimensional Scaling: Monographs on Statistics and Probability 88. Chapman and Hall/CRC, Boca Raton, Florida, p. 308.

Davis, S.M. and J.C. Ogden (eds.). 1994. Everglades: The Ecosystem and its Restoration. St. Lucie Press, Delray Beach, Florida.

Desikachary, T.V. 1959. Cyanophyta. Academic Press, New York.

Donar, C.M., K.W. Condon, M. Gantar and E.E. Gaiser. 2004. A new technique for examining the physical structure of Everglades floating periphyton mat. Nova Hedwigia 78: 107–119.

Dong, Q. 2006. Pulsing sheetflow and wetland integrity. Front. Ecol. Environ. 4: 9–10.

Douglas, M.S. 1947. The Everglades: River of Grass. Hurricane House, Coconut Grove, Florida.

Dow, C.L. and R.A. Zampella. 2000. Specific Conductance and pH as indicators of watershed disturbance in streams of the New Jersey pinelands. J. Environ. Manage. 26: 437–445.

Evans, J.H. 1959. The survival of freshwater algae during dry periods. Part II. Drying experiments. Part III. Stratification of algae in pond margin litter and mud. J. Ecol. 47: 55–81.

Firbank, L.G., C.J. Barr, R.G.H. Bunce, M.T. Furse, R. Haines-Young, M. Hornung, D.C. Howard, J. Sheail, A. Sier and S.M. Smart. 2003. Assessing stock and change in land cover and biodiversity in GB: an introduction to Countryside Survey 2000. J. Environ. Manage. 67: 207–218.

Gaiser, E. 2009. Periphyton as an early indicator of restoration in the Florida Everglades. Ecol. Indic. 6: S37–S45.

Gaiser, E.E., L.J. Scinto, J.H. Richards, K. Jayachandran, D.L. Childers, J.C. Trexler and R.D. Jones. 2004. Phosphorus in periphyton mats provides the best metric for detecting low-level P enrichment in an oligotrophic wetland. Water Res. 38: 507–516.

Gaiser, E.E., J.C. Trexler, J.H. Richards, D.L. Childers, D. Lee and A.L. Edwards. 2005. Cascading ecological effects of low-level phosphorus enrichment in the Florida Everglades. J. Environ. Qual. 34: 717–723.

Gaiser, E.E., J.H. Richards, J.C. Trexler, R.D. Jones and D.L. Childers. 2006. Periphyton responses to eutrophication in the Florida Everglades: cross-system patterns of structural and compositional change. Limnol. Oceanogr. 51: 617–630.

Gaiser, E.E., P. McCormick, S.E. Hagerthey and A.D. Gottlieb. 2011. Landscape patterns of periphyton in the Florida Everglades. Crit. Rev. Environ. Sci. Technol. 41(S1): 92–120.

Garcia-Pichel, F. and O. Pringault. 2001. Cyanobacteria track water in desert soils. Nature 413: 380–381.

Gleason, P.J. and W. Spackman. 1974. Calcareous Periphyton and Water Chemistry in the Everglades. Miami Geological Society, Coral Gables, Florida.

Gleason, P.J. and P. Stone. 1994. Age, origin, and landscape evolution of the Everglades peatland. pp. 149–197. *In*: S.M. Davis and J.C. Ogden (eds.). Everglades—The Ecosystem and its Restoration. St. Lucie Press, Delray Beach, FL.

Gottlieb, A. 2003. Short and Long-Hydroperiod Everglades Periphyton Mats: Community Characterization and Experimental Hydroperiod Manipulation, Dissertation. Florida International University, Miami, Florida, 253 pp.

Gottlieb, A., S. Hagerthey, D. Kilbane, R. Shufford and S. Newman. 2004. The effects of varying conductivity on Everglades periphyton community structure. Society of Wetlands Scientists, http://www.sws.org/archive/Seattle2004/program_contrib_schedule.htm.

Gottlieb, A., J.H. Richards and E.E. Gaiser. 2005. Effects of desiccation duration on the community structure and nutrient retention of short and long-hydroperiod Everglades periphyton mats. Aquat. Bot. 82: 99–112.

Gottlieb, A.D., J.H. Richards and E.E. Gaiser. 2006. Comparative study of periphyton community structure in long and short-hydroperiod Everglades marshes. Hydrobiologia 569: 195–207.

Gunderson, L.H. 1994. Vegetation of the Everglades: determinants of community composition. pp. 323–340. *In*: S.M. Davis and J.C. Ogden (eds.). Everglades: The Ecosystem and its Restoration. St. Lucie Press, Delray Beach, Florida, USA.

Hagerthey, S.E., S. Newman, K. Ruthey, E.P. Smith and J. Godin. 2008. Multiple regime shifts in a subtropical peatland: Community-specific thresholds to eutrophication. Ecol. Monogr. 78: 547–565.

Harvey, J., R. Schaffranek, G. Noe, L. Larsen, D. Nowacki and B. O'Connor. 2009. Hydroecological factors governing surface water flow on a low-gradient floodplain. Water Resour. Res. 45: W03421.

Ho, D.T., V.C. Engel, E.A. Variano, P.J. Schmieder and M.E. Condon. 2009. Tracer studies of sheet flow in the Florida Everglades. Geophys. Res. Lett. 36: L09401.

House, W.A. 2003. Geochemical cycling of phosphorus in rivers. Appl. Geochem. 18: 739–748.

Jarvie, H.P., C. Neal, R.J. Williams, E.J. Sutton, M. Neal and H.D. Wickham. 2002. Phosphorus sources, speciation and dynamics in the lowland eutrophic River Kennet, UK. Sci. Tot. Environ. 282: 175–203.

Johansson, C. 1982. Attached algal vegetation in running waters of Jiimtland, Sweden. Acta Phytogeogr. Suec. 71. Uppsala.

Jongman, R.H.G., C.J.F. Ter Braak and O.F.R. Van Tongeren. 1995. Data analysis in Community and Landscape Ecology. Cambridge University Press, Cambridge, UK, 299 pp.

Komárek, J. and A. Konstantinos. 1999. Sußwasserflora von Mitteleuropa. Gustav Fischer, Jena, p. 548. Krammer, K. 1992. Bibliotheca Diatomologica: Pinnularia eine Monographie der europaischen Taxa. J. Cramer, Berlin, 353 pp.

Krammer, K. and H. Lange-Bertalot. 1991. Bacillariophyaceae: Naviculaceae Sußwasserflora von Mitteleuropa. Fischer, Stuttgart, 876 pp.

La Hee, J.M. and E.E. Gaiser. 2012. Benthic diatom assemblages as indicators of water quality in the Everglades and three tropical karstic wetlands. Freshwater Sci. 31: 205–221.

Lamberti, G. 1996. The role of periphyton in benthic food webs. pp. 533–564. *In*: R.J. Stevenson, M.L. Bothwell and R. Lowes (eds.). Algal Ecology: Freshwater Benthic Ecosystems. Academic Press, San Diego, California, USA.

Lang, T.A., O. Oladeji, M. Josan and S. Daroub. 2010. Environmental and management factors that influence drainage water P loads from Everglades Agricultural Area farms of South Florida. Agr. Ecosyst. Environ. 138: 170–180.

Lange, O.L., J. Belnap and H. Reichenberger. 1998. Photosynthesis of the cyanobacterial soil-crust lichen *Collema tenax* from arid lands in southern Utah USA: role of water content on light and temperature responses of CO_2 exchange. Funct. Ecol. 12: 195–202.

Lange-Bertalot, H. 1993. Bibliotheca Diatomologica: 85 Neue Taxa unduber weitere neu definierte Taxaerganzend zurSuwasserflora von Mitteleuropa. J. Cramer, Berlin, 454 pp.

Larsen, L.G. and J.W. Harvey. 2010. How vegetation and sediment transport feedbacks drive landscape change in the everglades and wetlands worldwide. Am. Nat. 176(3): E66–E79.

Lee, Sylvia. 2014. Mechanisms of diatom assembly in a hydrologically-managed subtropical wetland. Dissertation. Florida International University. Miami, 181 pp.

Lee, Sylvia, E. Gaiser and J. Trexler. 2013. Diatom-based models for inferring hydrology and periphyton abundance in a subtropical karstic wetland: implications for ecosystem-scale bioassessment. Wetlands 33: 157–173.

Lee, S., E. Gaiser, B. Van de Vijver, M. Edlund and S. Spaulding. 2014. Morphology and typification of *Mastogloia smithii* and *M. lacustris*, with descriptions of two new species from the Florida Everglades and the Caribbean region. Diatom Res., doi:10.1080/0269249X.2014.889038.

Leland, H.V., L.R. Brown and D.K. Mueller. 2001. Distribution of algae in the San Joaquin River California, in relation to nutrient supply, salinity and other environmental factors. Freshwater Biol. 46: 1139–1167.

Liston, S.E. and J.C. Trexler. 2005. Spatial and temporal scaling of macroinvertebrate communities inhabiting floating periph-yton mats in the Florida Everglades. J. N. Am. Benthol. Soc. 24: 832–844.

McCormick, P., J. Harvey and E. Crawford. 2011. Influence of changing water sources and mineral chemistry on the Everglades Ecosystem. Crit. Rev. Environ. Sci. Technol. 41(S1): 28–63.

McCormick, P.V. and M.B. O'Dell. 1996. Quantifying periphyton responses to phosphorus in the Florida Everglades: A synoptic-experimental approach. J. N. Am. Benthol. Soc. 15: 450–468.

McCormick, P.V. and R.J. Stevenson. 1998. Periphyton as a tool for ecological assessment and management in the Florida Everglades. J. Phycol. 34: 726–733.

McCormick, P.V., P.S. Rawlik, K. Lurding, E.P. Smith and F.H. Sklar. 1996. Periphyton-water quality relationships along a nutrient gradient in the northern Florida Everglades. J. N. Am. Benthol. Soc. 15: 433–449.

McCormick, P.V., R.B.E. Shuford, III, J.G. Backus and W.C. Kennedy. 1997. Spatial and seasonal patterns of periphyton biomass and productivity in the northern Everglades, Florida, USA. Hydrobiologia 362: 185–208.

McCormick, P.V., M.B. O'Dell, R.B.E. Shuford, III, J.G. Backus and W.C. Kennedy. 2001. Periphyton response to experimental phosphorus enrichment in a subtropical wetland. Aquat. Bot. 71: 119–139.

McCune, B. and J.M. Medford. 1999. Multivariate Analysis of Ecological Data. MJM Software, Gleneden Beach, Oregon.

McVoy, C.W., W.P. Said, J. Obeysekera, J. Van Arman and T. Dreschel. 2011. Landscape and Hydrology of the Pre-drainage Everglades, University Press of Florida, 596 pp.

Miao, S.L., S. Newman and F.H. Sklar. 2000. Effects of habitat nutrients and seed sources on growth and expansion of *Typha domingensis*. Aquat. Bot. 68: 297–311.

Miao, S.L., P.V. McCormick, S. Newman and S. Rajagopalan. 2001. Interactive effects of seed availability, water depth, and phosphorus enrichment on cattail colonization in an Everglades wetland. Wetl. Ecol. Manag. 9: 39–47.

Morgan, M.D. 1987. Impact of nutrient enrichment and alkalinization on periphyton communities in the New Jersey Pine Barrens. Hydrobiologia 144: 233–241.

Newman, S. and S. Hagerthey. 2011. Water Conservation Area 1: A Case Study of Hydrology, Nutrient, and Mineral Influences on Biogeochemical Processes. Crit. Rev. Environ. Sci. Technol. 41(S1): 702–722.

Newman, S., J.B. Grace and J.W. Koebel. 1996. Effects of nutrients and hydroperiod on mixtures of *Typha*, *Cladium*, and *Eleocharis*: implications for Everglades restoration. Ecol. Appl. 6: 774–783.

Newman, S., P.V. McCormick, S.L. Miao, J. Laing, C. Kennedy and M. O'Dell. 2004. The effect of phosphorus enrichment on the nutrient status of a northern Everglades slough. Wetlands Ecol. Manag. 12: 63–79.

Noe, G.B., D.L. Childers and R.D. Jones. 2001. Phosphorus biogeochemistry and the impacts of phosphorus enrichment: Why is the Everglades so unique? Ecosystems 4: 603–624.

Noe, G.B., L.J. Scinto, J. Taylor, D.L. Childers and R.D. Jones. 2003. Phosphorus cycling and partitioning in an oligotrophic Everglades wetland ecosystem: a radioisotope tracing study. Freshwater Biol. 48: 1993–2008.

Odum, W.E., E.P. Odum and H.T. Odum. 1995. Nature's pulsing paradigm. Estuaries 18: 547–55.

Ogden, J.C., S.M. Davis, K.J. Jacobs, T. Barnes and H.E. Fling. 2005. The use of conceptual ecological models to guide ecosystem restoration in South Florida. Wetlands 25: 795–809.

Osborne, T.Z., S. Newman, P. Kalla, D.J. Scheidt, G.L. Bruland, M.J. Cohen, L.J. Scinto and L.R. Ellis. 2011. Landscape patterns of significant soil nutrients and contaminants in the Greater Everglades Ecosystem: Past, Present, and Future. Crit. Rev. Environ. Sci. Technol. 41(S1): 121–148.

Paerl, H.W., J.L. Pinckney and T.F. Steppe. 2000. Cyanobacterial-bacterial mat consortia: examining the functional unit of microbial survival and growth in extreme environments. Environ. Microbiol. 2: 11–26.

Patrick, R. and C.W. Reimer. 1975. The Diatoms of the United States, vol. (2) 1. Sutter House, Philadelphia, Pennsylvania, p. 213.

Potopova, M. and D.F. Charles. 2003. Distribution of benthic diatoms in US rivers in relation to conductivity and ionic composition. Freshwater Biol. 48: 1311–1328.

[RECOVER]. 2004. CERP Monitoring and Assessment Plan: Part 1. Monitoring and Supporting Research—January 2004. Comprehensive Everglades Restoration Plan, Restoration Coordination and Verification (RECOVER).

[RECOVER]. 2005. RECOVER Initial CERP Update Report, October 2005, http://www.evergladesplan.org/pm/recover/icu.aspx.

[RECOVER]. 2010. Planning Team, Technical Report on System-wide Performance of CERP 2015 Band 1 Projects, January 28, 2010, http://www.evergladesplan.org/pm/recover/band_1_report.aspx.

Rejmankova, E. and J. Komarkova. 2000. A function of cyanobacterial mats in phosphorus-limited tropical wetlands. Hydrobiologia 431: 135–153.

Richards, J.H. 2001. Bladder function in *Utricularia purpurea* (Lentibulariaceae): is carnivory important? Am. J. Bot. 88: 170–176.

Robinson, G.C., S.E. Gurney and L.G. Goldborough. 1997. Response of benthic and planktonic algal biomass to experimental water-level manipulation in a prairie lakeshore wetland. Wetlands 17: 167–181.

Ross, M.S., J.F. Meeder, J.P. Sah, P.L. Ruiz and G.J. Telesnicki. 2000. The Southeast Saline Everglades revisited: 50 years of coastal vegetation change. J. Veg. Sci. 11: 101–112.

Ryves, D.B., S. McGowan and N.J. Anderson. 2002. Development and evaluation of a diatom-conductivity model from lakes in West Greenland. Freshwater Biol. 47: 995–1014.

Sargeant, B.L., E.E. Gaiser and J.C. Trexler. 2010. Biotic and abiotic determinants of intermediate-consumer trophic diversity in the Florida everglades. Mar. Freshwater Res. 61: 11–22.

Slate, J.E. and R.J. Stevenson. 2000. Recent and abrupt environmental change in the Florida Everglades indicated from siliceous microfossils. Wetlands 20: 346–356.

[SFER]. South Florida Environmental Report. 2005. Ecology of the Everglades Protection Area, Chapter 6. 6-1 to 6-43, http://www.sfwmd.gov/portal/page/portal/pg_grp_sfwmd_sfer/portlet_prevreport/2005/volume1/chapters/V1_Ch6.pdf B. 2007, Appendix 2-2: Stage-Storage Relationship of Lakes and Impoundments.

Starmach, K. 1966. Cyanophyta-Sinice Glaucophyta-Glaukofity, Warsaw.

Stradling, D.A., T. Thygerson, J.A. Walker, B.N. Smith, D.L. Hansen, R.S. Criddle and R.I. Pendleton. 2002. Cryptogamic crust metabolism in response to temperature, water vapor, and liquid water. Thermochim. Acta 7121: 1–7.

Surratt, D., M.G. Waldon, M.C. Harwell and N.G. Aumen. 2008. Temporal and spatial trends of canal water intrusion into a northern Everglades marsh in Florida, USA. Wetlands 28: 173–186.

Swift, D.R. and R.B. Nicholas. 1987. Periphyton and water quality relationships in the Everglades Water Conservation Areas. Technical publication 87-2. South Florida Water Management District. West Palm Beach, Florida, 62 pp.

Townsend, P.A. 2001. Relationship between vegetation patterns and hydroperiod on the Roanoke River flood-plains, North Carolina. Plant Ecol. 156: 43–58.

Vymazal, J. 1995. Algae and Element Cycling in Wetlands. Lewis Publishers, Boca Raton, Florida.

Vymazal, J. and C.J. Richardson. 1995. Species composition, biomass, and nutrient content of periphyton in the Florida Everglades. J. Phycol. 31: 343–354.

Watts, D., M. Cohen, J. Heffernan and T. Osborne. 2010. Hydrologic modification and the loss of self-organized patterning in the ridge–slough mosaic of the Everglades. Ecosystems 13: 813–827.

Wetzel, R.G. 1983. Limnology. Saunders College Publishing, Philadelphia, PA, pp. 767.

Whitford, L.A. and G.J. Schumacher. 1984. A Manual of Fresh-Water Algae. Sparks Press, Raleigh, NC.

Wossenu, A., R.S. Huebner, C. Pathak and V. Ciuca. http://www.sfwmd.gov/portal/page/portal/pg_grp_sfwmd_sfer/portlet_prevreport/volume1/appendices/v1_app_2-2.pdf.

8

The Role of Periphyton Mats in Consumer Community Structure and Function in Calcareous Wetlands: Lessons from the Everglades

Joel C. Trexler,[1,a,*] *Evelyn E. Gaiser,*[2,b] *John S. Kominoski*[2,c] and *Jessica L. Sanchez*[1,d]

Introduction

Freshwater benthic microbial communities display a diversity of growth forms that shape food web structure and function through influencing availability of food and predation refuges of primary consumers. Periphyton, an assemblage of microflora growing on substrates (Wetzel 1983), includes a diversity of autotrophs and saprophytes (microscopic algae, bacteria, and fungi) that may form complex structures in floating mats, epiphyton, and benthic mats (Stevenson 1996). Through its dual influence on food resources and habitat structure, periphyton may play a controlling role in dynamics of aquatic communities (Power 1996). The constituents of periphyton display a

[1] Department of Biological Science, 3000 NE 151st Street, Florida International University, North Miami, FL 33181.
[a] Email: trexlerj@fiu.edu
[d] Email: jsanc318@fiu.edu
[2] Department of Biological Science, 11200 SW 8th Street, Florida International University, Miami, FL 33199.
[b] Email: gaisere@fiu.edu
[c] Email: jkominos@fiu.edu
* Corresponding author

range of palatability and nutrient content for consumers, including some taxa that are chemically defended with compounds that hamper digestibility or are toxic. Palatable and unpalatable biota may grow in a matrix of detritus, polysaccharide exudates, and mineral precipitates with the potential to interfere with consumer feeding (Paul and Hay 1986; Duffy and Paul 1992; Schupp and Paul 1994). Associational resistance, where neighboring species decrease the likelihood of or vulnerability to negative species interactions (Barbosa et al. 2009; Underwood et al. 2014), is an emergent property of periphyton mats with implications for shaping consumer community dynamics.

There is increasing theoretical evidence that the presence of inedible resources limits consumer abundance and diversity through food-web dynamics and stoichiometry (Hall et al. 2006). Inedible producers can compete with and control the nutrient content of edible producers, reducing the nutrient and energy available to grazers. Inedible producers can also control community assembly by effectively excluding specific grazer species that are weak competitors for the limited nutrients in edible producers (Fig. 1A). Grazer diversity can, in turn, shape the stoichiometry of producer communities both through selective removal and nutrient regeneration (Hillebrand et al. 2009; Hall 2009). Increasing grazer diversity may increase the likelihood that relatively efficient consumers are present that yield disproportionately large effects on stoichiometric patterns (Hillebrand et al. 2009). Thus, study of the linkages of algal growth forms to aquatic consumers may play a fundamental role in untangling the complexities of community structure and function in aquatic ecosystems.

Everglades wetlands have among the highest standing crops of periphyton compared to similar wetlands (Gaiser et al. 2012), and appear to support high rates of primary production (Ewe et al. 2006; Troxler et al. 2013). The remarkable amount of algal productivity in the Everglades would seem to support a large standing crop of aquatic consumers, but in fact the secondary productivity, as indicated by consumer standing stocks, is relatively low compared to other aquatic systems (Turner et al. 1999; Ruehl and Trexler 2011). This apparent energetic inefficiency is tied to

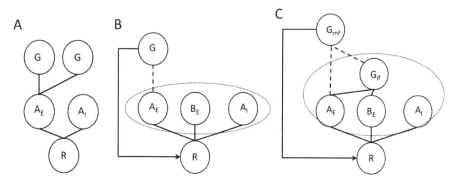

Figure 1. Conceptual models of food webs with nutrient resource (R), edible (A_E) and inedible (A_I) algal producers, and grazers (G). A. Two grazers competing to consume edible algae as in Hall et al. (2006). B. Edible bacterial decomposers (B_E), edible and inedible algae form a physically structured association (mat) that limits grazer consumption (dotted line); grazer excretion recirculates some of limiting nutrient resource (downward arrow). C. Infaunal grazers (G_{if}) have direct access to edible producers inside structured association, but are susceptible to macrofaunal grazers (G_{mf}) if their abundance gets too large to be confined in producer mat. See also Liston (2006) conceptual model.

the structural integrity of Everglades periphyton mats (Geddes and Trexler 2003; Chick et al. 2008). Mature periphyton mats in Everglades marshes are dominated by filamentous cyanobacteria that facilitate precipitation of calcium carbonate, and may reach thicknesses between 1.9 and 10 cm (Van Meter-Kasanof 1973; Browder et al. 1994). These mats host a complex community of algae, microbes, protozoans, and invertebrates (Browder et al. 1994; Donar et al. 2004; Gaiser et al. 2004; Liston and Trexler 2005), inhabiting copious detritus and extracellular polymeric substances (EPS) (Donar et al. 2004; Bellinger et al. 2010). The production of cyanotoxins (toxic secondary metabolites) has been shown for some common Everglades cyanobacteria (Bellinger and Hagerthey 2010). The Everglades is typical of other karstic wetlands in the Caribbean basin in supporting massive mats of seasonally accumulated periphyton (Rejmankova and Komarkova 2000; Rejmankova et al. 2004; La Hee and Gaiser 2012).

Periphyton mats include a diversity algal types that vary in their edibility. For example, common elements such as green algae and diatoms are generally considered more edible by animals than blue-green algae (Lamberti 1996; Steinman 1996; Sullivan and Currin 2000). Enrichment of Everglades wetlands by phosphorus dramatically changes periphyton communities, and at moderate to high levels leads to the complete disassociation of the mat complex followed by increased abundance of phytoplanktonic forms (McCormick et al. 2002; Gaiser et al. 2005). As P enrichment increases from background levels, the relative abundance of edible components of the periphyton mats increases, potentially making the mat more palatable, until the threshold level of enrichment when the mats dissociate (Gaiser et al. 2005). Geddes and Trexler (2003) suggested that edible algae (diatoms and green algae) inhabiting these mats gain an 'associational refuge' (Hay 1986; Duffy and Hay 1990) from grazers that results from physical and chemical defenses of cyanobacteria (Fig. 1B).

Geddes and Trexler (2003) demonstrated that destroying the physical structure of mature periphyton mats allowed more effective grazing of edible algae by omnivorous eastern mosquitofish (*Gambusia holbrooki*). Recent work with the herbivorous sailfin molly (*Poecilia latipinna*), one of only two herbivorous fish inhabiting the Everglades, provides additional support to the suggestion that periphyton mat structure limits grazer food selection (Fig. 2).

Algal mats can influence aquatic animal communities by providing food, but also habitat and predation risk (Fig. 1C). Everglades periphyton mats are home to a rich infauna that both consume it and use it as a refuge from predation (Liston and Trexler 2005; Liston 2006; Liston et al. 2008). Liston and Trexler (2005) observed that most Everglades macroinvertebrates inhabit periphyton mats as infauna. When these mats were eliminated by nutrient addition in field mesocosms, there was no evidence the displaced macroinvertebrates immigrated to the benthos; the macroinvertebrate density in the benthos was unchanged after mat dissolution, indicating a marked loss of these small animals from the mesocosms that was attributed to consumption by small fish and aquatic invertebrates (Liston et al. 2008). Path analysis of field data from a diversity of sites also supported consumption of enhanced macroinvertebrate production by small fishes and aquatic invertebrates along nutrient gradients (Sargeant et al. 2011). Phosphorus enrichment enhanced periphyton biomass and small fish and large macroinvertebrate density, but not periphyton-mat-dwelling macroinvertebrate density (chironomid larvae, amphipods, etc.). Periphyton mats do not provide a perfect

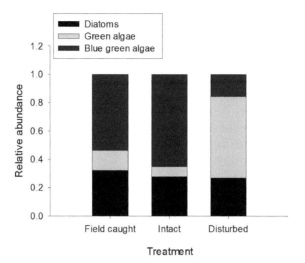

Figure 2. Diet selection by sailfin mollies (*Poecilia latipinna*). Relative abundance (proportion) of algal taxonomic categories from sailfin molly gut contents from field caught and experimental (intact mat and disturbed mat) treatments. Experimental methods described in Geddes and Trexler (2003) and gut contents from the esophagus to first bend of the intestines because this species lacks a defined stomach. Periphyton mats were collected from Water Conservation Area 3B and taxonomic composition was determined following Gottlieb et al. (2006). Sample sizes were: Field caught: N = 15; Intact mat (lab): N = 4; Disturbed mat (lab): N = 5.

refuge for invertebrate infauna, however. Dorn et al. (2006) observed a 50% reduction of periphyton mat infaunal invertebrates with a doubling of intermediate consumers (primarily grass shrimp, *Palaemonetes paludosus*). These authors suggested that some macro-consumers, particularly grass shrimp, are able to breach the refuge quality of mats to access invertebrates within; Geddes and Trexler (2003) noted that grass shrimp gut contents were less affected by mat disruption than eastern mosquitofish.

Food can vary in quality through its nutrient content (stoichiometry) or palatability (physical or chemical recalcitrance). Periphyton can be a dominant source of primary productivity in Everglades marshes (Ewe et al. 2006; Gaiser et al. 2012; Troxler et al. 2013), and formation of mature periphyton mats with their characteristic physical and biotic structure may be critical to trophic interactions and food-web dynamics there. It is likely that direct consumption of algae is not the primary route of energy flow in wetlands (Cebrian and Lartigue 2004; Moore et al. 2004), including those of the southern Everglades; analysis of stable isotopes of carbon and nitrogen suggests that detrital routes of energy flow dominate the energy budget of mesoconsumers inhabiting unenriched Everglades marshes (Williams and Trexler 2006). Lipid profile analysis documented an abundance of lipids only found in algae and prokaryotes, and paucity of lipids unique to vascular plants, in the tissues of a variety of Everglades consumers (Belika et al. 2012). Thus, the source of this "detritus" is heavily weighted toward EPS secreted by cyanobacteria, senescing components of algal mats, and bacteria that colonize these substrates. Suppression of grazing in the Everglades by chemical and physical means, important in controlling algal dynamics in most aquatic systems (Steinman 1996; Lamberti 1996), may contribute to the accumulation of periphyton found there.

Food web structure, as illustrated by variation in stable isotopic niche (Layman et al. 2007; Newsome et al. 2007), is affected by periphyton mat composition. Sargeant et al. (2010) found that isotopic niche area (an index of inter-individual variation in niche width) decreased with greater periphyton biomass and greater relative abundance of filamentous blue-green algae. Isotopic trophic position (range of δ^{15}N) and basal resource diversity (range of δ^{13}C) increased with increasing relative abundance of edible algal taxa and decreasing relative abundance of filamentous blue-green algae. At the population level, Abbey-Lee et al. (2012) found that isotopic niche area of eastern mosquitofish (*Gambusia holbrooki*) was best explained by periphyton algal composition, followed by days since last re-flooding of the site (hydrology). Isotopic niche area increased with greater relative abundance of edible taxa, after controlling for hydroperiod, availability of animal prey, and density of potential competitors; increasing hydroperiod was also positively correlated with increased isotopic niche breadth after controlling for food availability (Abbey-Lee et al. 2012).

In this chapter, we report new analyses evaluating the role of periphyton stoichiometry and algal taxonomic characteristics in supporting aquatic consumer standing crop. Everglades periphyton has very high C nutrient stoichiometric ratios (average C:P = 4,759; C:N = 21.8; N:P = 208.7; n = 159 compared to Redfield ratio of C:N:P = 106:16:1), indicative of very low quality food resources (King and Richardson 2007; Hagerthey et al. 2011) because of P limitation and a 'stoichiometric bottleneck' that limits trophic efficiency (Van Donk et al. 2008). We hypothesized that if nutrient availability limits aquatic consumer production, consumer standing crops would increase as a function of periphyton stoichiometry. However, if associational resistance resulting from periphyton mat structure limits energy flow from producers to consumers, mat taxonomic composition should be a better predictor of consumer standing stocks. These hypotheses are not mutually exclusive, because algal mat community structure is directly linked to nutrient availability, specifically phosphorus. However, the chain of causation, nutrient enrichment $\rightarrow$ algal production $\rightarrow$ consumer biomass versus nutrient enrichment $\rightarrow$ algal production $\rightarrow$ algal community structure $\rightarrow$ consumer biomass, will dictate the response of Everglades food-web structure to nutrient enrichment and is, therefore, worthwhile to evaluate.

Food limitation: Stoichiometric Quality or Palatability

Stoichiometric Analysis. We used data from the US Environmental Protection Agency's 2005 study of the Everglades (REMAP; Scheidt and Kalla 2007) to evaluate the relative power of algal mat stoichiometric composition and algal species composition to explain patterns of density of small omnivorous fishes and herbivorous macroinvertebrates and fish. REMAP is a system-wide sampling study that focuses on biogeochemical parameters and included throw-trap sampling of aquatic consumers in 2005. REMAP samples were collected between September 23 and December 7, 2005, using a stratified random protocol (Stevens and Olsen 2004). These data permitted an analysis of the relative impact of hydrological parameters and resources parameters, including stoichiometric ratios and relative abundance of edible algae, on aquatic consumers. We used multiple regressions that included parameters hypothesized to be informative for these dependent variables in other analyses. We focused on the N:P ratio because the accumulation of inorganic carbon in Everglades periphyton may complicate estimates of C:nutrient stoichiometric ratios. N:P is unaffected by partitioning of organic and inorganic components. We avoided estimating secondary productivity from our data because most such calculations require an assumption of samples from a closed system, while landscape-scale movements of animals tied to water level fluctuation is probably of great importance in this system (Trexler et al. 2002; Ruetz et al. 2005).

Periphyton N:P was linearly related to total phosphorus (TP) in periphyton in a semi-log plot (Fig. 3A) suggesting marked phosphorus limitation at low phosphorus

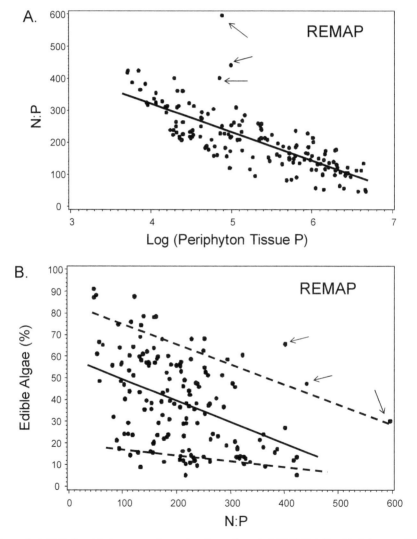

Figure 3. A. N:P of periphyton versus tissue phosphorus (ln ug/g + 1). B. Relationship between relative abundance of edible algae and N:P from REMAP data collected in wet season 2005. The solid line indicates a simple least-squares regression of these data, while the dashed line indicates a quantile regression slope from the highest and lowest 10% quantiles. Arrows indicate same three points from panel A.

availability or luxury uptake of phosphorus as it because more available relative to nitrogen (see Sterner and Elser 2002). There is a negative correlation between the relative abundance of edible algal taxa in periphyton tissues and the N:P ratio in the REMAP data (r = –0.447, n = 158, P < 0.001). While the relative abundance of edible algal types (green algae and diatoms) decreases with increasing N:P (Fig. 3B), a triangle pattern is apparent (Thomson et al. 1996). This is revealed by a quantile regression that fits a separate slope for selected quantiles of the data set (Terrell et al. 1996; Cade and Noon 2003), in this case, the highest 10th quantile of the data display a slope of –0.131 ($t_{1,15}$ = –3.64, P < 0.001) and the lowest 10th quantile displays a slope of –0.032 ($t_{1,15}$ = –1.87, P = 0.063). This pattern suggests that the maximum values of edible algae are more strongly set by phosphorus limitation (indicated by N:P) than are the lower values; other factors may be interacting with P availability to determine algal edibility when N:P is relatively low. Adding hydrological parameters, though improving the model fit, does not remove the changing slope with quantile of N:P as an explanatory variable for algal edibility.

The REMAP data permit us to ask if algal edibility or algal nutrient dynamics better explain patterns in density of aquatic consumers in the Everglades. This provides a test of the hypothesis that phosphorus limitation per se is transferred up the food web, or alternatively if food limitation is an outcome of changing algal composition (blue-green algae with grazer deterrents versus non-defended species). We determined the relationship of the density of herbivores and, separately, omnivorous fishes, to several independent variables that describe resource availability and hydrology (Table 1) to test these hypotheses. Herbivores were the summed density of all snails, grass shrimp, crayfish, and two species of herbivorous fishes (flagfish [*Jordanella floridae*] and sailfin molly). While crayfish and grass shrimp are omnivores, we included them because stable isotope and gut content analyses indicate plant matter is important in their diets (e.g., Loftus 1999). Omnivorous fishes were all fishes collected by throw trap except the two herbivorous species.

Both hydrological parameters and the relative abundance of edible algae contributed to explaining variation in herbivore density in the REMAP 2005 data (Table 1A). Interestingly, N:P was not a significant effect in this model that explained 49% of the variation in the dependent variable. When edible algae was dropped from the statistical model, the adjusted R^2 dropped to 0.263 and N:P was a significant parameter with a negative slope consistent with P limitation ($t_{1,154}$ = –2.89, P = 0.004). However, the $R^2_{adjusted}$ dropped to 0.227 when N:P was excluded, leaving only hydrological parameters. A similar weak effect of N:P was noted for analysis of omnivorous fishes (Table 1B). The $R^2_{adjusted}$ was 0.45 for the model of their density that included both resource and hydrological parameters. Dropping edible algae relative abundance yielded an $R^2_{adjusted}$ of 0.230, and a model with only hydrological parameters had $R^2_{adjusted}$ of 0.163, suggesting only a slightly better performance of N:P with this dependent variable.

These results indicate that food edibility is a factor limiting consumer density in the Everglades. Phosphorus limitation is an important factor in shaping Everglades periphyton mats. However, the effect of oligotrophy on the food web and consumer dynamics in general appears to be mediated through shaping the patterns of relative abundance of algal types in mats. Production of physical and chemical defenses against

Table 1. Regression analysis of data from the REMAP 2005 project relating resource availability and quality to consumer density in Everglades freshwater food webs. A. Herbivore density (log #/m²) B. Density of omnivorous fishes (log #/m²). See text for list of species.

Variable	Estimate	Error	t	P
A. Herbivore density	Adj R-Sq = 0.49; n = 155			
Intercept	-4.19	0.90	-4.67	<.0001
Edible algae (%)	0.04	0.004	8.21	<.0001
N:P molar	0.0005	0.001	0.51	0.612
Hydroperiod (days)	-0.01	0.002	-2.53	0.013
Days Since Reflooding (log days)	0.87	0.19	4.50	<.0001
Stem density (log #/m²)	0.11	0.09	1.23	0.220
B. Omnivorous fish density	Adj R-Sq = 0.45; n = 155			
Intercept	-2.25	0.86	-2.61	0.010
edible_algae	0.03	0.004	7.92	<.0001
N_P_mol	-0.00043	0.00091	-0.47	0.639
hydroperiod	-0.004	0.002	-2.03	0.045
LDSD	0.61	0.18	3.31	0.001
lstemden	0.12	0.09	1.44	0.152

grazers by blue-green algae are energetically costly (Bickel et al. 2000) and cause a trade-off between defense and other life functions for algae living with limited P availability. Presumably these costs render them inferior competitors to green algae and diatoms when P is even slightly less limited, leading to changes in Everglades periphyton mats that ultimately culminate in their breakdown as a cohesive matrix. At elevated P availability, green algae and diatoms appear to be able to reproduce at rates permitting them to outgrow consumption by grazers. These suggestions could be cast as hypotheses for tests in controlled experimental conditions.

Periphyton Edibility and Hydrological Drivers. Data collected for Comprehensive Everglades Restoration Plan (CERP) Monitoring and Assessment Plan (MAP) provide spatially extensive data from the Greater Everglades ecosystem (RECOVER 2009). CERP MAP provides independent information to test hypotheses about the linkage of Everglades consumers and periphyton edibility derived from the REMAP data. Aquatic consumer and periphyton samples were collected for the MAP in the wet season (late September through early December) of 2005 and 2006 using a spatially stratified sampling Generalized Random Tessellated (GRT) Grid design (Philippi 2003; Stevens and Olsen 2004). MAP periphyton data include information on algal species composition and algal tissue phosphorus, but not percent carbon and nitrogen, so stoichiometric analysis is not possible. However, analysis of REMAP data suggested that algal tissue phosphorus and relative abundance of edible taxa provide the best predictors for consumer dynamics. We focus on independent variables tied to periphyton volume and edibility, vascular plant density, and hydrology to evaluate the role of resource availability (mediated through algal productivity and edibility) and hydrology on consumer dynamics. We include emergent plant stem density and periphyton mat volume as indices of habitat structure.

We noted geographic patterns in the relative abundance of edible taxa, as well as in consumer communities such as fish collected by throw traps, in the MAP data. For example, edible algae relative abundance differed among water management areas such as Loxahatchee National Wildlife Refuge, WCA 2A, WCA 3A (Table 2; Region 2005: $F_{1,7} = 7.97$, $P < 0.001$; 2006 $F_{1,7} = 13.54$, $P < 0.001$), as did fish density (Table 2; Region 2005: $F_{1,7} = 8.76$, $P < 0.001$; Region 2006: $F_{1,7} = 7.18$, $P < 0.001$). Both relative abundance of edible algae and fish density generally decrease from north to south, though with exceptions (Fig. 4A,B; Table 2). We found that regional differences were confounded with differences in periphyton tissue TP and other possible environmental drivers, so we did not include spatial data in subsequent statistical analyses. We discuss this spatial confounding in our concluding section.

Prior to analysis of consumer data, we examined patterns of covariance of the independent variables to identify environmental gradients that could be treated as independent patterns in the search for potential drivers of community dynamics. The pattern of correlation among out independent variables was roughly similar in the two years we are examining (Table 3). However, every correlation that was significant in 2005 was weaker in 2006 (because of large sample sizes, relatively weak and possibly trivial correlations of $r = 0.15$ are 'significant' in these data); no correlation over $r = 0.4$ changed sign between the two years. Weaker coherence of the data between years was also revealed in principal components analyses (Table 4). In 2005, the first

Table 2. Regional means of edible algae (%) and fish density (log #/m²) in freshwater marsh ecosystems of the Everglades. Edible algae was estimated as the percent of all algae that was green or diatoms. Fish density is the sum of all species collected by throw trap.

Region	Edible algae (%)		Fish Density (log #/m²)	
	2005	2006	2005	2006
PM	54.2	41.6	2.08	1.96
LOX	50.9	53.3	2.88	3.02
2A	60.6	32.3	2.95	2.64
3A	44.4	34.1	2.81	2.35
3B	49.1	49.9	2.70	2.95
SRS	33.9	28.1	2.23	1.95
SMP	16.8	9.5	1.63	1.58
TS	10.2	9.6	1.47	0.44

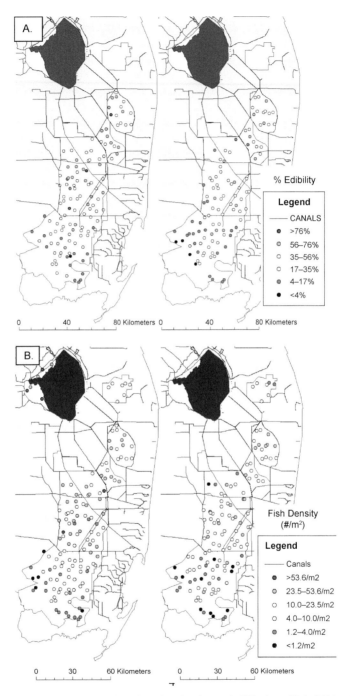

Figure 4. A. Maps illustrating distribution of relative abundance of edible algae (%) in 2005 and 2006. B. Maps illustrating distribution of fish density (#/m²) in 2005 and 2006.

Color image of this figure appears in the color plate section at the end of the book.

Table 3. Correlation matrix of independent variables used in this study. Units are listed on the left side of the table. Pearson correlations are listed in the first row for each variable, with P values on the second row, and sample size on the third. A. Data collected in wet season 2005. B. Data collected in wet season 2006.

	Stem Density	Hydro-period	Edible Algae	Periphyton AFDM	Periphyton Volume
A. 2005					
Periphyton tissue P (Log ug/g)	-0.088	0.456	0.397	-0.838	-0.708
	0.294	<.0001	<.0001	<.0001	<.0001
	143	126	143	144	144
Stem Density (Log #/m^2)	1	-0.035	0.115	0.102	0.032
		0.697	0.166	0.215	0.696
		128	146	148	148
Hydroperiod (d)		1	0.452	-0.466	-0.184
			<.0001	<.0001	0.037
			127	129	129
Edible Algae (%)			1	-0.432	-0.166
				<.0001	0.044
				147	147
Periphyton AFDM (Log g/m^2)				1	0.707
					<.0001
					149
B. 2006					
Periphyton tissue P (Log ug/g)	0.027	0.079	0.254	-0.712	-0.640
	0.761	0.394	0.003	<.0001	<.0001
	131	119	132	133	130
Stem Density (Log #/m^2)	1	0.069	0.122	0.017	0.011
		0.455	0.166	0.841	0.903
		120	131	134	133
Hydroperiod (d)		1	0.215	-0.058	0.091
			0.019	0.525	0.323
			119	122	119
Edible Algae (%)			1	-0.250	-0.098
				0.004	0.267
				133	130
Periphyton AFDM (Log g/m^2)				1	0.695
					<.0001
					133

Table 4. Principle Components Analysis of independent variables for 2005 and 2006 illustrating environmental gradients captured in this study. Correlations of factors to variables are reported. Total explained variance for each factor is reported at the bottom of each column, separately by year. Grey backgrounds indicate correlations exceeding 0.4.

Independent Variables	2005			2006		
	Factors			Factors		
	1	2	3	1	2	3
Periphyton Tissue P (log ug/g)	0.911	0.137	0.024	-0.867	0.073	-0.019
Periphyton Organic Matter (%)	0.86	0.111	0.042	-0.668	0.203	0.215
AFDM Periphyton Mat (log g)	-0.858	-0.169	0.129	0.882	0.031	0.023
Floating Mat Volume (log ml/m³)	-0.743	0.091	0.148	0.791	0.294	0.115
Chlorophyl a (ug/g)	0.577	0.068	0.168	-0.501	0.388	0.189
Average Hydroperiod (days)	0.319	0.881	-0.016	0.078	0.879	-0.064
Days Since Last Flooding (log days)	0.109	0.831	0.259	0.02	0.917	-0.077
Depth recession rate (cm/day)	-0.18	0.807	-0.201	0.057	0.039	-0.79
Depth (cm)	0.44	0.546	0.442	-0.298	0.681	0.46
Emergent Plant Stem Density (log #/m²)	0.098	-0.019	-0.885	0.173	-0.078	0.55
Edible Algae (%)	0.499	0.347	-0.294	-0.219	0.19	0.623
Total Variance Explained	39.5	19.1	10.9	27.5	21.9	14.8

factor explained 39.5% of the total variance and the first two factors explained 58.6%, while in 2006 the first factor explained only 27.5% and first two only 49.4%. In both years, the first factor from PCA was primarily tied to algal characteristics, though edibility (proportion of green algae and diatoms) was only strongly correlated to axis one (we arbitrarily define a strong correlation as r = 0.4) in 2005. Though reversed between years, the patterns of correlations to axis 1 were the same in both years (sign of correlations between variables and factors). Thus, tissue P, chlorophyll a, and edibility were inversely related to mat ash-free dry mass (AFDM) and volume. With the exception of depth at the time of sampling, which was positively correlated to tissue P, organic matter, etc. in both years, the hydrological parameters were primarily correlated with factor 2 in both years. Only factor 3 consistently revealed some common variability between algal and hydrological parameters; this is a surprising result because we generally expect periphyton volume and AFDM to be inversely correlated with hydroperiod.

In both years, periphyton edibility was correlated with periphyton tissue TP, hydroperiod, and stem density of emergent plants (Table 5). Thus, periphyton mats have a higher relative abundance of edible taxa in sites with more phosphorus, longer

Table 5. Regression results for analysis of periphyton edibility (%) reported separately by year. N = 124 in 2005 and N = 116 in 2006.

Variable	Estimate	Error	t	P
2005	$R^2_{adj} = 0.299$			
Intercept	-19.33	9.59	-2.01	0.046
Periphyton tissue TP (log ug/g)	5.52	1.77	3.11	0.002
Hydroperiod (days)	0.08	0.02	4.26	<.0001
Stem density (log #/m^2)	3.38	1.26	2.68	0.008
2006	$R^2_{adj} = 0.108$			
Intercept	-17.14	12.54	-1.37	0.175
Periphyton tissue TP (log ug/g)	5.53	2.01	2.75	0.007
Hydroperiod (days)	0.05	0.02	1.99	0.049
Stem density (log #/m^2)	2.84	1.44	1.97	0.051

hydroperiods, or higher stem density. Simple plots of the dependent variable against an explanatory variable can be misleading when models contain multiple independent variables because the effects of other independent variables have not been accounted for, as they are in the regression model. We correct for this by illustrating the results of our regression models with partial regression plots that display the residuals of the dependent variable against the residuals of each explanatory variable after it has been regressed on the other explanatory variables in the model (Belsley et al. 1980). Although the axes are not on the same scale as the original variables, the plots accurately show the relationships fit by regression (i.e., the slope equals the partial regression coefficient reported in our tables) and allow visual inspection of the scatter of points and possible influential cases and lack-of-fit. In both years, our statistical models predicted a 20% change in relative abundance of edible algae between the lowest and highest levels of algal tissue TP (Fig. 5). Though this relationship was significant, there was as much as a 75% range of in relative abundance of edible algae in both datasets (Fig. 5), which was only partially explainable by other independent variables we measured (R^2_{adj} was < 0.3 in both years, notably in 2006; Table 5).

Consumer density in general increased with increasing relative abundance of green algae and diatoms in periphyton mats. Herbivore density was correlated with algal edibility and hydroperiod in both 2005 and 2006 (Table 6). The amount of variation explained by our models were similar in both years (R^2_{adj} = 0.237 and 0.256 in 2005

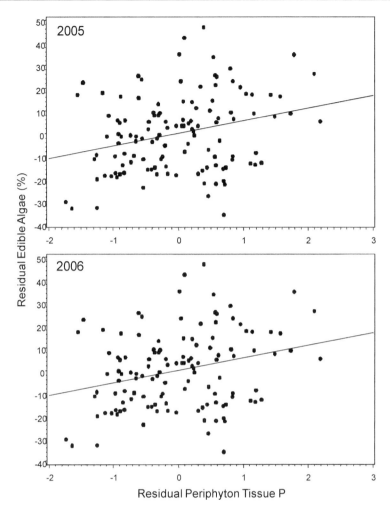

Figure 5. Partial regression plot of edible algae (%) versus periphyton tissue phosphorus reported separately for each year. The mean percentage of edible algae was 39.8 in 2005 and 32.6 in 2006. Residuals from a regression with hydroperiod (days) and stem density (#/m²) are plotted to indicate the unique pattern attributable to tissue phosphorus.

and 2006 respectively), though much variation in their density was unexplained. Edible algae relative abundance, hydroperiod, and emergent plant stem density were all positively related to herbivore density in 2005, but only edible algae relative abundance and hydroperiod in 2006. Though significant both years, the effect of algal relative abundance was stronger in 2006 than in 2007 (Fig. 6; partial regression slope was three times steeper in 2006 than in 2005). Omnivorous fish density also increased with relative abundance of edible algae in both years, and the partial regression of hydroperiod also yielded positive slopes both years though the effect was marginal in 2005 (Table 7). Stem density was significant in 2005 and not in 2006. The model explained more variation in 2006 than in 2005 (R^2_{adj} = 0.249 in 2005 and 0.321 in

Table 6. Regression results for analysis of herbivore density (#/m²) reported separately by year. N = 126 in 2005 and N = 117 in 2006.

Variable	Estimate	Error	t	P
2005	R^2_{adj} = 0.237			
Intercept	−0.75	0.40	−1.91	0.059
Edible algae (%)	0.01	0.01	2.18	0.031
Hydroperiod (days)	0.004	0.001	3.87	0.000
Stem density (log #/m²)	0.18	0.08	2.26	0.026
2006	R^2_{adj} = 0.256			
Intercept	−0.37	0.43	−0.86	0.389
Edible algae (%)	0.03	0.01	5.17	< .0001
Hydroperiod (days)	0.004	0.001	2.75	0.007
Stem density (log #/m²)	−0.01	0.09	−0.10	0.922

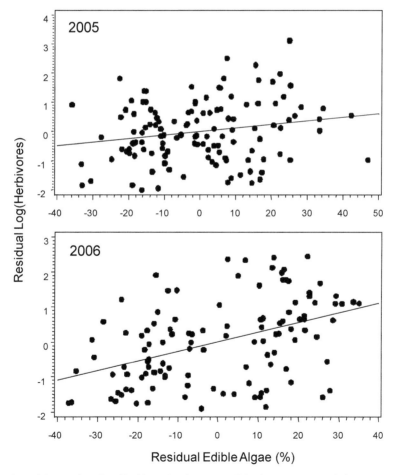

Figure 6. Partial regression plot of herbivore density versus edible algae (%). Residuals from a regression with hydroperiod (days) and stem density (#/m²) are plotted to indicate the unique pattern attributable to tissue phosphorus.

Table 7. Regression results for analysis of omnivorous fish density (#/m²) reported separately by year. N = 126 in 2005 and N = 117 in 2006.

Variable	Estimate	Error	t	P
2005	$R^2_{adj} = 0.249$			
Intercept	0.48	0.34	1.41	0.162
Edible algae (%)	0.02	0.00	3.79	0.001
Hydroperiod (days)	0.002	0.00	1.94	0.055
Stem density (log #/m²)	0.19	0.07	2.77	0.007
2006	$R^2_{adj} = 0.321$			
Intercept	0.63	0.34	1.85	0.067
Edible algae (%)	0.03	0.00	6.40	<.0001
Hydroperiod (days)	0.003	0.00	2.47	0.015
Stem density (log #/m²)	0.005	0.07	0.07	0.947

2006). The effect of algal relative abundance was 50% stronger in 2006 than in 2005 (Fig. 7; partial regression slope in Table 7). Note that both herbivore and omnivore density were log transformed, so that the effect would appear more marked and curvilinear on the arithmetic scale. Also, measures of periphyton mat volume and AFDM were not reported in these regression analyses because they never explained significant variation in either consumer dependent variable (neither in models with other periphyton and hydrology parameters nor alone).

We examined the two species of crayfish inhabiting Everglades wetlands in similar statistical models because past work demonstrated that their dynamics were controlled differently than fish (Dorn and Trexler 2007). In both 2005 and 2006, edible algae failed to explain a significant amount of variation (Table 8). Only depth-related hydrological parameters provided significant relationships, and these were most apparent for the Everglades crayfish (*Procambarus alleni*). Dorn and Trexler (2007) and Kellogg and Dorn (2012) provide data supporting the hypothesis that slough crayfish (*Procambarus fallax*) are more affected by the density of large aquatic predators than by physical parameters. Everglades crayfish are often the only crayfish in short-hydroperiod marshes or in long-hydroperiod wetlands that dried within the past year (Hendrix and Loftus 2000), but slough crayfish are the dominant (and only) species over much of the Greater Everglades ecosystem. Our statistical models fit the data better for both species in 2005 than in 2006 (Table 8), consistent with our data on periphyton but not other aquatic consumers.

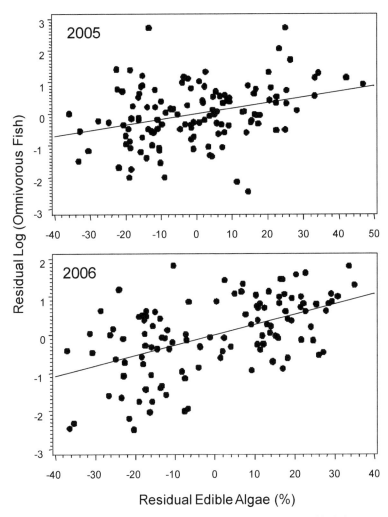

Figure 7. Partial regression plot of omnivore density versus edible algae (%). Residuals from a regression with hydroperiod (days) and stem density (#/m²) are plotted to indicate the unique pattern attributable to tissue phosphorus.

Discussion

The species composition and physical structure of periphyton yields emergent associational effects that are fundamentally important in shaping aquatic community structure and function, beyond those expected from simple stoichiometric relationships (Fig. 1). Luxuriant periphyton mats common to the Everglades and other karstic wetlands illustrate the consequences of associational resistance. King and Richardson (2007) proposed a subsidy-stress model (sensu Odum et al. 1979) for the effects of nutrient enrichment in the Everglades. They suggested that nutrient enrichment increased the quality and decreased the quantity of periphyton, stimulating

Table 8. Regression results for analysis of crayfish density (log #/m^2) reported separately by year. N = 126 in 2005 and N = 117 in 2006. A. Slough Crayfish. B. Everglades Crayfish.

Variable	A. Slough crayfish				B. Evergaldes crayfish			
	Estimate	Error	t	P	Estimate	Error	t	P
2005	$R^2_{adj} = 0.283$				$R^2_{adj} = 0.332$			
Intercept	-0.65	0.21	-3.15	0.002	1.60	0.24	6.71	<.0001
Edible algae (%)	0.003	0.003	1.05	0.294	-0.005	0.003	-1.59	0.113
Hydroperiod (days)	0.005	0.001	5.29	<.0001	-0.003	0.001	-3.12	0.002
Stem density (log #/m^2)	0.08	0.04	1.99	0.049	-0.02	0.05	-0.54	0.592
Depth (cm)	0.001	0.003	0.27	0.785	-0.01	0.003	-3.14	0.002
Average Depth past 60 days (cm)	0.001	0.01	0.25	0.803	0.01	0.01	2.02	0.046
Average Depth past 90 days (cm)	-0.01	0.005	-2.40	0.018	-0.01	0.01	-1.27	0.208
2006	$R^2_{adj} = 0.037$				$R^2_{adj} = 0.145$			
Intercept	0.26	0.21	1.25	0.215	1.00	0.23	4.40	<.0001
Edible algae (%)	0.002	0.003	0.71	0.478	0.001	0.003	0.43	0.671
Hydroperiod (days)	0.001	0.001	1.31	0.193	0.000	0.001	-0.09	0.932
Stem density (log #/m^2)	0.02	0.04	0.48	0.631	-0.03	0.05	-0.70	0.486
Depth (cm)	-0.01	0.003	-2.97	0.004	-0.01	0.003	-3.91	<.0001
Average Depth past 60 days (cm)	0.001	0.003	0.23	0.820	-0.001	0.003	-0.39	0.699
Average Depth past 90 days (cm)	-0.004	0.004	-1.16	0.248	0.0001	0.004	0.000	0.997

macroinvertebrate productivity at low to moderate levels and then stressing and decreasing it at high levels. This review supports the general model of subsidy-stress, providing additional information to the mechanism of the subsidy and stress phases. We report new research supporting published works suggesting that Everglades periphyton mats display associational resistance to grazers and mat species composition is a better predictor of invertebrate and fish density and biomass than stoichiometric quality (see also Gaiser et al. 2014). The mat structure creates a compartmentalized food web with infaunal invertebrates that experience a refuge from some predators while buried in the mat (Fig. 1C). However, when nutrient enrichment leads to mat degradation, these infauna are vulnerable to free-ranging mesoconsumers (Liston et al. 2008), whose densities and biomass increase until vascular plant density becomes so dense as to limit aquatic habitat and the animals that live there (Rader and Richardson 1994; King and Richardson 2007).

We found that the proportion of edible algal taxa in periphyton mats was a better predictor of consumer density than periphyton stoichiometric composition, even though our study included sites with low periphyton TP that could be expected to limit consumers (Sterner and Elser 2002; Hall 2009). Though the strength of some relationships varied across years and some of the weaker relationships were only significant in one or two of the data sets, we observed consistent results regarding the linkage of phosphorus in periphyton to changing periphyton mat composition, with implications for standing crops of herbivores and omnivores throughout the Greater Everglades ecosystem. Our analysis of REMAP data provided new support to the hypothesis that mat algal composition is a key driver of consumer dynamics, rather than stoichiometric relationships per se. To our knowledge, no other studies of food web function in the Everglades have been conducted at such a comprehensive scale. Past work (Trexler et al. 2002) indicates that the spatial separation of our sampling units (primary sampling units, PSUs) was appropriate to treat each as an independent sample and appropriate for inclusion in regression analyses of the type we report. There is pervasive spatial autocorrelation in the distribution of nutrient enrichment in the Everglades because the delivery of phosphorus is made through water flowing into the ecosystem from essentially point sources of canals and water-control structures. The most marked (and longest sustained) enrichment is in the northern portion of the ecosystem, with lower levels of P in the southern Everglades, particularly in Taylor Slough and the oligohaline ecotone (McCormick et al. 2002). This pattern of nutrient enrichment has undoubtedly shaped the spatial pattern of periphyton edibility in the ecosystem (Fig. 4A). We demonstrated that this oligotrophy has implications for the quality of periphyton as a food source, a conclusion that is consistent with past experimental studies. Adding phosphorus changes the periphyton community to favor more edible algae, leading to higher standing stocks of consumers. Interestingly, consumer density was not highly correlated with stoichiometric patterns unless algal relative abundance was excluded from the model. This suggests that algal defenses (physical or chemical) are more critical than stoichiometric considerations in limiting consumer productivity.

Turner et al. (1999) failed to find increased density of macroinvertebrates with nutrient enrichment, while Rader and Richardson (1994) observed increased macroinvertebrate density in enriched sites. Studying areas of the Everglades with low to moderate nutrient enrichment (max periphyton TP 1,278 μg g^{-1}), Liston (2006) observed that periphyton infaunal crowding (number g^{-1} AFDM periphyton) increased with periphyton P content in short-hydroperiod marshes, but not in long-hydroperiod ones. She proposed that re-current marsh drying in short-hydroperiod marshes limits the density of fish consumers (see also Trexler et al. 2001; 2005), releasing macroinvertebrates there from top-down control. In contrast, long-hydroperiod marshes are free from hydrological disturbance for extended periods, permitting fish and large predatory macroinvertebrate populations (notably crayfish) to increase in numbers and act more strongly as controlling factors for infaunal taxa such as nematodes, amphipods, midge larvae, and cladocerans. King and Richardson (2007) observed rapid recovery of macroinvertebrate biomass in enriched areas following drought, particularly of grass shrimp, *P. paludosus*. They attributed this recovery to a bloom of periphyton resulting from extra light reaching the substrate because of drought-induced senescence of macrophyte cover and possible incomplete drying of the flocculent organic layer of the marsh surface permitting high survival of the drying event. Short-hydroperiod marshes often produce very thick and dense benthic periphyton mats growing on karstic outcrops that provide additional refuge to invertebrates (see Fig. 17.7 in Gaiser et al. 2012). Hydrological conditions and seasonal patterns of periphyton mat senescence and growth interact with nutrient enrichment to determine the basis of aquatic community structure.

Future directions. Hairston et al. (1960) suggested that "the world is green", such that herbivore feeding is kept in check by predators. In reality, plants induce physical and chemical defenses (Stamp 2003) that keep herbivores from completely consuming all plant biomass in ecosystems. In addition, the majority of carbon fixed by plants is not consumed by herbivores but is catabolized and metabolized by heterotrophic microorganisms (Cebrian and Lartigue 2004; Moore et al. 2004). Heterotrophic microbes rapidly consume organic matter and mineralize inorganic nutrients that stimulate primary and secondary production in food webs, forming the microbial loop (Pomeroy and Weibe 1988; Batzer et al. 2006). The importance of detritus in maintaining food web interactions has been well understood for decades in ecology (e.g., Lindeman 1942; Moore et al. 2004; Rooney and McCann 2012), but the importance of algal-detrital interactions that drive the timing and magnitude of energy to microbial and higher-order metazoans is only recently being understood. Everglades periphyton mats illustrate the strong coupling of microbial autotrophic and heterotrophic production and nutrient cycling; here we propose that invertebrate infauna are a third important player forming a food-web compartment somewhat independent of water column dynamics. Geddes and Trexler (2003) reported evidence of nutrient regeneration stimulating growth of Everglades periphyton by manipulating grass shrimp and eastern mosquitofish density in field cages and Ruehl and Trexler (2013) reported similarly for ramshorn snails (*Planorbella duryi*) in mesocosms. To date, periphyton infauna density has proven difficult to manipulate without disrupting periphyton mat structure.

The microbial loop is characterized by the metabolic activities of bacteria, fungi, and protozoans, which are primed by the bioavailability of carbon and nutrients, and ultimately determine the trophic basis of food web production. The priming effect, whereby processing of recalcitrant organic matter is stimulated by the production of labile organic matter (Guenet et al. 2010; Danger et al. 2013), has important implications for carbon and nutrient constraints on the microbial loop. In terrestrial and aquatic ecosystems, the metabolic demands of heterotrophic microbes are subsidized by high-energy algal photosynthate, which enhances microbial production of enzymes associated with carbon and nutrients (Francouer et al. 2006; Rier et al. 2007; Mann et al. 2014). Priming by algae increases carbon flow to microbial biomass (Kuehn et al. 2014) and has important implications for how microbes drive ecosystem carbon flow and nutrient cycling. Reducing the recalcitrance of organic matter enhances nutrient uptake in ecosystems (Mann et al. 2014; Sinsabaugh 1994). A major uncertainty lies in how organic matter recalcitrance relative to lability influences microbial carbon- and nutrient-use efficiencies, and a stronger understanding of both would help explain how the trophic basis for food web production is interactively tied to the priming effect and the microbial loop. What is clear is that the priming effect plays a critical role in carbon flow and use by heterotrophic microbes and that photosynthetic stimulation by algae drive the microbial loop.

Plants have primacy in food webs, but the relative strength and directionality (top-down versus bottom-up) of resource-consumer dynamics is borne out across spatio-temporal scales (Batzer 1998) and explained by abiotic drivers playing out over large scales (Power 1992). For example, flow and hydrologic regimes determine the dynamic availability of basal resources. Extreme events, such as floods and droughts transport and retain organic matter, respectively, and antecedent hydrologic conditions can be predictors of the composition of available organic matter resources (Tank et al. 2010; Kominoski and Rosemond 2011). Future research should explore the linkage of microbial processes in Everglades periphyton mats to infaunal consumers and compartmentalization of the aquatic food web. Does this compartmentalization explain the accumulation of primary production typical of Caribbean karstic wetlands (Turner et al. 1999; Ruehl and Trexler 2011)?

Conclusions. Everglades periphyton mats are remarkable, but under-appreciated, contributors to function of a food web that historically supported high abundance of charismatic wading birds and alligators in a highly oligotrophic setting (Gaiser et al. 2012). Similar mats are found in karstic wetlands throughout the Caribbean basin, and beyond (Rejmankova and Komarkova 2000; Rejmankova et al. 2004; La Hee and Gaiser 2012). In this chapter, we discuss their role as food and structure for aquatic consumers living there. These mats display associational resistance that both protects edible components from consumption, permitting their biomass to accumulate, and provides a size-limited refuge for small invertebrate consumers protected from roving predators such as fish and large macroinvertebrates. Nutrient enrichment changes the balance of dynamics of the components in the mat, ultimately leading to loss of mat structure and related emergent effects of the association. This model characterizes the harmful effects of nutrient enrichment in the Everglades, helps explain patterns of consumer biomass observed, and provides relevant information for management solutions proposed to restore enriched portions of the ecosystem (Hagerthey et al. 2014). Future research should further explore the dynamics of this microbial loop, which will likely provide insight for similar periphyton mats in other aquatic systems.

Acknowledgements

We thank Peter Kalla, Daniel Scheidt, and Jennifer Richards for coordinated the REMAP research program, and Jana Newman and Andy Gottlieb for coordination of the CERP MAP research program. US EPA's Region 4, Science and Ecosystem Support Division and Water Management Division (EPA 904-R-07-001), and the Everglades National Park funded the REMAP project under cooperative agreement number H5297-05-0088 between FIU and ENP. The CERP MAP project was made possible by funding from Task Agreement CP040130 between the South Florida Water Management District and FIU. This research was enhanced by collaborations with the Florida Coastal Everglades Long-Term Ecological Research Program (funded by the National Science Foundation, DBI-0620409 and DEB-9910514). The research was conducted in accord with FIU animal care guidelines. This is contribution #670 of the Southeast Environmental Research Center.

References

Abbey-Lee, R., E.E. Gaiser and J.C. Trexler. 2013. Relative role of dispersal dynamics and competition in determining isotopic niche breadth. Freshwater Biol. 58: 780–792.

Alfaro, A.C., S.E. Dewas and F. Thomas. 2007. Food and habitat partitioning in grazing snails (Turbo smaragdus), northern New Zealand. Estuaries Coasts 30: 431–440.

Barbosa, P., J. Hines, I. Kaplan, H. Martinson, A. Szczepaniec and Z. Szendrei. 2009. Associational resistance and associational susceptibility: Having right or wrong neighbors. Annu. Rev. Ecol. Evol. S. 40: 1–20.

Batzer, D.P. 1998. Trophic interactions among detritus, benthic midges, and predatory fish in a freshwater marsh. Ecology 79: 1688–1698.

Batzer, D.P., R. Cooper and S.A. Wissinger. 2006. Wetland animal ecology. pp. 242–284. *In*: D.A. Batzer and R.R. Sharitz (eds.). Ecology of Freshwater and Estuarine Wetlands. University of California Press, Berkeley.

Belicka, L.L., E.R. Sokol, J.M. Hoch, R. Jaffe and J.C. Trexler. 2012. A Molecular and Stable Isotopic Approach to Investigate Algal and Detrital Energy Pathways in a Freshwater Marsh. Wetlands 32: 531–542.

Bellinger, B.J. and S.E. Hagerthey. 2010. Presence and diversity of algal toxins in subtropical peatland periphyton: The Florida Everglades, USA. J. Phycol. 46: 674–678.

Bellinger, B.J., M.R. Gretz, D.S. Domozych, S.N. Kiemle and S.E. Hagerthey. 2010. Composition of extracellular polymeric substances from periphyton assemblages in the Florida Everglades. J. Phycol. 46: 484–496.

Belsley, D.A., E. Kuh and R.E. Welsch. 1980. Regression Diagnostics. Identifying Influential Data and Sources of Collinearity. John Wiley & Sons, Inc., New York.

Bickel, H., S. Lyck and H. Utkilen. 2000. Energy state and toxin content—experiments on *Microcystis aeruginosa* (Chroococcales, Cyanophyta). Phycologia 39: 212–218.

Browder, J.A., P.J. Gleason and D.R. Swift. 1994. Periphyton in the Everglades: spatial variation, environmental correlates, and ecological implication. pp. 379–418. *In*: S.M. Davis and J.C. Ogden (eds.). Everglades, the Ecosystem and its Restoration. St. Lucie Press, Delray Beach.

Cade, B.S. and B.R. Noon. 2003. A gentle introduction to quantile regression for ecologists. Front. Ecol. Environ. 1: 412–420.

Cebrian, J. and J. Lartigue. 2004. Patterns of herbivory and decomposition in aquatic and terrestrial ecosystems. Ecol. Monogr. 74: 237–259.

Chick, J.H., P. Geddes and J.C. Trexler. 2008. Periphyton mat structure mediates trophic interactions in a subtropical wetland. Wetlands 28: 378–389.

Danger, M., J. Cornut, E. Chauvet, P. Chavez, A. Elger and A. Lecerf. 2013. Benthic algae stimulate leaf litter decomposition in detritus-based headwater streams: a case of aquatic priming effect. Ecology 94: 1604–1613.

Donar, C.M., K.W. Condon, M. Gantar and E.E. Gaiser. 2004. A new technique for examining the physical structure of Everglades floating periphyton mat. Nova Hedwigia 78: 107–119.

Dorn, N. and J.C. Trexler. 2007. Crayfish assemblage shifts in a large drought-prone wetland: the roles of hydrology and competition. Freshwater Biol. 52: 2399–2411.

Dorn, N.J., J.C. Trexler and E.E. Gaiser. 2006. Exploring the role of large predators in marsh food webs: Evidence for a behaviorally-mediated trophic cascade. Hydrobiologia 569: 375–386.

Duffy, J.E. and M.E. Hay. 1990. Seaweed adaptations to herbivory. Bio. Science 40: 368–375.

Duffy, J.E. and V.J. Paul. 1992. Prey nutritional quality and the effectiveness of chemical defenses against tropical reef fishes. Oecologia 90: 333–339.

Ewe, S.M.L., E.E. Gaiser, D.L. Childers, V.H. Rivera-Monroy, D. Iwaniec, J. Fourquerean and R.R. Twilley. 2006. Spatial and temporal patterns of aboveground net primary productivity (ANPP) in the Florida Coastal Everglades LTER (2001–2004). Hydrobiologia 569: 459–474.

Francoeur, S.N., M. Schaecher, R.K. Neely and K.A. Kuehn. 2006. Periphytic photosynthetic stimulation of extracellular enzyme activity in microbial communities associated with natural decaying wetland plant litter. Microb. Ecol. 52: 662–669.

Gaiser, E.E., L.J. Scinto, J.H. Richards, K. Jayachandran, D.L. Childers, J.D. Trexler and R.D. Jones. 2004. Phosphorus in periphyton mats provides best metric for detecting low level P enrichment in an oligotrophic wetland. Water Res. 38: 507–516.

Gaiser, E.E., J.C. Trexler, J.H. Richards, D.L. Childers, D. Lee, A.L. Edwards, L.J. Scinto, K. Jayachandran, G.B. Noe and R.D. Jones. 2005. Cascading ecological effects of low-level phosphorous enrichment in the Florida Everglades. J. Environ. Qual. 34: 717–723.

Gaiser, E.E., J.C. Trexler and P.R. Wetzel. 2012. The Florida Everglades. pp. 231–252. *In*: D.P. Batzer and A.H. Baldwin (eds.). Wetland Habitats of North America: Ecology and Conservation Concerns. Univ. California Press, Berkeley.

Gaiser, E., A. Gottlieb, S. Lee and J.C. Trexler. 2014. The importance of species-based microbial assessment of water quality in freshwater Everglades wetlands. *In*: J.A. Entry, A.D. Gottlieb, K. Jayachandran and A. Ogram (eds.). Microbiology of the Everglades Ecosystem. CRC Press, Boca Raton (This book).

Geddes, P. and J.C. Trexler. 2003. Uncoupling of omnivore-mediated positive and negative effects on periphyton mats. Oecologia 136: 585–595.

Guenet, B., M. Danger, L. Abbadie and G. Lacroix. 2010. Priming effect: bridging the gap between terrestrial and aquatic ecology. Ecology 91: 2850–2861.

Hagerthey, S.A., M.I. Cook, R.M. Kobza, S. Newman and B.J. Bellinger. 2014. Aquatic faunal responses to an induced regime shift in the phosphorus-impacted Everglades. Freshwater Biol., doi:10.1111/fwb.12353.

Hagerthey, S.E., B.J. Bellinger, K. Wheeler, M. Gantar and E. Gaiser. 2011. Everglades periphyton: a biogeochemical perspective. Crit. Rev. Environ. Sci. Technol. 41(S1): 309–343.

Hairston, N.G., F.E. Smith and L.B. Slobodkin. 1960. Community structure, population control, and competition. Am. Nat. 94: 421–425.

Hall, S.R. 2009. Stoichiometrically explicit food webs: feedbacks between resource supply, elemental constraints, and species diversity. Annu. Rev. Ecol. Evol. S. 40: 503–528.

Hall, S.R., M.A. Leibold, D.A. Lytle and V.H. Smith. 2006. Inedible producers in food webs: controls on stoichiometric food quality and composition of grazers. Am. Nat. 167: 628–637.

Hay, M.E. 1986. Associational plant defenses and the maintenance of species diversity: turning competitors into accomplices. Am. Nat. 128: 617–641.

Hendrix, A.N. and W.F. Loftus. 2000. Distribution and relative abundance of the crayfishes *Procambarus alleni* (faxon) and *P. fallax* (hagen) in southern Florida. Wetlands 20: 194–199.

Hillebrand, H., E.T. Borer, M.E.S. Bracken, B.J. Cardinale, J. Cebrian, E.E. Cleland, J.J. Elser, D.S. Gruner, W.S. Harpole, J.T. Ngai, S. Sandin, E.W. Seabloom, J.B. Shurin, J.E. Smith and M.D. Smith. 2009. Herbivore metabolism and stoichiometry each constrain herbivory at different organizational scales across ecosystems. Ecol. Lett. 12: 516–527.

Kellogg, C.M. and N.J. Dorn. 2012. Consumptive effects of fish reduce wetland crayfish recruitment and drive species turnover. Oecologia 168: 1111–1121.

King, R.S. and C.J. Richardson. 2007. Subsidy-stress response of macroinvertebrate community biomass to a phosphorus gradient in an oligotrophic wetland ecosystem. J. N. Am. Benthol. Soc. 26: 491–508.

Kominoski, J.S. and A.D. Rosemond. 2011. Conservation from the bottom-up: forecasting effects of global change on dynamics of organic matter and management needs for river networks. Freshwater Sci. 31: 51–68.

Kuehn, K.A., S.N. Francoeur, R.H. Findlay and R.K. Neely. 2014. Priming in the microbial landscape: periphytic algal stimulation of litter-associated microbial decomposers. Ecology 95: 749–762.

La Hee, J.M. and E.E. Gaiser. 2012. Benthic diatom assemblages as indicators of water quality in the Everglades and three tropical karstic wetlands. Freshwater Sci. 31: 205–221.

Lamberti, G.A. 1996. The role of periphyton in benthic food webs. pp. 533–564. *In*: R.J. Stevenson, M.L. Bothwell and R.L. Lowe (eds.). Algal Ecology. Academic Press, San Diego.

Layman, C.A., D.A. Arrington, C.G. Montana and D.M. Post. 2007. Can stable isotope ratios provide for community-wide measures of trophic structure? Ecology 88: 42–48.

Lindeman, R.L. 1942. The trophic-dynamic aspect of ecology. Ecology 23: 399–417.

Liston, S.E. 2006. Interactions between nutrient availability and hydroperiod shape macroinvertebrate communities in Florida Everglades marshes. Hydrobiologia 569: 343–357.

Liston, S.E. and J.C. Trexler. 2005. Spatio-temporal patterns in community structure of macroinvertebrates inhabiting calcareous periphyton mats. J. N. Am. Benthol. Soc. 24: 832–844.

Liston, S.E., S. Newman and J.C. Trexler. 2008. Macroinvertebrate community response to eutrophication in an oligotrophic wetland: An *in situ* mesocosm experiment. Wetlands 28: 686–694.

Mann, P.J., W.V. Sobczak, M.M. LaRue, E. Bulygina, A. Davydova, J.E. Vonk, J. Schade, S. Davydov, N. Zimov, R.M. Holmes and R.G.M. Spencer. 2014. Evidence for key enzymatic controls on metabolism of Arctic river organic matter. Glob. Change Biol. 20: 1089–1100.

McCormick, P.V., S. Newman, S. Miao, D.E. Gawlik, D. Marley, K.R. Reddy and T.D. Fontaine. 2002. Effects of anthropogenic phosphorus inputs on the Everglades. pp. 83–126. *In:* J.W. Porter and K.G. Porter (eds.). The Everglades, Florida Bay, and Coral Reefs of the Florida Keys. An Ecosystem Sourcebook. CRC Press, Boca Raton, FL.

Moore, J.C., E.L. Berlow, D.C. Coleman, P.C. de Ruiter, Q. Dong, A. Hastings, N.C. Johnson, K.S. McCann, K. Melville, P.J. Morin, K. Nadelhoffer, A.D. Rosemond, D.M. Post, J.L. Sabo, K.M. Scow, M.J. Vanni and D.H. Wall. 2004. Detritus, trophic dynamics and biodiversity. Ecol. Lett. 7: 584–600.

Newsome, S.D., C.M. del Rio, S. Bearhop and D.L. Phillips. 2007. A niche for isotopic ecology. Front. Ecol. Environ. 5: 429–436.

Odum, E.P., J.T. Finn and E.H. Franz. 1979. Perturbation theory and the subsidy-stress gradient. BioScience 29: 349–352.

Paul, V.J. and M.E. Hay. 1986. Seaweed susceptibility to herbivory—chemical and morphological correlates. Mar. Ecol. Prog. Ser. 33: 255–264.

Philippi, T. 2003. Final Report. CERP Monitoring and Assessment Plan: Stratified Random Sampling Plan. SFWMD Agreement C-C20304A.

Pomeroy, L.R. and W.J. Wiebe. 1988. Energetics of microbial food webs. Hydrobiologia 159: 7–18.

Power, M.E. 1992. Top-down and bottom-up forces in food webs: do plants have primacy? Ecology 73: 733–746.

Rader, R.B. and C.J. Richardson. 1994. Response of macroinvertebrates and small fish to nutrient enrichment in the northern Everglades. Wetlands 14: 134–146.

[RECOVER]. 2009. CERP Monitoring and Assessment Plan, Revised 2009. Comprehensive Everglades Restoration Plan, Restoration Coordination and Verification (RECOVER). U.S. Army Corps of Engineers Jacksonville District, Jacksonville, FL, and South Florida Water Management District, West Palm Beach, FL.

Rejmankova, E. and J. Komarkova. 2000. A function of cyanobacterial mats in phosphorus-limited tropical wetlands. Hydrobiologia 431: 135–153.

Rejmankova, E., J. Komarek and J. Komarkova. 2004. Cyanobacteria—a neglected component of biodiversity: patterns of species diversity in inland marshes of northern Belize (Central America). Diversity and Distributions 10: 189–199.

Rier, S.T., K.A. Kuehn and S.N. Francoeur. 2007. Enzymatic response of stream microbial communities colonizing inert and organic substrata to photosynthetically active radiation. J. N. Am. Benthol. Soc. 26: 439–449.

Rooney, N. and K.S. McCann. 2012. Integrating food web diversity, structure and stability. TREE 27: 40–46.

Ruehl, C.B. and J.C. Trexler. 2011. Comparisons of snail density, standing stock, and body size among freshwater ecosystems: A review. Hydrobiologia 665: 1–13.

Ruehl, C.B. and J.C. Trexler. 2013. A suite of prey traits determine predator and nutrient enrichment effects in a tri-trophic food web. Ecosphere 4: 75, doi.org/10.1890/ES13-00065.

Ruetz, C.R., III, J.C. Trexler, F. Jordan, W.F. Loftus and S.A. Perry. 2005. Population dynamics of wetland fishes: Spatio-temporal patterns shaped by hydrological disturbance? J. Animal Ecol. 74: 322–332.

Sargeant, B.L., E.E. Gaiser and J.C. Trexler. 2010. Biotic and abiotic determinants of intermediate-consumer trophic diversity in the Florida Everglades. Mar. Freshwater Res. 61: 11–22.

Sargeant, B.L., E.E. Gaiser and J.C. Trexler. 2011. Indirect and direct controls of macroinvertebrates and small fish by abiotic factors and trophic interactions in the Florida Everglades. Freshwater Biol. 56: 2334–2346.

Scheidt, D.J. and P.I. Kalla. 2007. Everglades ecosystem assessment: Water management and quality, eutrophication, mercury contamination, soils, and habitat. UPEPA Region 4, Athens GA. EPA 904-R-07-001. 98 pp. http://www.epa.gov/region4/sesd/reports/epa904r07001.html.

Schupp, P.J. and V.J. Paul. 1994. Calcium-carbonate and secondary metabolites in tropical seaweeds—variable effects on herbivorous fishes. Ecology 75: 1172–1185.

Sinsabaugh, R.L. and D.L. Moorhead. 1994. Resource allocation to extracellular enzyme production: a model for nitrogen and phosphorus control of litter decomposition. Soil Biol. Biochem. 26: 1305–1311.

Stamp, N. 2003. Out of the quagmire of plant defense hypotheses. Q. Rev. of Biol. 78: 23–55.

Steinman, A.D. 1996. Effects of grazers on freshwater benthic algae. pp. 341–373. *In*: R.J. Stevenson, M.L. Bothwell and R.L. Lowe (eds.). Algal Ecology. Academic Press, San Diego.

Sterner, R.W. and J.J. Elser. 2002. Ecological Stoichiometry: The Biology of Elements from Molecules to the Biosphere. Princeton University Press, NJ.

Stevens, D.L., Jr. and A.R. Olsen. 2004. Spatially balanced sampling of natural resources. J. Am. Stat. As. 99: 262–278.

Stevenson, R.J. 2006. An introduction to algal ecology in freshwater benthic habitats. pp. 3–30. *In*: R.J. Stevenson, M.L. Bothwell and R.L. Lowe (eds.). Algal Ecology: Freshwater Benthic Ecosystems. Academic Press, San Diego, CA.

Sullivan, M.J. and C.A. Currin. 2000. Community structure and functional dynamics of benthic microalgae in salt marshes. pp. 81–106. *In*: M.P. Weinstein and D.A. Kreeger (eds.). Concepts and Controversies in Tidal Marsh Ecology. Kluwer Academic Publishers, Dordrecht, South Holland.

Tank, J.L., E.J. Rosi-Marshall, N.A. Griffiths, S.A. Entrekin and M.L. Stephen. 2010. A review of allochthonous organic matter dynamics and metabolism in streams. J. N. Am. Benthol. Soc. 29: 118–146.

Terrell, J.W., B.S. Cade, J. Carpenter and J.M. Thompson. 1996. Modeling stream fish habitat limitations from wedge-shaped patterns of variation in standing stock. Trans. Am. Fish. Soc. 125: 104–117.

Thomson, J.D., G. Weiblen, B.A. Thomson, S. Alfaro and P. Legendre. 1996. Untangling multiple factors in spatial distributions: Lilies, gophers, and rocks. Ecology 77: 1698–1715.

Trexler, J.C., W.F. Loftus, C.F. Jordan, J. Chick, K.L. Kandl, T.C. McElroy and O.L. Bass. 2002. Ecological scale and its implications for freshwater fishes in the Florida Everglades. pp. 153–181. *In*: J.W. Porter and K.G. Porter (eds.). The Everglades, Florida Bay, and Coral Reefs of the Florida Keys: An Ecosystem Sourcebook. CRC, Boca Raton.

Trexler, J.C., W.F. Loftus and S. Perry. 2005. Disturbance frequency and community structure in a twenty-five year intervention study. Oecologia 145: 140–152.

Troxler, T., E.E. Gaiser, J.G. Barr, J.D. Fuentes, R. Jaffe, D.L. Childers, L. Collado-Vides, V.H. Rivera-Monroy, E. Castañeda-Moya, W.T. Anderson, R.M. Chambers, M. Chen, C. Coronado-Molina, S.E. Davis, V. Engel, C. Fitz, J.W. Fourqurean, T.A. Frankovich, J. Kominoski, C.J. Madden, S.L. Malone, S. Oberbauer, P.C. Olivas, J.H. Richards, C.J. Saunders, J. Schedlbauer, L.J. Scinto, F.H. Sklar, T.J. Smith, J.M. Smoak, G. Starr, R.R. Twilley and K.R.T. Whelan. 2013. Integrated carbon budget models for the Everglades terrestrial-coastal-oceanic gradient: Current status and needs for inter-site comparisons. Oceanography 26: 98–107.

Turner, A.M., J.C. Trexler, F. Jordan, S.J. Slack, P. Geddes and W. Loftus. 1999. Targeting ecosystem features for conservation: Standing crops in the Florida Everglades. Conserv. Biol. 13: 898–911.

Underwood, N., B.D. Inouye and P.A. Hambäck. 2014. A conceptual framework for associational effects: When do neighbors matter and how would we know? Q. Rev. Biol. 89: 1–19.

Van Donk, E., D.O. Hessen, A.M. Verschoor and R.D. Gulati. 2008. Re-oligotrophication by phosphorus reduction and effects on seston quality in lakes. Limnologica 38: 189–202.

Van Meter-Kasanof, N. 1973. Ecology of the microalgae of the Florida Everglades. Part I. Environment and some aspects of freshwater periphyton, 1959 to 1963. Nova Hedwigia 24: 619–664.

Wetzel, R.G. 1983. Attached algal-substrata interactions: Fact or myth, and when and how? Developments in Hydrobiology 17: 207–215.

Williams, A.J. and J.C. Trexler. 2006. A preliminary analysis of the correlation of food-web characteristics with hydrology and nutrient gradients in the southern Everglades. Hydrobiologia 569: 493–504.

9

Nitrogenase Activity in Everglades Periphyton: Patterns, Regulators and Use as an Indicator of System Change

Patrick W. Inglett

Introduction

As a major component of all living cells, nitrogen (N) is a fundamental element in the functioning of biological systems. Under N-limited conditions, certain prokaryotic organisms are able to convert atmospheric N_2 gas to ammonium a biologically-available N form, through the process of biological N fixation (Howarth et al. 1988a;b). The process of N fixation is catalyzed by the enzyme nitrogenase, which is only produced in specific prokaryotes including several genera of archaea, bacteria, and cyanobacteria. Cyanobacteria are a class of organisms exhibiting both algal morphologies and distinct bacterial features (Vymazal 1995).

Cyanobacteria and other N fixing microorganisms are important components of aquatic and wetland environments like the Everglades. The term 'periphyton' is used to refer to a complex of algae, bacteria, cyanobacteria, fungi and dead organic material. Periphyton is abundant in shallow aquatic systems and is commonly found in floating, epiphytic, and benthic forms (McCormick and Stevenson 1998; Gottlieb et al. 2006). In low nutrient areas periphyton biomass can reach levels > 6000 g AFDW m^{-2} (Iwaniec et al. 2006; Hagerthey et al. 2011). These assemblages are predominantly

Department of Soil and Water Science, University of Florida, 2181 McCarty Hall-A, Gainesville, FL 32611.
  Email: pinglett@ufl.edu

composed of filamentous cyanobacteria such as *Scytonema* and *Schizothrix* sp., while in short-hydroperiod systems, coccoids such as *Chroococcus* sp. can also be abundant (Gleason and Spackman 1974; Browder et al. 1994). In more eutrophic marsh areas, periphytic mats are much less abundant, but can still be dominated by cyanobacteria (e.g., *Lyngbya* sp. and *Oscillatoria* sp.), with other proportions of filamentous greens (e.g., *Spirogyra* sp. and *Mougeotia* sp.) (McCormick et al. 1996b; Gaiser et al. 2006).

Nitrogenase activity has been recorded in periphyton throughout the Everglades system including both oligotrophic and eutrophic conditions (Inglett et al. 2004; Liao and Inglett 2012). In eutrophic areas this is not surprising as P inputs have led to increasing N limitation of the system (TN:TP < 30) (Inglett et al. 2009; McCormick et al. 1996b). But in low nutrient areas, typical water column TP levels are below 10 mg P l^{-1} and water column TN:TP in excess of 250 (Inglett et al. 2004). Under such conditions, N$_2$ fixation would seemingly offer no benefit compared to the apparent need for P. Thus, measurable periphyton nitrogenase activity is somewhat of a paradox in many parts of the natural Everglades (Vitousek et al. 2002; Inglett et al. 2011).

Periphyton Nitrogenase Activity in Everglades Systems Ecosystem Components

Because of the widespread abundance of periphyton throughout the Everglades, N fixation from this community can occur on the surfaces of many submersed ecosystem components including benthic, periphytic (in particular associated with *U. purpurea*), and floating forms (Fig. 1; Inglett et al. 2004). Epiphytic and metaphytic forms dominate the deeper water areas, especially in Water Conservation Area (WCA) 2A and 3A, while benthic mats predominate in the more shallow or short-hydroperiod southern systems. Nitrogenase activity was also recorded for non-calcareous periphyton in the soft water system of WCA 1 (Inglett, unpublished results).

In contrast to the deeper water systems, in the southern Everglades areas, seasonally flooded systems such as the short hydroperiod marl prairies of Everglades National Park have periphyton mats that more closely resemble soil crusts of desert areas, especially during dry periods (Fig. 1; Gottlieb 2003; Gottlieb et al. 2005). Like other soil crusts (Belnap 2001), nitrogenase activity was recorded for these epibenthic forms (Liao and Inglett 2012; 2014), though at levels lower than those measured in Northern Everglades systems. Thick epibenthic, epiphytic and metaphytic growths also occur in southern systems like Shark River and Taylor Slough, but no measurements of nitrogenase activity have been conducted in these areas.

In nutrient impacted systems, periphytic forms are much less developed due to shading by abundant macrophytes and potentially reduced interaction between mat-forming genera. Mat communities on the soil surface are almost completely absent. However cyanobacterial epiphytes can still be found in eutrophic areas in association with the submersed plant stems and detritus, or as floating mats in cleared areas where light penetrates to the water surface (Inglett et al. 2004; Fig. 1). As stated earlier, these periphytic forms in eutrophic regions are dominated by heterocystous cyanobacteria, but because of reduced water column oxygen levels, there is a high potential for other heterotrophic N fixing genera.

Figure 1. Photos depicting the major types of periphyton assayed for nitrogen fixation activity in various Everglades systems as discussed in the text. Included are: epiphytic and floating mat periphyton from low nutrient areas (upper left), benthic mats lifting off from the soil surface and cross section of a floating mat showing thickness and layering (upper right), periphytic communities associated with detritus in eutrophic areas (lower left), and periphyton/soil crust during drought conditions in short hydroperiod marshes of the southern Everglades (lower right).

Color image of this figure appears in the color plate section at the end of the book.

Bacteria and archaea can also possess nitrogenase, but their presence in Everglades periphyton has only been assessed by the single study of Jasrotia and Ogram (2008). In their study they found a variety of groups represented in their *nifH* clone libraries from three sites along the WCA-2A nutrient enrichment gradient. Members of the δ-Proteobacteria were most abundant and present in eutrophic, transition and oligotrophic periphyton samples while members of the α-Proteobacteria and the γ-Proteobacteria were only found in the eutrophic and oligotrophic samples, respectively. Several recent studies have documented a significant role of methanotrophic bacteria of the α- and δ-Proteobacteria in the N fixation of northern bogs systems (Bragina et al. 2011; Larmola et al. 2014). Therefore, there is a distinct possibility of a similar function in Everglades periphyton communities because of the high methane production and emission rates observed throughout most of the Everglades system.

Measured Rates

Only a few studies have measured rates of nitrogenase activity in Everglades systems (Table 1). The first reported measurement appears to have been a student project by Goldstein (1980, cited in Davis and Ogden 1994) where acetylene reduction yielded estimated N fixation rates from 0.0961–0.593 mg N m^{-2} h^{-1}. This equates to approximately 2.6 g N m^{-2} yr^{-1}, but the exact location and periphyton composition of this measurement is unknown. Inglett et al. (2004) performed the first detailed examination of nitrogenase activity in floating periphyton from both the oligotrophic and eutrophic portions of WCA2A (Table 1). In that study, nitrogenase activity (measured as acetylene reduction) ranged from 19–105 nmol g dw^{-1} h^{-1} which considering areal biomass estimates equates to 23–213 nmol m^{-2} h^{-1} or 1.8–18 g N m^{-2} yr^{-1}.

Inglett et al. (2004) compared the fixation rates of floating mats with that of other system components (e.g., macrophyte detritus or soil). They found that higher rates of acetylene reduction occurred in the cyanobacterial mats (116 nmol g^{-1} h^{-1}) when compared on a weight basis with either macrophyte detritus (54 nmol g^{-1} h^{-1}), surficial floc (8.3 nmol g^{-1} h^{-1}), or surface soil (2.0 nmol g^{-1} h^{-1}). However, when normalized per meter squared (m^2), the fixation by mat communities is most significant only in areas where periphyton is abundant (i.e., the open water, low nutrient slough areas). In more eutrophic regions, macrophyte detritus and surficial soil are much more abundant

Table 1. Values of periphyton nitrogenase activity (as acetylene reduction) reported in the literature.

Everglades region	System type	Periphyton type	Units	Rate	Reference, notes
WCA2A	Eutrophic slough	Floating mat	nmol g^{-1} h^{-1}	116	1, a
	Mesotrophic slough	Floating mat	nmol g^{-1} h^{-1}	115	1, a
	Oligotrophic slough	Floating mat	nmol g OC^{-1} h^{-1}	147–240	1
	Oligotrophic slough with added P	Floating mat	nmol g OC^{-1} h^{-1}	200–2150	2, b
National Park	Marl prairie	Soil crust	nmol g^{-1} h^{-1}	1	3, c
	Marl prairie	Soil crust	nmol g^{-1} h^{-1}	1–7	4, d
	Restored site (2003 site)	Soil crust	nmol g^{-1} h^{-1}	0.2–62	4, e
	Restored site (2000 site)	Soil crust	nmol g^{-1} h^{-1}	BDL-60	4, e
	Mangroves	Pneumatophores	nmol g^{-1} h^{-1}	0–4.8	5

[1] Inglett et al. 2004
[2] Inglett et al. 2009
[3] Liao and Inglett 2012
[4] Liao and Inglett 2014
[5] Pelegri et al. 1997
[a] value for samples collected in August
[b] seasonal range based on samples assayed from the 0.4 g m^{-2} loading rate
[c] value reported for reference site during dry and wet seasons
[d] seasonal range for high elevation control site
[e] seasonal range for the low elevation restored sites

on an areal basis, and thus, their system-level rates of N fixation greatly exceed that of the periphytic communities.

Rates of N fixation in the soil-based periphyton in the marl prairies of The National Park are in general much lower than those of the Northern Everglades systems (Table 1; Liao and Inglett 2012; 2014). Rates of acetylene reduction in the unimpacted systems of these regions have been recorded as low as 0.3 nmol g^{-1} DW h^{-1} in the dry season (Liao and Inglett 2012) and as high as 7.5 nmol g^{-1} DW h^{-1} in the mid-wet season (September, Liao and Inglett 2014). Considering biomass estimates and measured seasonal patterns, Liao and Inglett (2014) arrived at an estimated annual fixation of 0.2 g N m^{-2} yr^{-1}. Periphyton from more nutrient enriched areas in the marl prairies (restored former agricultural areas) had much higher rates of fixation with peak rates > 60 nmol g^{-1} DW h^{-1} and annual estimates of 0.4 g N m^{-2} yr^{-1}. Based on these findings, it is likely to conclude that like many biogeochemical parameters, periphyton nitrogenase activity closely follows nutrient gradients with generally higher rates in the higher nutrient (P) Northern Everglades systems (or potentially downstream of discharge structures due to local concentrated loading effects).

In comparison with other ecosystems, the N fixation rates of unimpacted WCA-2A sloughs (9.7 g N m^{-2} yr^{-1}) are higher than similar estimates from other freshwater marshes (0.01–6.0 g N m^{-2} yr^{-1}) and within the ranges reported for cyanobacterial mats (1.3–76 g N m^{-2} yr^{-1}) (summarized by Howarth et al. (1988b)). Measurements of calcifying marine cyanobacterial mats dominated by *Schizothrix* sp. and *Calothrix* sp. show a range of acetylene reduction rates very similar to those observed in this study (28–561 mol m^{-2} h^{-1}) (Pinckney et al. 1995). Studies of cyanobacterial mats in Belize (50–175 mol m^{-2} h^{-1}) (Rejmánková and Komárková 2000) and Mexico (60–350 umol m^{-2} h^{-1} show fixation rates close to those of the Everglades (Vargas and Novelo 2006).

Apart from these reports, there are few studies documenting nitrogenase activity in other Everglades areas such as Shark River and Taylor Slough, and the Big Cypress National Preserve. Also, periphyton nitrogenase activity has been observed but not quantified in the soft water systems such as interior WCA1 and northern WCA2B. Aside from the marshes and prairies, other areas within the Everglades should also maintain epiphytic communities with nitrogenase activity. For example, epiphytes should readily colonize pneumatophores or prop roots of the mangrove species in portions of the lower Everglades National Park. In a study by Pelegri et al. (1999), however, pneumatophores in this region did not exhibit acetylene reduction rates indicative of cyanobacterial epiphytes with rates similar to sediments ranging from 0–31 nmol C_2H_4 g^{-1} DW hr^{-1}.

Temporal Patterns

Seasonal patterns

At least three studies have measured seasonal patterns of periphyton N fixation in the Everglades system. Of these, Inglett et al. (2004) observed a peak of N fixation in July and a somewhat rapid decline by September. This pattern could have been the result of senescence of the mats exposed to high summer irradiance at the water surface or species composition changes in the mat community as a function of growth and release from the benthic surface. In contrast, N fixation by periphyton in short hydroperiod marshes appears to be more closely tied with hydrology where there is a rapid increase in periphyton activity and N fixation following rehydration of the desiccated crust (Thomas et al. 2006; Liao and Inglett 2014).

These types of patterns with early summer peaks in N fixation have been observed in other wetland systems (e.g., Vargas and Novelo 2007). The exact cause for this seasonal pattern is unknown, but reduced light availability (resulting from increased water depth) and increased availability of N have been mentioned as factors for the low fixation rates observed in late summer (see Liao and Inglett 2014 for discussion). Another possibility could include erroneous assumptions that single time point measurements using the acetylene reduction encompass all organisms fixing N over a diel cycle (e.g., non-heterocystous cyanobacteria and other bacterial heterotrophs) (Paerl et al. 1996).

In the study by Inglett et al. (2004) the seasonal pattern was substantiated by measured isotopic ratios of N (δ^{15}N). However, δ^{15}N patterns did not coincide with seasonal measurements of acetylene reduction in periphyton from the marl prairies (Liao and Inglett 2014). A variety of reasons may explain this lack of agreement between isotopic ratios and acetylene reduction patterns in the marl prairie periphyton, such as the previously mentioned possibility of most N fixation occurring at times other than those sampled (e.g., night-time fixation), or confounding seasonal patterns of external N sources (Inglett and Reddy 2006). Based on these results, it appears that at least in some Everglades systems, seasonal patterns are characterized by high rates during wet season summer months, while in others, the exact seasonal pattern is unknown.

Diel patterns

Diel patterns are also frequently observed in cyanobacterial N fixation systems where changes in light levels control both the availability of reduced carbon (energy source) for fixation and levels of inhibitory oxygen (Stal 1995). This is true in response to environmental conditions, but also within a thick stratified periphyton mat which can have dramatic internal gradients (Paerl et al. 1989; Bebout and Paerl 1993). Thus, a variety of nitrogenase systems (e.g., heterocystous versus non-heterocystous cyanobacteria, or autotrophic versus heterotrophic organisms) have resulted to fill various niches created by diel cycles of photosynthesis and respiration (Paerl 1990; Fay 1992). In the Everglades, a range of N fixing metabolisms exist, with the most predominant form in the periphytic mats believed to be that of heterocystous cyanobacteria (Inglett et al. 2004; Jasrotia and Ogram 2008).

In single time point measurements, light to dark ratios of acetylene reduction consistently demonstrate a stimulation by light of N fixation in both floating mats of northern marshes and sloughs (Inglett et al. 2004; 2009) and periphyton on the soil surface of marl prairies (Liao and Inglett 2012; 2014). This type of pattern typically represents phototrophic N fixation (Fay 1992) as light stimulates the production of carbon substrates and energy used for nitrogenase activity. Both heterocystous and non-heterocystous cyanobacteria can produce this stimulation of nitrogenase under light conditions, but typically heterocystous cyanobacteria are believed to be most indicated by this pattern as they have protective cells (heterocysts) for isolating nitrogenase from photosynthetically generated oxygen.

To date, the only study of diel nitrogenase patterns in Everglades periphyton has been conducted by Inglett (Fig. 2) at the South Florida Water Management District experimental dosing site (mesocosms) in WCA2A. That study was based on methods from Inglett et al. (2004) and monitored nitrogenase activity of floating periphyton mat samples under ambient conditions over a single diel period in August. The results indicate that the majority of fixation in floating mats at this time of the year occurs in the early morning hours. This pattern closely resembles that of non-heterocystous cyanobacterial fixation (Fay 1992) and is a strong indication that the main N fixing component of the Everglades periphyton is non-heterocystous or heterotrophic in nature.

As mentioned earlier, natural abundance ^{15}N measurements have demonstrated that instantaneous acetylene reduction measurements do not apparently correlate with total N fixation occurring within a given system (Liao and Inglett 2014). For example, Liao and Inglett (2014) found a high correlation between N fixation rates and $\delta^{15}N$ between sites of differing nutrient composition, but poor correlations between these parameters for temporal patterns. Isotopic ratios integrate the fixation over a much longer time scale (ca weeks) than instantaneous acetylene reduction measurements (Gu and Alexander 1993; Rejmankova et al. 2004), therefore it is likely that seasonal patterns of species composition and environmental conditions give rise to expression of different nitrogenase forms at differing times of the year.

The use of molecular techniques based on *nif* expression have also revealed that other non-light stimulated mechanisms of N fixation exists in the Everglades periphyton forms (Jasrotia and Ogram 2008). Based on these studies, it appears that fixation rates

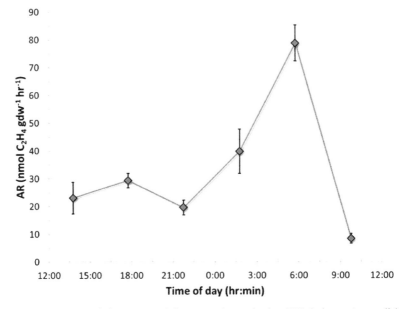

Figure 2. Measurements of nitrogenase activity as acetylene reduction (AR) during an August diel cycle in floating periphyton mats of Water Conservation Area 2A (see text for experimental details).

and patterns within the Everglades periphyton communities may be highly complex with the potential for fixation from several organism groups which can vary throughout seasonal cycles of light, temperature, and nutrient availability. It also appears that acetylene reduction or ^{15}N isotope incorporation observed under single, short duration light and dark incubations does not adequately assess all potential nitrogenase activity in the complex assemblage of periphyton organisms and may underestimate N fixation.

Regulators of Everglades Periphyton Nitrogenase Activity

Light

The seasonal patterns in nitrogenase activity illustrate the potential role of light and temperature as a regulator of N fixation Everglades periphyton. This is particularly important for deeper water systems such as sloughs in Northern Everglades systems where light availability is controlled by water depth or floating mat communities which are exposed to ambient light levels with little shading. Light availability is known to affect metabolism and species composition in periphytic communities (Kahn and Wetzel 1999), so it is reasonable to expect that light availability and its surrogate water level could directly influence N fixation. Everglades periphyton become light saturated at low levels of irradiance (less than 300 μmol m^{-2} s^{-1}, Hagerthey et al. 2011), and Inglett et al. (2004) speculated that high light intensity and UV exposure was responsible for suppressing nitrogenase activity in September. In contrast, Liao and Inglett (2014) hypothesized that increasing water depth and lower light penetration was responsible for lower N fixation rates in benthic periphyton. Therefore, while the exact mechanism or effect has yet to be supported, it is likely that light may act as both promoter (through photosynthetic stimulation) and inhibitor (via increased UV exposure of unshaded periphyton) of periphyton N fixation.

Nutrients

Phosphorus

By far, phosphorus (P) is the most highly influential nutrient parameter affecting observed N fixation rates for Everglades periphyton (Inglett et al. 2011). Studies in the northern water conservation areas as well as those in the National Park have observed strong correlations between periphyton acetylene reduction and P content (Inglett et al. 2004; 2009; Liao and Inglett 2012; 2014). N availability has also been shown to interact with P in many of these systems. For example, in WCA2A N content of periphyton was frequently observed as a covariate with N fixation and increased availability of available N (ammonium and nitrate) was believed to be the cause of both lower N fixation as well as green algal dominance of periphyton in moderately P enriched areas (Inglett et al. 2009). Similar observations were observed in soil-based periphyton in the marl prairies (Liao and Inglett 2014).

Nitrogen

Available inorganic N is also known to affect N fixation rates through inhibition of both heterocyst formation and overall nitrogenase expression (Horne et al. 1972; 1979). Spatial patterns of N availability have been shown to dictate N fixation rates in surface flow wetlands (Scott et al. 2007; 2008). Likewise, Inglett et al. (2004) reasoned that increased ammonium levels were responsible for lower N fixation in periphyton communities of the transition area of the WCA2A (areas with moderate P enrichment), and Liao and Inglett (2014) observed lowest nitrogenase activity in periphyton of marl prairies during flooded periods with high levels of water column nitrate. Similarly, N content of periphyton biomass has also shown a negative correlation with measured acetylene reduction rates (Inglett et al. 2009). For this reason, and as supported by stoichiometric theory, the availability of N relative to P (TN:TP) is often a better determinant of N limitation in algal communities than either N or P alone, and thus is generally a better predictor of observed N fixation rates in most systems (Smith 1983), including the Everglades (Inglett et al. 2009; Liao and Inglett 2014).

Hydrology

In addition to light and nutrients, other conditions may also influence periphyton growth or nutrient status, and thus also affect N fixation in these microbial communities. Some of these conditions may even be more influential within Everglades systems than nutrient inputs. The most obvious of these concerns the fact that periphyton growth is only found in aquatic environments, and the condition of adequate water availability may not always be met in Everglades systems. This is especially true for short hydroperiod marshes to the South where seasonal patterns of flooding, drying, and rewetting dictate metabolic activities, growth, and species composition of periphyton (Gottlieb et al. 2005; Iwaniec et al. 2006). Consequently, Liao and Inglett (2014) found that moisture content was the dominant factor controlling acetylene reduction rates in periphyton in the marl prairies of the National Park.

In adequately hydrated systems, water depth may also influence nitrogenase activity of periphyton by the previously stated effect on light availability. Deeper water conditions have been shown to affect metabolism of the periphytic community with the overall shift toward heterotrophic metabolism (Hagerthey et al. 2011) and lower levels of dissolved O_2. Under these conditions, oxidation processes would decrease within benthic periphyton mats and floc leading to higher levels of ammonium and likely a suppression of N fixation. The dominant fixers of these communities may also shift from more autotrophic to more heterotrophic organisms under the lower oxygen stressed conditions (Fay 1992). Another aspect of hydrology would include flow rate which could determine periphyton growth patterns (e.g., through floc redistribution) or nutrient supply rates. In higher flow areas, periphyton may have access to higher nutrient loads such as P (Gaiser et al. 2004) which could lead to altered species composition and biomass nutrient ratios which would result in altered nitrogenase activity.

Use of Periphyton Nitrogenase as an Ecosystem Indicator

Periphyton species composition is already a widely used indicator of hydrology and water quality and conditions in the Everglades system (e.g., McCormick et al. 1996b; Gaiser et al. 2011; Hagerthey et al. 2011). Periphyton activity and enzyme expression is also useful to illustrate environmental and nutrient related conditions in this and other aquatic systems (Newman et al. 2003; Sharma et al. 2005; Hagerthey et al. 2011). Similarly, periphyton nitrogenase which is regulated by both environmental and nutrient parameters, should also serve as a useful indicator of impact and ecosystem change in the Everglades.

Nutrient Inputs

The basic theory of why nitrogenase can be used as an indicator of nutrient conditions involves the production of this enzyme complex under conditions of N limitation. As stated earlier, the Everglades is generally thought of as a P limited system. However, there are also extremely low levels of available N throughout most of the Everglades system (Inglett et al. 2011). It is this condition of low available N which likely leads to the dominance of cyanobacteria in the periphytic forms.

In areas of increased P availability, N demand is even more elevated leading to continued dominance of the periphyton by cyanobacteria and enhanced rates of nitrogenase activity. This is frequently indicated by total P content or a reduced TN:TP ratio of the periphyton biomass (McCormick et al. 1996a). Strong correlations between periphyton TP and TN:TP and nitrogenase activity have been observed in every published study in the Everglades. Inglett et al. (2009) also demonstrated in a mesocosm dosing experiment that periphyton nitrogenase activity became enhanced relative to controls after receiving approximately 2–3 mg P m^{-2} of cumulative P dosing or achieving biomass TP content of 100–300 mg P kg^{-1}. The threshold sensitivity for detecting shifts in nitrogenase activity is currently unknown, however Inglett et al. (unpublished results) have observed elevated nitrogenase activity in response to P dosing in as little as 12 hours.

Hydrology

In addition to P, hydrology, including both water depth and flow velocity, may also affect periphyton composition and this may indirectly affect nitrogenase activity of this community. This possibility has not been investigated but it is potentially an important indicator for restoration within the Everglades system. With restoration of flow and likely changes in water depth, changes in light availability and light penetration to the benthic surface would undoubtedly result in altered photosynthetic activities leading to changes in N cycling via nitrogenase expression. The low light saturation point of benthic cyanobacterial mats suggests that even small reductions in water depth could have profound impacts on metabolism and thus N cycling in presently deeper water areas (Kahn and Wetzel 1999; Hagerthey et al. 2011).

Water depth is not the only hydrology related factor that could affect N fixation in periphyton communities of the Everglades. Changes in flow will also result in altered rates of nutrient delivery affecting nutrient balance in these sensitive microbial communities. Everglades periphyton have been shown to respond to the cumulative load of nutrients such as P (Gaiser et al. 2004) through the hypothesized calcium coprecipitation mechanism (Scinto and Reddy 2003). For this reason, increases in flow velocity following hydrologic restoration should also result in decreases in the N:P stoichiometry of periphyton and altered N demand leading to production of nitrogenase.

Ecosystem Restoration

Liao and Inglett (2012; 2014) demonstrated the utility of periphyton nitrogenase activity to indicate ecosystem restoration progress in the Hole-in-the-Doughnut project, Everglades National Park. In those studies, periphyton development and nutrient composition after clearing of prior agricultural lands was a good indicator of P from residual soil materials. As a result, recently cleared areas are more N limited than those for several years (Inglett and Inglett 2013). Likewise, periphyton of 7- and 10-year-old cleared plots had elevated P content and exhibited 3.3-fold and 2.6-fold higher rates of nitrogenase activity, respectively (Liao and Inglett 2014). Dynamics of nitrogenase activity were also found to be different for periphyton in the cleared areas as these communities were developing on very shallow soils (< 5 cm) which are more prone to desiccation than the deep (> 15 cm) marl soils of the native prairies (Smith et al. 2011; Inglett and Inglett 2013). As soils and plant communities of these systems continue to success, P levels become decreased, and periphyton nitrogenase activities should similarly decline (Inglett and Inglett 2014).

Summary and Conclusions

The observation of high rates of N fixation activity from periphyton of such a profoundly P limited ecosystem like the Everglades remains somewhat of an enigma. Rates of N fixation in the Everglades rival those recorded in coastal microbial mat communities (Howarth et al. 1988b). The identity of the N fixing organisms remains uncertain, but it is likely that the combination of non-heterocystous and heterocystous cyanobacterial species are the dominant fixers with a smaller contribution from bacteria and archaea.

An extreme diversity of conditions present within the Everglades system combined with the close tie between environmental conditions, periphyton composition, and N fixation suggests that much more work is needed to accurately assess and quantify nitrogenase activity within the system. Among the questions that warrant attention are the relative rates of N fixation between different periphytic forms (e.g., benthic versus epiphytic versus floating) and their response to changing environmental conditions (i.e., light/water depth, temperature, nutrient levels) in the various Everglades systems. To date, only two of the many Everglades systems (WCA2A and the marl prairies of the National Park) have received some attention, leaving several key Everglades habitats yet to be studied for their periphyton nitrogenase activity.

The sensitivity of periphyton (community structure and function) to the environmental conditions and the established coupling of nitrogenase activity to availability of N and P, suggests that periphyton is a very sensitive indicator of nutrient dynamics as well as hydrology. During the Everglades restoration process, restoration of flow and modification of existing water depth within the system will likely result in changes in composition of N fixers and measured rates within this ecosystem component. Periphyton nitrogenase activity could also be used to indicate P inputs or residual levels of P during restoration. For this potential to become reality, more studies evaluating the effects of environmental conditions on the relative abundance and identity of N fixing organisms, as well as documentation of related diel and seasonal patterns must be accomplished.

Based on the few studies conducted thus far, N fixation by periphytic communities in the Everglades is a significant process and an important contributor to the overall N cycle of the system. How periphyton N-fixation affects the N cycle of the Everglades remains unknown, but through the use of novel techniques such as molecular characterization of *nif* expression combined with studies to capture the spatial and temporal dynamics of periphyton and microbial mats, we will gain a better understanding of this dynamic process and its importance within the sensitive Everglades system. Combined with future studies tracing periphyton N through other system compartments (e.g., dissolved organic N, invertebrates, fish, etc.), the fate of N fixed may be established with larger scale implications for combined effects of P enrichment and hydrology on N cycling through the Everglades landscape.

References

Bebout, B.M., M.W. Fitzpatrick and H.W. Paerl. 1993. Identification of the sources of energy for nitrogen fixation and the physiological characterization of nitrogen-fixing members of a marine microbial mat communtiy. Appl. Environ. Microbiol. 59: 1495–1503.

Belnap, J. 2001. Factors influencing nitrogen fixation and nitrogen release in biological soil crusts. pp. 241–261. *In*: J. Belnap and O.L. Lange (eds.). Biological Soil Crusts: Structure, Function, and Management. Springer-Verlag, Berlin.

Bragina, A., S. Maier, C. Berg, H. Müller, V. Chobot, F. Hadacek and G. Berg. 2011. Similar diversity of Alphaproteobacteria and nitrogenase gene amplicons on two related Sphagnum mosses. Fronti. Microbio. 2: 275. doi: 10.3389/fmicb.2011.00275.

Browder, J.A., P.J. Gleason and D.R. Swift. 1994. Periphyton in the Everglades: spatial variation, environmental correlates, and ecological implications. pp. 343–348. *In*: S.M. Davis and J.C. Ogden (eds.). Everglades: The Ecosystem and its Restoration. St. Lucie Press, Delray Beach, Florida, USA

Davis, S.M. and J.C. Ogden (eds.). 1994. Everglades: The Ecosystem and its Restoration. St. Lucie Press, Delray Beach, Florida.

Fay, P. 1992. Oxygen relations of nitrogen fixation in cyanobacteria. Microbiol. Rev. 56: 340–373.

Gaiser, E., L. Scinto, J. Richards, K. Jayachandran, D. Childers, J. Trexler and R. Jones. 2004. Phosphorus in periphyton mats provides best metric for detecting low-level P enrichment in an oligotrophic wetland. Water Res. 38: 507–516.

Gaiser, E.E., D.L. Childers, R.D. Johns, J. Richards, L.J. Scinto and J.C. Trexler. 2006. Periphyton responses to eutrophication in the Florida Everglades: cross-system patterns of structural and compositional changes. Limnol. Oceanogr. 51: 617–630.

Gaiser, E.E., P.V. McCormick, S.E. Hagerthey and A.D. Gottlieb. 2011. Landscape patterns of periphyton in the Florida Everglades. Crit. Rev. Environ. Sci. Technol. 41: 92–120.

Gleason, P.J. and W. Spackman. 1974. Calcareous periphyton and water chemistry in the Everglades. pp. 225–248. *In*: P.J. Gleason (ed.). Environments of South Florida: Past and Present, Memoir No. 2. Miami Geological Society, Coral Gables, FL.

Gottlieb, A. 2003. Short and long hydroperiod Everglades periphyton mats: community characterization and experimental hydroperiod manipulation, Dissertation, Florida International University, Miami, FL.

Gottlieb, A., J. Richards and E. Gaiser. 2005. Effects of desiccation duration on the community structure and nutrient retention of short and long-hydroperiod Everglades periphyton mats. Aquatic Bot. 82: 99–112.

Gu, B. and V. Alexander. 1993. Estimation of N_2-fixation based on differences in the natural abundance of ^{15}N among freshwater N_2-fixing and non-N_2-fixing algae. Oecologia 96: 43–48.

Hagerthey, S.E., B.J. Bellinger, K. Wheeler, M. Gantar and E. Gaiser. 2011. Everglades periphyton: a biogeochemical perspective. Crit. Rev. Environ. Sci. Technol. 41: 309–343.

Horne, A.J., J.E. Dillard, D.K. Fujmita and C.R. Goldman. 1972. Nitrogen fixation in Clear Lake, California. 2. Synoptic studies on the autumn Anabaena bloom. Limnol. Oceanogr. 17: 693–703.

Horne, A.J., J.C. Sandusky and C.J.W. Carmiggelt. 1979. Nitrogen fixation in Clear Lake, California. 3. Repetitive synoptic sampling of the spring Aphanizomenon blooms. Limnol. Oceanogr. 24: 316–328.

Howarth, R.W., R. Marino and J.J. Cole. 1988a. Nitrogen fixation in freshwater, estuarine, and marine ecosystems. 2. Biogeochemical controls. Limnol. Oceanogr. 33: 688–701.

Howarth, R.W., R. Marino, J. Lane and J.J. Cole. 1988b. Nitrogen fixation in freshwater, estuarine, and marine ecosystems. 1. Rates and importance. Limnol. Oceanogr. 33: 669–687.

Inglett, P.W. and K.R. Reddy. 2006. Investigating the use of macrophyte stable C and N isotopic ratios as indicators of wetland eutrophication: patterns in the P-affected Everglades. Limnol. Oceanogr. 51: 2380–2387.

Inglett, P.W. and K.S. Inglett. 2013. Biogeochemical changes during early development of restored calcareous wetland soils. Geoderma 192: 132–141.

Inglett, P.W., K.R. Reddy and P.V. McCormick. 2004. Periphyton chemistry and nitrogenase activity in a northern Everglades ecosystem. Biogeochemistry 67: 213–233.

Inglett, P.W., E.M. D'Angelo, K.R. Reddy, P.V. McCormick and S.E. Hagerthey. 2009. Periphyton nitrogenase activity as an indicator of wetland eutrophication: spatial patterns and response to phosphorus dosing in a northern Everglades ecosystem. Wetlands Ecol. Manage. 17: 131–144.

Inglett, P.W., V.H. Rivera-Monroy and J.R. Wozniak. 2011. Biogeochemistry of nitrogen across the Everglades Landscape. Critical Reviews in Environmental Science and Technology 41: 187–216.

Iwaniec, D., D.L. Childers, D. Rondeau, C.J. Madden and C.J. Saunders. 2006. Effects of hydrologic and water quality drivers on periphyton dynamics in the southern Everglades. Hydrobiologia 569: 223–235.

Jasrotia, P. and A. Ogram. 2008. Diversity of nifH genotypes in floating periphyton mats along a nutrient gradient in the Florida Everglades. Curr. Microbiol. 56: 563–568.

Kahn, W.E. and R.G. Wetzel. 1999. Effects of microscale water level fluctuations and altered ultraviolet radiation on periphytic microbiota. Microb. Ecol. 38: 253–263.

Larmola, T., S.M. Leppänen, E. Tuittila, M. Aarva, P. Merilä, H. Fritze and M. Tiirola. 2014. Methanotrophy induces nitrogen fixation during peatland development. Proc. Nat. Acad. Sci. 111: 734–739.

Liao, X. and P.W. Inglett. 2012. Biological nitrogen fixation in periphyton of native and restored Everglades marl prairies. Wetlands 32: 137–148.

Liao, X. and P.W. Inglett. 2014. Dynamics of periphyton nitrogen fixation in short-hydroperiod wetlands revealed by high resolution seasonal sampling. Hydrobiologia 722: 263–277.

McCormick, P.V. and M.B. O'Dell. 1996. Quantifying periphyton responses to phosphorus enrichment in the Florida Everglades: a synoptic-experimental approach. J. N. Am. Benthol. Soc. 15: 450–468.

McCormick, P.V. and R.J. Stevenson. 1998. Periphyton as a tool for ecological assessment and management in the Florida Everglades. J. Phycol. 34: 726–733.

McCormick, P.V., P.S. Rawlik, K. Lurding, E.P. Smith and F.H. Sklar. 1996. Periphyton-water quality relationships along a nutrient gradient in the northern Florida Everglades. J. N. Am. Benthol. Soc. 15: 433–449.

Newman, S., P.V. McCormick and J.G. Backus. 2003. Phosphatase activity as an early warning indicator of wetland eutrophication: problems and prospects. J. Appl. Phycol. 15: 45–59.

Paerl, H.W. 1990. Physiological ecology and regulation of N_2 fixation in natural waters. Adv. Microb. Ecol. 11: 305–344.

Paerl, H.W., B.M. Bebout and L.E. Prufert. 1989. Naturally occurring patterns of oxygenic photosynthesis and N_2 fixation in a marine microbial mat: physiological and ecological ramifications. pp. 326–341. *In*: Y. Cohen and E. Rosenberg (eds.). Microbial Mats: Physiological Ecology of Benthic Microbial Communities. American Society for Microbiology, Washington, DC.

Paerl, H.W., M. Fitzpatrick and B.M. Bebout. 1996. Seasonal nitrogen fixation dynamics in a marine microbial mat: potential roles of cyanobacteria and microheterotrophs. Limnol. Oceanogr. 41: 419–427.

Pelegri, S.P., V.H. Rivera-Monroy and R.R. Twilley. 1997. A comparison of nitrogen fixation (acetylene reduction) among three species of mangrove litter, sediments, and pneumatophores in south Florida, USA. Hydrobiologia 356: 73–79.

Pinckney, J., H.W. Paerl, R.P. Reid and B. Bebout. 1995. Ecophysiology of stromatolithic microbial mats, Stocking Island, Exuma Cays, Bahamas. Microb. Ecol. 29: 19–37.

Rejmánková, E. and J. Komárková. 2000. A function of cyanobacterial mats in phosphorus-limited tropical wetlands. Hydrobiologia 431: 135–153.

Rejmánková, E., J. Komárková and M. Rejmánek. 2004. $\delta^{15}N$ as an Indicator of N_2 fixation by cyanobacterial mats in tropical marshes. Biogeochemistry 67: 353–368.

Scinto, L.J. and K.R. Reddy. 2003. Biotic and abiotic uptake of phosphorus by periphyton in a subtropical freshwater wetland. Aquatic Bot. 77: 203–222.

Scott, J.T., R.D. Doyle, J.A. Back and S.I. Dworkin. 2007. The role of N_2 fixation in alleviating N limitation in wetland metaphyton: enzymatic, isotopic, and elemental evidence. Biogeochemistry 84: 207–218.

Scott, J.T., M.J. McCarthy, W.S. Gardner and R.D. Doyle. 2008. Denitrification, dissimilatory nitrate reduction to ammonium, and nitrogen fixation along a nitrate concentration gradient in a created freshwater wetland. Biogeochemistry 87: 99–111.

Sharma, K., P. W. Inglett, K. R. Reddy and A. V. Ogram. 2005. Microscopic examination of photoautotrophic and phosphatase-producing organisms in phosphorus-limited Everglades periphyton mats. Limnol. Oceanogra. 50: 2057–2062.

Smith, C.S., L. Serra, Y.C. Li, P.W. Inglett and K. Inglett. 2011. Restoration of disturbed lands: the Hole-in-the-Donut restoration in the Everglades. Crit. Rev. Environ. Sci. Technol. 41: 723–739.

Smith, V. 1983. Low nitrogen-phosphorus ratios favor dominance by blue-green algae in lake phytoplankton. Science 221: 669–671.

Stal, L.J. 1995. Physiological ecology of cyanobacteria in microbial mats and other communities: Tansley Review No. 84. New Phytol. 131:1–32.

Thomas, S., E.E. Gaiser, M. Gantar and L.J. Scinto, 2006. Quantifying the responses of calcareous periphyton crusts to rehydration: A microcosm study (Florida Everglades). Aquatic Bot. 84: 317–323.

Vargas, R. and E. Novelo. 2007. Seasonal changes in periphyton nitrogen fixation in a protected tropical wetland. Biol. Fertil. Soils 43: 367–372.

Vitousek, P.M., K. Cassman, C. Cleveland, T. Crews, C.B. Field, N.B. Grimm, R.W. Howarth, R. Marino, L. Martinelli, E.B. Rastetter and J.I. Sprent. 2002. Towards an ecological understanding of biological nitrogen fixation. Biogeochemistry 57: 1–45.

Vymazal, J. 1995. Algae and Element Cycling in Wetlands. CRC Press, Inc., Boca Raton, FL.

Structure and Function of Cyanobacterial Mats in Wetlands of Belize

Jiří Komárek,[1] *Dagmara Sirová,*[2] *Jaroslava Komárková*[3] and
Eliška Rejmánková[4,*]

Introduction

Herbaceous wetlands of northern Belize are part of a group of phytogeographically related marshes that cover extensive areas on the Yucatan Peninsula, Caribbean islands, and the Florida Everglades (Fig. 1; Estrada-Loera 1991; Borhidi 1991; Chiappy-Jhones et al. 2001). These limestone-based wetlands are characterized by rather extreme conditions in terms of hydrology, nutrient availability and salinity, which limit growth of freshwater macrophytes (Borhidi 1991; Rejmánková et al. 1996). Extreme conditions are highly suitable for the development of cyanobacterial mats, formed by complex assemblages of cyanobacteria, eubacteria, and eukaryotic algae (Stal 2000; Paerl et al. 2002), which are the dominant primary producers in most of these wetlands. Although similar in many respects, these highly diverse and (relatively) isolated marshes have conditioned the adaptation and stabilization of cyanobacteria and algae, specifically diatoms, into numerous specialized eco- and morphotypes, with what appears a high degree of endemism (La Hée and Gaiser 2012).

[1] Institute of Botany ASCR and University of South Bohemia, Czech Republic.
  Email: Jiri.Komarek@ibot.cas.cz
[2] University of South Bohemia, Czech Republic.
  Email: dagmara_sirova@hotmail.com
[3] Hydrobiological Institute and Institute of Botany ASCR, Czech Republic.
  Email: komarkova@mujicin.cz
[4] University of California Davis, USA.
  Email: erejmankova@ucdavis.edu
* Corresponding author

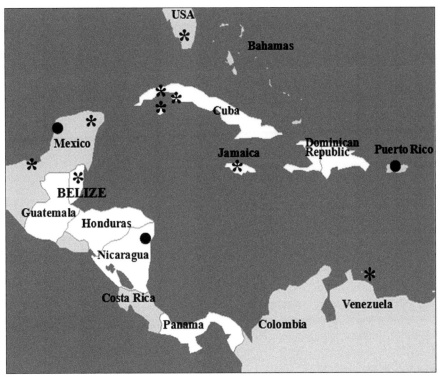

Figure 1. Caribbean and Central American regions with typical alkaline marshes (asterisks). The black points designate locations with habitually similar cyanobacterial vegetation, but different species composition.

In this chapter, we review and summarize what we have learned about the cyanobacterial mat (CBM) communities in Belizean marshes, with regards to both their uniqueness and similarity to the other limestone-based wetland types in the Caribbean region, especially the Everglades. We cover CBM structure and species composition, important biogeochemical processes, interaction with macrophytes, and sensitivity to eutrophication. We include materials that have already been published as well as hitherto unpublished data. Most of the results come from a long term nutrient addition experiment conducted in 15 marshes along the salinity gradient, where 10 x 10 m plots served as control, and N, P and N&P addition sampling sites (for details see Table 1 and Rejmánková et al. 2008). The field experiment results are complemented by numerous observational data from other marshes in the region, and data from mesocosm experiments (Rejmánková and Komárková 2005; Macek and Rejmánková 2007).

Study Site Description and History of Marsh Development in Belize

Belizean wetlands range in size from small < 1 ha marshes to large shallow inland lagoons and are strongly nutrient limited (Rejmánková et al. 1996). In many parts of this region, they are the only non-forest inland ecosystems still maintaining their natural composition and function. The flora of essentially all of these wetlands consist

Table 1. Physical and chemical characteristics of marshes of northern Belize. The values are means of five measurements available from these sites over the period August 2001 February 2003.

Marsh	Water depth range, cm	pH	Conductivity $\mu S\ cm^{-1}$	SRP $\mu g\ L^{-1}$	NH4-N $\mu g\ L^{-1}$	SO_4^{2-} mg L^{-1}	Cl^- mg L^{-1}
Frank	0–140	7.8	211	12.1	35.5	0.8	3.9
Big Snail	33–135	8.8	151	11.6	65.8	1.0	19.6
Cane	–3–95	7.5	121	10.9	68.9	2.9	4.6
Deep	7–96	8.1	112	14.7	85.8	3.8	8.1
Hidden	2–88	7.9	172	8.2	82.0	1.0	6.0
Buena Vista	4–70	7.8	1425	11.9	40.7	138.9	77.9
New	–3–140	7.9	2132	9.3	45.3	580.8	273.0
Quiet	–5–129	7.6	1153	15.5	23.8	318.7	154.7
Calabash	6–86	8.1	1915	9.3	41.5	652.8	248.5
Eli's	11–31	7.7	963	11.4	50.3	48.0	101.5
Doubloon	–5–190	8.1	5602	15.4	40.6	844.8	982.9
Little Belize	1–35	8.0	4420	14.5	81.9	45.3	808.0
LB road	15–57	8.4	2373	11.8	101.1	89.3	777.4
LB mangrove	0–30	8.1	3858	9.9	83.3	97.0	972.5
Chan Chen	15–127	8.4	3348	11.8	58.3	2438.4	273.0

of monodominant stands of *Eleocharis cellulosa, E. interstincta, Cladium jamaicense,* and *Typha domingensis*, with *Nymphaea ampla* and submerged species of the genera *Utricularia* and *Chara* as frequent co-dominants. Limestone geology is common to all the above mentioned areas (Fig. 1). The Yucatan Peninsula is a large carbonate platform consisting mostly of Tertiary limestone, marl, dolomite, and evaporites. The landscape is a tropical karst, with poorly developed surface drainage and many solution features. Regional variations in climate, bedrock lithology, and age and type of tectonic movements produce a varied landscape (Weidie 1985). The river courses that drain predominantly limestone terrain are tectonically controlled and are fed primarily by springs and only secondarily by surface runoff (Siemens 1978). Some of these karstic depressions are clay-lined and support a perched water table for part of the year. This varied karst landscape, with its variable hydrology, gives rise to a high diversity of wetland habitats. Karstic rivers transport mostly calcium carbonates and sulfates in solution and their water level fluctuation is usually less than 2 m due to the buffering effect of a well-developed subterranean drainage system (Siemens 1978). Perennial wetlands are found mostly below 100 m elevation and they have a water table that fluctuates 0.5–1 m between wet and dry seasons, reflecting regional changes in the water balance of coastal aquifers. The soils in these depressions are usually perennially moist haplaquolls: sticky montmorillonite-rich clays, often with gley, carbonate, gypsum, and peat horizons (Wright et al. 1959; Pope et al. 1989).

Recent paleoecological and archaeological researches indicate that lowland humid regions of Meso-America were as important as highland areas for development of early human societies (Pohl et al. 1996). Wetland agriculture in the Mayan Lowlands, specifically in Belize, has been well documented (Pohl 1990; Alcala-Herrera et al. 1994; Pope et al. 1996). Starting about 2500 BC, northern Belize was rapidly deforested, and the Mayan people probably settled around wetland margins where they found abundant plant, faunal and water resources (Pohl et al. 1996). By ca. 1000 BC, a rise in groundwater levels led farmers to construct drainage ditches. Several researchers have been exploring the possibility that the cyanobacterial mats could have functioned as a natural, renewable, and manageable source of agricultural fertilizer (Fedick and Morrison 2003; Morrison and Cózatl Manzano 2003; Novelo and Tavera 2003; Palacios Mayorga et al. 2003). Dried cyanobacteria could have been transported and applied as fertilizer in upland agricultural fields or home gardens. By the Classic period, AD 1–1000, wetland fields were flooded and mostly abandoned. As water levels rose, the lowland areas cultivated in the Maya Pre-classic period slowly changed into shallow lakes and mudflats characterized by a high content of minerals, namely gypsum and calcite. Through a process of natural siltation and plant succession, these shallow lakes slowly developed into their present state, i.e., herbaceous marshes (see Table 1 for marsh characteristics). For several centuries, the wetland ecosystems in northern Belize were influenced predominantly by natural processes until the middle of the 19th century, when sugarcane cultivation began. Many marshes are still relatively undisturbed by recent human activities and their present condition depicts a natural stage of wetland development as it has progressed since the termination of Maya cultivation.

Certain features make cyanobacteria particularly suitable for mat formation: sheaths and mucilage contribute to matrix building that supports other microbial and protozoan colonizers. The benthic mats at study sites (Fig. 2) have a cohesive, cotton-wool-like texture, range from few mm to 10 cm in thickness. They have a relatively smooth surface with a superficial golden-brownish layer composed mainly of cyanobacteria producing the photo-protective pigment scytonemin (Sirová et al. 2006). Below the surface there is a dark green layer dominated by filamentous cyanobacteria, under which a conspicuous pink layer of various phototrophic bacteria is often present, overlying the marl sediment (Fig. 2F).

Cyanobacteria, besides providing structural support, provide fixed N_2 which can be directly utilized by associated non-fixing microorganisms (Paerl et al. 1996; 2000). The CBM contribute to calcium carbonate and hence marl formation by consuming dissolved CO_2 during photosynthesis, thus displacing the bicarbonate—carbonate equilibrium in favor of carbonate (Hodell et al. 2007). They are also responsible for the input of organic matter to both the water column and sediments. Cyanobacterial mats are therefore of high importance in wetland ecosystem processes as sites of intense biochemical and ecological interactions. The importance of mats as a habitat and a source of food (fish feed on the diatoms and green algae component of the mats) has been stressed by Sargeant et al. (2010) and Lee et al. (2013) for the Everglades, no data are available for Belizean marshes.

Seasonal growth dynamics of CBM communities is driven mainly by the marsh flooding/drying cycles. Mat development starts at the beginning of the rainy season and, by the end of the dry season, many shallower marshes are completely dry (Fig. 3). The reflooding and/or increase in water level with the start of the rains stimulates rapid growth of epipelic algae that form a dense carpet of benthic mucilaginous communities. During the mid-rainy season, the water is usually quite deep with little metaphyton present. Cyanobacterial biomass increases slowly, with only occasional sloughing of bottom mats. The major increase in biomass happens at the end of rainy season/beginning of dry season when the water levels begin to decline. This is a time of massive periphytic growth on *Eleocharis* and other macrophytes, particularly

Figure 2. Examples of various types of cyanobacterial mats from marshes of Belize. 2A Aerial view of Belizean marshes; 2B Benthic mat (epipelon); 2C Floating mat (metaphyton); 2D Mat forming on plant stems (periphyton); 2E Dry mat; 2F Vertical cross-section of a benthic mat.

Color image of this figure appears in the color plate section at the end of the book.

Utricularia spp.; large areas of floating metaphyton form. Thus the rapid growth and development of algal mats with the highest biomass is achieved in the mid-dry season (February–March).

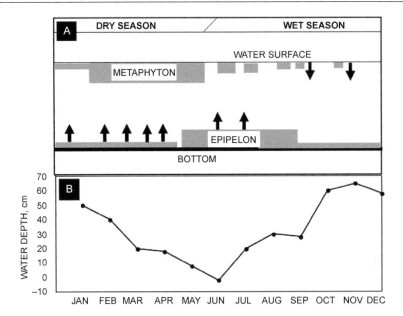

Figure 3. A. Scheme of mat development. Bold, upward pointing arrows indicate a process of intensive sloughing. Downward pointing arrows indicate sinking of floating mats due to heavy rains during the rainy season. B. A typical course of water depth during the year.

Species Composition of Cyanobacteria

CBM are not only important and functionally complex components of the marsh habitats, they also represent a significant reservoirs of biological diversity. The cyanobacterial vegetation from northern Belizean marshes was described in detail in papers of Rejmánková et al. (2004), Komárek et al. (2005), Komárek and Komárková-Legnerová (2007), Turicchia et al. (2009) and Komárek et al. (in press). The composition of these communities is often characterized by endemic morpho- and ecospecies, with distinct seasonal changes in species composition and their quantity, species life forms, and their specific function in the ecosystem. Species richness of cyanobacterial assemblages follows the conductivity gradient, being the highest in the marshes of medium conductivity ($\sim$1000–2000 μS cm^{-1}) and decreasing at low (< 500 μS cm^{-1}) and high (> 3000 μS cm^{-1}) conductivities (Rejmánková et al. 2004b). The richness and diversity of cyanobacterial species found in these habitats is unusually high, with a significant number of specific types and endemic or ecologically distinct species. Although the cyanobacterial mats found in the oligotrophic marshes of various regions of the Caribbean are ecologically similar and their character is comparable at all studied locations, some local deviations in species composition exist. For example, the very characteristic *Anabaena fuscovaginata* has been found only in the Florida Everglades (Mareš 2010). Several unique types are also known from Islas de los Aves near the Venezuelan coasts (Schiller 1956).

All main groups of cyanobacteria are present in Belizean CBM. The main proportion of biomass is formed by filamentous pseudanabaenacean and, in lesser frequency, oscillatorialean types without heterocytes. Dense filament clusters typically contain a surprising diversity of coccoid types. The coccoid cyanobacteria, with the exception of infrequent localized growth, do not occur in high abundance. The heterocytous types develop more sporadically, in distinct populations, mostly occurring seasonally, adapted to specific growth and ecological situations.

In total approximately 40 coccoid species, more than 30 filamentous cyanobacteria without heterocytes, and approximately 25 heterocytous species were identified to date from the Belizean marshes (Table 2). Several more types were found only in restricted number of samples or in non-identifiable developmental stages. However, from the review in Table 2, it is possible to conclude that the cyanobacterial flora from the Caribbean, particularly Belizean alkaline marshes is unique, highly specialized, containing numerous adapted forms, with distinct number of endemic species. The preliminary molecular studies of the Belizean communities confirm the specificity of phototrophic microbial flora. More than 100 traditional cyanobacterial taxa, which can mostly be designated as typical members of the Belizean shallow marshes, were identified (Komárek et al. 2005).

Coccoid Types

Coccoid cyanobacteria, which belong to several orders and families according to modern cyanobacterial systematics, are well diversified in Belizean marsh habitats. However, only the species *Aphanothece bacteriosa* (Fig. 4) belonging to the genus *Anathece* based on the revised system (Komárek et al. 2011), occurs in masses and forms very dense, greenish mucilaginous layers on the bottom of few marshes, mostly at the location Big Snail South (Table 1).

Other coccoid species occur sporadically, mostly dispersed among the dominant filamentous types. About 40 taxa (almost 80%) are characteristic for the Caribbean alkaline marshes only. About 15% are pantropical species and only a few can be designated as more or less cosmopolitan. Characteristic species include, for example, the endemic *Aphanothece bacilloidea, Aph. granulosa, Chlorogloea gardneri, Gloeothece aggregata, Gomphosphaeria semen-vitis, Lemmermanniella uliginosa,*

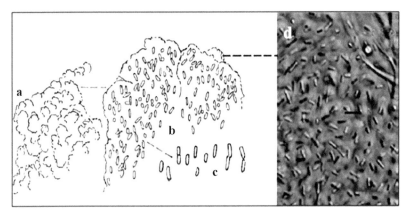

Figure 4. *Anathece bacteriosa*, an endemic coccoid cyanobacterium forming mass developments on the bottom of the mat Big Snail South (From Komárek and Komárková-Legnerová 2007).

Table 2. Summary of cyanobacteria from alkaline marshes of the Caribbean, comparing northern Belize and the Everglades. Category Common includes species occurring in both, Belizean and Everglades marshes, but not yet recorded from other regions. Category Endemic lists number of species endemic to the Caribbean.

Total number of genera	56
± cosmopolitan species	12
± pantropical species	17
Endemic:	
Species specific for the Caribbean r.	56
Species specific for Belize	11
Species specific for Everglades	6
Common species:	
coccoid	24
filamentous	20
heterocytous	7

	Generic name	Belize		Everglades		Common species	Endemic species
		Sp. No	End.	Sp. No.	End.		
Coccoid (21 genera, 44 species)	Aphanocapsa	3		2		intertexta, venezuelae	2
	Aphanothece	7		4		bacilloidea, granulosa, hardersii, variabilis	3
	Anathece	1		1		bacteriosa	1
	Asterocapsa	4	1	3		belizensis, nidulans, sp.	3
	Chlorogloea	2		2		gardneri, gessneri	2
	Chroococcidiopsis	1	1	-			1
	Chroococcus	7		8	1?	major, mediocris, cf. minutus, mipitanensis, occidentalis, pulcherimus	3
	Coelomoron	-		1			
	Coelosphaerium	-		1			
	Cyanodictyon	1		-			
	Cyanobacterium	-		1			
	Cyanokybus	-		1			1
	Cyanosarcina	1		1		sp.	1?
	Eucapsis	1		1	1?		1?
	Gloeothece	4	1	3		interspersa, opalothecata	3
	Gomphosphaeria	1		1		semen-vitis	1
	Johannesbaptistia	1		1		pellucida	
	Lemmermanniella	1		1		uliginosa	1
	Merismopedia	1		-			
	Onkonema	1	1	-			1?
	Rhabdogloea	1		-			1
		38	4	32	2(?)		22+3?

Table 2 contd....

Table 2 contd....

	Generic name	Belize		Everglades		Common species	Endemic species
		Sp. No	End.	Sp. No.	End.		
Filamentous (16 genera, 39 species)	Arthrospira	-		2	1		1
	Geitlerinema	3	2	4	1(?)	splendidum	
	Komvophoron	2	1	2		apiculatum	2
	Leibleinia	1		1		epiphytica	
	Leptolyngbya	5	1	4		angustissima, eliskae, mucosa, perelegans	3
	Limnothrix	1		1		unigranulata	
	Lyngbya	5	1	4		intermedia, martensiana, minor, splendens	2
	Oscillatoria	2		1			
	Phormidesmis	1		1		molle	
	Phormidium	3	1	4		granulatum, tergestinum	2
	Planktothrix	1	1(?)	-			1(?)
	Pseudanabaena	3		4		apiculato-flexuosa, belizensis, cf. galeata	2
	Schizothrix	2		1		sp.	1
	Spirulina	2		2		major, tenerrima	
	Symplocastrum	1		-		sp.	1
	Trichocoleus	1		-			
		33	6	31	1		14+1?
Heterocytous (19 genera, 37 species)	Anabaena	-		3			2
	Aulosira	1	1	1	1		2(?)
	Calothrix	-		3	(2?)		2(?)
	Chakia	1		1		ciliosa	1
	Cylindrospermum	2		1		breve	
	Dichothrix	-		1			1(?)
	Dictyophoron	1	1	-			1
	Fischerella	1		-			
	Fortiea	2		1			1
	Gloeotrichia	2		-			2
	Hapalosiphon	2		-			
	Hassallia	1	1	-			1
	Microchaete	2		-			2
	Nodularia	1		1			2(?)
	Nostoc	2		4			?
	Scytonema	2	1	1		belizensis	2
	Stigonema	1		-			1
	Tolypothrix	2		2		caribica, willei	2
	Trichormus	1		3		luteus	1
		24	4	22	1		16+7?

Asterocapsa spp. and others. Few genera are very diverse (*Aphanothece, Chroococcus*). Of these, *Ch. mipitanensis* and *Ch. pulcherrimus* belong to pantropical species, while *Ch. major, Ch. occidentalis* or *Ch. subsphaericus* are endemic for Belize or the alkaline marshes of Central America (Fig. 5).

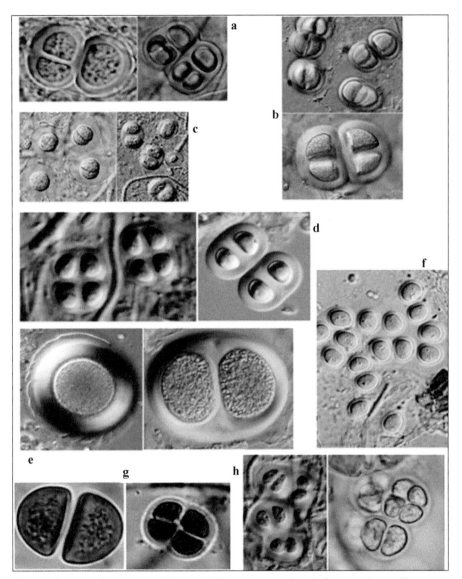

Figure 5. Examples of morphological diversity of *Chroococcus* types from Belizean marshes: a. *Chroococcus maior,* b. *C. mediocris,* c. *C. minutus,* d. *C. mipitanensis,* e. *C. occidentalis* f. *C. cf. occidentalis,* g. *C. pulcherrimus,* h. *C. subsphaericus.* From Komárek et al. 2005.

Color image of this figure appears in the color plate section at the end of the book.

Filamentous Types

Filamentous cyanobacteria without heterocytes have been recently classified in two distant orders (Pseudanabaenales, Oscillatoriales). Members of the genera *Leptolyngbya (L. mucosa, L. eliskae), Pseudanabaena*, and *Lyngbya* form the main biomass of the benthic mats (Fig. 2). They provide a suitable substrate for many other species, particularly from coccoid and heterocytous cyanoprokaryotes. While coccoid species are irregularly distributed in the clusters of filaments, heterocytous species usually colonize the floating mats only at the surface, although occasionally they can be found inside of other colonies. The simple filamentous types produce the main cyanobacterial biomass in the ecosystem. Endemic species contain several important dominants or very characteristic species such as *Geitlerinema serpens, Pseudanabaena apiculato-flexuosa* or *Komvophoron apiculatum. Geitlerinema splendidum, Lyngbya* cf. *martensiana, Spirulina tenerrima* are representatives of the cosmopolitan and widely distributed types. *Oscillatoria jenensis* or *Phormidesmis molle* represent the pantropical species. About 1/3 of species from about 30 taxa of this group belong to characteristic species occurring exclusively in alkaline habitats of the Caribbean marshes, several as endemic (Fig. 6).

Figure 6. Phylogenetic tree containing pseudanabaenacean and oscillatorialean strains isolated from alkaline marshes in northern Belize (in bold). (After Turicchia et al. 2009).

Heterocytous Types

The most morphologically complicated, multicellular cyanobacterial taxa with heterocytes and akinetes are not present in large amounts and also the number of species is distinctly lower compared to the more physiologically simple groups of cyanobacteria. However, their presence is very characteristic for distinct periods of CBM development and the group contains highly specific, probably endemic forms. They tend to colonize only distinct regions of CBM or certain marsh microhabitats and can form a dense biomass locally (*Aulosira doliispora, Nodularia aggregata, Scytonema belizensis, Tolypothrix willei*). Among these approximately 25 characteristic species none is known to have cosmopolitan distribution. *Cylindrospermum bourrellyi, C. breve, Fortiea bossei, F. monilispora, Hapalosiphon arboreus, H. luteolus* belong to pantropical species. Several commonly occurring species (*Tolypothrix willei, Trichormus luteus*) are known only from alkaline marshes of the Caribbean and numerous very interesting and distinct species must be considered as endemic (*Aulosira doliispora, Chakia ciliosa, Gloeotrichia aurantiaca, Hassallia maya, Scytonema belizensis* or *Stigonema eliskae*). As examples, we present two of these characteristic types: *Stigonema eliskae* and *Chakia ciliosa* (Fig. 7).

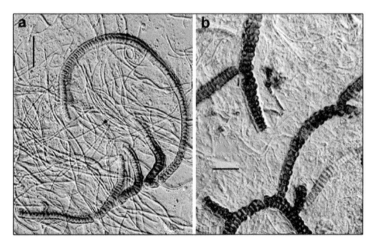

Figure 7. Examples of two heterocytous, probably endemic species from cyanobacterial mats from marshes of N Belize, a. *Chakia ciliosa* (with numerous thin filaments of *Leptolyngbya*), b. *Stigonema eliskae*; bars = 50 μm.

Color image of this figure appears in the color plate section at the end of the book.

There are not many data to compare the richness of Belizean cyanobacterial communities with biologically related or geographically close areas. The diversity of the cyanobacterial microflora of Cuba, Everglades, or Jamaica has not been studied in detail; certain similarities are known from calcareous locations in Puerto Rico, and the cenotes of the Yucatan (Gardner 1927). The extensive marshes of eastern Nicaragua (Mosquitia) are superficially similar, but more acidic with a dramatically different cyanobacterial species composition (Komárek, unpublished data). Alkaline marshes of

Caribbean, particularly those of northern Belize therefore represent a unique, species rich habitat that deserves protection and further studies.

Important Mat Biogeochemical Processes

The biogeochemical functioning of CBM is relatively well described from the Everglades (Browder et al. 1994; Vymazal and Richardson 1995; McCormick and O'Dell 1996; McCormick et al. 1996; 1998; Gottlieb et al. 2006; Chick et al. 2008; Geiser et al. 2011; Hagerthey et al. 2011; and others), but less is known about other places in the Caribbean, including Belize, Cuba, Mexico, and Jamaica (Vargas and Novelo 2006; Falcon et al. 2007; Gaiser et al. 2010; Beltran et al. 2012). In Belize, seasonal changes in mat development, together with temporal and spatial fluctuations in physico-chemical parameters determine the dynamics of biogeochemical processes. The mats play a crucial role in cycling of nitrogen, N, and phosphorus, P. The average mat N and P concentration in Belizean marshes is 11.5 (+/–0.7 SD) and 0.1 (+/–0.06) mg g^{-1} DW respectively, which corresponds to the other oligotrophic marshes in the Caribbean region (Grimshaw et al. 1993; McCormick et al. 1998; 2001; Scinto and Reddy 2003). In terms of stoichiometric ratios, the mats seem clearly P limited with molar N/P and C/P ratios 212 and 5240, respectively (see Hillebrand and Sommer 1999, and their suggested ratios of N/P > 22 and C/P > 180 indicating P limitation). Although the surrounding environment (water and sediments) are strongly P limited, the mat environment itself is capable of sequestering enough P to support significant N fixation.

Carbon

Despite the harsh conditions with regards to nutrient availability and hydrology, cyanobacterial mats grow to constitute substantial proportion of biomass in the ecosystem. The primary productivity determined by ^{14}C dark and light bottle technique varies according to location and season, usually between 1.5 and 7.5 μg C cm^{-2} h^{-1} (Rejmánková and Komárková 2000). The mat biomass at studied locations typically ranges between 400 and 800 g m^{-2} AFDW. (Note, local dense patches of macrophytes may occurr, so these values do not apply to the entire area of the marsh). The ash proportion correlates well with the increasing water conductivity up to about 2500 μS, and then it levels off and stays uniformly high at about 65%. Biomass of the mats (Fig. 8) is often higher than that of the sparse community of the most common emergent macrophyte, *Eleocharis cellulosa* (mean aboveground live biomass 120 g m^{-2}), and about equal to the higher limit of biomass range of the more robust macrophytes such as *Typha domingensis* and *Cladium jamaicense* (Rejmánková et al. 1996). These numbers are comparable to those reported for the Everglades (McCormick et al. 1998; Hagherthey et al. 2010); the seasonal patterns, however, are clearly different, as the highest biomass of Belizean mats is observed during the dry season.

Total carbon (TC) content of CBM typically ranges between 220–280 mg g^{-1} DW (mean 244 ± 43 SD), 30–40% of which is carbonate C. Carbon mineralization at night is completely anaerobic in the epipelon and even inside the floating mats (night redox measurements showed Eh values of < –150 mV; Rejmánková, unpublished data). The

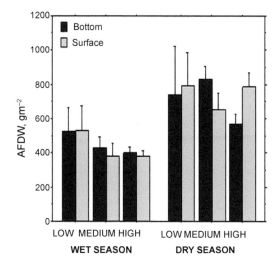

Figure 8. Biomass of bottom and floating CBM from the mid-wet season (August) and the end of the dry season (March) expressed as ash free dry weight (AFDW g m^{-2}). Each bar represents a mean of 5 marshes from low, medium and high salinity categories. Error bars indicate the standard error of mean.

main terminal electron acceptor is probably sulfate, especially in marshes with high sulfate concentration, although Fe^{3+} reduction and methanogenesis may contribute.

Phosphorus

It is clear that CBM are remarkably low in P, whose concentration often falls below 0.1 mg g^{-1} DW. But at the same time, they are capable of rapid growth from complete desiccation, following seasonal rewetting (D. Sirová, personal communication). How this is possible is one of the many questions regarding P cycling in CBM that cannot be answered without first gaining an understanding of the internal P forms and partitioning.

Although majority of the CBM volume is made up of pore water, over 90% of total P was shown to be associated with particulate matter (Borovec et al. 2010). In CBM, particulate matter is made up of cyanobacterial and bacterial cells, both living and dead, embedded in the matrix of extracellular polymeric substances, along with precipitated mineral ($CaCO_3$) crystals, and various detritus (Sirová et al. 2006). The process of P liberation from particles or complex organic molecules in CBM is mediated by extracellular enzymes, which break down high molecular weight compounds into smaller molecules available for uptake by cells (Sirová et al. 2006). The most important of these is alkaline phosphatase, which exhibited high activity at all studied locations. While visualizing its activity with enzyme-labeled fluorescence, we found that cyanobacteria, although dominant microorganisms, did not exhibit significant phosphatase activity. This lead us to presume that the main sources of the extracellular enzyme are different species of bacteria distributed in the mat and the mucilaginous sheaths of cyanobacteria. Moreover, most of the phosphatase activity appeared to be free activity, unbound to bacterial cells, but distributed in the matrix of extracellular polymeric substances which binds the CBM structure together. This may be the result of extracellular enzyme accumulation in the extracellular polymeric substance (EPS) matrix of the mat. It has been suggested that the matrix promotes accumulation of the more freely deployed enzymes by retarding losses (Lock et al. 1984), and several possible mechanisms that could affect a net accumulation (physical enmeshment, hindered diffusion, adsorption or cation bridging) have been proposed by Characklis and Marshall (1990). Extracellular enzymes are generally short-lived since they are susceptible to various forms of inactivation (Perez-Mateos et al. 1991). Mat EPS matrix provides photochemical protection of enzymes located within it, thereby increasing their longevity. Resources needed for enzyme synthesis, P for example, can thus be effectively conserved and directed into other processes such as microbial growth or nitrogen fixation. The EPS matrix seems to be able to conserve on average as much as 25% of the enzymatic activity in CBM samples that were completely sun-dried for a period of three months. It is possible that enzyme activity retained in dessicated samples contributes to the extremely high regeneration rates in CBM following flooding (Sirová et al. 2006).

While virtually no exchange of P with the water column was detected, the partitioning of different P forms within CBM clearly undergoes profound changes, following the large diurnal and vertical fluctuations of light, dissolved oxygen (DO) concentration, and pH on a microscopic scale (Fig. 9) (Borovec et al. 2010). Upon vertical cross-section, the CBM show three visually distinct zones (Fig. 2F) which differ in color (concentration of the dominant pigments such as cyanobacterial scytonemin, chlorophyll, and bacterial carotenoids) as well as in their physico-chemical characteristics. The bottom-most "red" layer is largely inert in terms of physicochemical fluctuations, possibly because it was found to contain many bacterial resting cysts (*Rhodocista* sp., unpublished data) and would therefore have low biological activity.

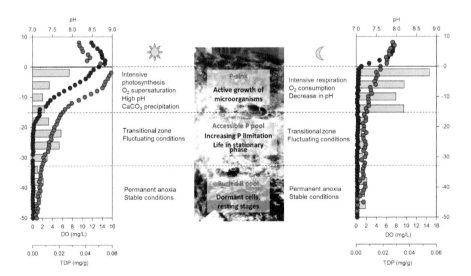

Figure 9. Diurnal changes in selected physico-chemical parameters of studied CBM at 5:00 pm (day) and 5:00 am (night).

Color image of this figure appears in the color plate section at the end of the book.

The top 0.5–1 cm of the mat, which includes the top "yellow" and middle "green" layers with photosynthetically active cyanobacteria, is supersaturated by oxygen during the day, when pH values rise above 9. During the night, the entire mat profile turns anoxic and pH values fall to the circum-neutral range. Such sharp changes in the top two layers have a profound effect on the redox-labile P, its forms as well as distribution. It has been commonly assumed that conditions in a photosynthetically active environment (high pH, high dissolved mineral content, and low CO_2 partial pressure) favor the co-precipitation of P with $CaCO_3$ or its adsorption onto $CaCO_3$ crystals (Otsuki and Wetzel 1972), thus making P less available for biotic uptake. We therefore expected the largest fraction of P to be associated with $CaCO_3$. This assumption, however, proved correct only for the bottom-most red layer, which experienced condition of permanent anoxia, stable pH and contained dormant bacterial cysts and cyanobacterial resting stages (D. Sirová, unpublished data). The most significant proportion of TP in the metabolically active layers of the mat was the exchangeable and loosely bound P (up to 52%, 55 µg g^{-1} DW), which we consider to be mainly associated with the EPS, and which is much more readily available both to the extracellular phosphatases and for uptake by microbial cells.

The night time decrease in pH seems to liberate significant amounts of exchangeable and loosely bound P, which moves, in the soluble reactive form, from the green to the yellow layer and is thus available for microbial uptake. Despite the large C:P ratios measured, this relatively large pool of readily available P in the metabolically active layers of CBM lead us to believe that the microorganisms may be primarily limited by other factors such as light availability rather than by P scarcity. The mats may possibly increase the limitation in macrophyte species to a certain extent, because they deplete the scarce P supply from the water column and prevent organic matter from reaching the sediment. However, the tight internal P cycling and dynamics seem to be one of the key structuring forces creating numerous ecological niches separated in both space and time that accommodate the large diversity of cyanobacterial species and render CBM highly susceptible to eutrophication.

Nitrogen

While Belizean marshes are P limited, the tissue composition of the cyanobacterial mats does not show any signs of N limitation. Yet in many marshes the N content in water is quite low. The mats are capable of sustaining high N concentrations mainly through a process of nitrogen fixation and are capable of providing the biologically available nitrogen to other components of the system. Marshes need this input of "new" N, because N sources are limited to precipitation and N recycling, and significant proportion of N can be lost through denitrification. N fixation is demanding energetically, but it also requires high P concentration, which often determines nitrogenase activity in aquatic systems (Inglett et al. 2009). The general consensus is that the N fixation becomes limited once the soluble reactive P (SRP) levels drop to less than 10 µg L^{-1} (Diaz et al. 2007). Since the SRP concentration in the majority of marshes is very close to 10 µg L^{-1} (Table 1) the high N fixation was something of a paradox until we learned about the very tight P recycling inside the mats (see above). Information on the amount of N contributed to the marsh ecosystems through N fixation by CBM is important for estimates of nutrient budgets and the understanding of ecosystem processes.

N fixation is related to spatial and temporal differences in mat taxonomic composition, namely the abundance of the heterocytous cyanobacteria. In our survey of over 20 marshes, the N fixation ranged from 0.52 to 4.13 nmol N_2 cm^{-2} h^{-1}, averaging 1.8 nmol N_2 cm^{-2} h^{-1} ($\pm$ 1.04 SD) (Note, the ratio of 4.31 based on acetylene reduction assay calibrated by $^{15}N_2$ reduction assay was used to convert moles of C_2H_4 to N_2 according to Šantrůčková et al. 2010). Differences in N fixation were best explained by the proportion of heterocyte-forming *Scytonema* ($R^2 = 0.699$; $P = 0.0001$). Nitrogen fixation was positively correlated with SRP ($R^2 = 0.46$; $P = 0.008$) and negatively with the concentration of inorganic N ($R^2 = 0.26$; $P = 0.05$).

The above values apply to floating mats. The N fixation of epipelon is generally lower averaging 0.53 nmol N_2 cm^{-2} h^{-1}. In our initial measurements of diurnal course of N fixation we did not detect any nitrogenase activity at night (Rejmánková and Komárková 2000). In subsequent measurements, we found dark N fixation from several marshes averaging 0.20 nmol N_2 cm^{-2} h^{-1} (Rejmánková, unpublished data). The dark N fixation was apparently carried out by non-heterocytous species of cyanobacteria (*Oscillatoria, Leptolyngbya*) or, possibly, heterotrophic bacteria.

The mesocosm experiment (Rejmánková and Komárková 2005) supported the data from the field. Nitrogen fixation changed following the shifts in functional groups of mat assemblages. It was most strongly negatively dependent on the level of N, but the effect of P was also very significant, i.e., N_2-fixation increased with increasing P especially at the end of the experiment. Nitrogen fixation reached up to 17.4 and 40.6 nmol N_2 cm^{-2} h^{-1}, at low nitrogen and medium or high P respectively. This high N_2-fixation coincided with a large quantity of cyanobacteria from the heterocytous genus *Nostoc* found in these treatments (Fig. 11).

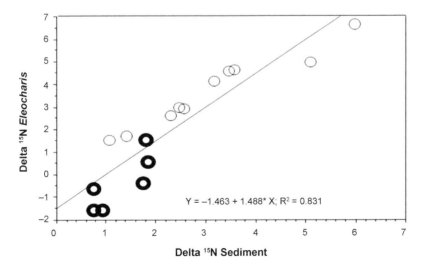

Figure 10. Delta ^{15}N of plant tissue, *Eleocharis cellulosa*, correlated with delta ^{15}N of the sediments. The black dots indicate marshes with dense cover of cyanobacteria. Sediment P availability in the selected marshes was comparable.

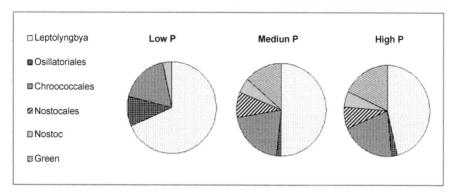

Figure 11. The relative proportions of the species groups in low, medium and high P treatments at the end of the experiment.

Nitrogen fixation by cyanobacteria is accompanied by relatively little isotopic fractionation and as a result, N-fixing cyanobacteria should have delta ^{15}N close to zero (Handley et al. 1998). The delta ^{15}N signatures of CBM have been shown to have a strong negative correlation with N-fixation (Rejmánková et al. 2004a; Inglett et al. 2009; Wozniak et al. 2012). There is little doubt that N-fixing cyanobacteria provide biologically available N (NH_4) for various components of mat communities, the question remains if this newly fixed N can also be used by the non-N-fixing vascular plants. Our correlative data (Fig. 11) indicate a close relationship between sediment and plant tissue delta [15]N (Rejmánková, unpublished data). However, experimental verification is needed because the natural abundance of ^{15}N in plant tissue is regarded as reliable but not precise indication of plants using N originating from N-fixation (Hopkins et al. 1998; Inglett et al. 2007).

Sensitivity to Eutrophication

Resource availability is one of the dominant structuring forces of communities and ecosystems (Goddard and Bradford 2003). Among examples of changes in resource availability is the accelerated land use change in the tropics resulting in increased nutrient loading to aquatic ecosystems (Downing et al. 1999). Increased nutrient input is resulting in species composition changes (Herbert et al. 1999) and shifts in species composition are known to change ecosystem functions such as primary production and nutrient cycling (Elser et al. 2000). Oligotrophic tropical and subtropical wetlands are among the ecosystems most vulnerable to impacts of increased nutrient input.

In the oligotrophic, P-limited marshes, the increased input of nutrients, especially P, profoundly changes the species composition and functions of cyanobacterial mats, often leading to their total elimination. This has been documented from the Everglades and, on a smaller scale, also from Belize. Throughout much of the Everglades, cyanobacterial communities have been either reduced or even completely eliminated due to exposure to high phosphorus loads and, in some areas, replaced by dense cattails (McCormick and O'Dell 1996; McCormick et al. 1996; Richardson et al. 2008). The majority of Belizean marshes is still in a relatively pristine state although nutrient enrichment and cyanobacteria elimination is apparent in areas of increased nutrient loading (marshes adjacent to agricultural fields and/or discharge of waste water) (Johnson and Rejmánková 2005). The response of CBM to changes in nutrient loading has been documented in Belize from the long term nutrient addition experiment (Rejmánková et al. 2008) and from a mesocosm experiment (Rejmánková and Komárková 2005).

Marsh Nutrient Enrichment Plots

The richness of morphospecies of cyanobacteria (Chroococcales, Oscillatoriales, Nostocales) greatly decreased in P-addition plots. At the same time, the common species of diatoms (*Cymbella*, *Gomphonema*, *Navicula*), green algae (Chlorococcales, Tetrasporales), filamentous green algae (*Oedogonium*, *Mougeotia*, *Spirogyra*) and desmids (*Cosmarium*, *Gonatozygon*, *Staurastrum*), i.e., indicators of more eutrophic conditions, appeared already 6 months after nutrient application. By the 3rd year of the project, CB mats from +P and +N&P plots have been almost completely eliminated, apparently due to large increase of macrophyte density and consequent shading of CB mats.

Mesocosms

The response of cyanobacterial mats to combinations of low, medium and high P, N, and salinity without the shading effect of macrophytes was evaluated in the mesocosm experiment (Rejmánková and Komárková 2005). Increasing concentrations of P and N correlated with the increased abundance of green algae and Chroococcales (Fig. 11), and high primary production. By the end of the 2-months experiment, we saw a clear differentiation: while *Leptolyngbya* was still the most abundant group followed by Chroococcales and Nostocales, a group of *Nostoc* appeared especially in medium and high P treatment. But the most dramatic was the increase in green algae, mostly Chlorococcales and Tetrasporales, some filamentous green algae (*Oedogonium, Mougeotia, Spirogyra, Bulbochaete*) and desmids (*Cosmarium, Gonatozygon, Staurastrum*, etc.), i.e., indicators of more eutrophic conditions. The proportion of diatoms was extremely low, less than 0.5% and by the end of the experiments, diatoms were almost undetectable.

Nitrogen fixation was negatively correlated with N concentration and positively correlated with the abundance of a group of heterocytous cyanobacteria from the genus *Nostoc*. At low N, increasing P concentrations supported higher N-fixation. The activity of alkaline phosphatase was negatively correlated with P and salinity, and positively with N. Phosphorus was an extremely important factor that had the potential to influence the structure (species composition) and function of CB mats. Nitrogen additions played less important role, while salinity exerted a significant effect on only a few processes. Even during the short duration of this experiment, we saw a decline in the richness of cyanobacterial morphospecies similar to what has been observed in the long term field experiment. At high P concentration, the mats were deteriorating as a consequence of heavier grazing.

Elimination of species capable of N-fixation can also have serious ecosystem consequences. Cyanobacterial mats in Belizean marshes contribute significant amount of N into the system. This is close to values reported for the unimpacted parts of the Everglades, i.e., in the range of 5–10 g N m^{-2} y^{-1} (Rejmánková and Komárková 2000; Inglett et al. 2004). Unless other types of N-fixers would replace cyanobacteria in contributing N into the system, N could eventually become limiting.

Is Oligotrophication Possible?

Cyanobacterial mats respond quickly (days to weeks) to changes in environmental conditions at a large spatial scale (meters to tens of kilometers; Gaiser et al. 2004; Rejmánková et al. 2008) and provide a good early indicator of ecosystem change. Of various metrics tested, the TP concentration in CBM tissue has been identified as one of the best measures of P load history (McCormick and Stevenson 1998; Gaiser et al. 2006; Gaiser 2009). Although increases in water and soil P are only detectable after years of enhanced P loading, effects upon periphyton TP concentration are immediate (Gaiser et al. 2004). The question is whether CBM can recover when P loading stops.

So far, no long term observations exist to provide reliable predictions for time frames for the recovery of CBM once they are degraded or eliminated from an area. However, there is anecdotal support for a relatively rapid recovery of cyanobacterial calcareous mats after the completion of the P dosing experiment in Everglades National Park (Gaiser, personal communication). After a complete desintegration of the calcareous mats following dosing of P above-ambient levels for 5 years, the CBM seemingly recovered about a year after dosing was terminated. CBM re-appeared quickly where cattails were not present, but the recovery was never fully documented. A necessary prerequisite for CBM regeneration in the areas that became dominated by tall dense macrophyte in response to increased P loading is to reduce macrophyte cover and decrease the internal loading from the sediments. This process will almost certainly require significant human intervention. As documented from the long-term nutrient addition plots in Belizean marshes, 8 years after the last P addition, the +P and +N&P plots are still dominated by *Typha domingensis* and/or dense *Eleocharis* spp. and nutrients are internally recycled. The Everglades experience confirmed that the restoration of P-impacted regions requires not only a reduction in P loads, but also active management efforts to reduce the resilience and resistance of *Typha* (Newman et al. 2013).

To conclude: Knowledge of species composition and ecosystem functions of cyanobacterial mats and understanding the consequences of their loss but even more importantly, the conditions of their regeneration, are necessary for management of these systems. The ecologically unique cyanobacterial microflora is very sensitive to nutrient enrichment, responding to eutrophication both directly and indirectly through shading by the expanding macrophytes. There is a whole suite of organisms dependent on these spatial marsh habitats as well as the CBM that grow within them. There are numerous interesting topics for future research, of which the most interesting and desirable should focus on studying if and under what conditions can once degraded marsh ecosystems be restored to the its original state dominated by diverse cyanobacterial communities.

Acknowledgments

Numerous students participated in various parts of research referred to in this chapter. Theirs, and field and lab technicians help is greatly appreciated. Parts of the research were supported by the NIH-NSF Grant # R01 AI49726 and NSF # 0089211.

References

Alcala-Herrera, J.A., J.S. Jacob and M.L. Machain Castillo. 1994. Holocene paleosalinity in a Maya wetland, Belize, inferred from the microfaunal assemblage. Quaternary Res. 41: 121–130.

Beltran, Y., C.M. Centeno, F. García-Oliva, P. Legendre and L.I. Falcón. 2012. N_2 fixation rates and associated diversity (nifH) of microbialite and mat-forming consortia from different aquatic environments in Mexico. Aquat. Microb. Ecol. 67: 15–24.

Borhidi, A. 1991. Phytogeography and Vegetation Ecology of Cuba. Akademia Kiado, Budapest.

Borovec, J., D. Sirová, P. Mošnerová, E. Rejmánková and J. Vrba. 2010. Spatial and temporal changes in phosphorus partitioning within a freshwater cyanobacterial mat community. Biogeochemistry 101: 323–333.

Browder, J.A., P.J. Gleason and D.R. Swift. 1994. Periphyton in the Everglades: spatial variation, environmental correlates, and ecological implications. *In:* S.M. Davis and J.C. Ogden (eds.). Everglades: The Ecosystem and its Restoration. St. Lucie Press, Delray Beach. pp. 379–417.

Characklis, W.G. and K.C. Marshall. 1990. Biofilms: a basis for an interdisciplinary approach. pp. 3–15. *In:* W.G. Characklis and K.C. Marshall (eds.). Biofilms. John Wiley and Sons, New York.

Chiappy-Jhones, C., V. Rico-Gray, L. Gama and L. Giddings. 2001. Floristic affinities between the Yucatan Peninsula and some karstic areas of Cuba. J. Biogeogr. 28: 535–542.

Chick, J.H., P. Geddes and J.C. Trexler. 2008. Periphyton mat structure mediates trophic interactions in a subtropical marsh. Wetlands 28: 378–389.

Diaz, M., F. Pedrozoa, C. Reynolds and P. Temporetti. 2007. Chemical composition and the nitrogen-regulated trophic state of Patagonian lakes. Limnologica 37: 17–27.

Downing, J.A., M. McClain, R. Twilley, J. Melack, J. Elser and N.N. Rabalais. 1999. The impact of accelerating land-use change on the N-cycle of tropical aquatic ecosystems: current conditions and projected changes. Biogeochemistry 46: 109–148.

Elser, J.J., R.W. Sterner, A.E. Galford, T.H. Chrzanowski, D.F. Findlay, K.H. Mills, M.J. Paterson, M.P. Stainton and D.W. Schindler. 2000. Pelagic C:N:P stoichiometry in a eutrophied lake: responses to a whole-lake food-web manipulation. Ecosystems 3: 293–307.

Estrada-Loera, E. 1991. Phytogeographic relationships of the Yucatan Peninsula. J. Biogeogr. 18: 687–697.

Falcón, L.I., R. Cerritos, L.E. Eguiarte and V. Souza. 2007. Nitrogen fixation in microbial mat and stromatolite communities from Cuatro Cienegas, Mexico. Microb. Ecol. 54: 363–373.

Fedick, S.L. and B.A. Morrison. 2003. Ancient use and manipulation of landscape in the Yalahau region of the northern Maya lowlands. Agr. Hum. Values 21: 207–219.

Gaiser, E. 2009. Periphyton as an indicator of restoration in the Florida Everglades. Ecol. Indic. 9: 37–45.

Gaiser, E., J.M. La Hee, F.A.C. Tobias and A.H. Wachnicka. 2010. *Mastogloia smithii* var *lacustris* Grun.: a structural engineer of calcareous mats in karstic subtropical wetlands. PANS 160: 99–112.

Gaiser, E., P.V. McCormick, S.E. Hagerthey and A.D. Gottlieb. 2011. Landscape patterns of periphyton in the Florida Everglades. Crit. Rev. Env. Sci. Tech. 41(S1): 92–120.

Gaiser, E.E., L.J. Scinto, J.H. Richards, K. Jayachandran, D.L. Childers, J.C. Trexler and R.D. Jones. 2004. Phosphorus in periphyton mats provides the best metric for detecting low-level P enrichment in an oligotrophic wetland. Water Res. 38: 507–516.

Gaiser, E.E., J.H. Richards, J.C. Trexler, R.D. Jones and D.L. Childers. 2006. Periphyton responses to eutrophication in the Florida Everglades: cross-system patterns of structural and compositional change. Limnol. Oceanogr. 51: 617–630.

Gardner, N.L. 1927. New Myxophyceae from Porto Rico. Mem. New York Bot. Garden 7: 1–144.

Goddard, M.R. and M.A. Bradford. 2003. The adaptive response of a natural microbial population to carbon and nitrogen limitation. Ecol. Lett. 6: 594–598.

Gottlieb, A.D., J.H. Richards and E.E. Geiser. 2006. Comparative study of periphyton community structure in long and short-hydroperiod Everglades marshes. Hydrobiologia 569: 195–207.

Grimshaw, H.J., M. Rosen, D.R. Swift, K. Rodberg and J.M. Noel. 1993. Marsh phosphorus concentrations, phosphorus content and species composition of Everglades periphyton communities. Arch. Hydrobiol. 128: 257–276.

Hagherthey, E.E., J.J. Cole and D. Kilbane. 2010. Aquatic metabolism in the Everglades: dominance of water column heterotrophy. Limnol. Oceanogr. 55: 653–666.

Hagherthey, S.E., B.J. Bellinger, K. Wheeler, M. Gantar and E. Gaiser. 2011. Everglades periphyton: a biogeochemical perspective. Crit. Rev. Env. Sci. Tech. 41(S1): 309–343.

Herbert, D.A., E.B. Rastetter, G.R. Shaver and G.I. Agren. 1999. Effects of plant growth characteristics on biogeochemistry and community composition in a changing climate. Ecosystems 2: 367–382.

Hillebrand, H. and U. Sommer. 1999. The nutrient stoichiometry of benthic microalgal growth: redfield proportions are optimal. Limnol. Oceanogr. 44: 440–446.

Hodell, D.A., M. Brenner and J.H. Curtis. 2007. Climate and cultural history of the Northeastern Yucatan Peninsula, Quintana Roo, Mexico. Clim. Chang. 83: 215–240.

Hopkins, D.W., R. E. Wheatley and D. Robinson. 1998. Stable isotope studies of soil nitrogen. pp. 75–88. *In*: H. Griffiths (ed.). Stable Isotopes. Integration of Biological, Ecological and Geochemical Processes. Bios Scientific Publishers, Oxford.

Inglett, P.W., K.R. Reddy and P.V. McCormick. 2004. Periphyton chemistry and nitrogenase activity in a northern Everglades ecosystem. Biogeochemistry 67: 213–233.

Inglett, P.W., K.R. Reddy, S. Newman and B. Lorenzen. 2007. Increased soil stable nitrogen isotopic ratio following phosphorus enrichment: historical patterns and tests of two hypotheses in a phosphorus-limited wetland. Oecologia 153: 99–109.

Inglett, P.W., E.M. D'Angelo, K.R. Reddy, P.V. McCormick and S.E. Hagerthey. 2009. Periphyton nitrogenase activity as an indicator of wetland eutrophication: spatial patterns and response to phosphorus dosing in a northern Everglades ecosystem. Wetl. Ecol. Manag. 17: 131–144.

Johnson, S. and E. Rejmánková. 2005. Impacts of land-use on nutrient distribution and vegetation composition of freshwater wetlands in northern Belize. Wetlands 25: 89–100.

Komárek, J. and J. Komárková-Legnerová. 2007. Taxonomic evaluation of cyanobacterial microflora from alkaline marshes of northern Belize. 1. Phenotypic diversity of coccoid morphotypes. Nova Hedwigia 84: 65–111.

Komárek, J., J. Komárková, E. Zapomělová and S.Ventura. Taxonomic evaluation of the cyanobacterial microflora from alkaline marshes of northern Belize. 3. Diversity of heterocytous genera. Nova Hedwigia (in prep.).

Komárek, J., S. Ventura, S. Turicchia, J. Komárková, C. Mascalchi and E. Soldati. 2005. Cyanobacterial diversity in alkaline marshes of northern Belize (Central America). Algolog. Stud. 117. Cyanobacterial Research 6: 265–278.

La Hée, J.M. and E.E. Gaiser. 2012. Benthic diatom assemblages as indicators of water quality in the Everglades and three tropical karstic wetlands. Fresh. Sci. 31: 205–221.

Lee, S.S., E.E. Gaiser and J.C. Trexler. 2013. Diatom-based models for inferring hydrology and periphyton abundance in a subtropical karstic wetland: Implications for ecosystem-scale bioassessment. Wetlands 33: 157–173.

Lock, M.A., R.R. Wallace, J.W. Costerton, R.M. Ventullo and S.E. Charlton. 1984. River epilithon—toward a structural-functional model. OIKOS 42: 10–22.

Macek, P. and E. Rejmánková. 2007. Response of emergent macrophytes to experimental nutrient and salinity additions. Funct. Ecol. 21: 478–488.

Mareš, J. 2010. *Anabaena fuscovaginata* (Nostocales), a new cyanobacterial species from periphyton of the freshwater alkaline marsh of Everglades, South Florida, USA. Fottea 10(2): 235–243.

McCormick, P.A. and R.J. Stevenson. 1998. Periphyton as a tool for ecological assessment and management in the Florida Everglades. J. Phycol. 4: 726–733.

McCormick, P.V. and M.B. O'Dell. 1996. Quantifying periphyton responses to phosphorus in the Florida Everglades: a synoptic-experimental approach. J. N. Amer. Benthol. Soc. 15: 450–68.

McCormick, P.V., R.S. Rawlick, K. Lurding, E.P. Smith and F.H. Sklar. 1996. Periphyton—water quality relationships along a nutrient gradient in the northern Florida Everglades. J. N. Amer. Benthol. Soc. 15: 433–449.

McCormick, P.V., R.B.E. Shuford, J.G. Backus and W.C. Kennedy. 1998. Spatial and seasonal patterns of periphyton biomass and productivity in the northern Everglades, Florida, U.S.A. Hydrobiologia 362: 185–208.

McCormick, P.V., M.B. O'Dell, R.B.E. Shuford, III, J.G. Backus and W.C. Kennedy. 2001. Periphyton responses to experimental phosphorus enrichment in a subtropical wetland. Aquat. Bot. 71: 119–139.

Morrison, B.A. and R. Cózatl Manzano. 2003. Initial evidence for use of periphyton as an agricultural fertilizer by the ancient Maya associated with the El Edén wetland, northern Quintana Roo, Mexico. pp. 401–413. *In*: A. Gómez-Pompa, M.F. Allen, S.L. Fedick and J.J. Jiménez-Osornio (eds.). The lowland Maya Area: Three Millennia at the Human-wildland Interface. Haworth Press, Binghamton, New York.

Newman, S., R. LeRoy, M. Manna, M. Cook, C. Coronado-Molina, M. Ross, C. Maddden, T. Troxler, S. Kelly, R. Bennett, D. Black and J.P. Sah. 2013. Ecosystem Ecology. South Florida Environmental report. pp. 6–37.

Novelo, E. and R. Tavera. 2003. The role of periphyton in the regulation and supply of nutrients in a wetland at El Eden, Quintana Roo. pp. 217–239. *In*: A. Gomez-Pompa, M.F. Allen, S.L. Fedick and J.J. Jimenez-Osornio (eds.). Lowland Maya Area: Three Millennia at the Human–Wildland Interface. The Haworth Press, Binghamton, New York.

Otsuki, A. and R.G. Wetzel. 1972. Co-precipitation of phosphate with carbonates in a marl lake. Limnol. Oceanogr. 17: 763–767.

Paerl, H.W., M. Fitzpatrick and B.M. Bebout. 1996. Seasonal nitrogen fixation dynamics in a marine microbial mat: Potential roles of cyanobacteria and microheterotrophs. Limnol. Oceanogr. 41: 419–427.

Paerl, H.W., J.L. Pinckney and T.F. Steppe. 2000. Cyanobacterial-bacterial mat consortia: examining the functional unit of microbial survival and growth in extreme environments. Environ. Microbiol. 2: 11–26.

Paerl, H.W., J. Dyble, L. Twomey, J.L. Pinckney, J. Nelson and L. Kerkhof. 2002. Characterizing man-made and natural modifications of microbial diversity and activity in coastal ecosystems. Anton Leeuw. Int. J. G. 81: 487–507.

Palacios-Mayorga, S., A.N. Anaya, E. Gonzales-Velazquez, L. Huerta-Arcos and A. Gomez-Pompa. 2003. Periphyton as a potential biofertilizer in intensive agriculture of the ancient Maya. pp. 389–400. *In*: A. Gomez-Pompa, M.F. Allen, S.L. Fedick and J.J. Jimenez-Osornio (eds.). Lowland Maya Area: Three Millennia at the Human–Wildland Interface. The Haworth Press, Binghamton, New York.

Perez-Mateos, M., M.D. Busto and J.C. Rad. 1991. Stability and properties of alkaline phosphatase immobilized by soil particles. J. Sc. Food Agr. 55: 229–240.

Pohl, M.D. 1990. Ancient Maya Wetland Agriculture: Excavations on Albion Island, Northern Belize. Westview Press, Boulder, CO.

Pohl, M.D., K.O. Pope, J.G. Jones, J.S. Jacob, D.R. Piperno, S.D. deFrance, D.L. Lentz, J.A. Gifford, M.E. Danforth and J.K. Josserand. 1996. Early agriculture in the Maya Lowlands. Lat. Am. Antiq. 74: 355–72.

Pope, K.O., B.H. Dahlin. 1989. Ancient Maya wetland agriculture: new insights from ecological and remote sensing. J. Field Archaeol. 16: 87–106.

Pope, K.O., M.D. Pohl and J.S. Jacob. 1996. Formation of ancient Maya wetland fields: natural and anthropogenic processes. pp. 165–176. *In*: S.L. Fedick (ed.). The Managed Mosaic: Ancient Maya Agriculture and Resource Use. University of Utah Press, Salt Lake City.

Rejmánková, E. and J. Komárková. 2000. Function of cyanobacterial mats in phosphorus-limited tropical wetlands. Hydrobiologia 431: 135–153.

Rejmánková, E. and J. Komárková. 2005. Response of cyanobacterial mats to nutrient and salinity changes. Aquat. Bot. 83: 87–107.

Rejmánková, E., K.O. Pope, R. Post and E. Maltby. 1996. Herbaceous wetlands of the Yucatan peninsula: communities at extreme ends of environmental gradients. Int. Rev. Ges. Hydrobio. 81: 223–252.

Rejmánková, E., J. Komárková and M. Rejmánek. 2004a. $\delta^{15}N$ as an indicator of N_2-fixation by cyanobacterial mats in tropical marshes. Biogeochemistry 67: 353–368.

Rejmánková, E., J. Komárek and J. Komárková. 2004b. Cyanobacteria—a neglected component of biodiversity: patterns of species diversity in inland marshes of northern Belize (Central America). Divers. Distrib. 10: 189–199.

Rejmánková, E., P. Macek and K. Epps. 2008. Wetland ecosystem changes after three years of phosphorus addition. Wetlands 28: 914–927.

Richardson, C.J. 2008. The Everglades Experiments: Lessons for Ecosystem Restoration. Springer-Verlag, New York, USA.

Šantrůčková, H., E. Rejmánková, B. Pivničková and J.M. Snyder. 2010. Nutrient enrichment in tropical wetlands: shifts from autotrophic to heterotrophic nitrogen fixation. Biogeochemistry 101: 295–310.

Sargeant, B.L., E.E. Gaiser, C. Joel and J.C. Trexler. 2010. Biotic and abiotic determinants of intermediate-consumer trophic diversity in the Florida everglades. Marine Freshw. Res. 61: 11–22.

Schiller, J. 1956. Die Mikroflora der roten Tümpel auf den Koralleninseln Los Aves "im Karibischen Meer. Ergebn. Deutsh. Limnol. Venezuela-Exped. 197–216.

Scinto, L.J. and K.R. Reddy. 2003. Biotic and abiotic uptake of phosphorus by periphyton in a subtropical freshwater wetland. Aquat. Bot. 77: 203–222.

Siemens, A.H. 1978. Karst and the pre-hispanic Maya in the southern low-lands. pp. 117–143. *In*: P.D. Harrison and B.L. Turner II (eds.). Pre-Hispanic Maya Agriculture. University of New Mexico Press, Albuquerque.

Sirová, D., J. Vrba and E. Rejmánková. 2006. Extracellular enzyme activities in benthic cyanobacterial mats: comparison between nutrient enriched and control sites in marshes of northern Belize. Aquat. Microb. Ecol. 44: 11–20.

Stal, L.J. 2000. Cyanobacterial mats and stromatolites. pp. 61–120. *In*: B.A. Whitton and M. Potts (eds.). The Ecology of Cyanobacteria. Kluver Academic Publishers, Dordrecht, The Netherlands.

Turicchia, S., S. Ventura, J. Komárková and J. Komárek. 2009. Taxonomic evaluation of cyanobacterial microflora from alkaline marshes of northern Belize. 2. Diversity of oscillatorialean genera. Nova Hedwigia 89: 165–200.

Vargas, R. and E. Novelo. 2006. Seasonal changes in periphyton nitrogen fixation in a protected tropical wetland. Biol. Fert. Soils 43: 367–372.

Vymazal, J. and C.J. Richardson. 1995. Species composition, biomass, and nutrient content of periphyton in the Florida Everglades. J. Phycol. 31: 343–354.

Weidie, A.E. 1985. Geology of the Yucatan platform. pp. 1–19. *In*: W.C. Ward, A.E. Weidie and W. Back (eds.). Geology and Hydrology of the Yucatan and Quternary Geology of Northeastern Yucatan Peninsula. New Orleans Geological Society.

Wozniak, J.R., W.T. Anderson, D.L. Childers, E.E. Gaiser, C.J. Madden and D.T. Rudnick. 2012. Potential N processing by southern Everglades freshwater marshes: are Everglades marshes passive conduits for nitrogen? Estuar. Coast. Shelf Sci. 96: 60–68.

Wright, A.C.S., D.H. Romney, R.H. Arbuckle and V.E. Romney. 1959. Land in British Honduras. Her Majesty's Stationery Office, London.

11

Biological Indicators of Changes in Water Quality and Habitats of the Coastal and Estuarine Areas of the Greater Everglades Ecosystem

Anna H. Wachnicka[1,]* and *G. Lynn Wingard*[2]

Introduction

Shallow estuarine and coastal ecosystems are home to a diverse array of benthic communities that are important trophic components of their food web (Wilson and Fleeger 2012). These ecosystems are often located in close proximity to highly urbanized areas, which exposes their biological communities to frequent changes in water quality. The estuaries and coastal zones of the Greater Everglades ecosystem, in close proximity to the metropolitan Miami-Dade County, Florida (Fig. 1) experience these stressors. The inter- and intra-annual water quality changes and climate variability cause alterations to the structure of the benthic assemblages (species composition and abundance). Records of the historic changes in the structure of benthic assemblages preserved in lake and ocean sediments have been successfully used in reconstructions of past water quality conditions, habitat types, and climate patterns around the world (Pickerill and Brenchley 1991; Smol et al. 2001; Smol and Stoermer 2010).

[1] Southeast Environmental Research Center, Florida International University, Miami, FL 33199, USA.
  Email: wachnick@fiu.edu
[2] U.S. Geological Survey, Reston, VA 20192, USA.
  Email: lwingard@usgs.gov
* Corresponding author

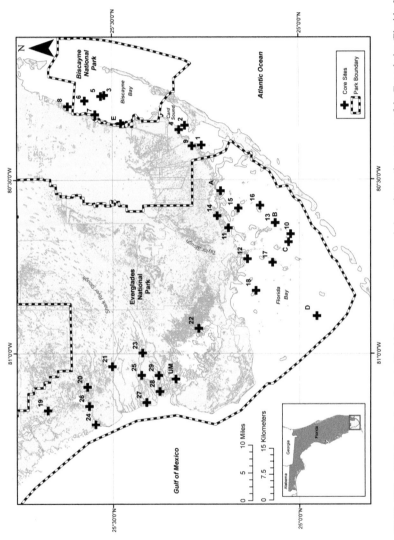

Figure 1. Map showing USGS, FIU and UM coring locations in Florida Bay, Biscayne Bay and along the southwest coast of the Everglades (Florida, U.S.A.). See Table 1 for detailed information about the core locations.

Color image of this figure appears in the color plate section at the end of the book.

Paleoenvironmental studies using different proxy organisms became especially popular since the late 1900s, after sediment coring and dating methods significantly improved. In addition, development of powerful numerical techniques enabled robust quantitative reconstructions of past environmental conditions based on historic records of subfossil organisms (i.e., remains of organisms whose fossilization process is not complete and that still contain original organic material; typically Holocene age).

Constituent remains of organisms in sediments are useful proxies of water quality in estuaries at the time of deposition. Commonly used bio-indicators include diatoms (Fig. 2), mollusks (Fig. 3), foraminifera (Fig. 4), ostracodes, seeds, pollen, and biomarkers. In order to be good bio-indicators, organisms must have strong relationships to water quality parameters that are of particular interest to scientists and typically must be well preserved in sediments, which enable reconstructions of environmental conditions over long time periods. Diatoms, which often represent a dominant component of the microbenthic algae assemblage in estuaries and coastal ecosystems, are especially well known as robust indicators of changing water quality in these environments. Their widespread distribution along the entire salinity gradient (Fig. 2) allows for accurate reconstructions of past salinity and sea level in estuaries and coastal regions, while high species richness (> 600 species recorded in south Florida estuaries) and diversity provide large amounts of ecological information and statistical power in complex data analyses (Smol and Stoermer 2010). Mollusks also occur in a wide range of environments, from terrestrial to freshwater, estuarine, and marine and are well preserved in most sediments (Fig. 3); however, many species are euryhaline and more tolerant of changes in water quality than diatoms. Because of their longer life spans (in south Florida estuaries typically 1.5 to 6 years), mollusks can be indicative of persistent conditions, while shorter-lived species such as diatoms, foraminifera, and ostracodes (from few days to up to 5 years; Reed 2009; EOL 2013) can provide snapshots of rapid change. Pollen grains are well preserved in peat deposits, which cause relatively rapid decay of calcareous species, and provide a linkage between estuarine and onshore environmental changes.

Paleoenvironmental reconstructions based on the historic changes in a single organism type that has a broad tolerance for environmental fluctuations do not always capture smaller fluctuations in water quality conditions. Consequently, due to the complex network of interactions that exist in the estuarine and coastal ecosystems, paleoenvironmental reconstructions based on combination of different biological and geochemical proxies, as described here, are highly desirable because no single proxy is capable of providing adequate information concerning the multiple, interrelated components of the ecosystem.

Multi-proxy paleoenvironmental studies are a pivotal component of the Comprehensive Everglades Restoration Plan (CERP). One of the primary goals of CERP is to restore the Everglades wetlands and estuaries to a more natural, pre-alteration (pre-1900) hydrology in terms of the quantity, quality, and timing of freshwater flow (U.S. Army Corps of Engineers 1999; 2006). Reliable modern records of water quality do not begin in this region until the 1980s (an already altered hydroscape); therefore, paleoenvironmental studies based on multiple biological and geochemical proxies provide the primary means for scientists to reconstruct pre-development water quality patterns that existed in the south Florida estuaries and

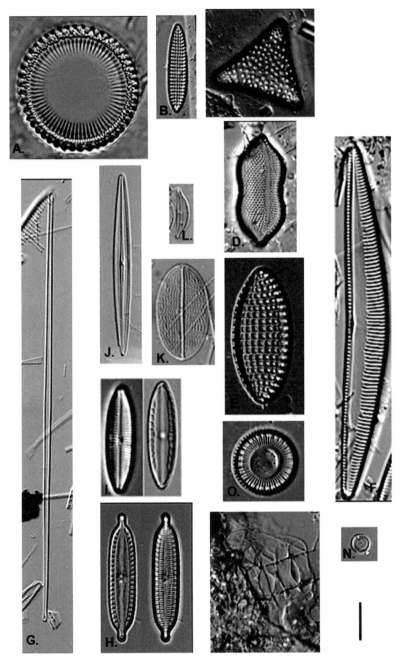

Figure 2. Examples of diatoms used as indicators of different salinity and nutrient regimes and seagrass habitats. Euryhaline species: A. *Paralia sulcata*, B. *Dimeregramma dubium*, C. *Triceratium reticulum*, D. *Psammodictyon panduriforme*; Mesohaline species: E. *Tryblionella granulata*, F. *Seminavis eulensteinii*; Epiphytic species: G. *Hyalosynedra laevigata* var. *angusta*, H. *Mastogloia corsicana*, I. *Mastogloia pusilla*, J. *Brachysira aponina*, K. *Cocconeis placentula*, L. *Amphora tenerrima*; High nutrients indicators: M. *Chaetoceros* sp., N. *Cyclotella choctawhatcheana*, O. *Cyclotella desikacharyi*.

Figure 3. Examples of mollusks used as indicators of different salinity regimes in south Florida's estuaries. Transition zone oligohaline fauna: A. *Polymesoda caroliniana*. B. *Cyrenoida floridana*. Nearshore euryhaline fauna: C. *Anomalocardia auberiana*. D. *Cerithidea costata*. Outer estuarine mesohaline to polyhaline fauna: E. *Nassarius vibex*. F. *Columbella rusticoides*. G. *Arcopsis adamsi*. Euryhaline fauna, ubiquitous throughout nearshore and central Florida Bay: I. *Cerithium muscarum*. J. *Modulus modulus*. L. *Chione elevata*. O. *Brachidontes exustus*. Outer estuarine polyhaline to euhaline fauna: H. *Astralium phoebium*. K. *Tegula fasciata*. M. *Cerithium eburneum*. N. *Turbo castanea*. All scale bars = 5 mm.

coastal regions. This information is essential in development of appropriate restoration targets and performance measures, assessment of ongoing restoration efforts and to distinguish natural from anthropogenic changes.

This chapter highlights the application of different biological indicators to studies on the impact of anthropogenic changes and climate variability on water quality and habitats in Florida Bay, Biscayne Bay, and adjacent coastal ecosystems.

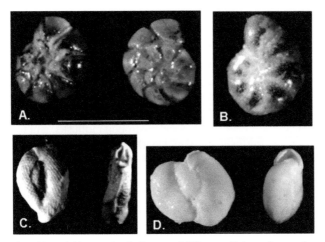

Figure 4. Examples of foraminifera used as indicators of different salinity regimes and seagrass habitats: Mesohaline species: A. *Ammonia tepida*, B. *Elphidium poeyanum*; Euryhaline species: C. *Triloculina bicarinata*; Epiphytic species: D. *Pateoris dilatata* (Source: Cheng 2010).

Geographical Setting

The low-lying coastal regions of south Florida include mangrove forests, dwarf mangroves, marshes, shallow lakes, embayments, and lagoons that provide diverse habitats for a unique fauna and flora. These habitats are connected via natural creeks (the largest of which are Shark River and Taylor River sloughs) and man-made canals with Biscayne Bay, Florida Bay, and the Gulf of Mexico (Fig. 1).

Biscayne Bay is a lagoonal estuary composed of barrier islands, shallow banks of carbonate sand and basins averaging 0.3 to 3.0 m deep (in areas that have not been dredged). The exchange of water between the bay and the Atlantic Ocean occurs primarily through the central opening called the "Safety Valve" and through the tidal creeks between the sandy barrier islands (Cantillo et al. 2000; Wang et al. 2003). The bay formed between 6,000 and 2,400 years ago, due to post-glacial Holocene transgression of the sea and filling of the late Pleistocene bedrock basin (Wanless 1976). The sediments of Biscayne Bay consist mostly of calcareous and siliceous skeletal benthic organisms that live in the bay, mixed with quartz-carbonate sand from the southern Appalachian Mountains and pure quartz Pleistocene Pamlico Sand of the mainland Atlantic Coastal Ridge in the northern part of the bay (Wanless 1976). Historically, freshwater flow from the Everglades into Biscayne Bay occurred via transverse glades, natural creeks, and groundwater seepage (Langevin 2003; Stalker 2008). The flow has been significantly reduced and in many places completely eliminated, and nutrient loading to the bay has increased significantly as a result of construction of a highly developed drainage system on the south Florida mainland in the mid-1900s (McIvor et al. 1994; Ross et al. 2001; Caccia and Boyer 2007; Wachnicka et al. 2013a).

Neighboring Florida Bay is a shallow estuary (between 0.3 to 2.0 m depth) composed of a series of small basins, separated from each other by carbonate mud banks and small mangrove islands (Tilmant 1989; McIvor et al. 1994; Fig. 1). The area occupied by the bay was gradually flooded by rising sea level between 4,500 and 1,500 years ago (Enos and Perkins 1979). The marine carbonate deposits that are mostly produced by calcareous algae, bivalve and gastropod mollusks, and foraminifera, lie on top of brackish and freshwater peat and marls that predate the marine transgression into Florida Bay (Bosence 1989; Wanless and Tagett 1989; Davies and Cohen 1989; Wachnicka 2009; Cheng 2010; Lodge 2010). Environmental conditions in Florida Bay have been significantly altered due to the construction of spoil islands along the Florida Keys associated with development of the Florida Overseas Railroad (now the Overseas Highway that carries U.S. Route 1 (US 1)) in the early 1900s. These constructed islands restricted the exchange of water between Florida Bay and the Atlantic Ocean, leading to alterations of water circulation patterns and residence time across the bay (Swart et al. 1996; 2001). Additionally, the quantity and quality of water being delivered to Florida Bay was significantly reduced by 1) construction of the Tamiami Trail between 1921 and 1926, which cuts across the Florida peninsula through the middle of the natural Everglades, and 2) conversion of large parts of the Everglades to agricultural lands and urban areas in the mid-1900s (Light and Dineen 1994; Sklar et al. 2002; McVoy et al. 2011).

The southwestern coast of Florida forms a complex network of natural creeks and channels, which interconnect shallow bays that are sandwiched between small mangrove islands and coastal marl prairies. This region receives large amounts of freshwater flowing in the southwest direction through Shark River Slough in Everglades National Park (ENP) from Lake Okeechobee (Fig. 1). Additionally, the area of Ten Thousand Islands, north of Lostmans River, receives freshwater from the Fakahatchee Strand and the Okaloacoochee Slough (Wingard et al. 2006). Sedimentary deposits in this region consist of recent marine and brackish-water sediments that overlie fresh-water calcitic mud that was deposited on bedrock or freshwater peat ~4,000 years ago (Scholl 1964; Perkinson 1989; Wanless 1994). This sequence of deposits reflects the history of marine transgression of the southwest Florida coast during the Holocene, which decelerated approximately 3,000 years ago when the rate of sea level rise decreased to < 10 cm/century from ~30 cm/century between 4,400 and 3,500 years ago, allowing build-up of the islands (Scholl 1964; Perkinson 1989; Wanless 1994). Hydrological modification of the Everglades south of Lake Okeechobee over the last century significantly reduced the amount of freshwater being delivered to the southwestern coastal regions of ENP (Light and Dineen 1994; Sklar et al. 2002; McVoy et al. 2011), affecting salinity patterns, habitats, and biological communities occupying the coastal bays and lagoons (Zarikian et al. 2001; Donders et al. 2008; Lammers et al. 2013). These changes are recorded in the assemblages of organisms preserved in the sediments.

Paleoenvironmental Reconstructions in South Florida Estuaries and Coastal Regions

Historic Changes in Salinity

Salinity in estuaries is a function of a number of interconnected natural processes, including freshwater surface runoff, groundwater, precipitation, sea level, tides, circulation, and evaporation. Alteration of any of these processes, whether anthropogenic or climate-driven, will change salinity patterns. As stated earlier, a primary goal of CERP is to restore more natural patterns of freshwater flow through the wetlands of the Everglades (U.S. Army Corps of Engineers 1999; 2006). Changes in the quantity and timing of freshwater flow in the Everglades throughout the 20th century have had the most impact on the estuaries of south Florida by altering the natural salinity patterns (McIvor et al. 1994). Determining the pre-alteration natural salinity patterns and the role of anthropogenic versus natural forces was the initial focus of the paleoecologic sediment core studies in the south Florida estuaries that began in the mid-1990s by researchers at Florida International University (Miami, FL), University of Miami (Miami, Florida), and the U.S. Geological Survey (Reston, Virginia) among others (Brewster-Wingard et al. 1998; 2001; Brewster-Wingard and Ishman 1999; Stone et al. 2000; Dwyer and Cronin 2001; Huvane and Cooper 2001; Ishman et al. 1998; Nelsen et al. 2002; Trappe and Brewster-Wingard 2001; Zarikian et al. 2001; Xu et al. 2007; Donders et al. 2008; Williams 2009; Wachnicka 2009; Cheng 2010; Cooper et al. 2010; Cheng et al. 2012; Wachnicka et al. 2013a;b;c; Lammers et al. 2013). Modern proxy data were gathered on the organisms' salinity tolerances and preferences to aid in interpretation of the historic salinity from the assemblages in the cores (Ishman 1997; Brewster-Wingard et al. 2001; Cronin et al. 2001; Huvane 2002; Ishman 2001; Carnahan et al. 2009; Wachnicka 2009; Wachnicka et al. 2010; 2011; Wingard and Hudley 2012).

Cores collected from near the terrestrial—estuarine margin contained some of the oldest records for Biscayne Bay and Florida Bay, dating back to approximately 500 to 4,600 years ago (Wingard et al. 2007; Wachnicka 2009; Cheng 2010). The lower portions of many of these cores contained freshwater and mangrove peats, and the mollusks, diatoms, ostracodes, and foraminifera indicate that fresh (< 0.5 practical salinity units (psu)) to oligohaline (0.5–5 psu) environments existed at the margins of Florida Bay and southern Biscayne Bay/Barnes Sound between 1,500 and 4,600 years ago (Ishman et al. 1998; Willard et al. 1997; Wingard et al. 2003; Wingard et al. 2004; Wachnicka 2009; Cheng 2010; Murray et al. 2010; Wingard et al. 2007; Wingard and Hudley 2012).

Sediment cores from the mudbanks of Biscayne Bay and Card Sound contained a 200 to 800 year old record (Wingard et al. 2007; Fig. 1; Table 1). Faunal and algal assemblages indicate a general trend of increasing salinity over time at most coring locations (Wingard et al. 2003; 2004; Wachnicka et al. 2013a). Analyses of ostracode, foraminifer, and mollusk assemblages from Manatee Bay in Barnes Sound (southern Biscayne Bay) and Middle Key Basin (adjacent to Manatee Bay; Fig. 1; Table 1) indicate that brackish water intrusion into this area began prior to the 20th century (Wingard et al. 2004; 2003; Ishman et al. 1998; Williams 2009). The same studies also showed that higher salinity tolerant faunal species became dominant in this region during the 20th century, implying development of more marine conditions that were interrupted by a few episodes of hypersalinity between 1940 and 1980. Additionally, Ishman et al. (1998) found that foraminifer species with lower-salinity preferences increased in abundance again in Manatee Bay in the post-1980 period, implying a lowering of salinity since that time. Wingard et al. (2004) found evidence of increasing fluctuations in salinity in Middle Key basin. The post-1980 decrease in salinity in the Manatee Bay region may be due to increased freshwater deliveries from the mainland via the drainage canals initiated by the South Florida Water Management District (SFWMD). Mollusk- and diatom-based studies conducted by Wingard et al. (2003; 2004) and Wachnicka et al. (2013a) in neighboring Card Sound in southern Biscayne Bay (Fig. 1; Table 1) showed that this region has been relatively isolated from direct marine influence over at least the last two centuries, but despite its isolation, the basin has fluctuated between more restricted upper estuarine environments and more open marine environments. The same studies also revealed that conditions in Card Sound, as well as Featherbed Bank and No Name Bank in central Biscayne Bay, became increasingly marine over the last century. Large salinity variations occurred in Card Sound and Featherbed Bank post-1950, while more stable salinity conditions were present at No Name Bank (Fig. 5a). Diatom-based quantitative salinity reconstructions showed that changes in average salinity conditions were larger in central Biscayne Bay (between ~7 psu and 13 psu at No Name Bank and Featherbed Bank, respectively) compared to southern Biscayne Bay (~2 psu in Card Sound) over the last few hundred years (Wachnicka et al. 2013a). Similar trends in the historic salinity levels at Featherbed Bank were discovered by Stone et al. (2000) based on analysis of foraminifer, mollusk, and ostracode assemblages. Analysis of subfossil mollusk assemblages in 36 sediment cores collected along a 0.5 to 2.0 km transect through remnant coastal wetlands between Princeton Canal, Mowry Canal, and L-31E Canal adjacent to southwestern Biscayne Bay, revealed that the average salinity levels in this region also increased significantly in the post-1950s period (from < 2 psu to 13.2 psu; Gaiser et al. 2006). Gaiser et al. (2006) suggested sea level rise and freshwater diversion between the 1940s and 1960s were the main causes of this change.

Analyses of faunal and algal assemblages in the 13 cores, listed in Table 1, recovered from the mudbanks in neighboring Florida Bay (Fig. 1) showed that northern and east-central parts of this estuary experienced larger magnitude variations in historic salinity values than south-central and southwestern parts of the bay. Cores from central and northeastern Florida Bay, which span a 150 to 1690-year time period, showed a significant decline or loss of many mollusk, diatom, and foraminifer species indicative of freshwater influx and mesohaline (15–18 psu) conditions around the onset of the 20th century (Brewster-Wingard et al. 1998; 2001; Trappe and Wingard 2001; Wingard et al. 2007; Wachnicka 2009; Cheng 2010; Cheng et al. 2012; Wingard and Hudley 2012; Wachnicka et al. 2013c). A quantitative mollusk-based salinity reconstruction performed by Wingard and Hudley (2012) on a core collected near the mouth of Taylor River in north-central Florida Bay (Figs. 1; 5b) showed a general trend toward higher salinity conditions over the last 1,600 years, which was especially pronounced in the post-1900 period. Similarly, the cores from mudbanks in Florida Bay show distinct shifts in patterns of assemblages throughout the 20th century, with species typical of

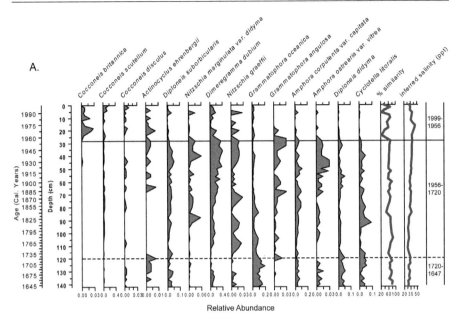

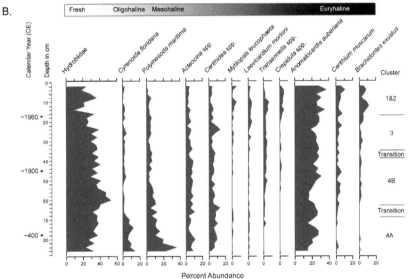

Figure 5. Changes in species composition and abundance: A. Changes in relative abundance of diatom taxa, percent similarity between consecutive samples in the core and diatom-inferred salinity at No Name Bank (core ID: GLW402-NNA) in Biscayne Bay (modified from Wachnicka et al. 2013a; figure 3 shows a few of the species listed here); B. Changes in percent abundance of mollusk species at the mouth of Taylor Creek (core ID: FB594 24) in Florida Bay (modified from Wingard and Hudley 2012; figure 2 shows some of the species listed here).

transitional or inner estuarine environments disappearing and species more typical of open marine environments appearing (Brewster-Wingard et al. 1998; 2001; Brewster-Wingard and Ishman 1999; Dwyer and Cronin 2001; Trappe and Brewster-Wingard 2001; Nelson et al. 2002). Diatom-based reconstructions of salinity at Russell Bank and Bob Allen Bank in east-central Florida Bay (Fig. 1; Table 1) also indicate that salinity was lower at these locations prior to the 20th century (Wachnicka et al. 2013c).

Increase in the baseline salinity levels (Wachnicka 2009; Wingard and Hudley 2012) since the beginning of the 20th century, suggest that freshwater deliveries from the mainland to the north-central and central regions of Florida Bay have decreased during this time as indicated by lower abundance and often complete disappearance of diatom and mollusk taxa capable of tolerating polyhaline (18–30 psu) and mesohaline (5–18 psu) conditions (Brewster-Wingard et al. 2001; Huvane and Cooper 2001; Wingard and Hudley 2012; Wachnicka et al. 2013b). Increasing salinity patterns over the 20th century were also observed by Huvane and Cooper (2001) and Brewster-Wingard et al. (2001) based on analysis of diatom and mollusk assemblages in a core retrieved from around Pass Key in east-central Florida Bay (Fig. 1; Table 1). Diatom assemblages examined by Huvane and Cooper (2001) indicated a significant increase in salinity at Pass Key and Russell Bank in east-central Florida Bay in the mid-1970s. Post-1900, the majority of the cores in Florida Bay were dominated by euryhaline (30–40 psu) diatom, foraminifer, and mollusk species, capable of tolerating wide swings in salinity (Brewster-Wingard et al. 2001; Murray et al. 2010; Trappe and Brewster-Wingard 2001; Huvane and Cooper 2001; Cheng et al. 2012; Wachnicka et al. 2013b). The dominance of these euryhaline assemblages implies that the magnitude of salinity variations increased in most of Florida Bay since that time.

Because of a greater distance from the mainland, the Ninemile Bank core assemblages in southwestern Florida Bay (Fig. 1) showed much smaller degree of restructuring (i.e., changes in species composition and abundance) than assemblages in cores from the rest of the bay (Wachnicka et al. 2013a). A post-1980s increase in abundance of taxa with wide salinity tolerances (also observed in the molluscan assemblages; Brewster-Wingard et al. 2001; Murray et al. 2010; Trappe and Brewster-Wingard 2001) and a decrease in average salinity was indicated by the diatom-based quantitative salinity reconstruction (Wachnicka et al. 2013a). These changes were contemporaneous with the initiation of the SFWMD's "Rainfall Plan" and the closing of the Buttonwood Canal in the southwest part of ENP. The "Rainfall Plan" adjusted the amount of freshwater flow to Shark River Slough based on the amount of rainfall and evaporation patterns present in south Florida and water levels in Water Conservation Area 3a (Light and Dineen 1994).

Cores recovered from around Bob Allen Keys and Ninemile Bank in south-central and southwestern Florida Bay (Fig. 1) contained mangrove peat deposits transected by freshwater marl deposits, dated back to approximately 4,190 years ago (Wachnicka 2009; Cheng 2010). High abundance of fresh and brackish water diatom and foraminifer species interrupted by sporadic occurrence of marine taxa in lower parts of these cores indicate presence of fresh to brackish water conditions during that time, similar to those that exist in the present-day coastal Everglades (Wachnicka 2009; Cheng 2010; Cheng et al. 2012; Wachnicka et al. 2013b;c).

Preliminary examination of mollusk assemblages in 11 sediment cores, covering between 1,000 to greater than 2,000 years of environmental history, along Shark, Harney, and Lostmans Rivers in the southwestern coastal area of ENP (Fig. 1; Table 1), revealed that larger freshwater species have disappeared in the upper levels of the cores collected from Shark and Harney Rivers, indicating a shift from freshwater toward more estuarine conditions (Wingard et al. 2005; 2006). Similar increase in salinity since about the turn of the century has been recorded by Zarikian et al. (2001) and Nelson et al. (2002) after examination of benthic ostracode and foraminifer assemblages and oxygen isotope levels in the core from Oyster Bay near Shark River outflow (Fig. 1; Table 1). A major departure in salinity values from the long-term linear trend that began in the mid-1980s was described by Nelson et al. (2002) as an "extreme excursion to more marine conditions" (p. 443), which coincided with record-low regional rainfall. The opposite trend in salinity since the early 1980s was recorded by Wachnicka et al. (2013c) based on an analysis of diatom assemblages and by Xu et al. (2007) based on the concentration of Teraxerol mangrove biomarker in the core collected from Ninemile Bank (southwest Florida Bay; Fig. 1), implying that the southwestern part of Florida Bay is being influenced by different factors than the southwest coastal region of ENP.

Impact of Sea Level Rise and Climate Change on Coastal Areas and Estuaries

Sea level rise and climate change have a profound influence on coastal and estuarine regions around the world. In south Florida, the challenge is attempting to understand the past history of changes and project how future change will impact the ecosystem and the planned 30 to 50 year restoration. Since the end of the last Ice Age approximately 21,000 years ago, global sea level rose by about 120 m and stabilized between 3,000 and 2,000 years ago (IPCC 2007). Geologic data suggest that sea level rose at an average rate of 0.5 mm/yr over the past 6,000 years and at an average rate of 0.1 to 0.2 mm/yr over the last ~3,000 years (IPCC 2007). A slow relative rise in sea level over the last ~3,000 years, allowed for establishment of a vast wetland ecosystem in south Florida (Wanless et al. 1994) that covered 1,166,000 ha of land (current cover ~618,000 ha; Davis et al. 1994). Since the beginning of the 20th century, many of the world's mountain glaciers have disappeared, and the ice sheets in both Greenland and Antarctica have lost mass due to global warming (NSIDC 2013; NASA 2013). Addition of water from the melting land-based ice masses and thermal expansion of the oceans are the two major factors responsible for the current increase in the rates of sea level rise to ~3 mm/yr (NOAA 2013). This rate is 6 to 10 times that of the past ~3,000 years and it is causing dramatic changes in the low-lying coastal wetland ecosystems of south Florida, which include accelerated coastal erosion, intrusions of salt into ground and surface waters of coastal areas, and replacement of freshwater sawgrass marshes with salt-tolerant mangroves (Wanless et al. 1994; Gaiser et al. 2006; Saha et al. 2011; Krauss et al. 2011; Langevin and Zygnerski 2012). Model results suggest that the global rate of sea level rise may accelerate further in the 21st century. The IPCC (2007) projections for worldwide sea level rise range from 20 to 60 cm during the 21st century; however, these rates do not include factors such as ice sheet flow dynamics that could significantly increase the rate. The more recent Copenhagen

Diagnosis (Allison et al. 2009) states that the IPCC (2007) report underestimated sea level rise and that it may be as much as twice what has been projected. The current average rate of sea-level rise in south Florida follows the global average rate, and it is ~2.24 mm/yr +/–0.16 mm/yr, according to the monthly mean sea level data from Key West, Florida, recorded between 1913 and 2006 (NOAA 2013a).

Three primary climate cycles influence interannual to multidecadal temperature and precipitation patterns in south Florida, and other regions of the United States (U.S.) and the world: El Niño-Southern Oscillation (ENSO), Pacific Decadal Oscillation (PDO), and Atlantic Multidecadal Oscillation (AMO). Because these oscillations are strongly interconnected, even small changes in the frequency and amplitude of one will have a profound effect on atmospheric patterns driven by the others (Gershunov and Barnett 1998; Enfield et al. 2001; McCabe et al. 2004; Hu and Feng 2012). South Florida usually experiences less than normal rainfall during cold phases of PDO that coincide with warm phases of AMO (McCabe et al. 2004). Additionally, warm phases of ENSO have greater influence on winter precipitation, while warm phases of AMO have greater influence on summer precipitation (Enfield et al. 2001; Moses et al. 2013). The mean climate over the equatorial Pacific Ocean and the North Atlantic Ocean will most likely undergo significant alterations under future climate change scenarios. As a result of these changes, the delicate energy balance between the oceans and atmosphere will be affected. It is unclear to what degree climate conditions over the equatorial Pacific Ocean and the North Atlantic Ocean will change over the next century, but scientists agree that global atmospheric teleconnections will be affected. Consequently, precipitation and temperature patterns in South Florida and other regions around the world will change, and these changes will affect natural ecosystems, agriculture, and management decisions related to water resources.

Changes in sea level and climatic patterns in south Florida are recorded in many of the sediment cores dating back to as old as 4,100 years ago, collected from Florida Bay, Biscayne Bay, and along the southwest coast of the Everglades (Stone et al. 2000; Zarikian et al. 2001; Nelson et al. 2002; Wachnicka 2009; Wingard et al. 2005; 2006; 2007; Cheng et al. 2012; Wachnicka et al. 2013a;b;c). The gradual inundation of the south Florida region by rising sea level during the last few thousand years has caused replacement of fresh and brackish water fauna and flora with marine organisms in the core assemblages. The 20th century salinity increase observed in many of the cores can be contributed to both natural and anthropogenic factors; however, these two factors cannot be easily separated. Anthropogenically caused changes in salinity regimes of the bays and coastal areas are related to development of a regional drainage and flood control system that diminished the quantity of freshwater deliveries from the mainland since the early 1900s, but these are superimposed on the natural climate-driven factors. Analyses of pollen assemblages in peat-dominated cores collected along the southwestern Florida coast by Willard and Bernhardt (2011) revealed a clear shift from freshwater marshes to transitional vegetation between ~3,000 and 2,200 years ago and to mangrove swamps between ~1,200 and 1,000 years ago, caused by rising sea level over the last ~3,600 years. Similar analyses of cores collected from Jim Foot Key, Crane Key, Spy Key, and Russell Key in Florida Bay, revealed a temporary shift from estuarine to mangrove sedimentation around 1,200–1,000 years ago, suggesting a still-stand or slowing of sea level rise before resumption of marine sedimentation and more rapid rates of sea level rise (Willard and Bernhardt 2011).

The natural variability of the regional climate causes natural seasonal and decadal-scale salinity fluctuations in the south Florida ecosystem (Dwyer and Cronin 2001; Cronin et al. 2001; Wachnicka et al. 2013a;c). Cronin et al. (2001) and Wachnicka et al. (2013a;c) discovered links between the timing of significant changes in the structure of ostracode and diatom assemblages in Florida Bay and Biscayne Bay cores and shifts in the intensity of ENSO, AMO and PDO. Wachnicka et al. (2013a) showed that the occurrence of severe and prolonged droughts associated with extreme cold phases of ENSO, AMO and PDO, and warm phases of AMO and cold phases of PDO coincided with many significant changes in diatom species composition and abundance in Biscayne Bay cores (Fig. 6). Additionally, switching of these oscillation patterns from one phase to another (most often from warm to cold phase) and to a lesser extent development of wet conditions during warm phases of ENSO and AMO were listed as possible causes of significant changes in the structure of diatom assemblages in Biscayne Bay and Florida Bay cores (Wachnicka et al. 2013a;c). Furthermore, reconstruction of the historic salinity levels at Russell Bank (central Florida Bay)

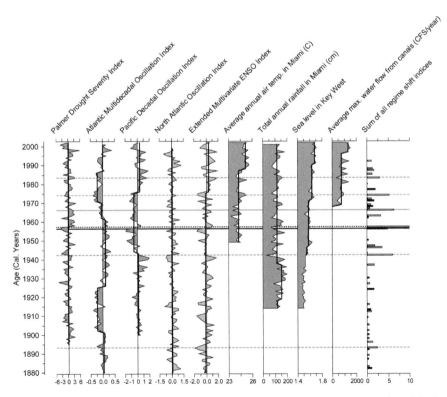

Figure 6. Links between historical changes in large-scale climate oscillation patterns, local precipitation and air temperature patterns, sea level and maximum water flow from major canals along the Biscayne Bay coast, and major shifts in diatom assemblage structure at No Name Bank (black bars and line), Featherbed Bank (gray bars and lines), and Card Sound Bank (stripped bars and dotted line) (Published in Wachnicka et al. 2013a).

based on analysis of Mg/Ca ratios in *Loxoconcha matagordensis* ostracode shells by Cronin et al. (2002) revealed cyclic oscillations in salinity values related to ENSO-driven rainfall variability.

Changes in Subaquatic Vegetation

Significant changes in water quality in the south Florida estuaries as a result of constantly changing climate and the 20th century alteration of the hydroscape of the south Florida mainland had a profound effect on subaquatic vegetation composition and density in these ecosystems. Numerous paleoenvironmental studies conducted in south Florida estuaries revealed major changes in seagrass abundance and density in Florida Bay and Biscayne Bay over the last few thousand years, with the largest changes recorded post-1950s (Cronin et al. 2001; Trappe and Wingard 2001; Cheng et al. 2012; Wachnicka et al. 2013a;b).

The well-documented late 1980s seagrass die-off and recovery event in Florida Bay was recorded in several studies conducted in this estuary. The Cheng et al. (2012) analyses of foraminifer assemblages in two cores collected from Russell Bank and Bob Allen Bank in central Florida Bay revealed a significant decrease in abundance of *Quinqueloculina polygona*—an epiphytic foraminifer species—in the late 1980s. Similarly, analyses of subfossil ostracode assemblages at several locations in central and eastern Florida Bay by Cronin et al. (2001) showed a decline in abundance of *Loxoconcha matagordensis, Xestoleberis* sp., and *Malzella floridana* epiphytic ostracode species during the 1970s and 1980s. Furthermore, biomarker analysis in cores collected from the same region by Xu et al. (2007) revealed a sharp drop in a *Paq* seagrass biomarker around 1988 to 1995. Analysis of epiphytic mollusks in the core from Whipray Basin in central Florida Bay by Trappe and Wingard (2001) revealed a > 30% decrease in their abundance between 1871 and 1913, suggesting that the seagrass die-off events are not new to the bay.

The highest values of seagrass biomarkers (*Paq*, C_{23}+ C_{25} n-alkanes and the C_{25}/C_{27} n-alkan-2-one ration) were reported at Trout Cove, Russell Bank and Ninemile Bank post-1950s. A high concentration of seagrass biomarkers, coinciding with a high concentration of microbial biomarkers since the 1950s, suggests a high rate of primary productivity at these sites. Xu et al. (2007) connected the post-1950 increases in seagrass and microbial abundance to anthropogenically induced nutrient enrichment in the bay. Cronin et al. (2001) also reported an increase in abundance of epiphytic species in central and eastern Florida Bay since the 1950s. Significant increase in abundance of epiphytic foraminifer and diatom species was also recorded by Cheng et al. (2012) and Wachnicka et al. (2013b) at Ninemile Bank in southwestern Florida Bay since the early 1980s. Additionally, Wachnicka et al. (2013a) showed that abundance and diversity of epiphytic diatom species also increased at Featherbed Bank in central Biscayne Bay since the mid-1970s and Card Sound since the mid-1950s, suggesting that subaquatic vegetation density at these locations also increased since the mid-1900s.

Changes in Nutrient Concentration

Cultural eutrophication (i.e., natural eutrophication accelerated by human activity) is a common problem facing coastal marine ecosystems around the world (Kelly 2008; Smith and Schindler 2009). Addition of nutrients to these ecosystems can increase the abundance of subaquatic vegetation and microbenthic communities attached to them, as well as those floating in the water column. These additional nutrients can cause algal blooms, which may reduce clarity and oxygen levels in the water as the algae and plants decompose, and decrease biodiversity.

Nitrogen and phosphorus concentrations in sediments of the south Florida estuaries have increased since the mid-1900s (Orem et al. 1999; Xu et al. 2007; Fourqurean et al. 2012; Wachnicka et al. 2013a;b) due to an increased influx of nutrient-rich waters from the drainage canals and submarine springs. Increases in seagrass and microbenthic community abundance in the last ~30 years of the 20th century are reflected in elevated concentrations of seagrass and algae biomarkers in sediment cores collected from Florida Bay; these increases were linked to the recent eutrophication of the south Florida estuaries (Xu et al. 2007). Wachnicka et al. (2013a) also linked the post-1950s increase in *Mastogloia* and *Cocconeis* (Figs. 2 and 5a) epiphytic diatom species in the Biscayne Bay cores to the post-1950s nutrient increase.

Changes in Freshwater Stage and Flow

One of the primary drivers of change in the south Florida coastal ecosystem is alteration of the natural flow of freshwater through the Everglades as a result of water management practices and land-use (Davis et al. 2005; Sklar and Browder 1998). Alterations in flow cause a cascade of changes to other key physical components of the ecosystems including water depth, salinity, nutrients, and dissolved oxygen, which in turn cause changes in biological components such as productivity, community structure, and species composition (Sklar and Browder 1998). Decreased freshwater inflow to coastal regions and adjacent estuaries and diversion of the flow from one region of the Everglades to another result in the encroachment of mangrove forests into freshwater marshes in ENP (Gaiser et al. 2006), the decline of mangrove forests in areas deprived of natural overland flow (McPherson and Halley 1996), and increases in the frequency of hypersalinity periods in central Florida Bay (McIvor et al. 1994). Paleoenvironmental investigations revealed that in the post-1950s period, when water stages and the rates and direction of freshwater flow throughout the Everglades changed, the ecological state of the south Florida coastal and estuarine ecosystems experienced unprecedented changes.

Analyses of subfossil diatom assemblages at several locations in Florida Bay and Biscayne Bay by Wachnicka et al. (2013a;b;c) showed that these basins experienced a distinct ecological regime shift during the 1950s and 1960s. Post-1950s, when flow of water throughout the Everglades had already been altered, the rate and magnitude of restructuring of the diatom assemblages increased significantly at most of the study locations within these estuaries (except for the southwest part of Florida Bay; Fig. 6).

Similarly, the amplitude of change in the structure of the foraminifer assemblages at several locations in Florida Bay increased since that time (Cheng et al. 2012).

Marshall et al. (2009) developed a three-part approach for deriving estimates of pre-alteration flow and stage in the wetlands of the Everglades based on paleoecology derived salinity estimates from cores collected in Florida Bay. First, paleosalinity is estimated from biotic assemblage information by comparing the core assemblages to modern analog data. To date, the method has utilized only molluscan faunal data, but any organism present in statistically significant quantities in the cores, and having an appropriate modern analog dataset could be utilized. Second, linear regression equations are developed that link stage and flow in the wetlands to salinity at different locations in Florida Bay. These equations were derived from hydrologic monitoring station data collected between 1991 and 2001. The final step is to insert the paleosalinity estimates into the linear regressions equations and obtain an estimate of what the pre-alteration stage and flow in the wetlands would have been. In a recent evaluation Marshall and Wingard (2012) used updated and improved linear regression models and improved paleosalinity estimates based on a statistical modern-analog approach (Wingard and Hudley 2012) applied to five cores. Results from the upgraded models estimated stage in the freshwater wetlands of the Everglades, which prior to alteration was 9 to 49 cm higher than the observed mean over the period of record of the hydrologic stations, and estimated flow into Shark River Slough at Tamiami Trail was 11.4 to 71.9 cubic meter per second ($m^3 s^{-1}$) higher than observed flow (Marshall and Wingard 2012). In Florida Bay, the combined average salinity for the northeastern bay stations prior to alterations was 14.7 psu less than observed hydrologic station data over the period of record (Marshall and Wingard 2012). Moving away from the sources of freshwater influx, this value drops to 1 psu less than observed for western Florida Bay near the transition zone with the Gulf of Mexico.

Summary

Paleoenvironmental studies are an essential first step in restoration of south Florida coastal and estuarine ecosystems. Restoration efforts are a significant component of national and regional land and water management strategies (U.S. Army Corps of Engineers 1999; 2006), but their success will depend heavily on the development of realistic targets and performance measurements for management and effective communication between policymakers, environmental managers, and scientists. The establishment of target conditions is an exceptionally complicated process, because stresses caused by anthropogenic alteration of the hydroscape need to be recognized and differentiated from stresses caused by climate variability. Since reliable instrumental data on water quality conditions date back only to the 1980s, and historical documents describing environmental conditions are incomplete, paleoenvironmental studies are the primary method for determining the pre-impact environmental conditions in the south Florida ecosystems.

Paleoenvironmental studies conducted in south Florida coastal and estuarine ecosystems successfully identified the initial point of change (environmental threshold), when the environmental conditions first departed from natural patterns of variability and revealed the responsiveness of the ecosystems to anthropogenic impacts. Multi-proxy analyses of sedimentary records obtained from numerous locations in Florida Bay, Biscayne Bay, and the adjacent coastal regions reveal that:

1. Fresh and brackish water faunal and algal assemblages in cores containing 1,500 to 4,600 year-old freshwater and mangrove peat deposits decreased in abundance over time, implying a gradual development of more marine conditions in the nearshore and coastal regions of south Florida due to progressing sea level rise and in the post-alteration period also due to decreased freshwater deliveries to the coastal regions.

2. The structure of faunal and algal assemblages and biochemical indicator values at most of the studied locations (except for those located in the more open marine regions of Florida Bay, e.g., Ninemile Bank) experienced the largest shift during the 1950s and 1960s. Additionally, the magnitude and rate of restructuring of faunal and algal assemblages has increased post 1950s, implying that the magnitude and rate of environmental changes in coastal and estuarine regions of south Florida increased since that time. During the 1950s and 1960s south Florida coastal and estuarine ecosystems experienced an ecological regime shift as a result of a combined effect of climate change intensification and intensive urbanization of the southeast Florida coast in the 1940s to 1960s, which decreased the ability of the ecosystems to absorb and respond to changes in the driving variables.

3. Salinity variations increased in central and southern Biscayne Bay and eastern and central Florida Bay in the post-1950s period due to changes in the quantity and timing of freshwater deliveries to the bays. Salinity remained stable in north-central Biscayne Bay and southwestern Florida Bay over the last few hundred years, in large part due to distance from the direct influx of freshwater.

4. The abundance of subaquatic vegetation and faunal and algal communities in central and eastern Florida Bay and southern Biscayne Bay increased since the 1950s as a result of nutrient enrichment. A significant drop in abundance of subaquatic vegetation (die-off events) in central Florida Bay was observed during 1870 to 1913 and 1970 to 1990.

5. Estimates of water stages and water flow rates in the freshwater wetlands of the Everglades in the pre-alteration period based on coupling of molluscan paleoecologic data to hydrologic regression models revealed that water stages were higher by 9–49 cm compared to recent stages, and flow of water to Shark River Slough at Tamiami Trail was 11.4–71.9 m^3 s^{-1} higher than observed flow.

Acknowledgments

These studies have been funded by grants from United States Geological Survey (USGS; Cooperative Agreement) # 800000961, Southeast Environmental Research Center's (SERC) Faculty Research Seed Funds Endowment, Everglades National Park and Biscayne National Park (National Park Service #5284-AP00-371), the National Science Foundation through the Division of Earth Science Geology and Paleontology Program (EAR-071298814), and by the USGS Greater Everglades Priority Ecosystems Science (GEPES) effort, G. Ronnie Best, Coordinator. The GIS map (Fig. 1) and the mollusk photographs (Fig. 3) were generated by Bethany Stackhouse (USGS), and the photographs of foraminifera (Fig. 4) were kindly provided to us by Laurel Collins (FIU). We would like to thank our USGS reviewers Christopher Bernhardt and Marci Robinson, and our anonymous reviewers for their insightful comments on the manuscript, which led to further improvement of its quality. Any use of trade names is for descriptive purposes only and does not imply endorsement by the U.S. Government. This is contribution number 637 from the Southeast Environmental Research Center at Florida International University.

References

Allison, I., N.L. Bindoff, R.A. Bindschadler, P.M. Cox, N. de Noblet, M.H. England, J.E. Francis, N. Gruber, A.M. Haywood, D.J. Karoly, G. Kaser, C. Le Quéré, T.M. Lenton, M.E. Mann, B.I. McNeil, A.J. Pitman, S. Rahmstorf, E. Rignot, H.J. Schellnhuber, S.H. Schneider, S.C. Sherwood, R.C.J. Somerville, K. Steffen, E.J. Steig, M. Visbeck and A.J. Weaver. 2009. The Copenhagen diagnosis, 2009: updating the world on the latest climate science. The University of New South Wales Climate Change Research Centre (CCRC), Sydney, Australia, 60 pp.

Aumann, H.H., A. Ruzmaikin and J. Teixeira. 2008. Frequency of severe storms and global warming. Geophys. Res. Lett. 35: L19805.

Bosence, D. 1989. Biogenic carbonate production in Florida Bay. Bul. Mar. Sci. 44: 419–433.

Brewster-Wingard, G.L. and S.E. Ishman. 1999. Historical trends in salinity and substrate in central and northern Florida Bay: a paleoecological reconstruction using modern analogue data: Estuaries. 22(2B): 369–383.

Brewster-Wingard, G.L., S.E. Ishman and C.W. Holmes. 1998. Environmental impacts on the southern Florida coastal waters: A history of Change in Florida Bay. J. Coast. Res. special issue 26: 162–172.

Brewster-Wingard, G.L., J.R. Stone and C.W. Holmes. 2001. Molluscan faunal distribution in Florida Bay, past and present: an integration of down-core and modern data. Bull. Am. Paleont. S361: 199–231.

Caccia, V.G. and J.N. Boyer. 2007. A nutrient loading budget for Biscayne Bay, Florida. Mar. Pollut. Bull. 54: 994–1008.

Cantillo, A.Y., K. Hale, E. Collins, L. Pikula and R. Caballero. 2000. Biscayne Bay: environmental history and annotated bibliography. NOAA Technical Memorandum NOS NCCOS CCMA 145.

Carnahan, E.A., M.M. Hoare, P. Hallock, B.H. Lidz and C.D. Reich. 2009. Foraminiferal assemblages in Biscayne Bay, Florida, USA: responses to urban and agricultural influence in a subtropical estuary. Mar. Pollut. Bull. 59: 221–233.

Cheng, J. 2010. Paleoenvironmental Reconstruction of Florida Bay, South Florida, Using Benthic Foraminifera: Unpublished Ph.D. Dissertation, Florida International University, Miami, 215 p.

Cheng, J., L.S. Collins and C. Holmes. 2012. Four thousand years of habitat change in Florida Bay, as indicated by benthic foraminifera. J. Foram. Res. 42: 3–17.

Cooper, S., E. Gaiser and A. Wachnicka. 2010. Estuarine paleoenvironmental reconstructions using diatoms. pp. 352–373. In: J. Smol and G. Stormer (eds.). Diatoms: Applications for the Environmental and Earth Sciences, Cambridge University Press, U.K.

Cronin, T.M., C.W. Holmes, G.L. Brewster-Wingard, S.E. Ishman, H. Dowsett, D. Keyser and N. Waibel. 2001. Historical trends in epiphytal ostracodes from Florida Bay: implications for seagrass and macro-benthic algal variability. Bull. Am. Paleont. 361: S159–S197.

Cronin, T.M., G.S. Dwyer, S.B. Schwede, C.D. Vann and H. Dowsett. 2002. Climate variability from the Florida Bay sedimentary record: possible teleconnections to ENSO, PNA, and CNP. Clim. Res. 19: 233–245.

Davis, S.M., L.H. Gunderson, W.A. Park, J.R. Richardson and J.E. Mattson. 1994. Landscape dimension, composition, and function in a changing Everglades ecosystem. pp. 419–444. In: S.M. Davis and J.C. Ogden (eds.). Everglades: The Ecosystem and its Restoration. St. Lucie Press, Delray Beach, FL.

Davis, S.M., D.L. Childers, J.J. Lorenz, H.R. Wanless and T.E. Hopkins. 2005. A conceptual model of ecological interactions in the mangrove estuaries of the Florida Everglades. Wetlands 25: 832–842.

Davies, T.D. and A.D. Cohen. 1989. Composition and significance of the peat deposits of Florida Bay. Bul. Mar. Sci. 44: 387–398.

Donders, T.H., P.M. Gorissen and F. Sangiorgi. 2008. Three-hundred-year hydrological changes in a subtropical estuary, Rookery Bay (Florida): Human impact versus natural variability. Geochem. Geophy. Geosy. 9: 1–15.

Dwyer, G.S. and T.M. Cronin. 2001. Ostracode shell chemistry as a paleosalinity proxy in Florida Bay. Bull. Am. Paleont. S361: 249–276.

Elsner, J.B., J.P. Kossin and T.H. Jagger. 2008. The increasing intensity of the strongest tropical cyclones. Nature 455: 92–95.

Enfield, D.B., A.M. Mestas-Nuñez and P.J. Trimple. 2001. The Atlantic multidecadal oscillation and its relation to rainfall and river flows in the continental. U.S. Geophys. Res. Lett. 28(10): 2077–2080.

Enos, P. and R.D. Perkins. 1979. Evolution of Florida Bay from island. Geol. Soc. Am. Bull. 1: 59–83.

EOL. 2013. http://eol.org/pages/4888/overview.

Fedorov, A.V. and S.G. Philander. 2000. Is El Niño Changing? Science 288: 1997–2002.

Fourqurean, J.W., G.A. Kendrick, L.S. Collins, R.M. Chambers and M.A. Vanderklift. 2012. Carbon, nitrogen and phosphorus storage in subtropical seagrass meadows: examples from Florida Bay and Shark Bay. Mar. Freshwater Res. 63: 967–983.

Gaiser, E., A. Zafiris, P.L. Ruiz, F.A.C. Tobias and M.S. Ross. 2006. Tracking rates of ecotone migration due to salt-water encroachment using fossil mollusks in coastal South Florida. Hydrobiologia 569: 237–257.

Gershunov, A. and T.P. Barnett. 1998. Interdecadal modulation of ENSO teleconnections. Bull. Am. Met. Soc. 79: 2715–2726.

Hu, Q. and S. Feng. 2012. AMO- and ENSO-driven summertime circulation and precipitation variations in North America. J. Clim. 25: 19, 6477–6495.

Huvane, J.K. 2002. Modern diatom distributions in Florida Bay: a preliminary analysis. pp. 479–496. *In*: J.W. Porter and K.G. Porter (eds.). The Everglades, Florida Bay, and Coral Reefs of the Florida Keys: An Ecosystem Sourcebook. CRC Press, Boca Raton, FL.

Huvane, J.K. and S.R. Cooper. 2001. Diatoms as indicators of environmental Change in sediment cores from northeastern Florida Bay. Bull. Am. Paleont. S 361: 145–158.

IPCC. 2007. Climate Change 2007: The Physical Science Basis. Intergovernmental Panel on Climate Change, Cambridge University Press.

Ishman, S.E. 1997. Ecosystem History of South Florida: Biscayne Bay Sediment Core Descriptions: U.S. Geologic Survey Open-File Report 97–437. 13 pp.

Ishman, S.E. 2001. Ecological controls on benthic foraminifera distributions in Biscayne Bay, Florida. Bull. Am. Paleont. S361: 233–248.

Ishman, S.E., T.M. Cronin, G.L. Brewster-Wingard, D.A. Willard and D.J. Verardo. 1998. A record of ecosystem Change, Manatee Bay, Barnes Sound, Florida: Proceedings of the International Coastal Symposium (ICS98). J. Coast. Res. Sci. 26: 125–138.

Kelly, J.R. 2008. Nitrogen effects on coastal marine ecosystems. pp. 271–332. *In*: J.L. Hatfield and R.F. Follett (eds.). Nitrogen in the Environment: Sources, Problems, and Management, Second edition. Elsevier Inc., Academic Press/Elsevier, Amsterdam, Boston. Konar, B. and K. Iken.

Krauss, K.W., A.S. From, T.W. Doyle, T.J. Doyle and M.J. Barry. 2011. Sea-level rise and landscape Change influence mangrove encroachment onto marsh in the Ten Thousand Islands region of Florida, USA. J. Coast. Conserv. 15: 629.

Lammers, J.M., E.E. van Soelen, T.H. Donders, F. Wagner-Cremer, J.S. Sinninghe Damsté and G.J. Reichart. 2013. Natural Environmental Changes versus Human Impact in a Florida Estuary (Rookery Bay, USA). Est. Coast. 36: 149–157.

Langevin, C.D. and M. Zygnerski. 2012. Effect of Sea Level Rise on Salt Water Intrusion near a Coastal Well Field in Southeastern Florida. Ground water.

Light, S.S. and J.W. Dineen. 1994. Water control in the Everglades: A historical perspective. pp. 47–84. *In*: S.M. Davis and J.C. Ogden (eds.). Everglades: The Ecosystem and its Restoration. St. Lucie Press, Delray Beach, FL. 826 pp.

Lodge, T.E. 2010. The Everglades Handbook–Understanding the Ecosystem, 3rd edn. CRC Press, Boca Raton.

Mann, M.E., J.D. Woodruff, J.P. Donnelly and Z. Zhang. 2009. Atlantic hurricanes and climate over the past 1,500 years. Nature 460: 880–885.

Marshall, F.E. and G.L. Wingard. 2012. Florida Bay salinity and Everglades wetlands hydrology circa 1900 CE: A compilation of paleoecology-based statistical modeling analyses: U.S. Geological Survey Open-File Report 2012-1054, 32 pp., available only online at http://pubs.usgs.gov/of/2012/1054.

Marshall, F.E., G.L. Wingard and P.A. Pitts. 2009. A simulation of historic hydrology and salinity in Everglades National Park: coupling paleoecologic assemblage data with regression models. Est. Coast. 32: 1: 37–53.

McCabe, G.J., M.A. Palecki and J.L. Betancourt. 2004. Pacific and Atlantic Ocean influences on multidecadal drought frequency in the United States. Proc. Nat. Acad. Sci. 101: 4136–4141.

McIvor, C.C., J.A. Ley and R.D. Bjork. 1994. Changes in freshwater inflow from the Everglades to Florida Bay including effects on biota and biotic processes: a review. pp. 117–146. *In*: S.M. Davis and J.C. Ogden (eds.). Everglades: The Ecosystem and its Restoration. St. Lucie Press, Delray Beach, FL.

McPherson, B.F. and R. Halley. 1996. The south Florida Environment. U.S. Geological Survey Circular 1134. 61 pp.

McVoy, C.W., W.P. Said, J. Obeysekera, J.A. VanArman and T.W. Dreschel. 2011. Landscapes and Hydrology of the Predrainage Everglades. University Press of Florida, Gainesville, FL.

Moses, C., W.T. Anderson, C.J. Saunders and F.H. Sklar. 2013. Regional climate gradients in precipitation and temperature in response to climate teleconnections in the Greater Everglades ecosystem of South Florida. J. Paleolimnol. 49: 5–14.

Murray, J.B., G.L. Wingard, T.M. Cronin, W.H. Orem, D.A. Willard, C.W. Holmes, C. Reich, E. Shinn, M. Marot, T. Lerch, C. Trappe and B. Landacre. 2010. Evidence of Environmental Change in Rankin Basin, Central Florida Bay, Everglades National Park: U.S. Geological Survey Open-File Report 2010-1125, 54 p.

NASA. 2013. http://climate.nasa.gov/key_indicators.

Nelsen, T.A., G. Garte, C. Featherstone, H.R. Wanless, J.H. Trefry, W.J. Kang, S. Metz, C. Alvarez-Zarikian, T. Hood, P. Swart, P. Blackwelder, L. Tedesco, C. Slouch, J.F. Pachut and M. O'Neal. 2002. Linkages between the south Florida peninsula and coastal zone: a sediment-based history of natural and anthropogenic influences. pp. 415–449. *In*: J.W. Porter and K.G. Porter (eds.). The Everglades, Florida Bay, and Coral Reefs of the Florida Keys: An Ecosystem Sourcebook. CRC Press, Boca Raton, FL.

NOAA. 2013. http://oceanservice.noaa.gov/facts/sealevel.html.

NOAA. 2013a. http://tidesandcurrents.noaa.gov/sltrends/sltrends_station.shtml?stnid=8724580.

NOAA. 2013b. http://www.srh.noaa.gov/mfl/?n=winteroutlookforsouthflorida.

NSIDC. 2013. http://nsidc.org/cryosphere/sotc/glacier_balance.html.

Orem, W.H., C.W. Holmes, C. Kendall, H.E. Lerch, A.L. Bates, S.R. Silva, A. Boylan, M. Corum and C. Hedgman. 1999. Geochemistry of Florida Bay sediments: nutrient history at five sites in eastern and central Florida Bay. J. Coastal Res. 15: 1055–1071.

Parkinson, R.W. 1987. Holocene Sedimentation and Coastal Response to Rising Sea Level Along a Subtropical Low Energy Coast, Ten Thousand Islands, Southwest Florida Ph.D. dissertation: Coral Gables, Fla., Univ. Miami, 224 p.

Parkinson, R.W. 1989. Decelerating Holocene sea-level rise and its influence in southwest Florida coastal evolution: A transgressive/regressive stratigraphy. J. Sed. Petrol. 59: 960–972.

Pickerill, R.K. and P.J. Brenchley. 1991. Benthic macrofossils as paleoenvironmental indicators in marine siliciclastic facies. Geosci. Can. 18: 119–138.

Reed, C. 2009. Twentieth-century Marine Science: Decade by Decade, 1st Edn. Facts on File Inc. (J), 298 pp.

Ross, M.S., E.E. Gaiser, J.F. Meeder and M.T. Lewin. 2001. Multi-taxon analysis of the "white zone", a common ecotonal feature of South Florida coastal wetlands. pp. 205–238. *In*: J.W. Porter and K.G. Porter (eds.). The Everglades, Florida Bay, and Coral Reefs of the Florida Keys. CRC Press, Boca Raton, FL, USA.

Saha, A.K., S. Saha, J. Sadle, J. Jiang, M.S. Ross, R.M. Price, L. Sternberg and K.S. Wendelberger. 2011. Sea level rise and South Florida coastal forests. Climatic Change 107: 81–108.

Scholl, D.W. 1964. Recent sedimentary record in mangrove swamps and rise in sea-level over the southwestern coast of Florida. Mar. Geol. 1: 344–367.

Sklar, F.H. and J.A. Browder. 1998. Coastal environmental impacts brought about by alterations to freshwater flow in the Gulf of Mexico. Environ. Manage. 22: 547–562.

Sklar, F.H., C. McVoy, R. VanZee, D.E. Gawlik, K. Tarboton, D. Rudnick and S. Miao. 2002. The effects of altered hydrology on the ecology of the Everglades. pp. 39–82. *In*: J.W. Porter and K.G. Porter (eds.). The Everglades, Florida Bay and Coral Reefs of the Florida Keys—An Ecosystem Sourcebook. CRC Press, Boca Raton, Florida, USA.

Smith, V.H. and D.W. Schindler. 2009. Eutrophication science: where do we go from here? Trends Ecol. Evol. 24: 201–207.

Smol, J.P. and E.F. Stoermer. 2010. The Diatoms: Applications for the Environmental and Earth Sciences, 2nd edn. Cambridge University Press, Cambridge.

Smol, J.P., H.J.B. Birks, W.M. Last, R.S. Bradley and K. Alverson. 2001. Tracking Environmental Change Using Lake Sediments. Volume 3: Terrestrial, Algal, and Siliceous Indicators. pp. 371.

Stone, J.R., T.M. Cronin, G.L. Brewster-Wingard, S.E. Ishman, B.R. Wardlaw and C.W. Holmes. 2000. A paleoecological reconstruction of the history of Featherbed Bank, Biscayne National Park, Biscayne Bay, Florida: USGS Open-File Report 00–191, 24 pp.

Swart, P.K., G. Healy, R.E. Dodge, P. Kramer, J.H. Hudson, R.B. Halley and M. Robblee. 1996. The stable oxygen and carbon isotopic record from a coral Florida Bay: A 160 year record of climatic and anthropogenic influence. Palaeogeogr. Paleocl. Palaeocl. 123: 219–237.

Swart, P.K., R. Price and L. Greer. 2001. The relationship between stable isotopic variations (O, H, and C) and salinity in waters and corals from environments in South Florida: implications for reading the paleoenvironmental record. Bull. Am. Paleont. 361: 17–30.

Tilmant, J.T. 1989. A history and an overview Bay. Bull. Mar. Sci. 44: 3–22.

Timmermann, A., J. Oberhuber, A. Bacher, M. Esch, M. Latif and E. Roeckner. 1999. Increased El Niño frequency in a climate model forced by future greenhouse warming. Nature 398: 694–697.

Trap, R.J., N.S. Diffenbaugh, H.E. Brooks, M.E. Baldwin, E.D. Robinson and J.S. Pal. 2007. Changes in severe thunderstorm environment frequency during the 21st century caused by anthropogenically enhanced global radiative forcing. Proc. Nat. Acad. Sci. 104: 19719–19723.

Trappe, C.A. and G.A. Brewster-Wingard. 2001. Molluscan fauna from Core 25B, Whipray Basin, Central Florida Bay, Everglades National Park: U.S. Geological Survey Open File Report 01–143, 20 p.

Trenberth, K.E. and T.J. Hoar. 1997. El Niño and climate change. Geophys. Res. Ltrs. 24: 3057–3060.

U.S. Army Corps of Engineers. 1999. Central and southern Florida comprehensive review study, final integrated feasibility report and programmatic environmental impact statement. Jacksonville, Florida. [Available at http://www.evergladesplan.org/].

U.S. Army Corps of Engineers. 2006. Comprehensive Everglades restoration plan system-wide performance measures, RECOVER Leadership Group draft.

Vecchi, G.A. and A.T. Wittenberg. 2010. El Nino and our future climate: where do we stand? Wiley Interdiscip. Rev. Clim. Change 1: 260–270.

Wachnicka, A. 2009. Quantitative diatom-based reconstruction of paleoenvironmental conditions in Florida Bay and Biscayne Bay, U.S.A. PhD dissertation, Florida International University, Miami, Florida, USA. http://etd.fiu.edu/ETD-db/available/etd-0506109-123854.

Wachnicka, A., E. Gaiser, L. Collins, T. Frankovich and J. Boyer. 2010. Distribution of diatoms and development of diatom-based models for inferring salinity and nutrient concentrations in Florida Bay and adjacent coastal wetlands (U.S.A.). Est. Coast. 33: 1080–1098.

Wachnicka, A., E. Gaiser and J. Boyer. 2011. Autecology and distribution of diatoms in Biscayne Bay, Florida: Implications for bioassessment and paleoenvironmental studies. Ecol. Indicat. 11: 622–632.

Wachnicka, A., E. Gaiser, L. Wingard, H. Briceño and P. Harlem. 2013a. Impact of the late Holocene climate variability and anthropogenic activities on Biscayne Bay (Florida, U.S.A.) environment: evidence from diatoms. Palaeogeogr. Palaeocl. Palaeoecol. 371: 80–92.

Wachnicka, A., L. Collins and E. Gaiser. 2013b. Response of diatom assemblages to 130 years of environmental Change in Florida Bay (U.S.A.). J. Pal. 49: 83–101.

Wachnicka, A., E. Gaiser and L. Collins. 2013c. Correspondence of historic salinity fluctuations in Florida Bay, U.S.A., to atmospheric variability and anthropogenic Changes. J. Pal. 49: 103–115.

Wang, J.D., J. Luo and J.S. Ault. 2003. Flows, salinity and some implications on larval transport in South Biscayne Bay. Fl. B. Mar. Sci. 72: 695–723.

Wanless, H.R. 1976. Geological setting and recent sediments of the Biscayne Bay region. pp. 32. In: A. Thorhaug and A. Volker (eds.). Biscayne Bay: Past, Present and Future, Special Report, University of Miami Sea Grant.

Wanless, H.R. and M.G. Tagett. 1989. Origin, growth and evolution of carbonate mudbanks in Florida Bay. B. Mar. Sci. 44: 454–489.

Wanless, H.R., R.W. Parkinson and L.P. Tedesco. 1994. Sea level control on stability of Everglades wetlands. pp. 199–224. In: S.M. Davis and J.C. Ogden (eds.). Everglades: The Ecosystem and its Restoration. St. Lucie Press, Delray Beach, Florida.

Willard, D.A. and C.E. Bernhardt. 2011. Impacts of past climate and sea level Change on Everglades wetlands: placing a century of anthropogenic Change into a late-Holocene context. Clim. Chan. 107: 59–80.

Willard, D.A., L. Brewster-Wingard, C. Fellman and S.E. Ishaman. 1997. Paleontological data from Mud Creek core 1, southern Florida. US Geological Survey Open-file Report 97–736.

Williams, R.A. 2009. Comparing reef bioindicators on benthic environments off southeast Florida. PhD dissertation, University of South Florida, St. Petersburg, Florida, USA.

Wilson, J.G. and J.W. Fleeger. 2012. Estuarine Benthos. *In*: J.W. Day, B.C. Crump, W.M. Kemp and A. Yáñez-Arancibia (eds.). Estuarine Ecology, Second Edition. John Wiley & Sons, Inc., Hoboken, NJ, USA. doi: 10.1002/9781118412787.ch12.

Wingard, G.L. and J.W. Hudley. 2012. Application of a weighted-averaging method for determining paleosalinity: a tool for restoration of south Florida's estuaries. Estuaries Coasts 35: 262–280. DOI: 10.1007/s12237-011-9441-3.

Wingard, G.L., T.M. Cronin, G.S. Dwyer, S.E. Ishman, D.A. Willard, C.W. Holmes, C.E. Bernhardt, C.P. Williams, M.E. Marot, J.B. Murray, R.G. Stamm, J.H. Murray and C. Budet. 2003. Ecosystem History of Southern and Central Biscayne Bay: Summary Report on Sediment Core Analyses: U.S. Geological Survey, OFR 03-375, 110 p. [Available online at http://sofia.usgs.gov/publications/ofr/03-375/].

Wingard, G.L., T.M. Cronin, C.W. Holmes, D.A. Willard, G.S. Dwyer, S.E. Ishman, W. Orem, C.P. Williams, J. Albeitz, C.E. Bernhardt, C. Budet, B. Landacre, T. Lerch, M.E. Marot and R. Ortiz. 2004. Ecosystem History of Southern and Central Biscayne Bay: Summary Report on Sediment Core Analyses—Year Two: U.S. Geological Survey, OFR 2004-1312, 109 p. [Available online at http://sofia.usgs.gov/publications/ofr/2004-1312/].

Wingard, G.L., T.M. Cronin, C.W. Holmes, D.A. Willard, C. Budet and R. Ortiz. 2005. Descriptions and preliminary report on sediment cores from the southwest coastal area, Everglades National Park, Florida: U.S. Geological Survey, OFR 2005-1360. 28 pp.

Wingard, G.L., C.A. Budet, R.E. Ortiz, J. Hudley and J.B. Murray. 2006. Descriptions and Preliminary Report on Sediment Cores from the Southwest Coastal Area, Part II: Collected July 2005, Everglades National Park, Florida. U.S. Geological Survey Open File Report 2006-1271.

Wingard, G.L., J.W. Hudley and C.W. Holmes. 2007. Synthesis of age data and chronology for Florida Bay and Biscayne Bay cores collected for the Ecosystem History of South Florida's Estuaries Projects. US Geological Survey Open-file Report 2007-1203. http://sofia.usgs.gov/publications/ofr/2007-1203/index.html.

Xu, Y., C.W. Holmes and R. Jaffe. 2007. Paleoenvironmental assessment of recent environmental changes in Florida Bay, USA: A biomarker based study. Estuar. Coast. Shel. S73: 201–210.

Zarikian, C.A.A., P.K. Swart, T. Hood, P.L. Blackwelder, T.A. Nelsen and C. Featherstone. 2001. A century of environmental variability in Oyster Bay using ostracode ecological and isotopic data as paleoenvironmental tools. B. Am. Pal. S361: 133–143.

12

Epiphytic Diatoms along Phosphorus and Salinity Gradients in Florida Bay (Florida, USA), an Illustrated Guide and Annotated Checklist

Thomas A. Frankovich[1,*] and *Anna Wachnicka*[2]

Introduction

Diatoms are important primary producers that are widespread in marine ecosystems. Their high species richness and diversity, and strong relationships to water quality, make them ideal organisms for bioassessments of ecosystems (Desrosiers et al. 2013). However, before these organisms can be used in bioassessments, their taxonomy must be better understood. The great number of individual diatom species increases the likelihood that a few of them will exhibit changes in relative abundance in response to changes in particular environmental characteristics. Salinity is one of the most influential variables affecting the spatial distribution of diatoms, with many taxa limited to either marine, brackish, or freshwater habitats (Frankovich et al. 2006; Saunders 2011; Haynes et al. 2011; Wachnicka et al. 2010; 2011). Nutrient availability has also been shown to influence the composition and abundance of diatom assemblages

[1] Department of Biological Sciences, Marine Science Program and Southeast Environmental Research Center. Florida Bay Interagency Science Center, 98630 Overseas Highway, Key Largo, Florida 33037.
Email: frankovich@virginia.edu
[2] Southeast Environmental Research Center, Florida International University, 11200 SW 8th St. OE 235, Miami, Florida 33199.
Email: wachnick@fiu.edu
* Corresponding author

in estuarine and coastal ecosystems (Armitage et al. 2006; Frankovich et al. 2006; 2009; Wachnicka et al. 2010; 2011; Arndt et al. 2011). Because estuaries are often adjacent to highly urbanized coastal areas or upstream agriculture, their biological communities are exposed to changes in nutrient concentrations and salinities. Water management in the upstream watersheds of Florida Bay has maintained water levels that are conducive for agriculture or residential development but these practices have often produced unnatural freshwater flows (i.e., water diversions and large pulses) into the Florida Bay estuary resulting in altered salinity regimes and pulsed nutrient inputs (Van Lent et al. 1993). The stress on estuarine biological communities will continue to increase due to escalating coastal development and rising sea-level.

The highly diverse nature of diatom communities is also a challenge to anyone who desires to use these organisms in bioassessments. A recent reference book published by Witkowski et al. 2000, describing benthic marine diatoms, identified 1183 "well-known" taxa representing 130 genera. Given that a majority of the species that investigators may encounter in their studies will be lesser-known or rare, many additional, sometimes obscure, and often difficult to obtain references will be required for correct species identification. Investigators can also expect to find species that are new to science. Species identifications are often difficult and highly subjective. They depend on the taxonomic skills of the observer, the quality of the diatom preparations, the availability of scanning electron microscopy (SEM), and the availability of suitable reference materials. This makes comparison of diatom species lists constructed by various authors at different times difficult. Detailed illustrations, valve morphometrics (i.e., descriptions of shape, size, striae densities, character of valve ornamentation), the referencing of the resources used in species identification, and comments on the key characters that differentiate similar-looking taxa, facilitate more accurate comparisons of species assemblages. A helpful first resource for anyone wishing to become familiar with a particular diatom community or perhaps initiate further study would be an illustrated guide of the locally common taxa that one is likely to find in their samples. Provided here is the first illustrated guide to the benthic diatom flora in Florida Bay.

Benthic diatom studies in Florida Bay and the Florida Keys are rare. The first list of 108 taxa from calcareous sands in the Florida Keys was published by Albert Mann in 1935. In 1978, two more lists were published by DeFelice and Lynts (list of 161 taxa) and Montgomery (list of 572 taxa) from the upper Florida Bay and adjoining sounds and from the coral reefs of the Florida Keys, respectively. With the exception of studies limited to newly described species or a specific genus (Wachnicka and Gaiser 2007), Montgomery (1978) is presently the only illustrated guide to the benthic diatoms in the shallow marine ecosystems of south Florida.

This chapter is the first illustrated guide to common epiphytic diatoms in Florida Bay. The abundances of the more common taxa and the locations of maximum relative abundance along existing salinity and nutrient gradients are also provided. We also report the response of some individual taxa to phosphorus fertilization.

Study Area

Florida Bay and the Florida Keys are the downstream end-members of the Greater Everglades. Florida Bay and the nearshore waters of the Florida Keys are shallow, seagrass-dominated and nutrient-limited ecosystems. Florida Bay is a large (approximately 2000 km²) sub-tropical coastal lagoon adjacent to the southern tip of Florida, USA (Fig. 1). It is bounded on the north by the coastal southern Everglades mangrove swamp, on the southeast by the Florida Keys archipelago, and on the west by the Gulf of Mexico. The bay bottom is covered by seagrasses, predominantly *Thalassia testudinum*, whose leaves provide most of the surface area available for benthic diatoms. A network of carbonate mud banks, some of which are often exposed during the winter dry season, divides the bay into 1–2 m deep basins. These mud banks and the Florida Keys act to limit oceanic water exchange, and provide the geography conducive for the presence of west to east salinity, salinity variability, and nutrient gradients in Florida Bay (Fourqurean and Robblee 1999). Salinities are relatively stable and marine (~35 psu) along the western edge of the bay and grade to more brackish salinities towards the northeast corner, except during droughts when low precipitation and high rates of evaporation create hypersaline conditions, with salinities occasionally reaching as high as 70 psu levels in the central and eastern parts of the bay. The Gulf of Mexico is the primary source of phosphorus (P), the limiting nutrient for most of Florida Bay primary production (Fourqurean et al. 1992). Limited P delivery from the freshwaters draining the upstream oligotrophic Everglades and P uptake by seagrass and associated epiphytes effectively reduce P availability in the eastern bay. As relatively P-rich Gulf of Mexico waters in the west are exchanged with eastern bay waters, a P-nutrient gradient from west to east is created (Boyer et al. 1997).

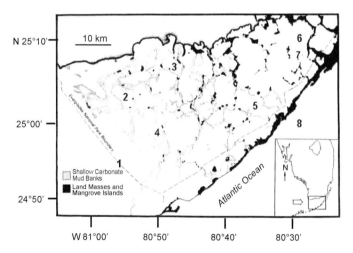

Figure 1. Florida Bay, USA, showing sampling locations for epiphytic diatoms. Distinct diatom community groups (Frankovich et al. 2006) are designated as Eastern Florida Bay (sites 5, 6, 7), Central Florida Bay (sites 2–4), Western Florida Bay (site 1), and Atlantic (site 8). Sites 4 and 6 were the locations of nutrient fertilization experiments (Frankovich et al. 2009).

Material and Methods

Diatom assemblages were examined from 96 permanent microscope slides prepared during two previously published investigations of epiphytic diatoms (Frankovich et al. 2006; 2009). Frankovich et al. (2006) identified four distinct diatom communities based on a surveys of epiphytic diatoms from seven sites within the bay and one site located east of the Florida Keys (Fig. 1), conducted during 6 sampling events between March 2000 and February 2001. Frankovich et al. (2009) enumerated epiphytic diatoms from fertilized and unfertilized seagrass plots at sites 4 and 6 in August 2004 and February 2005, and determined that addition of phosphorus changed the structure of diatom assemblages. The same methods were used to prepare slides in both of these studies (details provided in Frankovich et al. 2006; 2009). At least 500 diatom valves were counted along linear transects on each slide. Relative abundances of each taxa were determined based on enumerations performed using a Nikon Eclipse E600 light microscope (1,200 X magnification, oil immersion, numerical aperture = 1.40) equipped with differential interference contrast (DIC).

Taxa selected for illustrations in this chapter were those used in the ordination statistics to determine community groups and fertilization responses in Frankovich et al. 2006; 2009. In each of these studies, the taxa used were present in at least 5% of the samples and had a maximum abundance of > 1% in at least one of the samples. Although these taxa constituted only 28–39% of the total taxa observed in each of the studies, they represented 95–97% of the total valve count. All illustrated taxa were photographed using a Leica DFC425 digital camera and their morphometrics were determined using Leica Application Suite version 3.7 software. For each of the illustrated taxa, abundance is listed according to the maximum relative abundances observed from all 96 samples. Each taxon's abundance is defined as either abundant, common, or rare, which corresponds to maximum relative abundances of > 10%, 5–9%, and 1–5%, respectively. The very rare (< 1% maximum relative abundance) diatoms that were observed, and could be identified as established taxa, are not illustrated, but are listed in addition to the illustrated taxa in Table 1. The locations of maximum abundance for the illustrated taxa are also listed as either eastern Florida Bay, central Florida Bay, western Florida Bay, or Atlantic, and correspond to the four distinct community groups described in Frankovich et al. (2006). Western Florida Bay is site 1; Central Florida Bay consists of sites 2, 3, and 4; Eastern Florida Bay consists of sites 5, 6 and 7; and Atlantic is site 8. Mean and range of salinities and water column total phosphorus (TP) concentrations observed at these locations during sampling are listed in Table 2.

Although Frankovich et al. (2009) reported changes in community structure resulting from phosphorus enrichment at sites 4 and 6 in Florida Bay in August 2004, they did not report the responses of individual taxa to the enrichment. To determine individual species responses, we constructed data matrices of species relative abundances in P-enriched and P-unenriched seagrass plots and used Indicator Species Analysis (ISA; Dufrêne and Legendre 1997) to identify species that responded to P fertilization. ISA identifies indicator species based on the concentration of species abundance in a particular group and on the fidelity of occurrence in that group. Indicator values are calculated from the product of the proportional abundance of a particular species in a particular group relative to the abundance of that species in all groups, and the proportional frequency of the species in each group (Dufrêne and Legendre 1997; McCune and Grace 2002). The indicator values range from 0 (no indication) to 100 (perfect indication). The statistical significance of the indicator value is evaluated by

Table 1. List of observed epiphytic diatom taxa on *Thalassia testudinum.*

Class Coscinodiscophyceae
Actinocylus normanii (Gregory in Greville 1859) Hustedt 1957
Actinocylus octonarius var. tenellus (Brébisson 1854) Hendey 1954
Anaulus minutus Grunow in Van Heurck 1882
Ardissonea formosa Hantzsch in Rabenhorst 1863
Ardissonea fulgens (Greville 1827) Grunow in Cleve and Grunow 1880
Biddulphia biddulphiana (J.E. Smith 1807) Boyer 1900
Biddulphia rhombus (Ehrenberg 1839) W. Smith in Roper 1854
Cyclotella choctawatcheeana Prasad 1990
Cyclotella desikacharyi Prasad 2006
Cyclotella menighiniana Kützing 1844
Cymatosira belgica Grunow in Van Heurck 1880
Cymatosira lorenziana Grunow 1862
Cymatosirella cf. *capensis* (Giffen 1975) Dąbek, Witkowski and Archibald 2013
Glyphodesmis eximia Greville 1862
Hyalodiscus scoticus (Kützing 1844) Grunow 1879
Isthmia minima Harvey and J.W. Bailey 1854
Odontella aurita (Lyngbye 1819) Agardh 1832
Paralia sulcata (Ehrenberg 1838) Cleve 1873
Plagiogramma atomus Greville 1863
Plagiogramma pulchellum var. *pygmaeum* (Greville 1859) H. Peragallo and M. Peragallo 1901
Pleurosira laevis (Ehrenberg 1843) Compère 1982
Psammodiscus nitidus (Gregory 1857) Round and Mann 1980
Toxarium hennedyanum (Gregory 1857) Pelletan 1889
Toxarium undulatum Bailey 1854
Triceratium reticulum (Ehrenberg 1845)
Class Fragilariophyceae
Catacombas gaillonii (Bory 1844) Williams and Round 1986
Florella portoricensis Navarro 1982
Grammatophora angulosa Ehrenberg 1841 var. *angulosa*
Grammatophora hamulifera Kützing 1844
Grammatophora marina (Lyngbye 1819) Kützing 1844
Grammatophora oceanica (Ehrenberg 1854 pro parte) Grunow var. *oceanica*
Grammatophora serpentina (Ralfs1842) Ehrenberg 1844
Hyalosira interrupta (Ehrenberg 1838) Navarro 1991
Hyalosira tropicalis Navarro 1991
Hyalosynedra laevigata (Grunow 1877) Williams and Round 1986
Licmophora abbreviata Agardh 1831
Licmophora ehrenbergii (Kützing 1844) Grunow 1867 var. *ovata* (W. Smith 1853) Grunow 1867
Licmophora debilis (Kützing 1844) Grunow 1881
Licmophora gracilis (Ehrenberg 1838) Grunow 1867
Licmophora hastata Mereschkowsky 1901
Licmophora normaniana (Greville 1862) Wahrer 1985
Licmophora remulus Grunow 1867
Neodelphineis pelagica Takano 1982

Table 1. contd....

Table 1. contd.

Class Fragilariophyceae
Neosynedra provincialis (Grunow 1877) Williams and Round 1986
Neosynedra tortosa (Grunow 1877) Williams and Round 1986
Opephora cf. *horstiana* Witkowski 1994
Pteroncola inane (Giffen 1970) Round 1990
Reimerothrix floridensis Prasad 2001
Rhabdomena adriaticum Kützing 1844
Rhabdomena crozierii (Ehrenberg 1853) Grunow 1862
Rhabdomena minutum Kützing 1844
Striatella unipunctata (Lyngbye 1819) Agardh 1832
Synedra bacillaris (Grunow 1877) Hustedt 1932
Synedra fasciculata (C.A. Agardh 1812) Kützing 1844 f. *densestriata* (Møller 1950) Archibald 1983
Synedra lata (Giffen 1980) Witkowski 1990
Synedrosphenia gomphonema (Janisch and Rabenhorst 1863) Hustedt 1932
Tabularia tabulata (C. Agardh 1832) Snoeijs 1992
Thalassiothrix longissima Cleve and Grunow 1880
Class Bacillariophyceae
Achnanthes citronella (A. Mann 1925) Hustedt in Schmidt et al. 1937
Amphora abludens Simonsen 1960
Amphora acuta Gregory 1857
Amphora americana Wachnicka and Gaiser 2007
Amphora arcus Gregory 1854
Amphora arcus Gregory 1854 var. *sulcata* Schmidt 1875
Amphora bigibba Grunow 1875 var. *interrupta* (Grunow 1875) Cleve 1895
Amphora cingulata Cleve 1878
Amphora copulata (Kützing 1833) Schoeman and Archibald 1986
Amphora corpulenta var. *capitata*
Amphora crenulata Wachnicka and Gaiser 2007
Amphora decussata Grunow 1877
Amphora dubia Gregory 1857
Amphora floridae Wachnicka and Gaiser 2007
Amphora graeffei Cleve 1896
Amphora graeffeana Hendey 1973
Amphora hamata Heiden and Kolbe 1928
Amphora helenensis Giffen 1973
Amphora holsaticoides Nagumo and Kobayasi 1990
Amphora hyalina Kützing 1844
Amphora kolbei Aleem 1950
Amphora laevissima Gregory 1857
Amphora obtusa var. *oceanica* (Castracane 1886) Cleve 1895
Amphora ostrearia Brébisson in Kützing 1849 var. *vitrea* (Cleve 1868) Cleve 1894–1895
Amphora proteus Gregory 1857
Amphora pseudohybrida Simonsen 1960
Amphora pseudoproteus Wachnicka and Gaiser 2007

Table 1. contd....

Table 1. contd.

Class Bacillariophyceae
Amphora sulcata (Brébisson 1854) Peragallo 1897–1908
Amphora subtropica Wachnicka and Gaiser 2007
Auricula complexa (Gregory 1857) Cleve 1894
Auricula intermedia (Lewis 1865) Cleve 1894
Berkeleya scopularum (Brébisson in Kützing 1849) Cox 1979
Biremis circumtexta (Meister in Hustedt 1934) Lange-Bertalot and Witkowski 2000
Brachysira aponina Kützing 1836
Caloneis bicuneata (Grunow 1860) Boyer 1927
Caloneis excentrica (Grunow 1860) Boyer 1927
Caloneis liber (W. Smith 1853) Cleve 1894
Caloneis cf. *linearis* (Grunow 1860) Boyer 1927
Campylodiscus ralfsii W. Smith 1853
Campylodiscus subangularis Grunow in Cleve and Möller 1877–1882
Climaconeis colemaniae Prasad 2000
Climaconeis cf. *inflexa* (Brébisson in Kützing 1849) Cox 1982
Climaconeis koenigii Prasad 2000
Climaconeis lorenzii Grunow 1862
Climaconeis riddleae Prasad 2003
Cocconeis barleyi Frankovich and De Stefano 2005
Cocconeis britannica Naegeli in Kützing 1849
Cocconeis clandestina Schmidt 1894
Cocconeis coralliensis Riaux-Gobin and Compère 2008
Cocconeis dirupta Gregory 1857
Cocconeis discrepans Schmidt 1894
Cocconeis distantula Giffen 1967
Cocconeis euglypta Ehrenberg 1854
Cocconeis fraudulens Simonsen 1987
Cocconeis lineata Ehrenberg 1849
Cocconeis maxima (Grunow 1863) Peragallo and Peragallo 1897
Cocconeis pediculus Ehrenberg 1838
Cocconeis pseudodiruptoides Foged 1975
Cocconeis scutellum Ehrenberg 1838 var. *baldjikiana* (Grunow 1888) Cleve 1895
Cocconeis scutellum Ehrenberg 1838 var. *parva* (Grunow 1881) Cleve 1895
Cocconeis scutellum Ehrenberg 1838 var. *scutellum*
Cocconeis thalassiana Romero and López-Fuerte 2013
Cocconeis vitrea Brun 1891
Cocconeis woodii Reyes-Vasquez 1970
Cylindrotheca closterium (Ehrenberg 1839) Reimann and J.C. Lewin 1964
Diploneis crabro (Ehrenberg 1844) Ehrenberg 1854
Diploneis litoralis var. *litoralis* (Donkin 1871) Cleve 1894
Diploneis parca (Schmidt 1875) Boyer 1927
Diploneis smithii (Brébisson in W. Smith 1856) Cleve 1894
Diploneis splendida (Gregory 1857) Cleve 1894
Diploneis suborbicularis (Gregory 1856) Cleve 1894

Table 1. contd....

Table 1. contd.

Class Bacillariophyceae
Diploneis vacillans (A. Schmidt 1875) Cleve 1894
Entomoneis corrugata (Giffen 1963) Witkowski, Lange-Bertalot and Metzeltin 2000
Entomoneis pseudoduplex Osada and Kobayasi 1990
Fallacia florinae (Moeller 1950) Witkowski 1993
Gyrosigma balticum (Ehrenberg 1838) Rabenhorst 1853
Gyrosigma mediterraneum Cleve 1894
Gyrosigma obscurum (W. Smith 1852) Griffith and Henfrey 1856
Gyrosigma peisonis (Grunow 1860) Hustedt 1930
Gyrosigma tenuissimum (W. Smith 1853) Griffith and Henfrey 1856
Halamphora aequatorialis (Heiden 1928) Levkov 2009
Halamphora aponina (Kützing 1844) Levkov 2009
Halamphora cymbifera (Gregory 1857) Levkov 2009 var. *heritierarum* Wachnicka and Gaiser 2007
Halamphora tenerrima (Aleem and Hustedt 1951) Levkov 2009
Halamphora cf. *tenuissima* (Hustedt 1955) Levkov 2009
Halamphora cf. *tumida* (Hustedt 1956) Levkov 2009
Haslea ostrearia (Gaillon 1820) Simonsen 1974
Haslea spicula (Hickie 1874) Lange-Bertalot 1997
Mastogloia acutiuscula Grunow in Cleve 1883
Mastogloia adriatica Voigt 1963
Mastogloia adriatica Voigt 1963 var. *linearis* Voigt 1963
Mastogloia angulata Lewis 1861
Mastogloia angusta Hustedt 1933
Mastogloia apicululata W. Smith 1856
Mastogloia bahamensis Cleve 1893
Mastogloia barbadensis (Greville 1864) Cleve 1895
Mastogloia beaufortiana Hustedt 1955
Mastogloia binotata (Grunow 1863) Cleve 1895
Mastogloia biocellata (Grunow 1877) Novarino and Muftah 1991
Mastogloia cocconeiformis Grunow 1860
Mastogloia corsicana Grunow 1878
Mastogloia corsicana Grunow 1878 var. *constricta* Moller 1984
Mastogloia cribrosa Grunow 1860
Mastogloia crucicula (Grunow 1877) var. alternans Zanon 1948
Mastogloia crucicula var. *crucicula* (Grunow 1877) Cleve 1895
Mastogloia cuneata (Meister 1937) Simonsen 1990
Mastogloia cyclops Voigt 1942
Mastogloia decipiens Hustedt 1933
Mastogloia decussata Grunow in Cleve 1892
Mastogloia discontinua Kemp and Paddock 1990
Mastogloia elegans Lewis 1865
Mastogloia emarginata Hustedt 1925
Mastogloia erythraea Grunow 1860
Mastogloia erythraea Grunow 1860 var. *grunowii* Foged 1984
Mastogloia exilis Hustedt 1933

Table 1. contd....

Table 1. contd.

Class Bacillariophyceae
Mastogloia fimbriata (Brightwell 1859) Cleve 1895
Mastogloia foliolum Brun in Schmidt 1893
Mastogloia frickei Hustedt 1933
Mastogloia gibbosa Brun 1895
Mastogloia gibbosa Brun 1895 var. *orientalis* Voigt 1963
Mastogloia goesii (Cleve 1878) Cleve 1892
Mastogloia gracillima Hustedt 1955
Mastogloia grunowii Schmidt 1893
Mastogloia hainanensis Voigt 1952
Mastogloia horvathiana Grunow 1860
Mastogloia ignorata Hustedt 1933
Mastogloia jelineckii Grunow 1868
Mastogloia lacrimata Voigt 1963
Mastogloia laminaris Grunow 1883
Mastogloia lanceolata Thwaites 1856
Mastogloia lancettula Cleve 1892
Mastogloia laterostrata Hustedt 1933
Mastogloia lineata Cleve and Grove 1891
Mastogloia manokwariensis Cholnoky 1963
Mastogloia mauritiana Brun in Schmidt 1893 var. capitata Voigt 1942
Mastogloia mauritiana Brun in Schmidt 1893
Mastogloia meisteri Hustedt 1933
Mastogloia nebulosa Voigt 1952
Mastogloia cf. *neoborneensis* Pennesi and Totti 2011
Mastogloia neomauritania Paddock and Kemp 1988
Mastogloia ovalis Schmidt 1893
Mastogloia ovata Grunow 1860
Mastogloia paradoxa Grunow 1878
Mastogloia pisciculus Cleve 1893
Mastogloia pseudoelegans Hustedt 1955
Mastogloia pseudolacrimata Yohn and Gibson 1982
Mastogloia pseudolatecostata Yohn and Gibson 1982
Mastogloia pseudolatericia Voigt 1952
Mastogloia punctatissima (Greville 1857) Ricard 1975
Mastogloia punctifera Brun 1895
Mastogloia pusilla Grunow 1878
Mastogloia pusilla Grunow 1878 var. *linearis* Østrup 1910
Mastogloia rigida Hustedt 1933
Mastogloia rimosa Cleve 1893
Mastogloia robusta Hustedt 1959
Mastogloia rostellata Grunow 1877
Mastogloia stephensiana Yohn and Gibson 1982
Mastogloia subaffirmata Hustedt 1927

Table 1. contd....

Table 1. contd.

Class Bacillariophyceae
Mastogloia submarginata Cleve and Grunow in Cleve and Moller 1881
Mastogloia testudinea Voigt 1942
Mastogloia urveae Witkowski, Lange-Bertalot and Metzeltin 2000
Mastogloia varians Hustedt 1933
Navicula cf. *agnita* Hustedt 1955
Navicula cf. *duerrenbergiana* Hustedt 1931
Navicula flebilis Cholnoky 1963
Navicula halinae Witkowski 2000
Navicula cf. *klavsenii* Østrup 1897
Navicula longa (Gregory 1856) Ralfs in Pritchard 1861 var. *irregularis* Hustedt 1955
Navicula cf. *microdigitoradiata* Lange-Bertalot 1993
Navicula cf. *normalis* Hustedt 1955
Navicula perrhombus Hustedt in Schmidt et al. 1934
Navicula cf. *subrhyncocephala* Hustedt 1935
Nitzschia aequorea Hustedt 1939
Nitzschia angularis Smith 1853 var. *affinis* Grunow 1862
Nitzschia dubiiformis Hustedt 1939
Nitzschia insignis Gregory 1857
Nitzschia lanceola Grunow 1880
Nitzschia liebetruthii Rabenhorst 1864
Nitzschia longissima (Brébisson in Kützing 1849) Grunow 1862
Nitzschia lorenziana Grunow in Cleve and Möller 1879
Nitzschia macilenta Gregory in Greville 1859
Nitzschia marginulata Grunow 1880 var. *didyma* Grunow 1880
Nitzschia panduriformis Gregory 1857 var. *continua* Grunow in Cleve and Grunow 1880
Nitzschia pubens Cholnoky 1960
Nitzschia sigma (Kutzing 1844) var. *intercedens* Grunow 1878
Nitzschia spathulata Brébisson in W. Smith 1853
Nitzschia ventricosa Kitton 1873
Nitzschia vidovichii Grunow 1862
Nitzschia weissflogii Grunow 1878
Oestrupia powelli (Lewis 1861) Heiden in Hustedt 1935
Parlibellus cf. *plicatus* (Donkin 1873) Cox 1988
Parlibellus panduriformis John 1991
Petroneis granulata (Bailey 1854) D. Mann 1990
Petroneis plagiostoma (Grunow in Cleve and Möller 1881)
Plagiotropis cf. *baltica* Pfitzer 1871
Plagiotropis cf. *gibberula* Grunow 1880
Plagiotropis tayrecta Paddock 1988
Planothidium campechianum (Hustedt 1952) Witkowski and Lange-Bertalot 2000
Planothidium cf. *pericavum* (Carter 1966) Lange-Bertalot 1999
Pleurosigma formosum W. Smith 1852 var. *balearica* H. Peragallo 1891
Pleurosigma normanii Ralfs in Pritchard 1861

Table 1. contd....

Table 1. contd.

Class Bacillariophyceae
Proschkinia complanata (Grunow 1880) Mann 1990
Protokeelia cholnokyana (Giffen 1963) Round and Basson 1995
Rhopalodia pacifica Krammer 1987
Rhopalodia cf. *sterrenburgii* Krammer 1988
Seminavis basilica Danielidis 2003
Seminavis cyrtorapha Wachnicka and Gaiser 2007
Seminavis delicatula Wachnicka and Gaiser 2007
Seminavis eulensteinii (Grunow 1875) Danielidis, Ford and Kennett 2006
Seminavis robusta Danielidis and Mann 2002
Seminavis strigosa (Hustedt 1949) Danielidis and Economou-Amilli 2003
Stauroneis dubitabilis Hustedt 1959
Surirella fastuosa Ehrenberg 1840
Surirella scalaris Giffen 1967
Thalassiophysa hyalina (Greville 1865) Paddock and Sims 1981
Trachyneis aspera Ehrenberg 1843

Table 2. Mean and range of salinity and water column total phosphorus (TP) concentrations observed at Western, Central, and Eastern Florida Bay, and Atlantic sites during March 2000–February 2001.

	Salinity (mean, range)	TP (μM) (mean, range)
Western Florida Bay	(36.8, 34.6–39.3)	(0.43, 0.13–0.88)
Central Florida Bay	(37.0, 33.6–41.7)	(0.65, 0.24–1.75)
Eastern Florida Bay	(30.3, 23.7–39.9)	(0.30, 0.05–1.26)
Atlantic	(36.0, 35.1–36.6)	(0.25, 0.14–0.34)

a Monte-Carlo method that randomly assigns sampling units to groups and calculates the proportion of times that the indicator value from the randomized dataset equals or exceeds the indicator value from the actual dataset (Dufrêne and Legendre 1997; McCune and Grace 2002). ISA was performed using PC-ORD version 4.27 with 1000 permutations in the Monte-Carlo test. Taxa having a statistical significance of $p < 0.05$ were considered reliable indicators.

Results

A total of 284 taxa, including 272 established taxa and 13 previously undescribed or unnamed taxa, were observed on the leaves of *Thalassia testudinum*. Out of the total of 67 genera that were found in the samples, *Mastogloia*, *Amphora*, and *Cocconeis* were the most represented, with 76, 29 and 19 taxa, respectively.

Indicator Species Analysis determined that the species that increased in response to phosphorus enrichment ($p < 0.05$) at site 6 in Eastern Florida Bay were *Nitzschia panduriformis* var. *continua, Nitzschia aequorea, Amphora graeffeana, Halamphora tenerrima,* and *Halamphora aponina*, while *Brachysira aponina, Mastogloia punctifera, Mastogloia erythraea,* and *Mastogloia robusta* decreased in abundance

(Table 3). At site 4 in Central Florida Bay, *Mastogloia laminaris* was only present in P-enriched plots, while *Halamphora tenerrima* decreased in abundance following P addition (Table 3).

One hundred six taxa were used to determine community groups and fertilization responses in the ordination statistics in Frankovich et al. 2006; 2009. Photomicrographs and descriptions of these taxa are provided in the next section. Phosphorus indicator species and community group indicators (Frankovich et al. 2006) are denoted by the bolded letters "P" and "S", respectively, following the taxon name and authorities. RSV = raphe sternum valve, SV = sternum valve.

Table 3. Phosphorus indicator species by sampling event as determined by Indicator Species Analysis ($p <$ 0.05) (Dufrêne and Legendre 1997). +P, –P denotes with or without phosphorus fertilization, respectively; RA = relative abundance (%), % change = change in RA relative to –P plots.

Taxa Names	–P Mean RA	+P Mean RA	% Change	Indicator Value	Significance (*p*)
Site 6 August 2004					
Nitzschia panduriformis var. *continua*	0.07	0.55	690	89.3	0.012
Mastogloia erythraea var. *erythraea*	2.51	0.44	–82	84.6	0.024
Amphora graeffeana	zero	0.74	undefined	83.3	0.014
Halamphora tenerrima	1.64	5.07	210	79.3	0.008
Halamphora coffeaeformis var. *aponina*	0.90	3.19	250	77.0	0.005
Nitzschia aequoria	0.12	1.51	1200	77.0	0.022
Mastogloia robusta	0.99	0.32	–68	75.4	0.044
Mastogloia punctifera	13.87	4.83	–65	74.3	0.005
Brachysira aponina	13.24	6.47	–51	67.1	0.007
Site 4 August 2004					
Mastogloia laminaris	zero	0.73	undefined	100	0.002
Halamphora tenerrima	0.82	0.22	–73	78.9	0.013

Floristic Observations

Common taxa—Class Coscinodiscophyceae (centric diatoms)

Ardissonea fulgens (Greville 1853) Grunow in Cleve and Grunow 1880
References: Hustedt 1931–1959, as *Synedra fulgens*, pp. 211–212, fig. 717. Hein et al. 2008, p. 23, fig. 8: 1, 2
Plate 3: Fig. 20: length – 392 µm, width – 16.3 µm, striae density – 12/10 µm;
Plate 3: Fig. 23: Detail of valve apex, scale bar – 50 µm.
Abundance and Location of Maximum Abundance: rare, eastern Florida Bay

Cyclotella desikacharyi Prasad 2006
Reference: Prasad and Nienow 2006, figs. 1–17
Plate 1: Fig. 1: diameter – 20.0 µm, striae density – 16/10 µm.
Abundance and Location of Maximum Abundance: rare, central Florida Bay

Cyclotella choctawatcheeana Prasad 1990
References: Prasad et al. 1990, figs. 2–8; Prasad and Nienow 2006, figs. 47–49
Plate 1: Fig. 2: diameter–6.8 μm, striae density–27/10 μm.
Abundance and Location of Maximum Abundance: common, eastern Florida Bay

Cymatosira belgica Grunow in Van Heurck 1880, **S**
References: Hustedt 1931–1959, p. 121, fig. 649.
Plate 1: Fig. 5: length–11.2 μm, width–2.4 μm, striae density–13/10 μm.
Abundance and Location of Maximum Abundance: rare, western Florida Bay

Cymatosira lorenziana Grunow 1862, **S**
References: Hustedt 1931–1959, p. 120, fig. 648; Hein et al. 2008, p. 18, fig. 2: 4
Plate 1: Fig. 4: length–21.0 μm, width–10.2 μm, striae density–9/10 μm.
Abundance and Location of Maximum Abundance: rare, Atlantic

Toxarium undulatum Bailey 1854
References: Hustedt 1931–1959, as *Synedra undulata,* p. 207, fig. 714; Witkowski et al. 2000, pp. 83–84, figs. 31: 5, 6; Hein et al. 2008, pp. 33–34, fig. 15: 3
Plate 4: Fig. 28: Detail of middle of valve, scale bar–50 μm. Morphometrics of entire valve (not shown) are: length–709 μm, width–7.3 μm, striae density–9/10 μm
Abundance and Location of Maximum Abundance: rare, central Florida Bay

Common taxa—Class Fragilariophyceae (araphid, pennate diatoms)

Grammatophora oceanica (Ehrenberg 1854 pro parte) Grunow var. *oceanica*
References: Hustedt 1931–1959, pp. 44–45, fig. 573; Witkowski et al. 2000, p. 59, figs. 15: 13, 14, 16: 12, 17: 3, 4
Plate 2: Fig. 9: length–44.7 μm, striae density–28/10 μm
Abundance and Location of Maximum Abundance: rare, central Florida Bay

Hyalosira interrupta (Ehrenberg 1838) Navarro 1991
References: Witkowski et al. 2000, pp. 61–62, fig. 21: 6; Hein et al. 2008, p. 26, figs. 10: 3–8
Plate 2: Fig. 12: length–23.4 μm, striae density–28/10 μm
Abundance and Location of Maximum Abundance: common, central Florida Bay

Hyalosynedra laevigata (Grunow 1877) Williams and Round 1986
References: Witkowski et al. 2000, p. 62, figs. 17: 22, 29: 6–10, 30: 23; Hein et al. 2008, pp. 26–27, figs. 10: 11
Plate 3: Fig. 16: length–155 μm, width–4.0 μm, striae density–unresolved; Plate 3: Fig. 15: detail of valve apex, scale bar–20 μm
Abundance and Location of Maximum Abundance: abundant, central Florida Bay

Licmophora debilis (Kützing 1844) Grunow 1881
References: Hustedt 1931–1959, p. 72, fig. 602; Honeywill 1998, p. 248, figs. 15a–15e; Witkowski et al. 2000, p. 64, figs. 19: 16–19
Plate 4: Fig. 27: length–25.1 μm, striae density at both footpole and headpole–28/10 μm
Abundance and Location of Maximum Abundance: rare, Atlantic

Licmophora normaniana (Greville 1862) Wahrer 1985
References: Wahrer et al. 1985, figs. 10, 11; Hein et al. 2008, as *L. normanniana*, p. 27, figs. 11: 1, 2
Plate 4: Fig. 24: length–350 μm, width–5.2 μm, striae density at both footpole and headpole–17/10 μm
Abundance and Location of Maximum Abundance: rare, central Florida Bay

Licmophora remulus Grunow 1867
References: Hustedt 1931–1959, pp. 55–56, fig. 580; Witkowski et al. 2000, p. 68, figs. 19: 1, 2; Hein et al. 2008, pp. 27–28, figs. 11: 3–5
Plate 4: Fig. 26: length–207 μm, width–6.3 μm, striae density at footpole–29/10 μm, striae density at headpole–28/10 μm
Abundance and Location of Maximum Abundance: rare, eastern Florida Bay

Licmophora sp., **S**
Reference: Belando et al. 2012, pp. 284–285, figs. 26–35
Plate 4: Fig. 25: length–339 μm, width–22.7 μm, striae density at footpole–21/10 μm, striae density at headpole–20/10 μm
Remarks: The taxon depicted was erroneously identified as *L. grandis* in Frankovich et al. 2006. This large taxon has cuneate to spatulate valves and is similar to an incomplete description of *L. gigantea* Mereschkowsky 1901. We are unable to verify this identity because a drawing and striae densities were not provided in the original description. The *Licmophora* sp. described by Belando et al. 2012 is the same taxon described here. Septa on the valvocopulae are either lacking or are not well developed. Rimoportulae of the apical pole are located between the mantle and valve face and the basal rimoportulae are located near the sternum.
Abundance and Location of Maximum Abundance: rare, central Florida Bay

Neodelphineis pelagica Takano 1982
References: Round et al. 1990, pp. 412–413, fig. a; Prasad 1987, pp. 125–129, figs. 4, 5, 6.
Plate 2: Fig. 14: length–10.5, width–2.8, striae density–17/10 μm.
Abundance and Location of Maximum Abundance: rare, eastern Florida Bay

Neofragilaria sp., **S**
Plate 1: Fig. 3: length–11.7 μm, width–3.5 μm, striae density–15/10 μm
Remarks: This taxon was previously identified as *N. nicobarica* in Frankovich et al. 2006. The taxon depicted has finer striae densities than *N. nicobarica* Desikachary, Prasad and Prema (6–7/10 μm as listed in Witkowski et al. 2000) and also has opposite striae instead of alternate as in *N. nicobarica*.
Abundance and Location of Maximum Abundance: rare, Atlantic

Neosynedra tortosa (Grunow 1877) Williams and Round 1986
Reference: Witkowski et al. 2000, p. 69, fig. 29: 11
Plate 3: Fig. 19: length–116 μm, width–2.6 μm, striae density–31/10 μm
Abundance and Location of Maximum Abundance: common, eastern Florida Bay

Opephora cf. *horstiana* Witkowski 1994
References: Witkowski et al. 2000, pp. 70–71, figs. 25: 27–30; Hein et al. 2008, p. 28, figs. 11: 11
Plate 2: Fig. 13: length – 8.3 μm, width – 3.4 μm, striae density – 11/10 μm
Remarks: The taxon depicted agrees with *Opephora* cf. *horstiana* in Hein et al. 2008 and has coarser striae density than *O. horstiana* as listed in Witkowski et al. 2000. Morales 2002 suggests transfer of many similar taxa in the genus *Opephora* to *Pseudostaurosira*, *Staurosirella*, and as yet undescribed new genera.
Abundance and Location of Maximum Abundance: rare, western Florida Bay

Pteroncola inane (Giffen 1970) Round 1990
References: As *Fragilaria tenerrima* in Peragallo and Peragallo 1897–1908, pp. 325–326, fig. 81: 4; as *Fragilaria hyalina* in Hustedt 1931–1959, p. 141, fig. 664; Round et al. 1990, pp. 390–391, fig. a; Witkowski et al. 2000, pp. 50–51, figs. 2: 8–10
Plate 1: Fig. 6: length – 17.7 μm, width – 2.1 μm, striae density – unresolved
Abundance and Location of Maximum Abundance: rare, central Florida Bay

Reimerothrix floridensis Prasad 2001, **S**
Reference: Prasad et al. 2001, figs. 1–12
Plate 2: Fig. 11: length – 104 μm, width – 3.6 μm, striae density – unresolved
Abundance and Location of Maximum Abundance: abundant, central Florida Bay

Rhabdomena adriaticum Kützing 1844
References: Hustedt 1931–1959, pp. 21–23, fig. 552; Witkowski et al. 2000, p. 76, figs. 13: 10–12
Plate 4: Fig. 30: length – 52.3 μm, striae density – 10/10 μm
Abundance and Location of Maximum Abundance: rare, central Florida Bay

Striatella unipunctata (Lyngbye 1819) Agardh 1832
References: Hustedt 1931–1959, pp. 30–31, fig. 560; Witkowski et al. 2000, pp. 78–79, figs. 23: 5–7; Hein et al. 2008, p. 30, figs. 13: 1, 9
Plate 2: Fig. 8: length – 79.5 μm, width – 18.0 μm, striae density – 26/10 μm on diagonal
Abundance and Location of Maximum Abundance: rare, eastern Florida Bay

Synedra fasciculata (C.A. Agardh 1812) Kützing 1844 f. *densestriata* (Møller 1950) Archibald 1983
Reference: Archibald 1983, pp. 320–321, figs. 467–472. The taxon depicted was identified as *Tabularia tabulata* in Frankovich et al. 2006.
Plate 4: Fig. 29: length – 31.8 μm, width – 2.6 μm, striae density – 19/10 μm
Abundance and Location of Maximum Abundance: common, eastern Florida Bay

Synedra ? sp.
Plate 3: Fig. 18: length – 85.9 μm, width – 2.5 μm, striae density at footpole – 27/10 μm, striae density at headpole – 28/10 μm; Plate 3: Fig. 21: detail of valve apex, scale bar – 20 μm
Remarks: The taxon depicted has linear valves that widen at about 2/3 of the valve length. The apices are bluntly rounded. The sternum is very narrow. The central area

is indistinct and is only evidenced by a slight thickening of the valves that partly occlude the central striae as in the freshwater genus *Ulnaria*.
Abundance and Location of Maximum Abundance: rare, eastern Florida Bay

Tabularia sp.
Plate 3: Fig. 17: length – 105 μm, width – 4.4 μm, striae density – 28/10 μm; Plate 3: Fig. 22: detail of valve apex, scale bar – 20 μm
Remarks: Similar to *Synedra affinis* Kützing var. *hybrida* in Peragallo and Peragallo 1897–1908, pp. 318–319, fig. 80: 16–18, and *Synedra tabulata* var. *obtusa* in Witkowski et al. 2000, p. 82, fig. 30: 3. The taxon depicted has much higher striae densities than the 13–14/10 μm and the 10–11/10 μm listed for the previously mentioned similar taxa.
Abundance and Location of Maximum Abundance: rare, central Florida Bay

Thalassiothrix longissima Cleve and Grunow 1880
Reference: Hustedt 1931–1959, pp. 224–225, fig. 726
Plate 2: Fig. 10: length – 87.3 μm, width – 2.7 μm
Abundance and Location of Maximum Abundance: rare, Atlantic

Trachysphenia sp., **S**
Reference: The taxon depicted was earlier identified as *Dimerogramma dubium* in Frankovich et al. 2006. As *Trachysphenia* 1in Hein et al. 2008, p. 93, fig. 66: 5–7
Plate 1: Fig. 7: length – 13.7 μm, width – 6.2 μm, striae density – 11/10 μm.
Remarks: The taxon depicted differs from *T. australis* Petit in its smaller size and only slightly heteropolar to isopolar valves.
Abundance and Location of Maximum Abundance: common, Atlantic

Common taxa—Class Bacillariophyceae (raphid, pennate diatoms)

Amphora arcus Gregory 1854 var. *sulcata* Schmidt 1875, **S**
References: The taxon depicted was identified as *A. lineolata* in Frankovich et al. 2006. Schmidt et al. 1874–1959, figs. 26: 8, 9; Peragallo and Peragallo 1897–1908, p. 225, fig. 50: 5; Witkowski et al. 2000, p. 129, fig. 165: 15; As *A. sulcata* in Wachnicka and Gaiser 2007, pp. 418–419, figs. 113, 114
Plate 19: Fig. 100: Valve view, length – 47.0 μm, width – 9.6 μm, dorsal striae density – 20/10 μm, ventral striae density – 18/10 μm. Plate 19: Fig. 99: Girdle view of frustule, length – 48.5 μm, dorsal striae density – 20/10 μm, ventral striae density – 20/10 μm, girdle striae density – 21/10 μm.
Abundance and Location of Maximum Abundance: rare, central Florida Bay

Amphora cingulata Cleve 1878
References: Peragallo and Peragallo 1897–1908, p. 219, figs. 48: 3, 49: 5–7; Wachnicka and Gaiser 2007, p. 436, fig. 189.
Plate 19: Fig. 98: length – 114 μm, width – 17.3 μm, dorsal striae density – 21/10 μm, ventral striae density – 20/10 μm
Abundance and Location of Maximum Abundance: rare, eastern Florida Bay

Amphora crenulata Wachnicka and Gaiser 2007
Reference: Wachnicka and Gaiser 2007, p. 398, figs. 25–28

Plate 18: Fig. 96: length – 14.6 μm, dorsal striae density – 26/10 μm, ventral striae density – 31/10 μm.
Abundance and Location of Maximum Abundance: rare, eastern Florida Bay

Amphora graeffeana Hendey 1973, **P**
References: Witkowski et al. 2000, pp. 138–139, figs. 166: 24, 172: 6–9; As *Amphora* cf. *graeffeana* in Hein et al. 2008, p. 44, fig. 18: 13; As *Amphora graeffeana* var. F02 in Wachnicka and Gaiser 2007, p. 418, fig. 110
Plate 19: Fig. 101: length – 33.6 μm, width – 7.8, dorsal striae density – 20/10 μm, ventral striae density – 19
Abundance and Location of Maximum Abundance: rare, eastern Florida Bay

Amphora ostrearia Brébisson in Kützing var. *typica* (Cleve 1868) Cleve 1894–1895
References: Peragallo and Peragallo 1897–1908, p. 220, figs. 49: 13, 14; Wachnicka and Gaiser 2007, p. 411, figs. 82, 83; Hein et al. 2008, p. 45, figs. 21: 2, 7, 22: 1
Plate 19: Fig. 103: length – 33.6 μm, width – 7.8 μm, dorsal striae density – 20/10 μm, ventral striae density – 19/10 μm
Abundance and Location of Maximum Abundance: rare, eastern Florida Bay

Amphora proteus Gregory 1857
References: Schoeman and Archibald 1986, pp. 432–434, figs. 70–80; Wachnicka and Gaiser 2007, pp. 432–433, figs. 174–176; as *Amphora* cf. *marina* Smith 1857 in Hein et al. 2008, pp. 44–45, fig. 18: 15
Plate 17: Fig. 87: length – 33.7 μm, width – 7.0, dorsal striae density – 13/10 μm, ventral striae density – 12/10 μm
Remarks: Wachnicka and Gaiser 2007 report much larger valves of *A. proteus* (length – 52–81 μm), but the valve depicted shares the same striae densities, a lens-shaped central area on the dorsal side of the valve, and ventral striae interrupted by a longitudinal line which cuts them only up to the middle part of the valve. *A.* cf. *marina* depicted in Hein et al. 2008, fig. 18: 15 also shares these morphologies and is smaller (length – 43 μm) than the size range published for *A. proteus*. Though *A. marina* has been distinguished from *A. proteus* by its much smaller size (length – 24.0–46.5 μm) and finer dorsal and ventral striae densities, 15–20/10 μm and 12–16/10 μm, respectively, Cleve 1895 considered *A. marina* as a possible form of *A. proteus*. The shared morphologies of the valves observed in Florida Bay suggests a much larger size range for *A. proteus* than previously reported, possibly encompassing valves previously identified as *A. marina*.
Abundance and Location of Maximum Abundance: rare, Atlantic

Amphora cf. *tumida* Hustedt 1956
References: Sar et al. 2004, pp. 71–80, figs. 1–3; as the undescribed species *A.* sp. 31 in Montgomery 1978, fig. 20: D
Plate 17: Fig. 90: Valve view, length – 28.0 μm, frustule width – 4.9, dorsal striae density – 19/10 μm, ventral striae density – 29/10 μm; Plate 17: Fig. 89: Dorsal view of girdle band punctae, length – 22.7 μm, punctae density – 25/10 μm
Remarks: The depicted taxon is most similar to *A. tumida* (Sar et al. 2004), but exhibits coarser girdle punctae density than the 34–40/10 μm reported for *A. tumida*. It is also similar to *A. exigua* in valve outline and girdle band punctae, but differs from the

reported striae densities for *A. exigua* (dorsal striae density – 11–15/10 μm, ventral striae density – 18–22/10 μm in Witkowski et al. 2000).
Abundance and Location of Maximum Abundance: rare, eastern Florida Bay

Amphora sp. 1
Reference: As the undescribed species *A.* B35 in Wachnicka and Gaiser 2007, p. 412, fig. 88.
Plate 19: Fig. 102: length – 40.0 μm, width – 7.0, dorsal striae density – 14/10 μm, ventral striae density – 26/10 μm
Remarks: Depicted taxon has slightly finer ventral striae density than *Amphora* B35 in Wachnicka and Gaiser 2007.
Abundance and Location of Maximum Abundance: rare, eastern Florida Bay

Amphora sp. 2
Reference: As the undescribed species *A.* sp. L06 in Wachnicka and Gaiser 2007, pp. 400–403, fig. 43
Plate 18: Fig. 95: length – 9.5 μm, frustule width – 4.8, dorsal striae density – 34/10 μm, ventral striae density – unresolved
Remarks: Depicted taxon differs from other small *Amphora* in Florida Bay by coarser dorsal striae density in *A. tenerrima* (Plate 15: Fig. 93) and more elliptical valves that exhibit easily discernible dorsal striae in LM than *A.* cf. *tenuissima* (Plate 15: Fig. 92).
Abundance and Location of Maximum Abundance: common, central Florida Bay

Brachysira aponina Kützing 1836. **S, P**
References: Round et al. 1990, p. 540, fig. a; Witkowski et al. 2000, p. 160, figs. 134: 5, 6
Plate 16: Fig. 80: length – 35.2 μm, width – 4.2 μm, striae density – 37/10 μm
Abundance and Location of Maximum Abundance: abundant, eastern Florida Bay

Climaconeis colemaniae Prasad 2000
References: Prasad et al. 2000, pp. 204–207, figs. 18–26; Hein et al. 2008, p. 49, fig. 22: 7
Plate 16: Fig. 79: length – 194 μm, width – 6.6 μm, striae density – 26/10 μm
Abundance and Location of Maximum Abundance: common, eastern Florida Bay

Cocconeis barleyi Frankovich and DeStefano 2005, **S**
Reference: De Stefano and Romero 2005, pp. 16–18, figs. 7: 1, 8: 1
Plate 5: Fig. 32: RSV, length – 16.1 μm, width – 10.0 μm, striae density – 23/10 μm;
Plate 5: Fig. 33: SV, length – 15.7 μm, width – 9.3 μm, striae density – 24/10 μm
Abundance and Location of Maximum Abundance: common, western Florida Bay

Cocconeis britannica Naegeli in Kützing 1849, **S**
References: Witkowski et al. 2000, p. 102, figs. 39: 21–23; De Stefano and Romero 2005, pp. 18–20, figs. 10: 1, 11: 1; Hein et al. 2008, p. 36, fig. 16: 13, 14
Plate 6: Fig. 42: RSV, length – 27.9 μm, width – 18.1 μm, striae density – 10/10 μm;
Plate 6: Fig. 43: SV, length – 21.2 μm, width – 12.1 μm, striae density – 14/10 μm
Abundance and Location of Maximum Abundance: rare, central Florida Bay

Cocconeis coralliensis Riaux-Gobin et Compère 2008
Reference: Riaux-Gobin et al. 2011, p. 22, figs. 3: 12–14, 33: 1–7
Plate 5: Fig. 38: RSV, length – 16.0 μm, width – 8.2 μm, striae density – 29/10 μm; Plate 5: Fig. 39: SV, length – 16.0 μm, width – 7.3 μm, striae density – 28/10 μm
Abundance and Location of Maximum Abundance: abundant, central Florida Bay

Cocconeis euglypta Ehrenberg 1854
References: As *Cocconeis placentula* var. *euglypta*. Hustedt 1931–1959, p. 308, fig. 802c; Navarro 1982, p. 29, figs. 18: 4, 5; Hein et al. 2008, p. 38, fig. 17: 8; Romero and Jahn 2013. pp. 6–7, figs. 9–11
Plate 5: Fig. 36: RSV, length – 23.3 μm, width – 13.4 μm, striae density – 23/10 μm; *Plate 5*: Fig. 37: SV, length – 19.8 μm, width – 10.7 μm, striae density – 21/10 μm
Abundance and Location of Maximum Abundance: abundant, eastern Florida Bay

Cocconeis scutellum Ehrenberg 1838
References: Hustedt 1931–1959, pp. 298–301, fig. 790; Witkowski et al. 2000, p. 114, figs. 36: 1–7, 38: 11; De Stefano et al. 2008, pp. 508–523, figs. 19–35
Plate 5: Fig. 34: RSV, length – 23.8 μm, width – 15.0 μm, striae density – 13/10 μm; *Plate 5*: Fig. 35: SV, length – 22.0 μm, width – 13.0 μm, striae density – 14/10 μm
Remarks: Vars. parva, obliqua, and baldjikiana are also present in the flora.
Abundance and Location of Maximum Abundance: abundant, western Florida Bay

Cocconeis woodii Reyes-Vasquez 1970, **S**
References: Reyes-Vasquez 1970, pp. 119–120, fig. 3; As *C. scutellum* var. *clinoraphis* in De Stefano et al. 2008, p. 528, figs. 54–70
Plate 6: Fig. 40: RSV, length – 46.0 μm, width – 26.5 μm, striae density – 12/10 μm; *Plate 6*: Fig. 41: SV, length – 37.2 μm, width – 24.4 μm, striae density – 11/10 μm
Remarks: The taxon depicted is the taxon described in De Stefano et al. 2008 as *C. scutellum* var. *clinoraphis*. De Stefano et al. 2008 assigned the variety name based on the description and a single drawing of the SV in the original publication of Zanon 1948. Zanon's original drawing reproduced in De Stefano et al. 2008 resembles *C. scutellum* var. *obliqua* Brun 1895. Brun's original drawings in Tempère 1893–1896 depict elliptic to broadly elliptic valves with a slightly oblique raphe and raphe-sternum. Zanon's taxon may be conspecific with *C. scutellum* var. *obliqua*. *C. scutellum* var. *clinoraphis* as described by De Stefano et al. 2008 exhibits valves with more sigmoid raphes and raphe-sternums in better agreement with the description and drawings of *C. woodii*. Florida Bay specimens exhibit broadly elliptic valves with slightly cuneate apices and sigmoid raphes and raphe-sternums. The combination of these features lead us to the species determination of *C. woodii*, though future analysis may indicate the taxon to be conspecific with *C. scutellum* var. *obliqua*.
Abundance and Location of Maximum Abundance: rare, western Florida Bay

Diploneis vacillans (A. Schmidt 1875) Cleve 1894
References: Hustedt 1931–1959, pp. 555–557, fig. 1060a–d; Witkowski et al. 2000, p. 196, figs. 89: 14, 90: 11, 12, 91: 9, 10
Plate 16: Fig. 82: length – 14.6 μm, width – 5.9 μm, striae density – 25/10 μm
Abundance and Location of Maximum Abundance: rare, eastern Florida Bay

Entomoneis pseudoduplex Osada and Kobayasi 1990
References: Osada and Kobayasi 1988, p. 165, figs. 4, 5, 32–42; Witkowski et al. 2000, p. 199, figs. 173: 9, 10; Hein et al. 2008, pp. 54–55, fig. 27: 10
Plate 22: Fig. 119: length – 54.8 μm, striae density – 26/10 μm
Abundance and Location of Maximum Abundance: rare, central Florida Bay

Halamphora aponina (Kützing 1844) Levkov 2009, **P**
References: As *Amphora aponina* in Kützing 1844, p. 108, fig. 5: 33; Archibald and Schoeman 1984, p. 95, figs. 70–75; Wachnicka and Gaiser 2007, p. 396, figs. 9–14; Hein et al. 2008, p. 42, fig. 18: 2
Plate 18: Fig. 92: length – 16.6 μm, dorsal striae density – 25/10 μm, ventral striae density – unresolved
Abundance and Location of Maximum Abundance: abundant, eastern Florida Bay

Halamphora cymbifera (Gregory 1857) Levkov 2009 var. *heritierarum* Wachnicka and Gaiser 2007
References: Wachnicka and Gaiser 2007, pp. 394–395, figs. 6–8; Hein et al. 2008, p. 43, fig. 20: 8
Plate 17: Fig. 91: length – 33.2 μm, width – 4.9, dorsal striae density – 19/10 μm, ventral striae density – 30/10 μm.
Remarks: The taxon depicted (Plate 17: Fig. 91) has higher ventral striae density than the 18–26/10 μm reported by Wachnicka and Gaiser 2007.
Abundance and Location of Maximum Abundance: common, central Florida Bay

Halamphora tenerrima (Aleem and Hustedt 1951) Levkov 2009, **S**
References: Clavero et al. 2000, pp. 199–202, figs. 1–14; Witkowski et al. 2000, p. 152, fig. 164: 20; Wachnicka and Gaiser 2007, p. 400, figs. 41, 42
Plate 18: Fig. 93: length – 16.1 μm, frustule width – 5.6 μm, dorsal striae density – 24/10 μm, ventral striae density – unresolved; Plate 18: Fig. 94: Dorsal view of girdle band punctae, length – 15.7 μm, punctae density – 26/10 μm
Remarks: Clavero et al. 2000 stated that the relatively coarse striation (26–32/10 μm) of the girdle bands of *A. tenerrima* (visible in LM, Fig. Plate 18: Fig. 94) is a distinctive feature for distinguishing this taxon from other similar small *Amphora* taxa.
Abundance and Location of Maximum Abundance: abundant, central Florida Bay

Halamphora cf. *tenuissima* (Hustedt 1955) Levkov 2009, **S**
References: As *Amphora* sp. 1 in Frankovich et al. 2006. Hustedt 1955, p. 39, fig. 14: 16; Clavero et al. 2000, p. 202, figs. 26–33; Hein et al. 2008, p. 46, fig. 20: 5
Plate 18: Fig. 97: length – 11.8 μm, frustule width – 2.9 μm, dorsal striae density – 36/10 μm, ventral striae density – unresolved.
Remarks: The depicted taxon shares characteristics visible in LM that fit with descriptions of *A. tenuissima*. However, similar small *Amphora* spp. exist, and SEM studies on this collection have not yet been performed to identify distinguishing features only visible using SEM.
Abundance and Location of Maximum Abundance: abundant, western Florida Bay

Mastogloia angusta Hustedt 1931–1959
Reference: Hustedt 1931–1959, p. 437, fig. 940

Plate 10: Fig. 54: length – 39.8 µm, width – 7.5 µm, striae density – 20/10 µm
Abundance and Location of Maximum Abundance: rare, eastern Florida Bay

Mastogloia binotata (Grunow 1863) Cleve 1895
References: Hustedt 1931–1959, pp. 404–405, fig. 889; Witkowski et al. 2000, p. 240, figs. 75: 15–18
Plate 9: Fig. 50: length – 30.2 µm, width – 17.2 µm, striae density – 15/10 µm
Abundance and Location of Maximum Abundance: rare, Atlantic

Mastogloia biocellata (Grunow 1877) Novarino and Muftah 1991
References: Hustedt 1931–1959, as *Mastogloia erythraea* var. *biocellata* p. 448, fig. 959 d; Witkowski et al. 2000, pp. 240–241, figs. 77: 19, 20; Hein et al. 2008, p. 60, figs. 33: 8, 39: 3
Plate 12: Fig. 62: length – 35.9 µm, width – 8.2 µm, striae density – 25/10 µm
Abundance and Location of Maximum Abundance: rare, eastern Florida Bay

Mastogloia corsicana Grunow 1878, **S**
References: Hustedt 1931–1959, pp. 454–455, fig. 966; Witkowski et al. 2000, p. 242, figs. 77: 15–18; Hein et al. 2008, pp. 61–62, figs. 34: 3, 4
Plate 10: Fig. 52: length – 37.9 µm, width – 11.5 µm, striae density – 14/10 µm
Abundance and Location of Maximum Abundance: abundant, eastern Florida Bay

Mastogloia cribrosa Grunow 1860, **S**
References: Hustedt 1931–1959, p. 403, fig. 887; Hein et al. 2008, p. 62, figs. 36: 1–3
Plate 8: Fig. 46: length – 47.1 µm, width – 35.9 µm, striae density – 9/10 µm
Abundance and Location of Maximum Abundance: abundant, Atlantic

Mastogloia crucicula var. *crucicula* (Grunow 1877) Cleve 1895, **S**
References: Hustedt 1931–1959, pp. 408–409, fig. 894; Witkowski et al. 2000, p. 242, fig. 75: 3; Hein et al. 2008, p. 63, figs. 36: 6, 7
Plate 8: Fig. 47: length – 16.1 µm, width – 8.4 µm, striae density – 19/10 µm
Remarks: Var. alternans is also present in flora.
Abundance and Location of Maximum Abundance: abundant, Atlantic

Mastogloia cuneata (Meister 1937) Simonsen 1990
References: Montgomery 1978, as *Mastogloia inaequalis*, figs. 140: A - H, 202: C, D; Simonsen 1990, pp.134–138, figs. 76–111; Witkowski et al. 2000, p. 243, figs. 74: 19–26, 81: 15–18; Hein et al. 2008, p. 63, figs. 35: 2, 36: 4
Plate 12: Fig. 64: length – 19.3 µm, width – 5.0 µm, striae density – 37/10 µm
Abundance and Location of Maximum Abundance: rare, Atlantic

Mastogloia cyclops Voigt 1942
References: Voigt 1942, p. 8, fig. 8; Stephens and Gibson 1980, pp. 144–146, figs. 8–14; Witkowski et al. 2000, p. 243, figs. 77: 7, 8
Plate 11: Fig. 58: length – 27.4 µm, width – 10.8 µm, striae density – 23/10 µm
Abundance and Location of Maximum Abundance: common, Atlantic

Mastogloia discontinua Kemp and Paddock 1990
References: Kemp and Paddock 1990, p. 319–320, figs. 48–58; Hein et al. 2008, p. 64, fig. 40: 4

Plate 11: Fig. 59: length – 34.2 µm, width – 10.5 µm, striae density – 12/10 µm
Abundance and Location of Maximum Abundance: common, eastern Florida Bay

Mastogloia erythraea Grunow 1860, **P**
References: Hustedt 1931–1959, p. 448, fig. 959a–c; Witkowski et al. 2000, p. 246, figs. 76: 2–7; Hein et al. 2008, p. 65, fig. 39: 1, 2, 4
Plate 14: Fig. 69: length – 60.3 µm, width – 14.1 µm, striae density – 20/10 µm
Abundance and Location of Maximum Abundance: abundant, central Florida Bay

Mastogloia fimbriata (Brightwell 1859) Cleve 1895
References: Hustedt 1931–1959, pp. 400–402, fig. 884; Witkowski et al. 2000, pp. 247–248, figs. 77: 1–4, 83: 1, 2; Hein et al. 2008, p. 66, fig. 38: 1, 3
Plate 7: Fig. 44: length – 38.6 µm, width – 26.1 µm, striae density – 7/10 µm
Abundance and Location of Maximum Abundance: rare, Atlantic

Mastogloia hainanensis Voigt 1952
References: Voigt 1952, p. 443, fig. 1: 9; Hein et al. 2008, as *Mastogloia* 13, p. 99, figs. 70: 4, 5
Plate 14: Fig. 71: length – 50.4 µm, width – 14.4 µm, striae density – 16/10 µm
Remarks: Voigt 1952 states that *M. hainanensis* is similar to the lanceolate form of *M. braunii*, differing only in the uniform shape of the partectae in *M. hainanensis*. Valves depicted by Hein et al. 2008 and Voigt 1952 have slightly higher striae density – 18–21/ 10 µm.
Abundance and Location of Maximum Abundance: rare, central Florida Bay

Mastogloia ignorata Hustedt 1933
References: Hustedt 1931–1959, p. 433, fig. 932; Witkowski et al. 2000, p. 250, figs. 76: 8–11; 77: 13, 14; 81: 5–8
Plate 11: Fig. 57: length – 30.0 µm, width – 9.7 µm, striae density – 21/10 µm
Remarks : This taxon is very similar to *M. decipiens*, but differs in the structure of the areolae. The areolae of *M. ignorata* are irregularly sized quadrangular to round in the median area becoming smaller and rounder towards the valve margin. The areolae of *M. decipiens* are uniform on the valve face.
Abundance and Location of Maximum Abundance: common, western Florida Bay

Mastogloia lacrimata Voigt 1963, **S**
References: Voigt 1963, pp. 116–117, fig. 24: 6; Hein et al. 2008, pp. 67–68, fig. 41: 4; Lobban et al. 2012, p. 276, fig. 30: 6–8
Plate 12: Fig. 61: length – 37.3 µm, width – 17.0 µm, striae density – 17/10 µm
Abundance and Location of Maximum Abundance: rare, Atlantic

Mastogloia laminaris Grunow 1883, **P**
Reference: Hustedt 1931–1959, p. 432, fig. 931a
Plate 10: Fig. 53: length – 21.5 µm, width – 6.1 µm, striae density – 24/10 µm
Abundance and Location of Maximum Abundance: rare, eastern Florida Bay

Mastogloia lanceolata Thwaites 1856
References: Hustedt 1931–1959, p. 425, fig. 922; John 1980, as *M. halophila*, pp. 853–854, figs. 3: 7, 8; Stephens and Gibson 1980, pp. 146–149, figs. 15, 16; Witkowski et al. 2000, p. 251, figs. 73: 6–9

Plate 11: Fig. 56: length – 58.2 μm, width – 14.8 μm, striae density – 18/10 μm

Remarks: The taxon depicted shares characteristics with *M. halophila* John 1980, *M. lanceolata* Thwaites 1856, and *M. aquilegiae* Grunow 1893. John 1980 differentiates *M. halophila* from *M. lanceolata* by a doubly twisted raphe in *M. halophila* versus an only slightly bent raphe in *M. lanceolata*, slightly higher striae density in *M. lanceolata* (20/10 μm) versus *M. halophila* (16–17/10 μm), and smaller size of *M. lanceolata* (40–50 μm) versus *M. halophila* (45–93 μm). John 1980 mainly differentiates *M. halophila* from *M. aquilegiae* by only one deflection in the raphe of *M. aquilegiae*. Differences in striae densities and size ranges are small. Hustedt 1931–1959 differentiates *M. aquilegiae* from *M. lanceolata* by slightly coarser striae density in *M. aquilegiae* and the wider partectae (3.5 um) in *M. aquilegiae* versus that of *M. lanceolata* (2 um). The width of the partectae of *M. halophila* are listed as 2.7–3 um. The taxon depicted has characteristics intermediate between these three taxa with striae density of 18 most similar to *M. aquilegiae*, a doubly twisted raphe in agreement with *M. halophila*, and partectae width (2.4 um) most similar to *M. lanceolata*. Hustedt 1931–1959 stated that Cleve 1895 would have combined *M. lanceolata* and *M. aquilegiae*, but disagreed and maintained separate species. Based on these descriptions, we are inclined to agree with Cleve 1895, and also believe that the three species may represent morphological variation of the same taxon. The oldest valid description is that of *M. lanceolata* Thwaites 1856.

Abundance and Location of Maximum Abundance: common, central Florida Bay

Mastogloia manokwariensis Cholnoky 1963, **S**
References: Cholnoky 1963b, p. 173, fig. 26: 52–55; Witkowski et al. 2000, pp. 253–254, fig. 80: 11; Hein et al. 2008, p. 68, fig. 38: 2
Plate 10: Fig. 55: length – 15.9 μm, width – 5.7 μm, striae density – 23/10 μm
Abundance and Location of Maximum Abundance: abundant, western Florida Bay

Mastogloia ovalis Schmidt 1893, **S**
References: Hustedt 1931–1959, p. 408, fig. 893; Witkowski et al. 2000, p. 255, figs. 75: 11–13; Hein et al. 2008, p. 69, fig. 40: 6
Plate 8: Fig. 48: length – 17.2 μm, width – 11.8 μm, striae density – 19/10 μm
Abundance and Location of Maximum Abundance: common, Atlantic

Mastogloia ovata Grunow 1860
References: Hustedt 1931–1959, pp. 409–410, fig. 895; Witkowski et al. 2000, p. 255, fig. 82: 1
Plate 7: Fig. 45: length – 29.6 μm, width – 20.5 μm, striae density – 18/10 μm
Abundance and Location of Maximum Abundance: rare, eastern Florida Bay

Mastogloia pseudoelegans Hustedt 1955
References: Hustedt 1955, p. 19–20, fig. 10; Hein et al. 2008, p. 70, fig. 42: 5
Plate 13: Fig. 67: length – 42.2 μm, width – 16.7 μm, striae density – 22/10 μm
Abundance and Location of Maximum Abundance: rare, central Florida Bay

Mastogloia pseudolatecostata Yohn and Gibson 1982
References: Yohn and Gibson 1982, pp. 43–44, figs. 20–25; Hein et al. 2008, p. 70, fig. 45: 1, 3; Lobban et al. 2012, p. 281, figs. 35: 1–2

Plate 9: Fig. 49: length–46.3 µm, width–30.8 µm, striae density–11/10 µm
Abundance and Location of Maximum Abundance: common, eastern Florida Bay

Mastogloia punctifera Brun 1895, **P**
References: Hustedt 1931–1959, pp. 419–420, fig. 914; Hein et al. 2008, p. 71,
fig. 43: 2, 5
Plate 12: Fig. 63: length–47.3 µm, width–17.4 µm, striae density–18/10 µm
Abundance and Location of Maximum Abundance: common, eastern Florida Bay

Mastogloia pusilla Grunow 1878 var. *linearis* Ostrup 1910, **S**
Reference: Hustedt 1931–1959, p. 481, fig. 1002 d
Plate 13: Fig. 68: length–31.5 µm, width–7.0 µm, striae density–17/10 µm
Abundance and Location of Maximum Abundance: abundant, western Florida Bay

Mastogloia rimosa Cleve 1893
References: Yohn and Gibson 1981, p. 644, figs. 24–30; Hein et al. 2008, p. 72,
fig. 46: 3, 5
Plate 11: Fig. 60: length–37.8 µm, width–14.3 µm, striae density–10/10 µm
Abundance and Location of Maximum Abundance: rare, central Florida Bay

Mastogloia robusta Hustedt 1931–1959, **P**
References: Hustedt 1931–1959, p. 443, fig. 951; Witkowski et al. 2000, p. 261,
figs. 79: 3–6; Lobban et al. 2012, p. 282, figs. 36: 3–6
Plate 10: Fig. 51: length–49.0 µm, width–13.4 µm, striae density–16/10 µm
Abundance and Location of Maximum Abundance: rare, eastern Florida Bay

Mastogloia subaffirmata Hustedt 1927
Reference: Hustedt 1931–1959, pp. 449–450, fig. 960a
Plate 14: Fig. 70: length–42.1 µm, width–10.8 µm, striae density–20/10 µm
Abundance and Location of Maximum Abundance: rare, eastern Florida Bay

Mastogloia sp. 1
Reference: Hustedt 1931–1959, p. 419, fig. 913
Plate 12: Fig. 65: length–19.7 µm, width–6.3 µm, striae density–20/10 µm
Remarks: The taxon depicted is the same as that shown as *Mastogloia* spec. (cf. *rimosa*)
in Witkowski et al. 2000 (Plate 77: figs. 21, 22).
Abundance and Location of Maximum Abundance: rare, eastern Florida Bay

Mastogloia sp. 2
Reference: As *Mastogloia* cf. *braunii* in López-Fuerte et al. 2013
Plate 13: Fig. 66: length–50.9 µm, width–21.3 µm, striae density–14/10 µm
Remarks: The depicted taxon belongs to the Group Sulcatae and shares features with
M. foliolum, *M. braunii*, and *M. neomauritiana*. The taxon differs from *M. foliolum*
in having a more sinuous raphe and lacks the transverse fascia that is present in
M. folioloum. The partectae are also structured differently. In contrast to the gradual
decrease in partectae width from the middle of the partectae ring toward the apices
that is observed in both *M. foliolum* and *M. braunii*, the partectae abruptly decrease in
width on either side of the larger central partectae. The partectae are most similar to that
of *M. neomauritiana* with larger central partectae and smaller surrounding, uniform,

and transapically widened partectae extending towards the apices. The depicted taxon differs from *M. neomauritiana,* in the former having only slightly drawn out apices, as opposed to produced almost capitate poles in the latter. The depicted taxon also possesses an H-shaped hyaline area in the center of the valve, a feature that is common in the Sulcatae and is shared by both *M. foliolum* and *M. braunii.*
Abundance and Location of Maximum Abundance: rare, central Florida Bay

Navicula cf. *agnita* Hustedt 1955
References: Similar taxa. *N. agnita* in Hustedt 1955, p. 27, figs. 9: 13–16 and Witkowski et al. 2000, p. 266, figs. 136: 21, 142: 10. *N. lagunae* in Seddon et al. 2011, pp. 871–872, figs. and in Hein et al. 2008 as *N.* cf. *flagellifera*, p. 75, figs. 49: 9, 10
Plate 15: Fig. 77: valve view, length – 31.8 μm, width – 6.6 μm, striae density – 15/10 μm, lineolae/poroid density – 33/10 μm; Fig. ?: girdle view, length – 31.0 μm, striae density – 14/10 μm
Remarks: We distinguish *N. agnita* from the similar *N. flagellifera* by the raised apices, most prominent in girdle view, in the latter. The taxon depicted is also very similar to *N. lagunae* but the striae densities are slightly larger, the lineolae are more distinct, and the central raphe endings are less sinuous.
Abundance and Location of Maximum Abundance: abundant, eastern Florida Bay

Navicula cf. *duerrenbergiana* Hustedt 1934
References: Hustedt 1934 in Schmidt et al. 1874–1959, Tafel 393, figs. 8, 9; Witkowski et al. 2000, p. 276, figs. 116: 8–14, 145: 14
Plate 15: Fig. 75: length – 24.8 μm, width – 5.1 μm, striae density – 17/10 μm, lineolae/poroid density – 29/10 μm
Remarks: In Florida Bay samples, valves are delicate and lightly silicified and often deformed as seen in Witkowski et al. 2000, fig. 116: 10, possibly from the drying process. Striae density is slightly higher than the 14–15/10 μm listed in Witkowski et al. 2000.
Abundance and Location of Maximum Abundance: abundant, eastern Florida Bay

Navicula flebilis Cholnoky 1963a
References: Cholnoky 1963a, p. 57, fig. 49; Witkowski et al. 2000, p. 278, figs. 130: 13, 28; Hein et al. 2008, p. 75, figs. 49: 6, 7
Plate 15: Fig. 73: length – 53.4 μm, width – 9.9 μm, striae density – 12/10 μm, lineolae/poroid density – 32/10 μm
Abundance and Location of Maximum Abundance: rare, central Florida Bay

Navicula halinae Witkowski et al. 2000
References: Witkowski et al. 2000, p. 276, figs. 116: 1–4
Plate 15: Fig. 74: length – 52.3 μm, width – 6.8 μm, striae density – 14/10 μm, lineolae/poroid density – 28/10 μm
Abundance and Location of Maximum Abundance: rare, central Florida Bay

Navicula cf. *microdigitoradiata* Lange-Bertalot 1993
Reference: Witkowski et al. 2000, pp. 290–291, figs. 126: 1–7
Plate 15: Fig. 78: length – 18.0 μm, width – 4.2 μm, striae density – 22/10 μm, lineolae/poroid density – unresolved

Remarks: The taxon depicted differs from *N. microdigitoradiata* in having higher striae density than the 14–17/10 μm listed in Witkowski et al. 2000.
Abundance and Location of Maximum Abundance: rare, eastern Florida Bay

Navicula cf. *normalis* Hustedt 1955
References: Hustedt 1955, p. 29, fig. 9: 3; Witkowski et al. 2000, p. 292, figs. 121: 15, 145: 30
Plate 15: Fig. 72: length – 59.3 μm, width – 9.9 μm, striae density – 11/10 μm, lineolae/poroid density – 33/10 μm
Remarks: Taxon depicted is most similar to *N. normalis* but exhibits undulating striae in the center of the valves similar to *N. johannrossii* (see figs. 137: 1–10 in Witkowski et al. 2000, pp. 284–285). The taxon differs from *N. johannrossii* in having more distinct lineolae and evenly spaced striae throughout.
Abundance and Location of Maximum Abundance: rare, western Florida Bay

Navicula sp.
Plate 15: Fig. 76: length – 15.3 μm, width – 3.2 μm, striae density – 17 center – 21 apices/10 μm, lineolae/poroid density – unresolved
Remarks: This taxon was most often seen as frustules in girdle view. This taxon is most similar to *N. salinicola* but differs in having valves that are narrow lanceolate to sub-rhombic in outline with acutely rounded apices, and radial striae throughout.
Abundance and Location of Maximum Abundance: abundant, central Florida Bay

Nitzschia aequorea Hustedt 1939, **P**
Reference: Witkowski et al. 2000, p. 367, fig. 210: 14–15
Plate 21: Fig. 113: length – 26.6 μm, width – 4.3 μm, striae density – 34/10 μm, fibulae density – 16/10 μm
Abundance and Location of Maximum Abundance: rare, eastern Florida Bay

Nitzschia angularis Smith 1853 var. *affinis* Grunow 1862
References: Peragallo and Peragallo 1897–1908, p. 284, fig. 73: 8; Witkowski et al. 2000, p. 368, figs. 199: 5, 6; Hein et al. 2008, p. 83, fig. 58: 7
Plate 20: Fig. 109: length – 51.5 μm, width – 5.8 μm, striae density – unresolved, fibulae density – 6/10 μm
Abundance and Location of Maximum Abundance: common, central Florida Bay

Nitzschia dubiiformis Hustedt 1939
References: Witkowski et al. 2000, p. 379, fig. 179: 8–10; as *N. pseudohybrida* in Hein et al. 2008, pp. 86–87, fig. 60: 3
Plate 20: Fig. 111: length – 30.0 μm, width – 2.5 μm, striae density – unresolved, fibulae density – 18/10 μm
Remarks: We were unable to distinguish *N. dubiiformis*, *N. thermaloides*, and *N. pseudohybrida* from each other. As stated in Hein et al. 2008, Krammer and Lange-Bertalot 1988 (p. 60) suggest that *N. dubiiformis* and *N. pseudohybrida* are conspecific. The earliest described species is *N. dubiiformis*.
Abundance and Location of Maximum Abundance: common, eastern Florida Bay

Nitzschia liebetruthii Rabenhorst 1864, **S**
Reference: Witkowski et al. 2000, p. 390, fig. 209: 21

Plate 21: Fig. 117: length – 14.5 μm, width – 2.9 μm, striae density – 25/10 μm, fibulae density – 15/10 μm

Remarks: Fibulae density in the depicted taxon is slightly larger than that reported by Witkowski et al. 2000. We distinguish *N. liebetruthii* from the very similar *N. frustulum* by the unevenly spaced middle fibulae in the latter.

Abundance and Location of Maximum Abundance: abundant, eastern Florida Bay

Nitzschia marginulata Grunow 1880 var. *didyma* Grunow 1880
References: Peragallo and Peragallo 1897–1908, p. 270, fig. 70: 16; Witkowski et al. 2000, p. 393, figs. 183: 4, 5
Plate 20: Fig. 107: length – 35.9 μm, width – 9.2 μm, striae density – 31/10 μm, fibulae density – 14/10 μm
Remarks: Striae density of the depicted taxon is slightly finer than that reported in Witkowski et al. 2000, p. 392 for the nominate variety (19–28/10 μm).
Abundance and Location of Maximum Abundance: rare, Atlantic

Nitzschia panduriformis Gregory 1857 var. *continua* Grunow in Cleve and Grunow 1880, **P**
References: As *Psammodictyon panduriforme* var. *continua* in Snoeijs and Balashova 1998, p. 88, fig. 476; Witkowski et al. 2000, p. 398, fig. 183: 6
Plate 20: Fig. 110: length – 21.0 μm, width – 8.3 μm, striae density – 19/10 μm, fibulae density – 16/10 μm
Remarks: Fibulae density for the depicted taxon is smaller than nominate variety; fibulae densities for the var. *continua* is not reported in either Snoeijs and Balashova 1998 or Witkowski et al. 2000.
Abundance and Location of Maximum Abundance: common, eastern Florida Bay

Nitzschia pubens Cholnoky 1960
Reference: Archibald and Schoeman 1983, p. 285, fig. 307
Plate 21: Fig. 116: length – 15.6 μm, width – 4.1 μm, striae density – 37/10 μm, fibulae density – 16/10 μm
Remarks: We distinguish *N. pubens* from the similar *N. microcephala* by the broad to elliptically lanceolate valves in the former versus the linearly lanceolate valves in the latter.
Abundance and Location of Maximum Abundance: common, central Florida Bay

Nitzschia sigma (Kutzing 1844) var. *intercedens* Grunow 1878
References: Peragallo and Peragallo 1897–1908, p. 290, fig. 74: 7; as *N. sigma* in Witkowski et al. 2000, p. 404, fig. 206: 6 and Hein et al. 2008, pp. 86–87, fig. 60: 3; Lobban et al. 2012, pp. 303–304, fig. 61: 4
Plate 20: Fig. 104: length – 112 μm, striae density – 34/10 μm, fibulae density – 11/10 μm
Abundance and Location of Maximum Abundance: rare, central Florida Bay

Nitzschia ventricosa Kitton 1873
References: Witkowski et al. 2000, p. 408, fig. 204: 8; Cleve and Grunow 1880, p. 100, no figures; Hustedt 1955, p. 48, no figures.

Plate 21: Fig. 112: Detail of the valve central area, scale bar – 50 µm. Morphometrics of entire valve (not shown) are : length – 303 µm, width – 7.1 µm, striae density – 34/10 µm, fibulae density – 6/10 µm

Abundance and Location of Maximum Abundance: rare, central Florida Bay

Nitzschia vidovichii Grunow 1862
References: Peragallo and Peragallo 1897–1908, p. 283, fig. 72: 13; Witkowski et al. 2000, p. 409, fig. 201: 2, 3; as *N. scalpelliformis* in Hein et al. 2008, p. 86, fig. 60: 2
Plate 20: Fig. 105: length – 148 µm, width – 4.9 µm, striae density – 29/10 µm, fibulae density – 10/10 µm; Plate 20: Fig. 106: Detail of the valve central area, scale bar – 20 µm
Remarks: We were unable to distinguish *N. vidovichii* from *N. scalpelliformis*. Witkowski et al. 2000 states that the "complex of forms and different established taxa around *N. scalpelliformis* needs a general taxonomic revision". The earliest described of the two species is *N. vidovichii*.
Abundance and Location of Maximum Abundance: rare, central Florida Bay

Nitzschia sp.
Reference: As *Nitzschia* 5 in Hein et al. 2008, p. 103, fig. 73: 10
Plate 20: Fig. 108: length – 79.7 µm, width – 6.0 µm, striae density – 40/10 µm, fibulae density – 17/10 µm
Remarks: The taxon depicted is most similar to *N. cursoria* (Donkin 1858) Grunow 1880 but differs in the latter having less dense fibulae (6–10/10 µm) as reported in Cleve and Grunow 1880 and Peragallo and Peragallo 1897–1908.
Abundance and Location of Maximum Abundance: rare, central Florida Bay

Planothidium cf. *pericavum* (Carter 1966) Lange-Bertalot 1999
Reference: Witkowski et al. 2000, pp. 122–123, fig. 49: 4–7
Plate 5: Fig. 31: RSV, length – 12.6 µm, width – 3.7 µm, RSV and SV striae density – 21/10 µm; SV, length – 12.6 µm, width – 3.7 µm, RSV and SV striae density – 21/10 µm
Remarks: The taxon depicted has slightly higher striae densities than the 16–18/10 µm reported in Witkowski et al. 2000.
Abundance and Location of Maximum Abundance: rare, eastern Florida Bay

Pleurosigma sp.
References: As *P.* (*delicatulum* var. ?) *kariana* Grunow in Cleve and Grunow 1880, p. 50, fig. 3: 69; *P. elongatum* W. Smith 1852 var. *kariana* (Grunow 1860) Grunow 1880. Similar taxa – *P. salinarum* Grunow 1878 in Cleve 1894, p. 39, no figures and Peragallo and Peragallo 1897–1908, pp. 166–167, fig. 33: 16. *P. delicatulum* Smith 1852 var. *salinarum* Grunow 1878 in H. Peragallo 1890–1891, p. 13, no figures. *P. elongatum* Smith 1852 var. *gracilescens* Grunow 1880 in Cleve 1894, p. 38, no figures. *P. gracilescens* Grunow 1880 in H. Peragallo 1890–1891, p. 7, fig. 3: 9.
Plate 21: Fig. 114: length – 148 µm, width – 15.6 µm, transverse and oblique striae density – 21/10 µm, crossing angle of oblique striae – 57°, raphe angle – 4°; Plate 21: Fig. 115: Detail of valve central area, scale bar – 20 µm.
Remarks: The taxon depicted appears to be most similar to *P.* (*delicatulum* var. ?) *kariana* Grunow depicted in in Cleve and Grunow 1880, p. 50, fig. 3: 69. The currently accepted name of that taxon is *P. elongatum* (W. Smith 1852) var. *kariana* Grunow

1860. The depicted taxon has similar striae densities and valve outline to *P. elongatum* var. *kariana* but differs in having a smaller central area. It is not possible to compare the stria angles because these have not been found in the literature for *P. elongatum* var. *kariana*. It is also similar, yet different from, *P. salinarum* and *P. gracilescens* Grunow 1880. *P. salinarum* in Cleve 1894 and Peragallo and Peragallo 1897–1908 has slightly higher transverse and oblique striae densities, 22–25/10 μm and 25–28/10 μm, respectively. *P. gracilescens* has more similar transverse and oblique striae densities, both 16.5–18.5/10 μm but significantly differs in that the apices are acute as opposed to blunt in the depicted taxon. Frithjof Sterrenburg has examined the depicted taxon and type material of *P. salinarum* and *P. elongatum* and determined that the depicted taxon is neither of these taxa because of differences in stria angles and the sizes of the central areas. (F. Sterrenburg, personal communication).

Abundance and Location of Maximum Abundance: rare, central Florida Bay

Proschkinia complanata (Grunow 1880) Mann 1990
Reference: Witkowski et al. 2000, p. 341, figs. 60: 29–32, 147: 8–11.
Plate 16: Fig. 81: length–32.2 μm, width–3.0 μm, striae density in middle–26/10 μm, striae density at apices– 30/10 μm.
Abundance and Location of Maximum Abundance: rare, central Florida Bay

Rhopalodia pacifica Krammer 1987, **S**
References: Krammer 1988, p. 168, figs. 129–138; Witkowski et al. 2000, p. 411, figs. 214: 3–4.
Plate 21: Fig. 118: length–34.5 μm, width–11.2 μm, striae density–20/10 μm, primary costae density–10/10 μm, punctae density–28/10 μm.
Remarks: The depicted taxon is similar to *Rhopalodia guettingeri* but differs in its larger size and coarser punctae density easily visible under LM.
Abundance and Location of Maximum Abundance: common, central Florida Bay

Seminavis basilica Danielidis 2003
References: Danielidis and Mann 2003, pp. 22–27, figs. 1–19; Wachnicka and Gaiser 2007, p. 440, figs. 219–220.
Plate 17: Fig. 84: length–72.3 μm, width–10.4 μm, dorsal and ventral striae density–31/10 μm.
Abundance and Location of Maximum Abundance: rare, central Florida Bay

Seminavis cyrtorapha Wachnicka and Gaiser 2007
Reference: Wachnicka and Gaiser 2007, pp. 444–446, figs. 237–240.
Plate 17: Fig. 86: length–41.4 μm, width–7.6 μm, dorsal striae density–26/10 μm, ventral striae density–25/10 μm.
Abundance and Location of Maximum Abundance: common, central Florida Bay

Seminavis delicatula Wachnicka and Gaiser 2007, **S**
References: As *Amphora obtusiuscula* in Sullivan 1990, pp. 251–258, figs. 1: 1–6 and Witkowski et al. 2000, pp. 145–146, figs. 164: 6–8; Wachnicka and Gaiser 2007, pp. 442–444, figs. 228–234.
Plate 17: Fig. 85: length–31.4 μm, width–5.3 μm, dorsal striae density–37/10 μm, ventral striae density–35/10 μm.

Abundance and Location of Maximum Abundance: common, Atlantic

Seminavis robusta Danielidis and Mann 2002
References: Danielidis and Mann 2002, pp. 440–443, figs. 40–47; Wachnicka and Gaiser 2007, p. 442 figs. 221–222; Hein et al. 2008, pp. 81–82, figs. 57: 3, 5.
Plate 17: Fig. 83: length – 84.8 µm, width – 12.2 µm, dorsal striae density – 16/10 µm, ventral striae density – 15/10 µm.
Abundance and Location of Maximum Abundance: rare, eastern Florida Bay

Seminavis strigosa (Hustedt 1949) Danielidis and Economou-Amilli 2003
References: Danielidis and Mann 2003, pp. 30–32, figs. 23–26; as *Amphora strigosa* in Witkowski et al. 2000, p. 151, figs. 164: 3, 4, 166: 5–7; Wachnicka and Gaiser 2007, p. 439, figs. 210–212
Plate 17: Fig. 88: length – 28.9 µm, width – 4.8 µm, dorsal striae density – 20/10 µm, ventral striae density – 17/10 µm.
Abundance and Location of Maximum Abundance: common, eastern Florida Bay

Surirella fastuosa Ehrenberg 1840
References: Witkowski et al. 2000, pp. 414–415, figs. 215: 1–3; Hein et al. 2008, p. 90, figs. 62: 2, 3, 64: 1.
Plate 22: Fig. 121: length – 72.1 µm, width – 48.4 µm, dorsal striae density – 16/10 µm, striae density – 16/10 µm, costae density – 2/10 µm.
Abundance and Location of Maximum Abundance: rare, central Florida Bay

Thalassiophysa hyalina (Greville 1865) Paddock and Sims 1981
References: As *Auricula insecta* in Peragallo and Peragallo 1897–1908, p. 194, figs. 42: 16–18; Round et al. 1990, p. 606, fig. a; Hein et al. 2008, p. 87, figs. 60: 6, 61: 1.
Plate 22: Fig. 120: length – 66.8 µm, striae density – unresolved, fibulae density – 10/10 µm.
Abundance and Location of Maximum Abundance: rare, central Florida Bay

Discussion

The spatial distribution of diatom species (current study) and community groups (Frankovich et al. 2006) in Florida Bay and the adjacent oceanside back-reef suggest that the diatom species assemblages are structured either by salinity and nutrient availability. The four distinct community groups determined by Frankovich et al. 2006 and designated here as the eastern Florida Bay, central Florida Bay, western Florida Bay, and Atlantic groups are aligned along existing salinity (Frankovich and Fourqurean 1997) and nutrient availability (Fourqurean et al. 1992) gradients. Long-term average salinities increase towards marine and less variable salinities from eastern to western Florida Bay and the Atlantic Ocean environments (Frankovich and Fourqurean 1997; Boyer et al. 1999). The salinity gradient observed during the species observations was similar in direction, but small and occurred over a polyhaline range (24–42 psu) from Eastern Florida Bay to Central and Eastern Florida Bay (Table 2). The salinity at the Atlantic site (site 8) was stable and marine around 36 psu. The availability of the limiting nutrient phosphorus generally increases from eastern towards western Florida Bay and is lowest in the Atlantic environment (Fourqurean

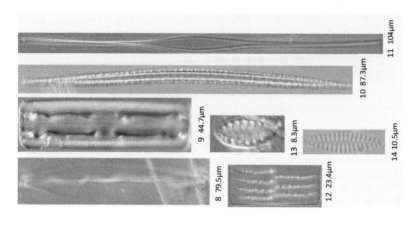

PLATE 2. Figure 8. *Striatella unipunctata.* Figure 9. *Grammatophora oceanica.* Figure 10. *Thalassiothrix longissima.* Figure 11. *Reimerothrix floridensis.* Figure 12. *Hyalosira interrupta.* Figure 13. *Opephora* cf. *horstiana.* Figure 14. *Neodelphineis pelagica.*

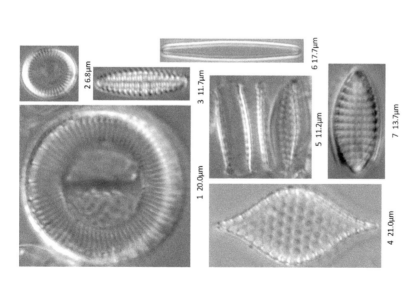

PLATE 1. Figure 1. *Cyclotella desikacharyi.* Figure 2. *Cyclotella choctawatcheeana.* Figure 3. *Neofragilaria* sp. Figure 4. *Cymatosira lorenziana.* Figure 5. *Cymatosira belgica.* Figure 6. *Pteroncola inane.* Figure 7. *Trachysphenia* sp.

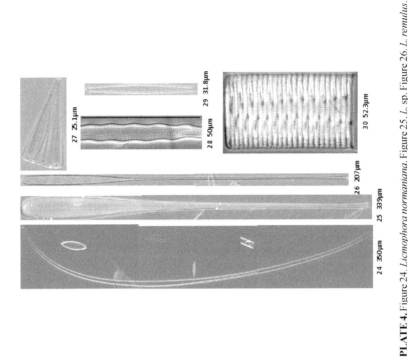

PLATE 4. Figure 24. *Licmophora normaniana*. Figure 25. *L.* sp. Figure 26. *L. remulus*. Figure 27. *L. debilis*. Figure 28. *Toxarium undulatum*, detail of valve center. Figure 29. *Synedra fasciculata* f. *densestriata*. Figure 30. *Rhabdomena adriaticum*.

PLATE 3. Figures 15, 16. *Hyalosynedra laevigata*. Figure 15. Detail of valve apex. Figure 16. Entire valve. Figure 17. *Tabularia* sp. Figure 18. *Synedra* sp. Figure 19. *Neosynedra tortosa*. Figure 20. *Ardissonea fulgens*. Figure 21. *Synedra*. sp., detail of valve apex. Figure 22. *Tabularia* sp., detail of valve apex. Figure 23. *A. fulgens*, detail of valve apex.

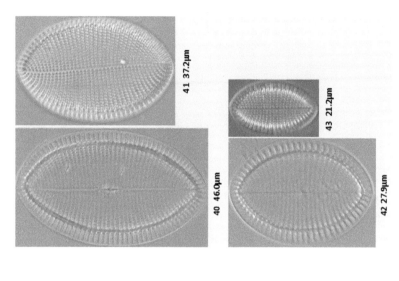

PLATE 6. Figures 40, 41. *Cocconeis woodii*. Figure 40. RSV. Figure 41. SV. Figures 42, 43. *Cocconeis britannica*. Figure 42. RSV. Figure 43. SV.

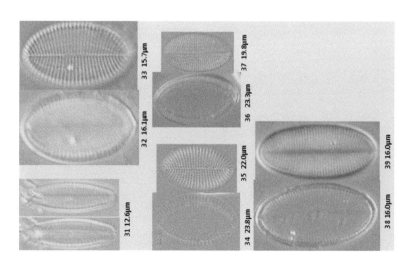

PLATE 5. Figure 31. *Planothidium* cf. *pericavum*, frustule, RSV (left), SV (right). Figures 32, 33. *Cocconeis barleyi*. Figure 32. RSV. Figure 33. SV. Figures 34, 35. *C. scutellum*. Figure 34. RSV. Figure 35. SV. Figures 36, 37. *C. euglypta*. Figure 36. RSV. Figure 37. SV. Figures 38, 39. *C. coralliensis*. Figure 38. RSV. Figure 39 SV.

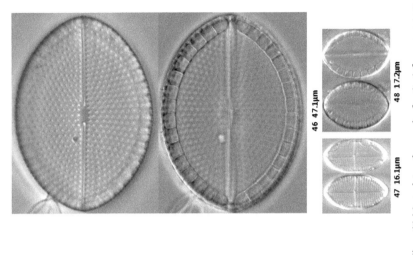

PLATE 8. Figure 46. *Mastogloia cribrosa*, valve face (top), focus on partectae (below). Figure 47. *M. cruicicula* var. *cruicicula*, valve face (left), focus on partectae (right). Figure 48. *M. ovalis*, valve face (left), focus on partectae (right).

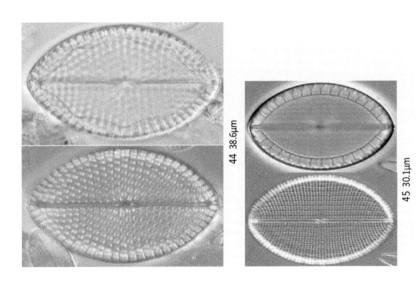

PLATE 7. Figure 44. *Mastogloia fimbriata*, valve face (left), focus on partectae (right). Figure 45. *M. ovata*, valve face (left), focus on partectae (right).

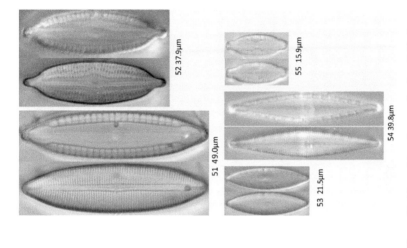

PLATE 10. Figure 51. *Mastogloia robusta*, valve face (left), focus on partectae (right). Figure 52. *M. corsicana*, valve face (left), focus on partectae (right). Figure 53. *M. laminaris*, valve face (left), focus on partectae (right). Figure 54. *M. angusta*, valve face (left), focus on partectae (right). Figure 55. *M. manokwariensis*, valve face (left), focus on partectae (right).

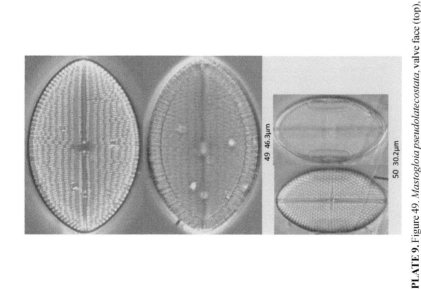

PLATE 9. Figure 49. *Mastogloia pseudolatecostata*, valve face (top), focus on partectae (below). Figure 50. *M. binotata*, valve face (left), focus on partectae (right).

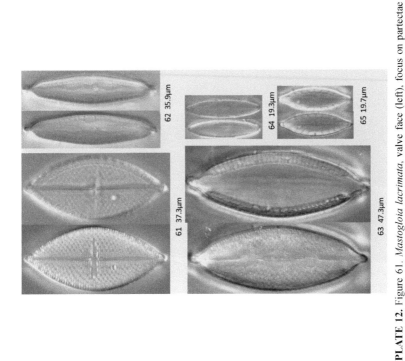

PLATE 12. Figure 61. *Mastogloia lacrimata*, valve face (left), focus on partectae (right). Figure 62. *M. biocellata*, valve face (left), focus on partectae (right). Figure 63. *M. punctifera*, valve face (left), focus on partectae (right). Figure 64. *M. cuneata*, valve face (left), focus on partectae (right). Figure 65. *M.* sp. 1, valve face (left), focus on partectae (right).

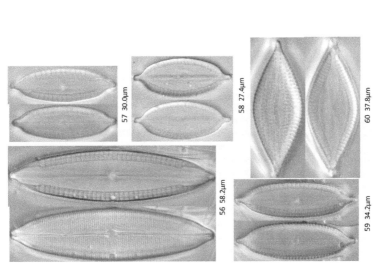

PLATE 11. Figure 56. *Mastogloia lanceolata*, valve face (left), focus on partectae (right). Figure 57. *M. ignorata*, valve face (left), focus on partectae (right). Figure 58. *M. cyclops*, valve face (left), focus on partectae (right). Figure 59. *M. discontinua*, valve face (left), focus on partectae (right). Figure 60. *M. rimosa*, valve face (top), focus on partectae (bottom).

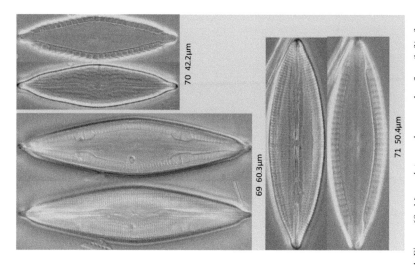

PLATE 14. Figure 69. *Mastogloia erythraea*, valve face (left), focus on partectae (right). Figure 70. *M. subaffirmata*, valve face (left), focus on partectae (right). Figure 71. *M. hainanensis*, valve face (top), focus on partectae (bottom).

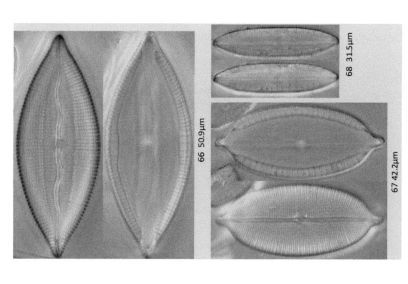

PLATE 13. Figure 66. *Mastogloia* sp. 2, valve face (top), focus on partectae (bottom). Figure 67. *M. pseudoelegans*, valve face (left), focus on partectae (right). Figure 68. *M. pusilla* var. *linearis*, valve face (left), focus on partectae (right).

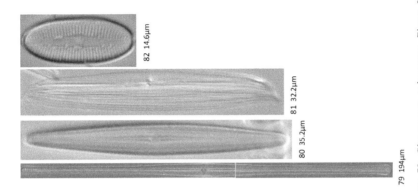

PLATE 16. Figure 79. *Climaconeis colemaniae.* Figure 80. *Brachysira aponina.* Figure 81. *Proschkinia complanata.* Figure 82. *Diploneis vacillans.*

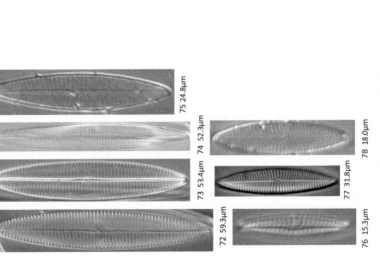

PLATE 15. Figure 72. *Navicula* cf. *normalis.* Figure 73. *N. flebilis.* Figure 74. *N. halinae.* Figure 75. *N.* cf. *duerrenbergiana.* Figure 76. *N.* sp. Figure 77. *N.* cf. *agnita.* Figure 78. *N.* cf. *microdigitoradiata.*

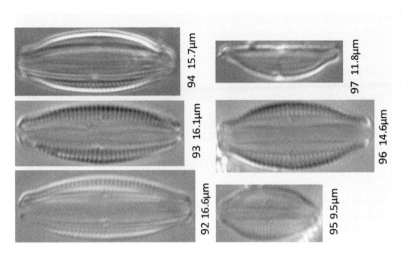

94 15.7μm

93 16.1μm

92 16.6μm

97 11.8μm

96 14.6μm

95 9.5μm

PLATE 18. Figure 92. *Halamphora aponina*. Figures 93, 94. *H. tenerrima*. Figure 93. Frustule, focus on ventral surface. Figure 94. Focus on dorsal girdle bands. Figure 95. *Amphora* sp. 2. Figure 96. *A. crenulata*. Figure 97. *H. tenuissima*.

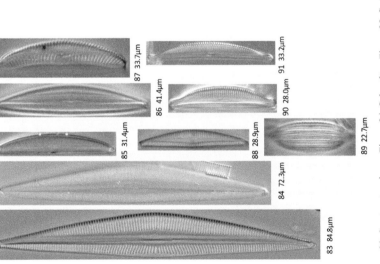

87 33.7μm

86 41.4μm

85 31.4μm

84 72.3μm

83 84.8μm

91 33.2μm

90 28.0μm

88 28.9μm

89 22.7μm

PLATE 17. Figure 83. *Seminavis robusta*. Figure 84. *S. basilica*. Figure 85. *S. delicatula*. Figure 86. *S. cyrtorapha*. Figure 87. *Amphora proteus*. Figure 88. *S. strigosa*. Figures 89, 90. *A.* cf. *tumida*. Figure 89. Frustule, focus on dorsal girdle bands. Figure 90. Focus on valve face. Figure 91. *Halamphora cymbifera* var. *heritierarum*.

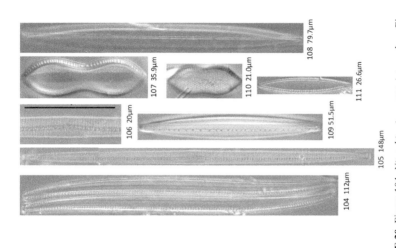

PLATE 20. Figure 104. *Nitzschia sigma* var. *intercedens*. Figures 105, 106. *N. vidovichii*. Figure 105. Entire valve. Figure 106. Focus on central area. Figure 107. *N. marginulata* var. *didyma*. Figure 108. *N.* sp. Figure 109. *N. angularis* var. *affinis*. Figure 110. *N. panduriformis* var. *continua*. Figure 111. *N. aequoria*.

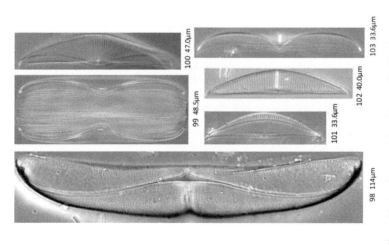

PLATE 19. Figure 98. *Amphora cingulata*. Figures 99, 100. *A. arcus* var. *sulcata*. Figure 99. Frustule, focus on ventral surface. Figure 100. Focus on dorsal striae. Figure 101. *A. graeffeana*. Figure 102. *A.* sp. 1. Figure 103. *A. ostrearia* var. *typica*.

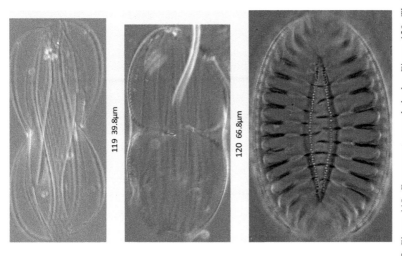

PLATE 22. Figure 119. *Entomoneis pseudoduplex.* Figure 120. *Thalassiophysa hyalina.* Figure 121. *Surirella fastuosa.*

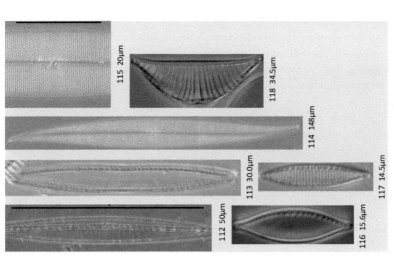

PLATE 21. Figure 112. *Nitzschia ventricosa,* focus on central area. Figure 113. *N. dubiiformis.* Figures 114, 115. *Pleurosigma* sp. Figure 114. Entire valve. Figure 115. Focus on central area. Figure 116. *N. pubens.* Figure 117. *N. liebetruthii.* Figure 118. *Rhopalodia pacifica.*

et al. 1992; Boyer and Jones 2002). Water column total phosphorus concentrations observed during the species observations exhibited a similar pattern with the lowest concentrations in the Atlantic and Eastern Florida Bay and higher concentrations in the Central and Western Florida Bay (Table 2).

A comparison of the diatom indicator species of the community groups (Frankovich et al. 2006) and the results of the Indicator Species Analysis of the responses of individual taxa to the phosphorus (P) enrichment (present study) reveals that only *Brachysira aponina*, a brackish to oligohaline species (Witkowski et al. 2000), was an indicator of both P fertilization and spatial distribution. The decrease in abundance of *B. aponina* following P enrichment may suggest that this species is a poor competitor under nutrient replete conditions. This response to P addition and its dominance in eastern Florida Bay is consistent with the lower salinity and low nutrient environment in eastern Florida Bay. Despite the significant but small changes in diatom community structure resulting from P addition, the nutrient enriched diatom communities retained greater similarity to control communities from the same site than to the diatom communities of other spatial groups located at different sites along the nutrient and salinity gradients (Frankovich et al. 2006). Montgomery 1978 argued that benthic diatom communities are likely not limited by nutrient availability because of their close proximity to rich benthic sources of nutrients and the large and diverse sympatric bacterial populations that provide a stable input of nutrients. He also reported negative correlations of species richness, diversity, and evenness with community density, indicating the presence of interspecific competition, and suggested that space competition rather than nutrient competition is a dominant factor for epiphytic diatoms. The limited number of nutrient indicator species (10) and the small changes in community structure (Frankovich et al. 2009) further suggest that nutrient availability is only a minor factor structuring these communities. However, Snoeijs 1999 hypothesized that the overwhelming influence of salinity variability may mute the response to nutrient availability.

The number of established taxa observed (272, Table 1) greatly exceeds the number of established taxa identified and listed as epiphytes in previous studies from the Gulf of Mexico and the Caribbean Sea. Though Frankovich et al. 2006 report 332 taxa from the same material, that number includes 40 additional taxa that were very rare (< 1% maximum relative abundance) and remain undescribed. Therefore these taxa are not included in the present listing. Many of the taxa observed in the present study were also noticed in previous diatom studies from south Florida, the Gulf of Mexico and the Caribbean Sea, suggesting similar diatom floras. The number of taxa in common is roughly proportional to the number of taxa reported by the previous authors. 122 common taxa are illustrated in Hein et al. 2008 that represent 32% of the taxa they observed from macroalgal, stromatolite, *Thalassia testudinum* seagrass leaves, sand, and glass slide substrates and from plankton in the Bahamas. Montgomery 1978 provides SEM photomicrographs of 102 shared taxa (many undescribed at the time) that represent 18% of the taxa they list from sediment, coral and *Thalassia* substrates in the Florida Keys. The low percentage of taxa in common with Montgomery 1978 results from the unique groups of taxa observed by Montgomery among the sediment, coral, and seagrass substrates and the much greater species richness from the sediment and coral substrates which he attributed to the greater structural complexities of these substrate types. DeFelice and Lynts 1978 list 161 diatom species from surface sediments and *Thalassia* leaves in northeast Florida Bay (eastern Florida Bay community group of present study). Despite the lack of mounting media in their prepared diatom slides, which likely limited taxonomic resolution, 48 shared taxa were reported, representing 34% of their epiphytic taxa. Huvane 2002 also surveyed diatoms (260 taxa) from surface sediments of Florida Bay (eastern, central, and western Florida Bay community groups of present study) but a list of the observed taxa was not provided preventing further comparison.

The greatest percentages of shared taxa with the present study were observed from diatom surveys focused similarly on the epiphytes of *Thalassia testudinum* (Reyes-Vasquez 1970; Corlett and Jones 2007; Lopez-Fuerte et al. 2013). 42 diatom species were described from *Thalassia* leaves in Biscayne Bay, Florida, USA (located to the immediate north of Florida Bay) of which 67% were shared taxa (Reyes-Vasquez 1970). Corlett and Jones 2007 list 61 taxa from the Cayman Islands with 50% of their identifiable taxa in common with the present study. *Thalassia* epiphytes from the Mexican Caribbean were also similar to the epiphytic diatoms in Florida Bay, with 50% of the 106 taxa in common (Lopez-Fuerte et al. 2013).

The dominance of *Mastogloia* species observed in the present study (27% of the 284 taxa) is a common characteristic of benthic diatom surveys from many tropical shallow marine habitats (DeFelice and Lynts 1978; Montgomery 1978; Reyes-Vasquez 1970; Frankovich et al. 2006; Corlett and Jones 2007; Lopez-Fuerte et al. 2013). The percentage of *Mastogloia* taxa in the species lists ranged from 21% (Reyes-Vasquez 1970) to 35–71% from *Thalassia* leaves in Montgomery 1978. Montgomery 1978 reported much lower *Mastogloia* percentages from sand and coral substrates (range = < 1–20.1%). The complete lack of *Mastogloia* taxa in a 40 cm sediment core sampled from an unvegetated site in Rookery Bay, FL, USA (Cremer et al. 2007; V. McGee personal communication) is a notable exception to their prevalence in sediments from other subtropical South Florida estuaries such as Florida Bay (Wachnicka et al. 2013) and Biscayne Bay (Wachnicka et al. 2011), where seagrass coverage is much greater (Yokel 1975; Zieman et al. 1989; Lirman and Cropper 2003). A lower range of salinities is also experienced in Rookery Bay due to the direct influence of freshwater discharge from Henderson Creek (Cremer et al. 2007) may also influence the lack of *Mastogloia* taxa. A similar paucity of *Mastogloia* taxa was also reported in the seagrass epiphyte flora in Mississippi Sound where average salinity is 30 psu (Sullivan 1979), suggesting that the prevalence of seagrass-associated *Mastogloia* taxa may be limited to tropical/subtropical marine environments with dense seagrass coverage.

Concluding Remarks

This study provides the first illustrated guide and a preliminary checklist to the epiphytic diatom flora of Florida Bay. This will hopefully serve as a reference and source for future investigations of biogeography, environmental change, and benthic ecology.

The environmental preferences of individual diatom taxa or community groups, as indicated by their spatial distribution and response to nutrient additions, suggest how these communities may change as environments are altered by future and on-going water management practices, continuing sea-level rise and future unknown changes to the greater Everglades ecosystem. The existence of at least 40 additional undescribed taxa suggests that descriptions of the Florida Bay and the Florida Keys diatom flora is far from complete, and that much taxonomic work is still needed to fully characterize these communities and further determine relationships with similar ecosystems. The scarcity of taxonomic expertise currently limits benthic diatom ecological studies; therefore, knowledge of the varied community compositions will advance future investigations of the functional ecology (e.g., productivity, biogeochemical cycling) of these important ecosystem components.

Acknowledgments

The authors thank Dr. Jay Zieman for the use of University of Virginia boats and microscope facilities. The authors thank Drs. Evelyn Gaiser, Frithjof Sterrenburg, Mike Sullivan, Chiara Pennesi, Mike Hein, Chris Lobban, Matt Ashworth and Andrzej Witkowski for assistance in diatom taxa identification. The authors also thank Drs. Mike Sullivan and Evelyn Gaiser for reviewing earlier drafts of the manuscript. P.J. Walker and Dr. Dave Rudnick (Everglades National Park, ENP) aided permit issuance and use of ENP facilities. Funding for this work was provided by the George Barley Scholars Program of the University of Virginia. This study was conducted as part of the Florida Coastal Everglades Long-Term Ecological Research Program (FCE-LTER) National Science Foundation grant DEB-9901514. This is contribution 695 of the Southeast Environmental Research Center at Florida International University.

References Cited

Archibald, R.E.M. and F.R. Schoeman. 1983. The diatoms of the Sundays and Great Rivers in the eastern Cape Province of S. Afr. Bibliotheca Diatomologica 1: 1–362.

Archibald, R.E.M. and F.R. Schoeman. 1984. *Amphora coffeaeformis* (Agardh) Kützing: a revision of the species under light and electron microscopy. S. Afr. J. Bot. 3: 83–100.

Armitage, A.R., T.A. Frankovich and J.W. Fourqurean. 2006. Variable responses within epiphytic and benthic microalgal communities to nutrient enrichment. Hydrobiologia 569: 423–35.

Arndt, S., G. Lacroix, N. Gypens, P. Regnier and C. Lancelot. 2011. Nutrient dynamics and phytoplankton development along an estuary–coastal zone continuum: a model study. J. Mar. Syst. 84: 49–66.

Belando, M.D., A. Marín and M. Aboal. 2012. *Licmophora* species from a Mediterranean hypersaline coastal lagoon (Mar Menor, Murcia, SE Spain). Nova Hedwigia, Beiheft. 141: 275–288.

Boyer, J.N. and R.D. Jones. 2002. A view from the bridge: external and internal forces affecting the ambient water quality of the Florida Keys National Marine Sanctuary. pp. 609–628. *In*: J.W. Porter and K.G. Porter (eds.). The Everglades, Florida Bay, and Coral Reefs of the Florida Keys, an Ecosystem Sourcebook. CRC Press, Boca Raton, Florida.

Boyer, J.N., J.W. Fourqurean and R.D. Jones. 1997. Spatial characterization of water quality in Florida Bay and Whitewater Bay by multivariate analyses: zones of similar influence. Estuaries 20: 743–758.

Boyer, J.N., J.W. Fourqurean and R.D. Jones. 1999. Seasonal and long-term trends in the water quality of Florida Bay. Estuaries 22: 417–430.

Cholnoky, B.J. 1963a. Beiträge zur kenntnis des marinen litorals von Sudafrika. Botanica Marina 5: 38–83.

Cholnoky, B.J. 1963b. Ein Beitrag zur kenntnis der diatomeenflora von Holländischen Neuguinea. Nova Hedwigia 5: 157–198.

Clavero, E., J.O. Grimalt and M. Hernandes-Marine. 2000. The fine structure of two small *Amphora* species. *A. tenerrima* Aleem and Hustedt and *A. tenuissima* Hustedt. Diatom Res. 15: 195–208.

Cleve, P.T. 1894. Synopsis of the Naviculoid Diatoms, Part 1. Kongliga Svenska Vetenskaps-Academiens Handlingar 26: 1–194.

Cleve, P.T. 1895. Synopsis of the Naviculoid Diatoms, Part 2. Kongliga Svenska Vetenskaps-Academiens Handlingar 27: 1–220.

Cleve, P.T. and A. Grunow. 1880. Beiträge zur kenntnis der arktischen Diatomeen. Kongliga Svenska Vetenskaps-Academiens Handlingar 17: 1–121.

Corlett, H. and B. Jones. 2007. Epiphyte communities on *Thalassia testudinum* from Grand Cayman, British West Indies: their composition, structure, and contribution to lagoonal sediments. Sed. Geol. 194: 245–262.

Cremer, H., F. Sangiorgi, F. Wagner-Cremer, V. McGee, A.F. Lotter and H. Visscher. 2007. Diatoms (Bacillariophyceae) and dinoflagellate cysts (Dinophyceae) from Rookery Bay, Florida, USA. Caribbean J. Sci. 43: 23–58.

Danielidis, D.B. and D.G. Mann. 2002. The systematics of *Seminavis*: the lost identities of *Amphora angusta, A. ventricosa* and *A. macilenta*. Eur. J. Phycol. 37: 429–448.

Danielidis, D.B. and D.G. Mann. 2003. New species and new combinations in the genus *Seminavis* (Bacillariophyta). Diatom Res. 18: 21–39.

DeFelice, D.R. and G.W. Lynts. 1978. Benthic marine diatom associations: upper Florida Bay (Florida) and associated sounds. J. Phycol. 14: 25–33.

Desrosiers, C., J. Leflaive, A. Eulin and L. Ten-Hage. 2013. Bioindicators in marine waters: benthic diatoms as a tool to assess water quality from eutrophic to oligotrophic coastal ecosystems. Ecol. Indic. 32: 25–34.

De Stefano, M. and O.E. Romero. 2005. A survey of alveolate species of the diatom genus *Cocconeis* (Ehr.) with remarks on the new section Alveolatae. Bibliotheca Diatomologica 52: 1–133.

De Stefano, M., O.E. Romero and C. Totti. 2008. A comparitive study of *Cocconeis scutellum* Ehrenberg and its varieties (Bacillariophyta). Botanica Mar. 51: 506–536.

Dufrêne, M. and P. Legendre. 1997. Species assemblages and indicator species: the need for a flexible asymmetrical approach. Ecol. Monogr. 67: 345–366.

Fourqurean, J.W. and M.B. Robblee. 1999. Florida Bay: A history of recent ecological changes. Estuaries 22: 345–357.

Fourqurean, J.W., J.C. Zieman and G.V.N. Powell. 1992. Phosphorus limitation of primary production in Florida Bay: evidence from C:N:P ratios of the dominant seagrass *Thalassia testudinum*. Limnol. Ocean. 37: 162–71.

Frankovich, T.A. and J.W. Fourqurean. 1997. Seagrass epiphyte loads along a nutrient availability gradient, Florida Bay, USA. Mar. Ecol. Progr. Ser. 159: 37–50.

Frankovich, T.A., E.E. Gaiser, J.C. Zieman and A.H. Wachnicka. 2006. Spatial and temporal distributions of epiphytic diatoms growing on *Thalassia testudinum* Banks ex König: relationships to water quality. Hydrobiologia 569: 259–71.

Frankovich, T.A., A.R. Armitage, A.H. Wachnicka, E.E. Gaiser and J.W. Fourqurean. 2009. Nutrient effects on seagrass epiphyte community structure in Florida Bay. J. Phycol. 45: 1010–1020.

Haynes, D., R. Skinner, J. Tibby, J. Cann and J. Fluin. 2011. Diatom and foraminifera relationships to water quality in The Coorong, South Australia, and the development of a diatom-based salinity transfer function. J. Paleolimnol. 46: 543–560.

Hein, M.K., B.M. Winsborough and M.J. Sullivan. 2008. Bacillariophyta (Diatoms) of the Bahamas. A.R.G. Gantner Verlag Kommanditgesellschaft, Ruggell, Germany.

Honeywill, C. 1998. A study of the British Licmophora species and a discussion of its morphological features. Diatom Res. 13: 221–271.

Hustedt, F. 1927–1930. Die Kieselalgen Deutsschlands, Österreichs und der Schweiz unter Berücksichtigung der übrigen Länder Europas sowie der angrenzenden Meeresgebiete. Band 7, Teil 1. *In*: L. Rabenhorst (ed.). Kryptogamen-Flora von Deutschland, Österreich und der Schweiz. Akademische Verlagsgesellschaft, Geest und Portig K.-G., Leipzig, Germany.

Hustedt, F. 1955. Marine Littoral Diatoms of Beaufort, North Carolina. Duke University Marine Station Bulletin No. 6. Duke University Press, Durham, North Carolina.

Hustedt, F. 1931–1959. Die Kieselalgen Deutsschlands, Österreichs und der Schweiz unter Berücksichtigung der übrigen Länder Europas sowie der angrenzenden Meeresgebiete. Band 7, Teil 2. *In*: L. Rabenhorst (ed.). Kryptogamen-Flora von Deutschland, Österreich und der Schweiz. Akademische

Verlagsgesellschaft, Geest and Portig K.-G., Leipzig, Germany. English translation—The Pennate Diatoms (1985), Koeltz Scientific Books, Koenigstein, West Germany.

Huvane, J.K. 2002. Modern diatom distributions in Florida Bay: a preliminary analysis. pp. 479–495. *In*: J.W. Porter and K.G. Porter (eds.). The Everglades, Florida Bay, and Coral Reefs of the Florida Keys. CRC Press, Boca Raton, Florida.

John, J. 1980. Two new species of the diatom Mastogloia from Western Australia. Nova Hedwigia 33: 49–858.

Kemp, K.D. and T.B.B. Paddock. 1990. A description of two new species of the diatom genus *Mastogloia* with further observations on *M. amoyensis* and *M. gieskesii*. Diatom Res. 5: 311–323.

Krammer, K. 1988. The Gibberula-group in the genus *Rhopalodia* O. Müller (Bacillariophyceae) II. Revision of the group and new taxa. Nova Hedwigia 47: 159–205.

Krammer, K. and H. Lange-Bertalot. 1988. Bacillariophyceae, Teil 2: Bacillariaceae, Epithemiaceae, Surirellaceae. *In*: A. Pascher (ed.). Süßwasserflora von Mitteleuropa. Band 2/2. Gustav Fischer Verlag, Stuttgart, West Germany.

Lirman, D. and W.P. Cropper. 2003. The influence of salinity on seagrass growth, survivorship, and distribution within Biscayne Bay, Florida: field, experimental, and modeling studies. Estuaries 26: 131–141.

Lobban, C.S., M. Schefter, R.W. Jordan, Y. Arai, A. Sasaki, E.C. Theriot, M. Ashworth, E.C. Ruck and C. Pennesi. 2012. Coral-reef diatoms (Bacillariphyta) from Guam: new records and preliminary checklist, with emphasis on epiphytic species from farmer-fish territories. Micronesica 43: 237–479.

López-Fuerte, F.O., D.A. Sisqueiros-Beltrones and O.U. Hernández-Almeida. 2013. Epiphytic diatoms of *Thalassia testudinum* (Hydrocharitaceae) from the Mexican Caribbean. Mar. Biodiv. Rec. 6: 1–11.

Mann, A. 1935. Diatoms in bottom deposits from the Bahamas and the Florida Keys. Papers from the Tortugas Laboratory 29: 121–128.

McCune, B. and J.B. Grace. 2002. Analysis of Ecological Communities. MJM Software Design, Gleneden Beach, Oregon.

Mereschkowsky, C. 1901. Diagnoses of new Licmophorae. Nouva Notarisia 12: 141–153.

Montgomery, R.T. 1978. Environmental and ecological studies of the diatom communities associated with the coral reefs of the Florida Keys. Ph.D. Dissertation. Florida State University, Tallahassee, Florida.

Morales, E.A. 2002. Studies in selected fragilarioid diatoms of potential indicator value from Florida (USA) with notes on the genus *Opephora* Petit (Bacillariophyceae). Limnologica 32: 102–113.

Navarro, J.N. 1982. Marine diatoms associated with the mangrove prop roots in the Indian River, Florida, U.S.A. Bibliotheca Phycologica 61: 1–151.

Osada, K. and H. Kobayasi. 1988. Observations on the forms of the diatom Entomoneis paludosa and related taxa. pp. 161–172. *In*: H. Simola (ed.). Proceedings of the Tenth International Diatom Symposium, Joensuu, Finland, August 28–September 2, 1988. Koeltz Scientific Books, Koenigstein, Germany.

Peragallo, H. 1890–1891. Monographie du genre *Pleurosigma* et des genres alliès. Le Diatomiste 1: 1–35.

Peragallo, H. and M. Peragallo. 1897–1908. Diatomées Marines de France et des Districts Maritimes Voisins. À Grez-sur-Loing (S.-et-M.). Reprinted in 1984 by Koeltz Scientific Books, Koenigstein, West Germany.

Prasad, A.K.S.K. 1987. Marine diatoms of St. George Sound, the northeastern Gulf of Mexico: *Neodelphineis pelagica* Takano (Diatomaceae, Bacillariophyceae). Northeast Gulf Science. 9: 125–129.

Prasad, A.K.S.K. and J.A. Nienow. 2006. The centric diatom genus *Cyclotella* (Stephanodiscaceae: Bacillariophyta), from Florida Bay, USA with special reference to *Cyclotella choctawatcheeana* and *Cyclotella desikacharyi*, a new species related to the *Cyclotella striata* complex. Phycologia 45: 127–140.

Prasad, A.K.S.K., J.A. Nienow and R.J. Livingston. 1990. The genus *Cyclotella* with special reference to *C. striata* and *C. choctawatcheeana* sp. nov. Phycologia 29: 418–436.

Prasad, A.K.S.K., K.A. Riddle and J.A. Nienow. 2000. Marine diatom genus *Climaconeis* (Berkeleyaceae, Bacillariophyta): two new species *Climaconeis koenigii* and *C. colemaniae*, from Florida Bay, USA. Phycologia 39: 199–211.

Prasad, A.K.S.K., J.A. Nienow and K.A. Riddle. 2001. Fine structure, taxonomy and systematics of *Reimerothrix* (Fragilariaceae: Bacillariophyta), a new genus of synedroid diatoms from Florida Bay, USA. Phycologia 40: 35–46.

Reyes-Vasquez, G. 1970. Studies on the diatom flora living on *Thalassia testudinum* König in Biscayne Bay, Florida. Bull. Mar. Sci. 20: 105–34.

Riaux-Gobin, C., O.E. Romero, P. Compère and A.Y. Al-Handal. 2011. Small-sized Achnanthales (Bacillariophyta) from coral sands off Mascarenes (western Indian Ocean). Bibliotheca Diatomologica 57: 1–234.

Romero, O.E. and R. Jahn. 2013. Typification of *Cocconeis lineata* and *Cocconeis euglypta* (Bacillariophyta). Diatom Res. 28: 175–184.

Round, F.E., R.M. Crawford and D.G. Mann. 1990. The Diatoms: Biology and Morphology of the Genera. Cambridge University Press, Cambridge, England.

Saunders, K.M. 2011. A diatom dataset and diatom-salinity inference model for southeast Australian estuaries and coastal lakes. J. Paleolimnol. 46: 525–542.

Schmidt, A., M. Schmidt, F. Fricke, H. Heiden, O. Müller and F. Hustedt. 1874–1959. Atlas der Diatomaceen-Kunde. R.Reisland, Leipzig, Germany.

Schoeman, F.R. and R.E.M. Archibald. 1986. Observations on *Amphora* species (Bacillariophyceae) in the British Museum (Natural History). V. Some species from the subgenus *Amphora*. S. Afr. J. Bot. 52: 425–437.

Seddon, A.W.R., C.A. Froyd and A. Witkowski. 2011. Diatoms (Bacillariophyta) of isolated islands: new taxa in the genus *Navicula* sensu stricto from the Galapágos Islands. J. Phycol. 47: 861–879.

Simonsen, R. 1987. Atlas and Catalogue of the Diatom Types of Friedrich Hustedt. J. Cramer, Berlin, West Germany.

Simonsen, R. 1990. On some diatoms of the genus *Mastogloia*. Nova Hedwigia Beiheft 100: 121–142.

Snoeijs, P. 1993. Intercalibration and distribution of diatom species in the Baltic Sea, Volume 1. The Baltic Marine Biologists Publication No. 16a. Opulus Press, Uppsala, Sweden.

Snoeijs, P. 1999. Diatoms and environmental change in brackish waters. pp. 298–333. *In*: E.F. Stoermer and J.P. Smol (eds.). The Diatoms: Applications for the Environmental and Earth Sciences. Cambridge University Press, New York.

Snoeijs, P. and N. Balashova. 1998. Intercalibration and distribution of diatom species in the Baltic Sea, Volume 5. The Baltic Marine Biologists Publication No. 16e. Opulus Press, Uppsala, Sweden.

Stephens, F.C. and R.A. Gibson. 1980. Ultrastructural studies of some *Mastogloia* (Bacillariophyceae) species belonging to the groups Undulatae, Apiculatae, Lanceolatae and Paradoxae. Phycologia 19: 143–152.

Sullivan, M.J. 1979. Epiphytic diatoms of three seagrass species in Mississippi Sound. Bull. Mar. Sci. 29: 459–464.

Sullivan, M.J. 1990. A light and scanning electron microscope study of the marine epiphytic diatom *Amphora obtusiuscula* Grunow. *In*: M. Ricard and M. Coste (eds.). Ouvrage dédié à la mémoire du Prof. H. Germain 1903–1989. Koeltz Scientific Books, Koenigstein, West Germany.

Tempère, J.A. 1893–1896. Le Diatomiste, Volume 2. Paris.

Van Lent, T.J., R. Johnson and R. Fennema. 1993. Water managemnet in Taylor Slough and effects on Florida Bay. South Florida Research Center, Everglades National Park, National Park Service. Homestead, Florida, USA.

Voigt, M. 1942. Contribution to the knowledge of the diatom genus *Mastogloia*. J. R. Microsc. Soc. 62: 1–20.

Voigt, M. 1952. A further contribution to the knowledge of the diatom genus *Mastogloia*. J. R. Microsc. Soc. 23: 440–449.

Voigt, M. 1963. Some new and interesting *Mastogloia* from the Mediterranean area and the Far East. J. R. Microsc. Soc. 82: 111–121.

Wachnicka, A.H. and E.E. Gaiser. 2007. Characterization of *Amphora* and *Seminavis* from South Florida, USA. Diatom Res. 22: 387–455.

Wachnicka, A., E. Gaiser, L. Collins and J. Boyer. 2010. Distribution of diatoms and development of diatom-based models for inferring salinity and nutrient concentrations in Florida Bay and adjacent coastal wetlands of south Florida (USA). Est. Coast. 33: 1080–1098.

Wachnicka, A., E. Gaiser and J. Boyer. 2011. Ecology and distribution of diatoms in Biscayne Bay, Florida (USA): implications for bioassessment and paleoenvironmental studies. Ecol. Indic. 33: 1080–1098.

Wachnicka, A., L.S. Collins and E.E. Gaiser. 2013. Response of diatom assemblages to 130 years of environmental change in Florida Bay (USA). J. Paleolimnol. 49: 83–101.

Wahrer, R.J., G.A. Fryxell and E.R. Cox. 1985. Studies in pennate diatoms: valve morphologies of *Licmophora* and *Campylostylus*. J. Phycol. 21: 206–217.

Witkowski, A.H., H. Lange-Bertalot and D. Metzeltin. 2000. Diatom Flora of Marine Coasts I. *In*: H. Lange-Bertalot (ed.). Iconographia Diatomologica Annotated Diatom Micrographs, Volume 7. A.R.G. Gantner Verlag Kommanditgesellschaft, Ruggell, Germany.

Yohn, T.A. and R.A. Gibson. 1981. Marine diatoms of the Bahamas I. *Mastogloia* Thw. ex Wm. Sm. species of the groups Lanceolatae and Undulatae. Botanica Mar. 24: 641–655.

Yohn, T.A. and R.A. Gibson. 1982. Marine diatoms of the Bahamas II. *Mastogloia* Thw. ex Wm. Sm. species of the groups Decussatae and Ellipticae. Botanica Mar. 25: 41–53.

Yokel, B.J. 1975. Rookery Bay Land Use Studies: Environmental Planning Strategies for the Development of a Mangrove Shoreline. Study No. 5 Estuarine Biology. The Conservative Foundation, Washington, D.C.

Zanon, D.V. 1948. Diatomee marine di Sardegna e pugillo di alghe marine della stessa. Acta Pontiffical Academy of Sciences 12: 202–246.

Zieman, J.C., J.W. Fourqurean and R.L. Iverson. 1989. Distribution, abundance, and productivity of seagrasses and macroalgae in Florida Bay. Bull. Mar. Sci. 44: 292–311.

CHAPTER

13

Pigment-Based Chemotaxonomy and its Application to Everglades Periphyton

J.W. Louda,[1,*] *C. Grant,*[1,2] *J. Browne*[1,3] *and S. Hagerthey*[4,5]

Introduction

This chapter has several goals. First, we cover the concept of pigment-based chemotaxonomy and its use of internal pigment ratios to derive taxon-specific chlorophyll-*a* contributions to a microalgal, here periphyton, community. Second, we introduce and review methods and a variety of real or potential problems associated with pigment-based chemotaxonomy. Third, we hope that this chapter can serve as a source for those studying photosynthetic pigments in the various microalgal communities and ecotones of the Everglades and other subtropical ecosystems.

[1] Organic Geochemistry Group, Department of Chemistry and Biochemistry and the Environmental Sciences Program, Florida Atlantic University, Boca Raton, Florida, USA 33431.
Email: blouda@fau.edu
[2] Presently at Palm Beach Atlantic University, 901 South Flagler Drive, West Palm Beach, Florida, U.S.A., 33401.
Email: CIDYA_GRANT@pba.edu
[3] Presently at University of North Carolina, Chapel Hill, North Carolina, U.S.A., 27523.
Email: jbrowne8@live.unc.edu
[4] South Florida Water Management District, 3301 Gun Club Road, West Palm Beach, Florida, U.S.A., 33406.
Email: Hagerthey.scot@epa.gov
[5] Presently at the United States Environmental Protection Agency, 1200 Pennsylvania Avenue Northwest, Washington, D.C., U.S.A., 20460.
* Corresponding author

Periphyton is a ubiquitous component of the Everglades landscape and is a complex mixture of oxygenic micro-photoautotrophic (algae and cyanobacteria) and heterotrophic (bacteria, fungi, larvae, etc.) organisms (Browder et al. 1981; 1984; Gaiser et al. 2011; Hagerthey et al. 2011; 2012). These mélanges can also provide non-oxygenated microhabitats (aka niches) that are conducive for the growth of anoxygenic phototrophs such as the purple-sulfur bacteria and, when the oxidation potential (pE, aka Eh) becomes low enough, the brown- and/or green-sulfur bacteria (Krabbenhoft 1996; Cleckner et al. 1999). Several chapters in this volume provide detailed descriptions, often to the species level, of the various taxa of aerobic phototrophs that are present in the numerous forms of Everglades periphyton (see Gaiser et al.; Gottlieb et al.; Frankovitch and Wachnika; this volume).

"Benthic algae (periphyton or microphytobenthos) are primary producers and an important foundation of many stream food webs. These organisms also stabilize substrata and serve as habitat for many other organisms. Because benthic algal assemblages are attached to substrate, their characteristics are affected by physical, chemical, and biological disturbances that occur in the stream reach during the time in which the assemblage developed" (Stevenson and Bahls 1999). We included this quotation noting that not all 'periphyton is attached or even benthic but rather to link the idea of "disturbances in the stream" to the long held description of the Everglades as "The River of Grass" (Stoneman-Douglas 1947). As such, these phototrophic communities are obviously controlled by and respond to alterations in water quality (viz. nutrients, toxicants) and quantity (viz. hydroperiod).

Periphyton is routinely utilized to assess the ecological status of the Everglades because the species that comprise periphyton are sensitive to changes in environmental conditions (McCormick and Stevenson 1998). Long-term monitoring by the South Florida Water Management District (SFWMD), the Long-Term Ecological Research (LTER) project at Florida International University's (FIU) Southeast Ecological Research Center (SERC) and others have established strong relationships between individual algal species with specific water quality conditions such as low and high total phosphorous (TP) and salinity (S_{psu}) (see Browder et al. 1981; 1994; McCormick et al. 2002; Gaiser et al. 2006; 2011; Hagerthey et al. 2011). However, taxonomic identification of algae requires highly skilled individuals, and may have lengthy turn-around times depending on sample number and to what taxonomic resolution is requested. Cost and time constraints associated with periphyton taxonomic study, as well as the need to assess ecological changes over short time periods and large spatial expanses of the Everglades, support the development of alternative metrics of community structure that can be derived from pigment-based methods. This becomes especially important when a key aspect of the Comprehensive Everglades Restoration Plan (CERP), notably its 'monitoring and assessment function', is "Adaptive Feedback". That is, primary producer community changes due to alteration in water management processes need to be made in a timely fashion in order to notify managers and allow for the appropriate adjustments in water quality and/or quantity and delivery schedules to be made. Microscopy and chemotaxonomy each have advantages and disadvantages in addressing microalgal community structure (Louda 2008; Havskum et al. 2004).

There are two main disadvantages to pigment-based chemotaxonomy. First, except for *Karenia* sp. (gyroxanthin as biomarker) marine red tide organisms and the CYANO group 1 organisms such as *Oscillatoria* sp. and *Tricodesmium* sp. (oscillaxanthin as biomarker), pigment-based evaluations only go to Class or, more commonly, to the Division level (cf. Jeffrey et al. 2011). Secondly, determination of the proper biomarker coefficients for the calculation (~estimation) of taxon-specific chlorophyll-*a* (CHL*a*) and then its relation to that taxon's biomass is difficult given the known heterogeneity of periphyton communities, nutrient stressors and unknown or highly complex light fields. These drawbacks are discussed in detail later in text.

Advantages of pigment-based community evaluation relative to microscopy include: (a) Sample size. That is, pigment extractions are performed on much larger samples than are microscopic examinations; (b) Turnaround times on large spatial/ temporal sample sets are lower, as are (c) Costs per sample.

Higher level photosynthetic taxa do appear to be generally associated with environmental conditions (see Browder et al. 1994; Gaiser et al. 2006; 2011; Hagerthey et al. 2012; Iwaniec et al. 2006). For example, cyanobacteria (aka blue-green algae) are abundant in nutrient poor, slightly alkaline environments such as the Everglades National Park while desmids (green algae) are often common in nutrient poor, slightly acidic environments typified by Water Conservation Area 1 (WCA-1). This suggests that periphyton methods other than identification to lowest taxonomic unit (species) might be used to rapidly assess changing ecological conditions. Chemotaxonomy is commonly used to assess algal composition in marine, estuarine, and freshwater ecosystems. The method is based on the principle that different groups of algae have unique and diagnostic chlorophyll and carotenoid pigments that can be used to assess biomass and phylogenetic composition (Millie et al. 1993; Mackey et al. 1996; 1998; Havens et al. 1999; Veldhuis and Kraay 2004; Wright and Jeffrey 2005; Louda 2008; Mineeva 2011; Chapters in Roy et al. 2011). These principles have been used to assess phytoplankton and periphyton composition in certain ecosystems of South Florida, notably Lake Okeechobee (Winfree et al. 1997; Steinman et al. 1998; Havens et al. 1999), Florida Bay coastal waters (Louda 2008) and the microphytobenthos and epiphytes of Florida Bay (Louda et al. 2000). The analyses of 'higher level' taxonomic groupings or "functional groups" is routine with marine phytoplankton (Paerl et al. 2003) including satellite-based remote sensing (e.g., SeaWifs. See Nair et al. 2008).

Periphyton pigment extraction protocols were known to give highly variable results depending upon the solvent choice (cf. Biggs and Kilroy 2000). An exhaustive study by the authors (Hagerthey et al. 2006) using Everglades periphyton grab and periphytometer samples resulted in a reliable extraction protocol for Everglades periphyton as well as other microalgal samples. Subsequent pigment quantification using high-performance liquid chromatography (HPLC) chemotaxonomy was then examined as a potential rapid, low cost alternative for the assessment the periphyton communities of the Everglades.

In addition to allowing for a rapid, theoretically unbiased between analysts, method for assessing the major taxa (Divisions, Classes) of periphyton, pigment analyses can also easily detect purple- and green-/brown-sulfur bacteria. This later fact then allows for the identification of anoxic microzones and the potential for mercury methylation by coincident Archaea (Cleckner et al. 1999). The presence, absence or abundance of the photoprotective scytonemin pigments and especially the recently identified scytonemin-imine (Grant and Louda 2013) can also be utilized to infer light intensity history. However, all of these pigment-based methods are not without inherent constraints and these shortfalls are the main thrust of the current chapter. Identification of these issues will hopefully guide future research to more highly reliable and rapid pigment-based chemotaxonomic evaluation of Everglades periphyton and other wetland microalgal communities. Rapid evaluation of these communities as they respond to natural and anthropogenic alterations of hydroperiod, nutrients and other changes with 'Adaptive Feedback' to managers is the ultimate goal.

Periphyton Samples and Standard Cultures

Samples were provided by the South Florida Water Management District (SFWMD) and the LTER project at Florida International University (E. Gaiser, J. Trexler).

Figure 1 is a map of southern Florida with the areas (WCA1, WCA2A, WCA3A, Taylor and Shark River Sloughs) and sites labeled.

Figure 1. Location of periphyton monitoring sites in the Florida Everglades. The major transects are identified as WCA-1, transects X, Y & Z; WCA-2A, transects E, F & U; WCA-3A, transects 3AW &#AE; WCA-3B, transects 345B & 345C; Shark River Slough, transects S12A, S12C, S355A & S355B; Taylor Slough, transects 1-7. Experimental mesocosms are marked with a bold X.

Samples were frozen to liquid nitrogen temperature and placed under vacuum, in the dark for 18–24 hours, on a standard freeze-drying unit (Labconco Corp, Kansas City, Mo, USA). Upon completion the dry weight and water content was determined. To allow for direct comparison between freeze-dried and fresh/frozen samples we applied a conversion factor to the fresh/frozen samples to convert the units to $\mu g \cdot g^{-1}$ dry weight. The conversion was obtained by determining the average bulk wet/dry weight ratio of the filtered mélange.

Calcification was qualitatively (+++/++/+/–) gauged by a very simple microscopic exam during which a few drops of 0.2N hydrochloric acid was added to an aliquot of sample and any evolution of CO_2 noted.

Periphyton was removed from periphytometers by scrapping the material from the glass slide and rinsing it into a small aluminum pan with deionized water. This material was then filtered through a 47 mm Whatman GF/F glass fiber filter which was then folded in half, blotted between paper towels, refolded (quarters), reblotted and rapidly frozen until freeze dried as above.

Pigments were extracted in a 10 mL all glass/Teflon tissue grinder at ice bath temperatures using the MADW solvent (methanol/acetone/dimethylformamide/water; 30:30:30:10; v/v/v/v). This method is a significant improvement over conventional pigment extractions methods such as 90–100% acetone, methanol or mixtures of those solvents (Hagerthey et al. 2003; 2006). Ground samples in the glass mortar of the tissue grinder were next sonicated at ice-bath temperature in short spurts, centrifuged, decanted and syringe filtered (0.45 μm) to obtain the crude extract. 1.00 mL of the extract was mixed with 0.125 mL (125 μL) of an ion-pairing (aka ion-suppression) reagent consisting of 15.0 g tetrabutylammonium acetate (TBAA) and 77.0 g ammonium acetate in 1.00L water (cf. Mantoura and Llewellyn 1983; Louda et al. 1998) to prepare the injectate solution. Extracts were kept cold (ice-bath or frozen) and under nitrogen whenever possible.

Typically, 100 μL of the injectate solution was loaded into the HPLC system. Separation and identification of pigments was by reverse phase HPLC and photodiode array detection (PDA) as detailed previously (Louda 2008; Louda et al. 1998; 2000; 2002; Hagerthey et al. 2006).

HPLC was performed on Waters NovaPak C18 reversed phase (RP) columns (3.9 x 150 or 3.9 x 300 mm; 4 μm particle size, 7% carbon load, end-capped, 60Å pore size, 120 $m^2 g^{-1}$). Column storage was in 85% aqueous methanol.

The solvent profile used throughout these studies was that of Louda et al. (1998; 2002) as also detailed in Hagerthey et al. (2006) and Louda (2008) and is essentially a time lengthened variation of that given by Mantoura and Llewellyn (1983). The extra "time" was found to be necessary to allow for more "room" on the column for pigment separation, especially important for the resolution of lutein (LUT) from zeaxanthin (ZEA). Solvents were; (A) 0.5 M ammonium acetate in methanol:water (85:15, v/v), (B) 90% acetonitrile, and (C) ethyl acetate. Linear gradients were used with the following solvent makeup (A/B/C) at each end time (minutes) over the following schedule: 0(60/40/0), 5(60/40/0), 10(0/100/0), 40(0/30/70), 45(0/30/70), 45(0/0/100), 47(0/100/0), 48(60/40/0). A minimum of 10 minutes re-equilibration of the 60/40/0 initial conditions solvent was incorporated between injections.

Detector response and column degradation were monitored with the use of an internal standard (IS). Here, we used copper mesoporphyrin-IX-dimethyl ester (Cu meso-IX-DME) which has an intense and narrow Soret absorption at 394 nm and does not interfere with the 'routine' monitoring wavelengths of 440 nm (Chlorophylls and carotenoids) and 410 nm (pheopigments). Other internal standards have been discussed by Mantoura and Repeta (1997) and attention to procedural detail and mathematical considerations are given by Hooker and Van Heukelem (2011). Bacteriochlorophyll-*a* and its pheo-pigments were monitored in our system at 360 nm.

In addition to the use of the internal standard (IS) to monitor column degradation (retention time), its percent recovery (%R = (IS$_{rec}$/IS$_{inj}$) x 100; where rec and inj = recovered and injected) is monitored for every HPLC analysis. The extractant (MADW) contains the internal standard in known concentration and then becomes a complete procedural standard encompassing extraction through analysis. Periodic checks with known amounts of pure chlorophylls and carotenoids in similar fashion (rec/inj) also provide 'Data Quality Indicators' (DQIs: see Hanrahan 2009).

Post-run integrations were performed at 440, 410, 394 and 360 nm. Millimolar or specific extinction coefficients were taken from the literature (e.g., Davies 1965; Foppen 1971; Holden 1976; Jeffrey et al. 1997; Egeland et al. 2011). Given that we utilize integration at 440 nm (for chlorophylls-*a*/-*b*/-*c$_1$*/-*c$_2$*/*c$_3$* and the carotenoids), most extinction coefficients required adjustment for not being used at the exact wavelength for which they reported. For example, the "recommended specific absorption coefficient" (α in L• g^{-1} •cm^{-1}) for chlorophyll-*a* is 129 at 428 nm in 90% acetone (Egeland et al. 2011). However, since we integrated at 440 nm that value would need to be adjusted down (X ~ 0.526) to reflect the relative absorption of chlorophyll-*a* at 440 nm. We utilized published E$^{1\%}_{1\,cm}$ values that were adjusted in similar fashion and then ran through QA/QC with known pigments. That is, spectrophotometrically formulated solutions of pure known pigments were ran on the HPLC system and the integrated peak areas converted to 'recovered' values which were then compared to the known injectate quantity. Alternately, one may integrate at each specific wavelength for each pigment. Currently we integrate at 440 nm (chlorophylls-*a*/-*b*/-*c*s & carotenoids), 410 nm (pheopigments-*a*), 394 nm (internal standard), 360 nm (bacteriochlorophylls-/ bacteriopheopigments-*a*), and often 666 nm (CHL*a*') and 460 nm (echinenone) in order to quantify these last two, each in the presence of the other (viz. coelution).

Obtaining excellent to 'adequate' separation of pigments is paramount to being able to perform pigment-based chemotaxonomy. The chromatogram shown here as Fig. 2 is for the separation of the DHI mixed pigment standards (Batch "Mix-103) using our system as described above. It should be noted that 'baseline monitoring' subroutines to force a flat baseline are not used here in order to better address system run-to-run variations. The 'peak' at about 49–50 minutes is in response to the column flush with 100% ethyl acetate.

A very important point here is that lutein and zeaxanthin need to be fully separated and are also separated from diatoxanthin. That is, as zeaxanthin and lutein are both biomarkers that are utilized in various chemotaxonomic equations, they must be well separated. Often there are reports of "L+Z" (=lutein plus zeaxanthin) in the literature.

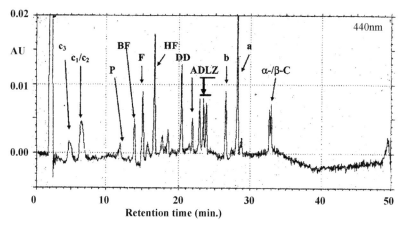

Figure 2. RP-HPLC chromatogram (λ = 440 nm) of DHI mixed pigments. *Code*: c_3 = chlorophyll-c_3, $-c_1/c_2$ = chlorophylls-$c_1/-c_2$, P = peridinin, BF = 19'-butanyloxy-fucoxanthin, F = fucoxanthin, HF = 19'-hexanyloxy-fucoxanthin, DD = diadinoxanthin, ADLZ = alloxanthin/Diatoxanthin/Lutein/Zeaxanthin, b = chlorophyll-b, a = chlorophyll-a, α-/β-C = α-/β-carotenes.

Such data of little, if any, use in dissecting pigment signatures of a mixed community. The importance of quality assurance (QA) is also the topic of a recent book chapter by Van Heukelem and Hooker (2011).

Figure 3 reveals the base-line separation, or nearly so due to run-to-run slight variations, of lutein (LUT) from zeaxanthin (ZEA) in an injectate with about a five-fold higher amount of 'L+Z' than shown above in Fig. 2. The separation of lutein from zeaxanthin by HPLC has also been found to be extremely important in the study of the human eye and HPLC methods, in addition to ours as given above, for that and other purposes are known (cf. Bone and Landrum 1992; Craft 1992). Therefore, pigment-based geological or biological studies should stress the complete separation of lutein and zeaxanthin which then allows valid individual quantification!

A previous HPLC method used by our group incorporated a 3.9 x 150 mm NovaPak column to separate chlorophyll-a-epimer (CHLa') from echinenone (ECHIN). Our latest system, using a 300 mm long column optimized for the separation of both the polar scytonemins and lutein from zeaxanthin led to these 2 pigments (CHLa'/ECHIN) partially co-eluting. Therefore we modified our integration methodology to allow for the quantitation of these 2 pigments in the presence of each other. Figure 4 is an overlay of the UV/Vis absorption spectra Chlorophyll-a' and echinenone showing but

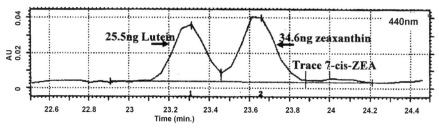

Figure 3. Partial RP-HPLC chromatogram revealing the separation of lutein from zeaxanthin.

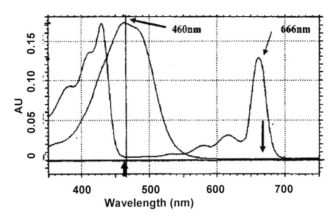

Figure 4. Absorption spectral overlay of chlorophyll-*a*-epimer and echinenone revealing the analytical wavelengths used for their quantitation and the lack of interference.

a 2–3% interference between in the quantitation of echinenone at 460 nm by CHLa' and no interference when CHLa' is measured at 466 nm. As one becomes more major or minor this alters a "bit". When echinenone is present, it usually vastly dominates over chlorophyll-*a*' and there is much less (< 1%) interference at 460 nm for the quantitation of echinenone.

Analyzing each pigment individually would be very time and effort consuming. For example α-carotene (448 nm), β-carotene (454 nm), myxoxanthophyll (476 nm), CHLa (430 or 662 nm), CHLb (458 or 646 nm) *et cetera* would all require individual integrations. Adjusting extinction coefficients and using grouped integration wavelengths, as given above, speeds syn- and post-HPLC data processing.

Three HPLC systems were used during these studies. System #1 consisted of a ThermoSeparations Products (TSP) Mdl. 4100 quaternary solvent pump, a Rheodyne Mdl. 7120 injector with a 100 μL sample loop, a Waters MDl. 990 photodiode array detector (PDA, aka Diode Array Detector or DAD) with Waters 990 software. System #2 was similar to system #1 but utilized a Waters 996 PDA and Waters Millenium software. System #3 was a Thermo Separations Spectra System consisting of an AS3000 autosampler with Peltier chilled sample storage, a TSP P4000 quaternary HPLC pump, and detection by either a TSP UV1000 single wavelength detector or, more recently, a Waters Mdl. 490E dual wavelength detector. The analog output of either detector was processed by PeakSimple A-D converters and software.

Over 75 individual known pigments plus the commercial pigment mixtures were utilized to standardize the chromatographic system (Louda et al. 2002). In order to better dissect pigment arrays and classify periphyton contributors, we purchased several additional known compounds (aphanizophyll, alloxanthin, diadinoxanthin, lutein, *inter alia*), of interest in this project, for RP-HPLC/PDA standardization.

The firm of DHI Water and Environment (Denmark) provides authentic known biomarker chlorophylls and carotenoids. We utilized 15 of these standards, to augment the 90+ pigment standards in the OGG lab. For the purpose of routine QA/QC, we also used the DHI certified "mixed" pigment standard set from DHI for use in all HPLC protocols.

In order to process large numbers of samples during the CERP-MAP and -RECOVER programs, it was necessary to use the autosampling HPLC system described above. Cross standardization of the autosystem and the Waters 990 and 996 PDAs (ng pigment from the in-house Excel© program we call 'PIGCALC' using photodiode array detector data) has proven facile and resulted in (Autosytsem {mV*min} versus PDA {AU*min}) peak area regressions with $R^2 \gg 0.9$ in all cases. This allowed the "routine" running of samples on the Autosystem with 1 out 6–8 samples being ran on both systems as QA/QC cross checks. Chromatographic data and chemotaxonomic formulae have been standardized to weights per square centimeter (periphytometers) or per gram dry weight (grab samples). This made it easier to link the SFWMD MetaData and chemotaxonomic regression formulae could then be adjusted in accord with ground-truthing, light field and other ancillary data.

Studies on the effect of light on microalgal pigment quantities, ratios and relationships to biomass (cell number, cell volume, organic carbon, proteins and carbohydrates) were performed as portions of both the Master's and Ph.D. research of co-author Grant (Grant 2008; 2010).

The above preliminary Everglades-specific equation was derived to compare relative differences in periphyton phylogenetic composition collected from a variety of environmental conditions.

We have previously demonstrated that the performance of chemotaxonomic assessment is highly dependent upon extraction methods, especially for mixed communities such as those found in the Everglades (Hagerthey et al. 2006).

The Master's thesis research of co-author Browne (Browne 2010) extended pigment-based chemotaxonomic estimation of these sample using SLE (Simultaneous Linear Equations) variants, the CHEMTAX algorithm (Higgins and Mackey 2000; Latasa 2007; Mackey et al. 1996; 1997; 1998) and the Bayesian Community Estimator (BCE) running under the "R" routine (Van der Meersche et al. 2008; 2009).

Microscopic identification of a subset of the field (periphytometer) samples (n = 211) was conducted by the Florida Department of Environmental Protection (FDEP) laboratory in Tallahassee and supplied by the South Florida Water Management District. This included periphytometer cell counts (which count all cells and 10-micron filament lengths individually) and unit counts (which count each colony or filament as one) for 17 categories of algal groups. These groups were combined here to provide relative abundance estimates for chlorophytes, cryptophytes, cyanobacteria, diatoms, and dinoflagellates. For the grab samples (n = 122), relative abundance in relationship to biovolume, rather than cell/unit counts, was provided by the FIU FCE-LTER group. These estimates were based on microscopic identification and listed the relative biovolume per sample for seven categories of algae (diatoms, desmid and non-desmid greens, and four categories of cyanobacteria), which were then combined into total cyanobacteria, chlorophytes, and diatoms. Cryptophytes and dinoflagellates were not reported in the microscopic exam data set.

A "Statistical Analysis of Everglades Algal-Water Quality Relations" was also conducted by M. Cohen and S. Lamsal (2008) of the School of Forest Resources and Conservation, University of Florida, in Gainesville. The resultant Classification And Regression Tree analysis (CART) is aimed at predicting water quality conditions (TP, TKN, pH, temperature, specific conductance, and dissolved oxygen) from periphytometer periphyton composition. Paired water quality data was available for 245 of the periphytometer samples. CART provides a classification/regression tree, through recursive partitioning that can be utilized to predict water quality based on the relative abundance of different algal groups (McCune and Grace 2002).

Results-pigment Background and Everglades Periphyton

Data presented herein was collected during the ongoing pigment studies of the Organic Geochemistry Group at Florida Atlantic University. This includes pigment analyses of unialgal cultures over several decades by the senior author and studies during the Masters' or Masters' plus Ph.D. research of co-authors Browne or Grant, respectively.

The Everglades studies reported herein were conducted from July 2003 through July 2009. The first series of studies were aimed at developing, testing and evaluating the best methods with which to reliably extract and analyze the chlorophylls, carotenoids and scytonemin pigments in Everglades periphyton (Hagerthey et al. 2003; Louda et al. 2003; 2006; Mongkhonsri et al. 2007). The final procedure and recommendations were published in the Journal of Phycology (Hagerthey et al. 2006). Preliminary application of pigment-based chemotaxonomy with Everglades periphyton was next (2004–2005). This was followed by the main study (2006–2009) during which the majority of data given herein was collected.

In the pages to follow, we hope to give the reader an overview of pigment distributions in the major taxa involved in periphyton and to introduce the basic concepts of taxon-specific pigment analyses as well as both the inroads and shortcomings of its application to the chemotaxonomy of periphyton.

Examples of Taxon-specific Pigment Arrays

This section is a general introduction to the pigments found within certain taxa in order to lay the groundwork for the dissection of mixed taxon natural algal communities. It is that numerical dissection of complicated pigment arrays which is aimed at estimating the taxonomic composition of the source community. Additionally, this overview of taxon-specific pigment arrays is also meant to serve as background information for the reader.

It needs to be pointed out that not all of the RP-HPLC chromatograms given in this chapter were obtained with exactly the same column and/or gradient. That is, the reader may notice that chlorophyll-*a*, for example, is shown with retention times of about 27 (cf. Fig. 5b) or 30 (Fig. 6b) minutes and retention times for other pigments will be shifted as well. However, each system was verified versus known mixtures, as given above (Fig. 2) as part of our routine QA/QC standard operating procedures (SOPs).

Cyanobacteria

In essence, we deal with three major sub-types or functional groups of cyanobacteria. These are the unicellular coccoidal, filamentous and diazotrophic groupings.

The unicellular elongated (baciliform) cyanobacterium (Cyanobacteria {Cyanophyta}, Chroococcaceae) *Anacystis nidulans*, now *Synechococcus elongatus* (see http://www.uniprot.org /taxonomy/1140), gives the pigment distribution shown in Fig. 5a when grown in nutrient replete culture under low to moderate light (e.g., 30–100 μmol quanta•m^{-2}•s^{-1}). Here the non-taxon-specific pigments chlorophyll-*a* (CHL*a*) and β-carotene are evident. The peaks to the right and left of CHL*a* are CHL*a*' (epimer) and the pair of CHL*a* allomers, normal and epimeric forms, repectively. Zeaxanthin, a dihydroxy derivative of β-carotene (viz. 3, 3'-dihydroxy-β-carotene; see Appendices A and B2) is highly abundant in the cyanobacteria and also, in lower amounts, in chlorophytes. Often, coccoidal cyanobacteria such as *S. elongatus* in natural communities/blooms will have only CHL*a*, β-carotene and ZEA (Louda et al. 2000), as shown in Fig. 5b, and the more highly oxidized derivatives caloxanthin and nostoxanthin are absent (below detection limits) or in trace amounts only. Therefore, pigment-based chemotaxonomy is often if not always forced to use only zeaxanthin as a marker for (coccoidal) cyanobacteria (Louda 2008).

The filamentous cyanobacterium *Anabaena flos-aquae* (Nostocaceae) is a common filamentous form in the waters and periphyton of south Florida. When grown or found in moderate light conditions, the pigment distribution is relatively simple, as given in Fig. 6a, with all pigment peaks on scale. Here CHL*a*, echinenone and β-carotene are the major pigments.

In the y-axis (AU) expanded chromatogram (Fig. 6b), a few minor pigments can be better noticed. An additional carotene, tentatively δ–(ε, ε-carotene), was detected and eluted between echinenone and β-carotene (t$_r$ = 32.4 min. Fig. 6b). Canthaxanthin, known to be a photo-protectorant pigment (Grant and Louda 2010) is present in a small amount in this moderate light grown culture. Zeaxanthin always accompanies echinenone but is in much lower abundance in the filamentous relative to coccoidal types. This coincidence of zeaxanthin in both coccoidal and filamentous cyanobacteria, as well as small amounts in chlorophytes, is a strong complicating factor for the pigment-based estimation of coccoidal cyanobacteria. Filamentous forms can be estimated by echinenone and/or myxoxanthophyll contributions (Winfree et al. 1997; Neto et al. 2006; Louda 2008). Myxoxanthophyll(s), various glycosides of the carotenol myxol (refs. in App. A), are found in a wide variety of filamentous and some coccoidal cyanobacteria. *Anabaena variablis* is reported as either containing several variants of myxol-fucosides or just myxol and 4-hydroxy-myxol, depending upon the strain (Takaichi et al. 2005). The exact myxoxanthophyll, namely myxol glycoside, present is therefore variable, shifting with the exact sugar moiety present, and HPLC retention times are then obviously variable as well. The myxoxanthophyll(s)

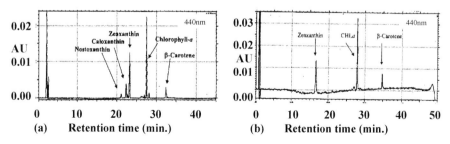

Figure 5. RP-HPLC chromatogram of the pigments in cultures of (a) *Anacystis nidulans* in lab culture and (b) a *Synechococcus elongatus* bloom in the northern Florida Bay/Everglades fringe.

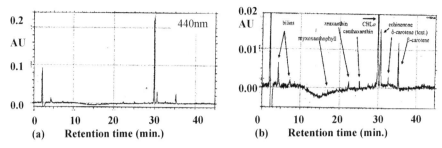

Figure 6. RP-HPLC chromatogram of the pigments extracted from *Anabaena flos-aquae*. (a) Y-axis adjusted to have all pigment peaks on scale. (b) Approximate 10-fold expansion of Fig. 5a.

can also be utilized to estimate cyanobacteria (Winfree et al. 1997) and appear to be related to cell wall integrity in certain cyanobacteria (references in West and Louda 2011). Lastly, in Fig. 6b, we found 2 bilins to be present in the lipophilic extract of this *A. flos-aquae* culture. The bilins are the straight-chain tetrapyrrole prosthetic groups of the phycobiliproteins, accessory photosynthetic pigments (Rowan 1989; Falkowsi and Raven 2007) and were most likely released by solvent denaturation of the protein.

Certain filamentous nitrogen-fixing (diazotrophic) cyanobacteria contain aphanizophyll (Hertzberg and Liaaen-Jensen 1966b; 1971) and studying the use of pigment to estimate 'nitrogen-fixing phototrophs' is suggested here. Additionally, the myxoxanthophyll relative oscilloxanthin (App. B2) is found only in certain genera, such as *Oscillatoria* and *Athrospira* (Hertzberg and Liaaen-Jensen 1966a; Roy et al. 2011).

The filamentous cyanobacterium *Scytonema hofmanni* (Scytonemataceae) is prevalent in Everglades periphyton (Gleason and Spackman 1974; Browder et al. 1994). When cultured in nutrient replete conditions under moderate light (e.g., ~150 µmol quanta•m^{-2}•s^{-1}) it exhibits the pigment distribution shown in Fig. 7a. The primary accessory carotenoids, the myxol glycosides, known collectively as myxoxanthophylls (Takaichi et al. 2001; Mohamed et al. 2005; Graham and Bryant 2009), echinenone and β-carotene are present. Canthaxanthin is also present and is the only indication of the generation of photoprotectorant pigments, as detailed by Grant and Louda (2010). However, in the many of cases in the Everglades were *S. hofmanni* is present, it is found in low hydroperiod areas with intense photic flux density (> 1,500 µmol quanta•m^{-2}•s^{-1}). In these cases, as well as cultures grown in high to intense light

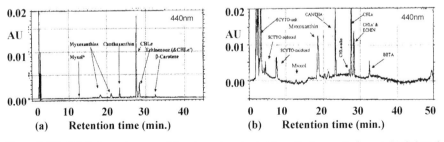

Figure 7. RP-HPLC chromatograms of the pigments extracted from *Scytonema hofmanni*. (a) Cultured in moderate (~150 µmol quanta•m^{-2}•s^{-1}) light. (b) Cultured in intense (~1,500 µmol quanta•m^{-2}•s^{-1}) light. "Scyto-unkn" is now known to be scytonemin-3a-imine (Grant and Louda 2013).

(350– 1,500+ µmol quanta•m^{-2}•s^{-1}), the pigment distribution changes to that found when *S. hofmanni* is cultured under high-to-intense photon flux density, as shown in Fig. 7b.

That is, under high to intense light conditions *S. hofmanni* contains greatly increased amounts of canthaxanthin. Additionally, the scytonemin UVR sunscreen pigments, in oxidized and reduced forms, are significant. In the most intense light conditions, such as the very low hydroperiod portions of the Shark River Slough, scytonemin-3a-imine (Grant and Louda 2013) is present, sometimes in very high abundance, as is obvious from the HPLC of the extract from the culture given in Fig. 7b.

Chlorophytes

The conceptual model of Browder et al. (1994), based on water quality parameters (e.g., hardness, pH, phosphorous, hydroperiod, etc.), emphasized the opposite ends of the Everglades periphyton community spectrum as being the calcareous periphyton, dominated by the cyanobacterial types covered above, and the desmid plus filamentous green-rich associations occurring in low nutrient low carbonate waters.

Figures 8a-b are the RP-HPLC chromatograms of two of the desmids (Chlorophyta, Zygnematales, Desmidaceae) common (Vymazal and Richardson 1995) to Everglades periphyton. *Cosmarium* sp. has been given as a desmid common to high-conductivity waters while others fall into low conductivity sites (Gleason and Spackman 1978; Swift 1984).

As with other green algae, the pigments found were chlorophyll-*a* (CHL*a*), chlorophyll-*b* (CHL*b*), β-carotene, lutein (LUT) and the 'xanthophyll-cycle' pigments neoxanthin, violaxanthin antheraxanthin and minor amounts of zeaxanthin depending upon time of day algae are harvested (see Demmig-Adams 1990; Muller et al. 2001; Demmig-Adams and Adams 2006; Falkowski and Raven 2007). Zeaxanthin is not identified in Figs. 8a-b but traces were noticeable in highly y-axis (AU) expanded chromatograms eluting immediately after lutein (see Fig. 3). The epimeric forms of CHL-*a* and -*b* were identified also. These are the small peaks eluting about one minute after CHL*a* and CHL*b* respectively (Figs. 8a-b).

Figure 9 presents the interconversions within the chlorophyte 'xanthophyll cycle'.

Minor amounts of chlorophyllide-*a* were also observed (Figs. 8a-b) and its abundance somewhat correlates with culture age. The small peaks occurring

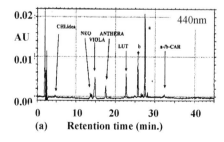
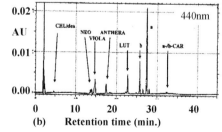

Figure 8. RP-HPLC chromatograms (λ = 440 nm) of the desmid chlorophytes (a) *Closterium acerosum* and (b) Cosmarium *turpini.* Pigment codes follow Appendix A.

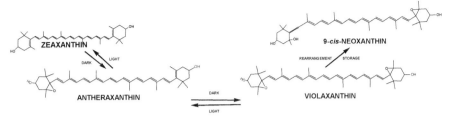

Figure 9. The chlorophyte and higher plant xanthophyll cycle pigments and their interconversions (after Demmig-Adams 1990; Demmig-Adams and Adams 2006; Falkowski and Raven 2007).

immediately after CHLs-a and -b are the epimeric forms (epimers: App. A) of each. The Master's thesis and Doctoral dissertation studies of co-author Grant (2008; 2010) greatly added to our understanding of chlorophyte pigments as they pertain to pigment-based chemotaxonomy. Namely, lutein was shown to be a photoprotectorant pigment (PPP) rather than strictly being a photosynthetic accessory pigment (PAP). That is, the CHLa/LUT ratio decreased significantly with increasing photon flux density. This indicated that lutein was being preferentially synthesized and/or retained as CHLa per cell decreased (Grant and Louda 2010). Pigment ratios which drastically change with light conditions are of less utility in chemotaxonomy than are pigments with more stable relationships to CHLa, the biomass proxy in chemotaxonomy. CHLa/CHLb ratios were found to be much more stable, at about 2.5:1 (molar) and became the chemotaxonomic ratio of choice for estimating chlorophyte abundance (Grant and Louda 2010). The xanthophyll-cycle pigments change rapidly in response to light conditions (cf. Demmig-Adams 1990). The changes in the proportions of the xanthophyll cycle pigments occur on the scale of minutes in response to changing light conditions (Muyller et al. 2001).

Similar pigment arrays were found during on studies for a wide variety of chlorophytes, including many filamentous forms (e.g., *Platymonas* sp., *Chlorella* spp., *Cladophora* sp., *Spirogyra* sp., etc.).

In figure 9, we labeled the conversion of violaxanthin to 9-cis-neoxanthin as 'storage'. While 9-cis-neoxanthin is indeed formed from violaxanthin, it is not part of the 'xanthophyll cycle' per se. That is, it does not enter into the epoxidation/deepoxidation reactions associated with the non-photochemical quenching of excess light energy (see Takaichi and Mimuro 1998; Bouvier et al. 2000; Muller et al. 2001; Takaichi 2011; Dembek et al. 2012). In higher plants, 9-cis-neoxanthin is a putative precursor for absisic acid (ABA) biosynthesis (Li and Walton 1990; Takaichi and Mimuro 1998).

In the 'brown line' algae, neoxanthin serves as an intermediate in the biosynthesis of fucoxanthin, a primary photosynthetic accessory pigment, and diadinoxanthin, a photoprotectorant pigment involved in the 'xanthophyll cycle' (Dambeck et al. 2012), as shown in Fig. 10.

Lutein, shown to increase in high light environments and therefore appear to be functioning as a photoprotectorant pigment (PPP: Grant and Louda 2011), has also been shown to form lutein-5, 6-epoxide via reactions analogous to the formation of violaxanthin (Niyogi et al. 1997; Muller et al. 2001), forming yet another 'xanthophyll cycle' pair. Figure 11 contains a structural comparison of lutein and its epoxide.

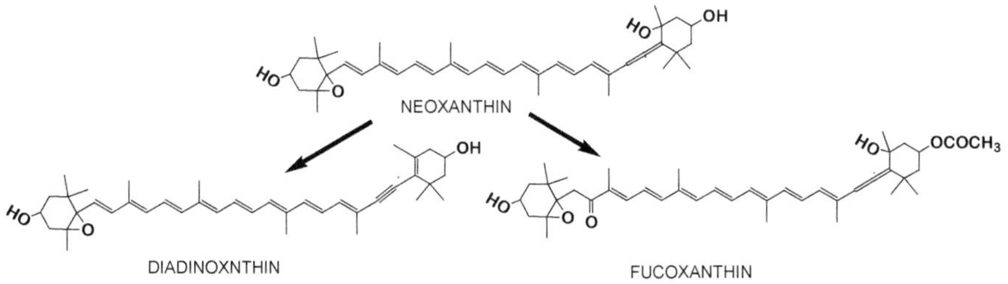

Figure 10. Structural comparisons of neoxanthin as a precursor to diadinoxanthin and fucoxanthin.

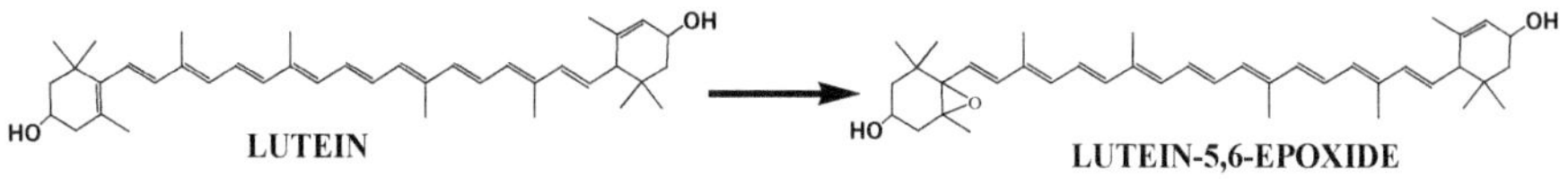

Figure 11. Epoxidation of lutein yielding lutein-5, 6-epoxide.

Chromophytes, specifically diatoms (Bacillariophyceae)

One of the major Everglades periphyton communities is given as being 'diatom-rich calcareous mat forming' (Browder et al. 1994). Diatoms have been extensively studied and their assemblages across the Everglades have been dissected to provide a "Disturbance Index" (Cooper et al. 1999). A prime driver for diatom species selection is given as the weighted-average periphyton total phosphorous (TP) within the contexts of optima and tolerances (McCormick et al. 1996; Gaiser et al. 2006). Diatom diversity at the species level is determined microscopically and, as detailed throughout this chapter, pigment-based chemotaxonomy strives only to estimate major taxonomic differences (e.g., percentages of cyanobacteria, diatoms and chlorophytes). However, changes in bulk periphyton composition have also been shown to reflect perturbations in water quality and quantity (Gaiser et al. 2011; Hagerthey et al. 2011).

In addition to CHLa, the vast majority of diatoms have chlorophylls-c_1-c_2 (see Baker and Louda 1986; Louda and Baker 1986; Scheer 1991; Graham and Wilcox 2000; Zapata et al. 2006), though a few have CHLs-c_2-c_3 or CHLs-c_1-c_2-c_3.

Carotenoids characteristic of the chromophyte taxa, notably diatoms here, include β-carotene, fucoxanthin, and the chromophyte 'xanthophyll cycle' pigments diatoxanthin and diadinoxanthin (Bjørnland and Liaaen-Jensen 1989; Graham and Wilcox 2000; Falkowski and Raven 2007).

Here we present the RP-HPLC chromatogram (Fig. 12) of the pigments in the fresh-water diatom *Navicula* sp. CHLa and fucoxanthin stand out as the major photosynthetic pigments in diatoms. This was to be expected as these two pigments exist in rather stoichiometric relationships in the antenna. That is, CHLa plus CHLs-c plus fucoxanthin exist in protein-bound antenna complexes, also known as fucoxanthin-chlorophyll-a/-protein (FCP) complexes (Glidenhoff et al. 2010). The ratio of CHLa to fucoxanthin has been reported as 0.84 (Fujii et al. 2012) to 1.0 (Emmanouil et al. 2005) in the FCP complexes. Grant and Louda (2010) reported the CHLa to FUCO molar ratio as 0.85:1 in total pigment extracts of both the marine diatom *Phaeodactylum tricornutum* and the fresh water *Navicula* species across a large range of photon-flux densities.

In addition to CHL*a* and FUCO, diatoms (see Fig. 12) contain β-carotene and the chromophyte xanthophyll cycle pigments (Fig. 13) diadinoxanthin and diatoxanthin.

Diatoxanthin (DIATO) is formed from diadinoxanthin in the light (see Hager 1980; Falkowski and Raven 2007). The presence of violaxanthin (see Fig. 9 for structure) in chromophytes has been known for a while but both its sporadic occurrence and function were unclear.

The presence of violaxanthin, zeaxanthin and a few other carotenoids such as β-cryptoxanthin, in addition to the better known and more prevalent 'brown line' xanthophyll cycle pigments (DIATO, DIADINO), are known but occur in very limited

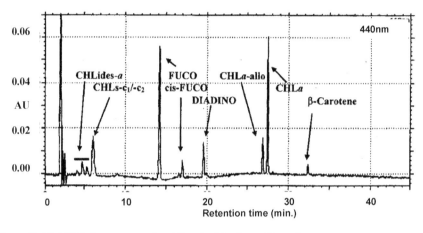

Figure 12. RP-HPLC chromatogram (λ = 440 nm) of the diatom *Navicula* sp.

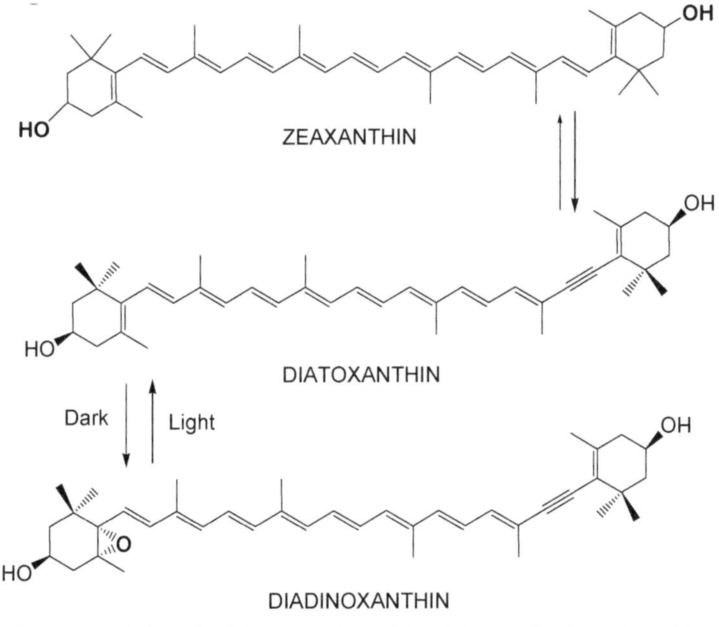

Figure 13. The xanthophyll cycle of diatoms and a variety of algae in the 'brown' line (chrysophytes, xanthophyceae, etc.) and dinoflagellates (see Hager 1980; Falkowsi and Raven 2007).

quantities and these may only become apparent in high-light stress situations (see Lohr and Wilhelm 2001).

In Fig. 11, *cis*-fucoxanthin, a geometric isomer of FUCO, is shown. Figure 14 has the structural comparison of fucoxanthin (FUCO) and one of the known cis-fucoxanthins (viz. 9') as well as fucoxanthinol (FUCOL), the deacetylated product often found in senescent/dead diatoms (see Louda et al. 2002). "Cis-fucoxanthin" has also been previously reported in phytoplankton (Breton et al. 2000; Louda et al. 2008) from various locales.

Diatoms have highly active chlorophyllase pools and the enzymatically mediated hydrolytic scission of phytol often occurs just with the physical disruption of cellular integrity during the extraction of viable diatoms. Chlorophyllase activity in diatoms is species specific leading to considerable, some or no conversion of chlorophyll-*a* to chlorophyllide-*a* (CHLide-*a*: Jeffrey and Hallegraeff 1987).

The chlorophyllides-*a* are highlighted in Fig. 12. Often, only CHLide-*a* and pyro-CHLide*a* are found (see Louda et al. 2002; 2011). In the present case, this *Navicula* sp. (Fig. 12) produced CHLide-*a* (4.8 min.) plus its allomer (aka 13^2-oxy derivative: 4.1 min) and epimer (aka CHLide-*a*': 5.2 min). These designations are covered in Appendix A.

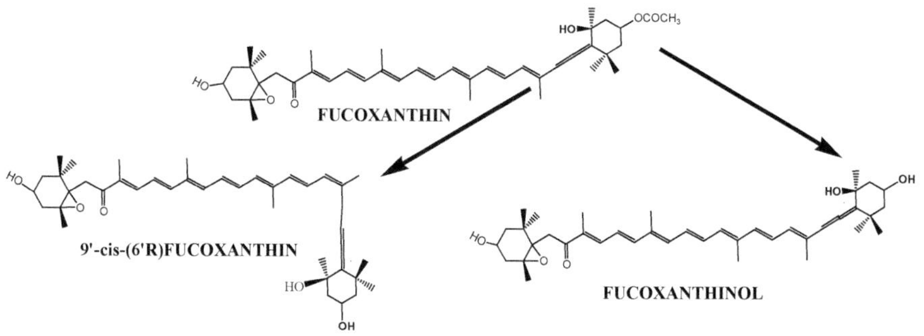

Figure 14. Alteration of fucoxanthin in senescent, dead, grazed and diagenetic stages.

Dinoflagellates (Pyrrophyta/Dinophyceae): About 1700 species of dinoflagellates are marine and 220 species are freshwater (Taylor et al. 2008). Additionally, dinoflagellates can be photoautotrophic, heterotrophic or mixotrophic, often in response to light or prey availability (Morden and Sherwood 2002; Taylor et al. 2008). The dinoflagellates have been described as containing at least six sub-groups based on their pigment complements (Zapata et al. 2012: cf. Van Heukelem and Hooker 2011) are divided into many sub-groups. Herein only the peridinin (PERI) and non-peridinin containing dinoflagellates will be considered. Complicating the estimation of dinoflagellates is the group's extreme variability. The peridinin-containing dinoflagellates, as that given in Fig. 15, if shown by microscopy to be the only or main type present in a community could be relatively well estimated using CHL*a* to PERI ratios as such ratios are 'relatively' stable across within low to moderate (30–300 µmol quanta $m^{-2}s^{-1}$) light regimes and increasing only slightly under intense (> 1,600 µmol quanta $m^{-2}s^{-1}$) illumination (Grant and Louda 2010). In addition to the chlorophylls, peridinin

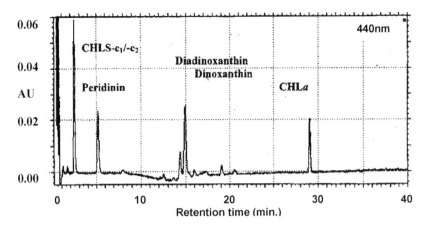

Figure 15. RP-HPLC chromatograms ($\lambda = 440$ nm) of a Type-1 (peridinin) dinoflagellate (*Peridinium* sp.). Type designation after Zapata et al. (2012).

and the 'xanthophyll cycle' pigments (dinoxanthin, diadinoxanthin), the carotenoid glycosides P468 and P457 (t = 1–2 minutes) are present in Fig. 15.

However, there are at least six dinoflagellate pigment groupings (Jeffrey et al. 2011; Zapata et al. 2012) Group #1 is the peridinin containing species that lack other primary biomarker carotenoids.

To date, we have found only very small signals from peridinin-containing dinoflagellates in Everglades periphyton and, without culturing individual dinoflagellates from living material or cysts, cannot offer any insight as to the presence or absence of non-peridinin containing autotrophic or mixotrophic dinoflagellates (viz. types 2–5) in Everglades periphyton. Very few reports of dinoflagellates in Everglades periphyton exist though McCormick et al. (1996) reported a few species under the heading "Other Taxa".

Cryptophytes

Though only present in very low amounts, the presence of alloxanthin in some Everglades periphyton samples mandates consideration of the Cryptophyta. Alloxanthin, the penultimate pigment biomarker for cryptophytes (Jeffrey and Vesk 1997; Breton et al. 2000; Pandolfini et al. 2000; Wright and Jeffrey 2005) has also been observed in minor amounts in certain marine dinoflagellates (Schnepf and Elbrachter, 1999; Zapata et al. 2012).

In the present example, *Cryptomonas* sp. (Fig. 16), alloxanthin is accompanied by crocoxnthin and monadoxanthin, two other cryptophyte specific carotenoids. In place of β-carotene, the cryptophytes have α-carotene, a useful qualitative marker. Chlorophyll-c_3 was also identified as an accessory photosynthetic pigment in this sample. It is reported that the cryptophytes will have any one of the 3 main chlorophylls-c (viz. -c_1, -c_2, -c_3: O'Kelley 1993).

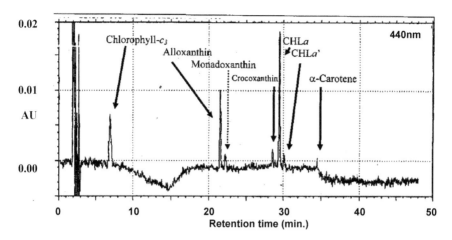

Figure 16. HPLC chromatogram of a typical cryptophyte (*Cryptomonas* sp.).

In vitro *mixed lab cultures*

In order to assess our ability to both chromatographically separate and quantify the biomarker pigments required for pigment-based chemotaxonomy and to assist in developing equations for the estimation of taxon-specific chlorophyll-*a* contributions, we prepared various mixtures with known algal cultures. Figure 17 is an example of one of these mixtures. In this case, we mixed the five major taxa of interest herein. That is, this is a mixture of the taxa described above (cyanobacteria, chlorophytes {desmid}, diatom, dinoflagellate {peridinin-containing} and cryptophytes).

In the methods section, we described the integration and quantification of individual pigments. To reiterate, pigment identity was based on retention time and UV/Vis spectra in comparison to authentic knowns. Currently the only coelution that is somewhat problematic is that of chlorophyll-*a*-epimer (CHL*a*') and echinenone and overcoming that quantitatively was described in the text associated with Fig. 4.

Aside from the example given here (Fig. 17), numerous mixes were made using 2, 3, 4 or all 5 of the taxa covered above. These were prepared in order to test (ground-truth) the use internal pigment ratios to back-calculate community structure. Again we point out the fact that we use *molar* ratios of CHL*a* to the taxon-specific biomarker pigment(s). For example, in the case of diatoms, the CHL*a* to FUCO ratio found by Grant and Louda (2010) is 0.85:1. That is, for each mole of fucoxanthin one would 'expect' 0.85 moles of diatom sourced CHL*a*. Other studies by our group have found CHL*a* to FUCO up to 1.1 or 1.2 to 1 (see Louda 2008; Louda et al. 2002; Winfree et al. 1997). As will be shown in the following sections, pigment ratios need to be determined and verified for each individual environment and community since they change with various parameters such as light, nutrient and species-specificity.

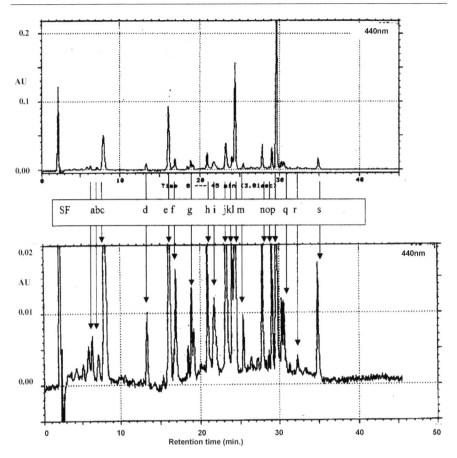

Figure 17. HPLC Chromatogram of the extract of mixed phytoplankton cultures: SF = Solvent Front; a = CHLide-*a*; b = CHLide-a'; c = CHLs-*c*/-*c*,; d =PERI; e = FUCO; f = NEO; g = *cis*-FUCO (VIOLA to right); h = DIADINO; i = ALLO; j = DIATO; k = LUT; l = ZEA; m = CANTH; n = CHL-*b*; o = CHL*a*-allo; p = CHL-*a*; q = CHL-*a*-epimer & ECHIN; r = meso-CHL-*a* (tent-?); s = β-CAR.

Pigment-based Chemotaxonomic Estimation of Periphyton Community Structure

Pigment-based chemotaxonomy (Mantoura and Llewellyn 1983; Millie et al. 1993; Mackey et al. 1996; 1997; 1998; Hagerthey et al. 2006; Louda 2008) depends on the ability to adequately extract, separate and identify the photosynthetic and non-photosynthetic (aka photoprotectorant) pigments from a mixed taxa microalgal community. Then, having that analytical data, the use of mathematical relationships of pigment biomarker quantities to the taxon-specific chlorophyll-*a* contributions allows an estimate of each contributing taxon to be made. This, to this point, assumes that chlorophyll-*a* alone is a valid biomass indicator and with that assumption comes a certain level of uncertainty. For now, we will proceed using the time-honored tradition of using CHL*a* as a microalgal biomass indicator (see Kreps and Verbinskaya 1930;

Harvey 1934; Atkins and Parke 1951; Richards and Thompson 1952; Parsons and Strickland 1968; Hambrook-Berkman and Canova 2007: *inter alia*).

Simultaneous Linear Equations (SLE)

Taxon-specific CHL*a* contributions of cyanobacteria, green algae, diatoms, dinoflagellates and cryptophytes were made using simultaneous linear equation (SLE) equations such as the following:

ΣCHL*a* = [1.1(ZEA-ECHIN)+11.1(ECHIN)]+[3.2(CHL*b*)]+[1.2(FUCO)]+[1.5(PERI)]+[3.8(ALLO)]   Eq. #1

Where ZEA = zeaxanthin (cyanobacteria), ECHIN = echinenone (filamentous cyanobacteria), CHL*b* = chlorophyll-*b* (chlorophytes), FUCO = fucoxanthin (chrysophytes, esp. diatoms), PERI = peridinin (dinoflagellates), and ALLO = alloxanthin (cryptophytes) (Hagerthey et al. 2006). Myxoxanthophyll containing cyanobacteria can also be estimated separately (= 7.5xMYXO), provided that ZEA is first corrected for contributions from the MYXO-containing population (ZEA$_{corrected}$ = ZEA – 1.6xMYXO: see Winfree et al. 1997; Louda et al. 1998; 2004).

That is, total estimated community CHL*a* is derived from the summation of the taxon-specific CHL*a* derived from such equations. Regression coefficients were derived from CHL*a* - to - diagnostic pigment ratios obtained from regional field and laboratory data studies of representative taxa (> 30) conducted over several years by the Organic Geochemistry Group at FAU. Taxonomic analysis of long-term datasets was also used in the development of the equation. One test that we employed to estimate the accuracy of the equation was to compare the chemotaxonomic estimation (CHL*a*$_{est}$) *of* total CHL*a* with the sum of all HPLC measured CHL*a* forms (CHL*a*$_{HPLC}$). That is, CHL*a* itself plus its epimer, allomers and phytol free (chlorophyllide) forms, including any pyro (C13^2-H-decarbomethoxy) forms. The closer CHL*a*$_{est}$/CHL*a*$_{HPLC}$ is to unity (1.0), the better the estimate appears. Often, we found that this measure fell between 0.9–1.1.

More recent *in vitro* studies have (re-) emphasized the fact that much more attention needs to be given to the photon flux density (PFD) received by periphyton communities. That is, chlorophyll-*a* per cell and the relationship of several but not all taxon-specific biomarker pigments to chlorophyll-*a* vary in relation to felt PFD (Grant and Louda 2010 and references therein). For example, we now feel that 'routine' estimation of chlorophytes are better using a 2.5 (CHL*b*) factor rather than the 3.2 (CHL*b*) given above. Thus, our previous estimations of chlorophyte biomass may be elevated be a factor of about 1.3. An adjusted SLE equation is as follows:

ΣCHL*a* = [1.1(ZEA-ECHIN)+11(ECHIN)]+[2.5(CHL*b*)]+[1.2(FUCO)]+[1.5(PERI)]+[3.8(ALLO)] Eq. #2

In both equations #1 and #2, the term [1.1(ZEA-ECHIN) +11(ECHIN)] encompasses the estimation of coccoidal [viz. 1.1(ZEA-ECHIN)] and filamentous (viz. 11.0(ECHIN)] cyanobacteria. Albeit, this is a first approximation at teasing apart these two separate cyanobacterial types and we have utilized other pigments, such as myxoxanthophyll (see Winfree et al. 1997; Steinman et al. 1998; Havens et al. 1999; Louda et al. 2004) for the estimation of cyanobacteria in south Florida waters.

A third type of cyanobacteria, the diazotrophic (nitrogen fixing) cyanobacteria may also be able to be estimated using the carotenoid aphanizophyll. An example of a pigment array which includes diazotrophic cyanobacteria is given here as Fig. 18. Here, the microphyto-benthos from Paurotis Pond in the southern Everglades/Mangrove fringe was found to contain diatoms (fucoxanthin), chlorophytes (CHL*b* and lutein) and nitrogen-fixing cyanobacteria (aphanizophyll). The presence of zeaxanthin in an amount exceeding aphanizophyll, plus the co-occurrence of echinenone, likely indicates the presence of non-nitrogen fixing cyanobacteria as well. Teasing apart all three cyanobacteria types (coccoidal, filamentous and nitrogen-fixing) will require significant additional study.

The choice of biomarker to taxon-specific chlorophyll-*a* conversion factor needs to be ground-truthed for each environment/community and such drivers as light field, nutrient level and growth/death stages all need to be considered. These points are detailed in following sections.

Above, in Equations 1 and 2, we utilized molar ratios, though others (Millie et al. 1992; Mackey et al. 1997; see References in Roy et al. 2010) utilize weight-based calculations and this is routinely given as the inverse ratio (viz. biomarker to CHL*a*). The inter-conversion of moles to mass is obviously facile.

Illustrating the use of weight versus molar pigment values, we point to the study of phytoplankton chemotaxonomy in the Neuse Estuary of North Carolina by Pinckney and others (1997). They arrived at the regression formula based upon weight ratios.

$$\Sigma CHLa = (0.27 \times ZEA) + (3.8 \times CHLb) + (1.02 \times FUCO) + (0.85 \times PERI) + (2.06 \times ALLO) \qquad \text{Eq. \#3}$$

Converting the equation of Pinckney et al. (1997) to a molar relationship, one arrives at:

$$\Sigma CHLa = (0.42 \times ZEA) + (3.9 \times CHLb) + (1.4 \times FUCO) + (1.2 \times PERI) + (3.2 \times ALLO) \qquad \text{Eq. \#4}$$

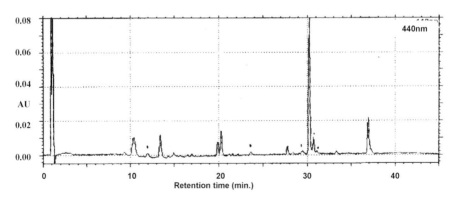

Figure 18. HPLC chromatogram of the microphytobenthos from Paurotis Pond in the southern Everglades/Mangrove fringe. APHAN (10.5 min), FUCO (15 min), MYXO (13.2 min), Lut/Zea (19.9/20.3 min), CANTH (23.5 min), CHL*b* (23.7 min), CHL*a* (30.1 min), ECHIN/CHL*a*` (30.9 min) and βCAR (37.0 min). N$_2$-fixing and filamentous cyanos, diats and chloros.

When compared to the present authors' formula (partial Eq. #2), differences are found for the estimators of all divisions. This likely represents geographical differences as well as the manner(s) in which the ratios were determined.

$$\Sigma CHLa = (1.1xZEA) + (2.4xCHLb) + (1.2xFUCO) + (1.5xPERI) + (3.8xALLO)$$
$$\text{Eq. #5}$$

or, on a wt/wt basis;

$$\Sigma CHLa = (1.7xZEA) + (2.4xCHLb) + (1.6xFUCO) + (2.1xPERI) + (6.0xALLO)$$
$$\text{Eq. #6}$$

Equations such as the above (Eq. #6) can be utilized as a first approximation for the CHEMTAX (see: Mackey et al. 1996; 1997; 1998) algorithm by taking the inverse of the weight-based equation. For example, the inverse of Eq. #6 yields the following;

$$\Sigma CHLa = (0.59xZEA)+(0.42xCHLb)+(0.62xFUCO)+(0.48xPERI)+(0.17xALLO)$$
$$\text{Eq. #7}$$

The protocol for recording the SFWMD sample analyses was set up with the lab's current chemotaxonomic method—simultaneous equations (SLE)—in mind. In contrast to factor analysis methods such as CHEMTAX, which use a large set of samples in a data matrix, SLE is effectively only applied to one sample at a time. Although batch calculations can be set up which provide an answer for many samples at once, the actual calculation is performed independently on each sample.

Aside from the SLE method(s) of estimating taxon-specific contributions to mixed microalgal communities, there are two additional methodologies used in pigment-based chemotaxonomy. These are the CHEMTAX algorithm and the Bayesian Community Estimator (BCE) methods covered below.

CHEMTAX

Given that the SLE methodology discussed above uses only one pigment per Division or Class, it has the potential of taxonomic overlap. For instance, zeaxanthin, often the only biomarker available for coccoidal cyanobacterial, also occurs in chlorophytes and sporadically in other taxa. The CHEMTAX algorithm attempts to circumvent this ambiguity by using more than one pigment per Class or Division.

The CHEMTAX steepest decent algorithm was developed by investigators studying the Southern Ocean at the CSIRO (Commonwealth Scientific and Industrial Research Organisation) in Australia (Mackey et al. 1996; 1997; 1998). It is a factor analysis program developed in 1996 (Mackey et al. 1996), licensed through CSIRO Marine Laboratories and was written to run inside MATLAB (The MathWorks, Inc. 2008). Although revised versions of CHEMTAX (v. 1.95, v. 2) have been referenced by the program's developers (Wright and Jeffrey 2006; Wright et al. 2009), they have not become available and the original 1997 release (v. 1.0) was tested by the current authors.

CHEMTAX works by evaluating groups of samples, with pigment data arranged in matrix form; biomarker ratios to chlorophyll-*a* are also arranged in a matrix. Using the (unknown) algal class composition of the samples as a third matrix, this forms a linear inverse problem which is solved by matrix factorization, using a straightforward algorithm to provide the least-squares solution (Mackey et al. 1996). Unlike SLE, which takes only sample data and biomarker ratios as input, CHEMTAX allows input as to how much and which ratios are allowed to vary and how the data is weighted; several other input options control more specifically how the calculations are run. Results include not only the algal class composition of each sample, but also a revised ratio matrix, residuals (from the least-squares calculations), and, if requested, breakdowns of pigments assigned to each algal class within each sample and more information regarding the iterative calculation process.

A priori choices must also be made prior to running the analysis and include: which algal classes should be included, which pigments will be chosen for biomarkers of specific algal classes, what ratios of pigment to chlorophyll-*a* should be used for the calculation coefficients, how should sample groups be constructed, and how to set the program parameters. Equation #7 above, served as one of our starting points for testing the algorithm with Everglades periphyton.

Bayesian Community Estimator (BCE)

The Bayesian Compositional Estimator (BCE) is a chemotaxonomic program which was developed in 2007 by researchers at the Netherlands Institute of Ecology (Van den Meersche et al. 2008). BCE is implemented as a package (Van den Meersche and Soetaert 2009) in the open source software R (R Development Core Team 2009). Periodic updates of the package are available by downloading the updated package from the R website; this study used BCE version 1.4 and R version 2.9.1.

Similarly to CHEMTAX, BCE uses a ratio matrix, a data matrix, and an unknown sample composition matrix to compose a linear inverse problem. However, BCE uses Bayesian methods to fit a probability distribution to the data and find a maximum likelihood solution for the problem. BCE first finds the least-squares solution (although this methodology can be altered by specifying different parameters when starting the program) and uses it a starting point for a Markov Chain Monte Carlo (MCMC) simulation. This version of BCE (1.4) exhibits some problems with the MCMC mixing which the authors of the method intend to improve through inclusion of a new algorithm (Van den Meersche, pers. comm. to J.B. 2009).

A priori choices to be made prior to running BCE are similar to choices in CHEMTAX: which algal classes should be included, which pigments will be chosen for biomarkers of specific algal classes, what ratios of pigment to chlorophyll-*a* should be used for the calculation coefficients, how should sample groups be constructed, and how should the program parameters be set. Choices should also be made as to when the results of the random walk will be considered acceptable. For example, what should be a minimum for the number of runs accepted, and what will constitute acceptable mixing?

Again, all pigment-based chemotaxonomic 'estimations' hinge on the proper choice of pigment ratios. As will be detailed, it is the living/dead, light, pE and nutrient conditions within the periphyton community which very much complicates this process. Attention to each detail must be made and additional study is certainly required to 'perfect' the methodology (-ies) with periphyton.

Complicating Factors

This section is presented in order to address a few of the factors which potentially diminish the accurate application of pigment-based chemotaxonomy to microalgal community assessment.

Cellular Senescence/Death and Early (Biotic/Abiotic) Diagenesis

The studies of Louda et al. (1998; 2002; 2011; Baker and Louda 1986; 2002) have detailed cellular senescence and the alteration of chlorophylls and carotenoids in many marine and fresh-water microalgae. Szymczak-Zyla et al. (2008) investigated the influence of microbial activities on the alteration of chlorophylls. Given that periphyton is an accumulation of living, senescing, dead and heterotrophically altered microalgae, it is important to consider these products and how their presence might skew chemotaxonomic interpretations of community structure.

The essential tenet of pigment-based chemotaxonomy holds that we can estimate taxon-specific biomass, usually using CHLa as its proxy, from the pigments present in a sample. However, if senescent and/or dead microalgae are present, the ratios required for this mathematically related estimation can be skewed. This may even extend to seasonal differences in community dynamics. While exacting studies on "viable" versus senescent/dead and/or mixed 'health' periphyton communities are lacking, we can offer some insight as to how pigment ratios are likely to be skewed.

Carotenols, such as zeaxanthin from cyanobacteria and alloxanthin from cryptophytes are found to be more stable with time during senescence death than are fucoxanthin from diatoms/chrysophytes and peridinin from dinoflagellates (cf. Repeta and Gagosian 1982; Louda et al. 2002; Louda 2008; Yacobi and Ostrovsky 2012). In the case of both fucoxanthin and peridinin, the first step in loss is deesterification giving fucoxanthinol and peridinol, respectively. These, however, can be added with the parent compound for a total. It is the further degradation, often going to small non-pigment molecules called loliolides (Repeta and Gagosian 1982), that removes these biomarkers from chemotaxonomic utility.

Figure 19 contains some of the known carotenoid alterations which complicate pigment-based chemotaxonomy when the sample contains senescent/dead and/or reworked materials.

The instability of fucoxanthin and peridinin are found to selectively remove their signals, diatoms and dinoflagellates respectively, from the pool of biomarkers available for typing a microalgal community. Even the secondary marker diadinoxanthin, widely used in the CHEMTAX algorithm (cf. Mackey et al. 1996; 1997; 1998), is fraught with problems. That is, diadinoxanthin (see Fig. 13) reverts to diatoxanthin in illuminated cells (see Hager 1980; Falkowsi and Raven 2007) and is also oxidatively cleaved to loliolides (Repeta 1989).

Aside from CHLb and CHLc_3, it is the carotenoids that serve as the prime taxon-specific biomarkers for the chemotaxonomic processes. Thus, carotenoid alterations can and do have profound effects on the outcome of these analyses. In Figs. 13 and 14, and accompanying text, we pointed out several of the main alterations. However, it can be even more complicated than shown above. For example, taking just the case

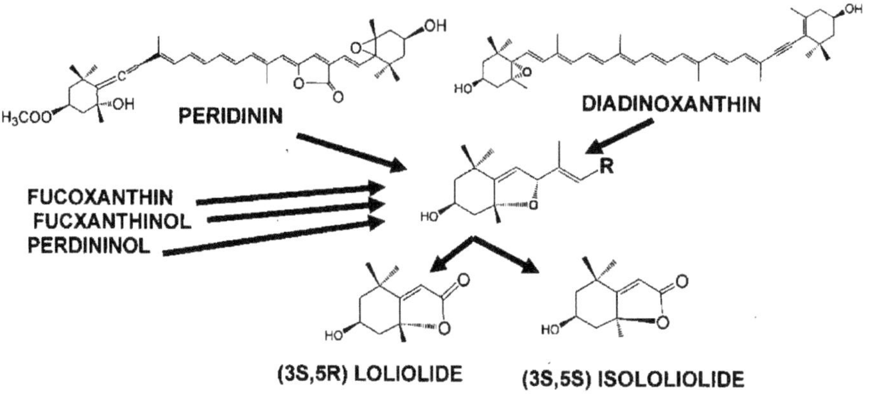

Figure 19. Selected carotenoid alterations (after Repeta 1989).

of fucoxanthin, the following products have been reported as forming during diatom senescence, death and early diagenesis: fucoxanthin, cis-fucoxanthin, isofucoxanthin, fucoxanthin dehydrate, fucoxanthin-3,3'-didehydrate, isofucoxanthin dehydrate, fucoxanthinol, isofucoxanthinol, cis-isofucoxanthinol, and 2 forms of loliolides (Repeta and Gagosian 1982; 1987).

Diadinoxanthin (DIADINO) has been used as a secondary or modifying marker for diatoms (fucoxanthin), dinoflagellates (peridinin), chrysophytes (fucoxanthin + 19'-butanoyloxy-fucoxanthin), euglenophytes (chlorophyll-*b*), and haptophytes (fucoxanthin + 19'-butanoyloxy-fucoxanthin + 19'-hexanoyloxyfucoxanthin) in CHEMTAX (Mackey et al. 1998). In degraded samples, such as sediments, diadinoxanthin has also been used to indicate the total diatom plus dinoflagellate contributions (Reuss 2005). Given both the breakdown of DIADINO (Fig. 19) and the light/dark related conversion of DIADINO to diatoxanthin (DIATO: Fig. 13) and the conversion of DIATO to 9-cis-zeaxanthin in senescent/dead degrading samples (Louda et al. 2002), the use of DIADINO appears highly speculative.

The concept of 'selective losses' (Hurley and Armstrong 1991; Leavitt 1993; Louda et al. 2002) reveals that the more stable carotenols, such as zeaxanthin (cyanobacteria), lutein (chlorophytes) and alloxanthin (cryptophytes) often last longer after senescence/death (Louda et al. 1998; 2002; 2001) and grazing (Louda et al. 2008).

Therefore, an emphasis needs to be placed upon the study of fresh viable communities collected and immediately frozen from a known light field and as free as possible from senescent/dead and reworked cells. This is the conundrum of pigment-based chemotaxonomy when applied to periphyton.

Similar problems are likely with the chlorophylls. Chlorophyll degradation has been split into Type-I and Type-II reactions. Type-I reactions involve retention of pigment character while Type-II reactions leads to colorless but often fluorescent catabolites and smaller molecules (Hendry et al. 1987; cf. Baker and Louda 2002). The differences in the disappearance trends of CHL*a* has been linked to accompanying differences in oxygen fugacity and light conditions, plus other factors (Baker and Louda 1986, 2002; Bale et al. 2011; Louda et al. 2011).

Aging experiments also revealed that CHL*b* disappeared faster than CHL*a* as evidenced by increasing CHL*a*/CHL*b* ratios (Louda et al. 1998), likely linked to differences in macrocyclic ring opening kinetics yielding the 19-formyl-1[21H,23H]-bilinones, akin to that reported for CHL*a* catabolism (Engel et al. 1996; Gossauer 1996; Matile et al. 1996; Doi et al. 2001). Sediment catalysis aside, it is often found that pigments such as the chlorophylls-c, fucoxanthin and peridinin disappear with age. Such was the case in Florida Bay sediments wherein pigment-based chemotaxonomy could 'prove' only a good history of cyanobacteria (zeaxanthin) and purple-S bacteria (bacteriochlorophylls-*a*/bacteriopheophytin-*a*) despite the strong and obvious co-occurrence of an abundance of diatom frustules (Louda et al. 2000).

Effects of Light and Nutrient Drivers on Pigment Contents and Ratios

Despite the fact that chlorophyll-*a* is a long accepted proxy for microalgal biomass (see Kreps and Verbinskaya 1930; Harvey 1934; Atkins and Parke 1951; Richards and Thompson 1952; Parsons and Strickland 1968; Hambrook-Berkman and Canova 2007: *inter alia*), the relation of taxon-specific CHL*a* to biomass under variant light and nutrient drivers is less well studied. General relationships of carbon to CHL*a* have indeed been modeled (Zonneveld 1998) but the number of species or even classes studied is low. There are, however, various reports on certain drivers of pigment contents and ratios. Nitrogen (N) or phosphorus (P) limitation/starvation is reported to lead to drastic decrease in all pigment (CHL*a*, CHL*b*, lutein, canthaxanthin, β-carotene) contents, except astaxanthin, per cell in the chlorophyte *Haematococcus pluvialis* (Boussiba et al. 1999). CHL*a* contents are reported to respond in a non-systematic manner with total phosphorus (TP; Watson et al. 1992; 1997). That is, their data revealed that "—the general trend followed—is actually opposite to that which would be expected if the relationship between TP and algal biomass were primarily influenced by this ratio" (Watson et al. 1992). CHL*a* per cell has also been reported as decreasing with increasing cell size (Agusti 1991).

The effect of light is perhaps the most widely studied aspect of cellular CHL*a* variability.

The carbon-to-CHL*a* ratio (θ^{-1}) is reported to increase in concert with light levels for nutrient-replete cultures (Gelder 1987). The effect of light levels on cellular CHL*a* contents has been more widely studied and, in all cases, is reported to decrease with increasing irradiance (e.g., Bautista and Jimenez-Gomez 1996; Woitke et al. 1997; Grant and Louda 2010). Cellular CHL*a* contents are so closely linked to light that changes can be detected within the diel light/dark cycle (Owens et al. 1980). Self-shading in dense algal populations also leads to light induced changes in CHL*a* contents (Agusti 1991).

Input of Redeposited Flocculent Organic Matter (aka floc)

Flocculent organic matter (OM), known as floc, is a complex mixture of degrading OM, living microorganisms and inorganic constituents (Droppo 2001).

In the Everglades, floc is reported to be a critical component of detrital energy cycles (Belicka et al. 2012; Bellinger et al. 2012). The presence of diatom materials in Everglades floc has been shown by certain polyunsaturated fatty acid (PUFA) distributions, including the $C_{20:5}$ and $C_{22:6}$ (Neto et al. 2006). The studies of Neto et al. (2006) and Pisani et al. (2013) also revealed cyanobacterial, diatom and chlorophyte pigment signatures in addition to higher plant materials. In their study of Everglades floc microbial biomarkers Bellinger et al. (2012) also reported fatty acid methyl ester (FAME) distributions showing the presence of chlorophytes and diatoms. Therefore, Everglades floc, often up to 20 cm in thickness (Bruland et al. 2007), includes microalgal OM and pigments in addition to higher plant and microbial materials.

As its name implies, floc is only mildly denser than water and is easily resuspended easily and redeposited downstream (Bruland et al. 2007). Periphyton sloughing during active growth (periphytometer) has been reported (Childers et al. 2002) and it is easy to assume that similar sloughing occurs from epiphytic growths, such as the so-called 'sweaters' on macrophytes such as *Utricularia* sp. and *Typha* sp., and dense epipelic accumulations. Floc entrainment into flow and redeposition downstream is known (Larsen et al. 2009) and yields redistribution from slough to ridges with significant capture in dense *Eleocharis* stands (Larsen et al. 2008). Presence/absence of floc adds yet another confounding aspect to the valid interpretation of pigment-based chemotaxonomy from periphyton accumulations. That is, aside from short term periphytometer samples, all periphyton in the Everglades can be assumed to be a collection of living, senescing, dead and redeposited microalgae and higher plant materials.

Pigment-based Chemotaxonomy Applied to Everglades Periphyton

In this section, we present some of the results derived from a six-year study (2003–2008) of several aspects of the utilization of pigment analyses for the description of Everglades periphyton communities.

Examples of Everglades Periphyton Pigment Arrays

Figure 20 is the RP-HPLC chromatogram ($\lambda = 440$ nm) of the pigments extracted from a periphyton sample (TS-3-3) from Taylor Slough in the southeastern Everglades. Integrated peak areas (AU•min) were then entered into Beer-Lambert calculations (A= Elc; A= absorbance, E = an extinction coefficient $\{\varepsilon_{mM}, \alpha, E^{1\%}_{1cm}\}$, l = light path {usually 1cm}, c = concentration {dictated by units of E, α or ε}). Appropriate extinction coefficients were taken from the literature, adjusted for A at wavelengths used for integration (440 nm, 410 nm, etc.) and used to calculate the molar amount of each pigment and those values inserted into Equation #2, described earlier. From that exercise, we determined that the periphyton community here (TS-3-3; Fig. 20) was 59% cyanobacteria, 27% chlorophytes and 14% diatom, based on their individual (taxon-specific) contributions of CHL*a* to the total community CHL*a* (Figs. 3 and 21).

Results can be presented in tabular fashion or as a community histogram, as shown for TS-3-3 in Fig. 21.

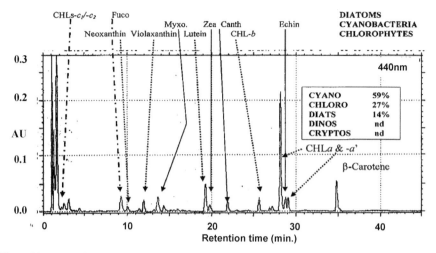

Figure 20. RP-HPLC Chromatogram of pigments from Taylor Slough (#TS-3-3) periphyton.

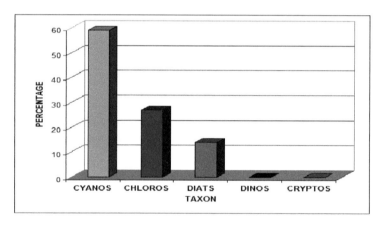

Figure 21. Histogram of the chemotaxonomic estimate of Taylor Slough (#TS-3-3) periphyton.

Pigment-based Chemotaxonomy Compared to Microscopic Evaluations

Figures 22a-c are the cross-plots of chematoxomic (y) versus microscopically (cell number, x) determined percent composition for 207 periphyton samples from a variety of Everglades areas. Regressions (R^2) are not great, going from ~ 0.04 to 0.30 and the slopes (m) in the equation y = mx, forcing the regression to the zero intercept reveal both over (chlorophytes) and under (cyanobacteria) chemotaxonomic estimation of those taxa contributions. Diatoms (m = 1.14) were relatively better estimated. Examining the slope (m) in the equation "HPLC% = m x Cell#", one sees that the HPLC estimates, on the regression line trend are about 0.6, 2.0 and 1.1 times that obtained by cell number. This would indicate correction of the SLE formula used (Equation #2) by the inverse of these factors as follows; $\Sigma CHLa = [(1/0.6)1.1(ZEA)]$ + $[(1/2.0)2.5(CHLb)]$ + $[(1/1.1)1.2(FUCO)]$ to obtain $\Sigma CHLa = [(1.83(ZEA))]$ +

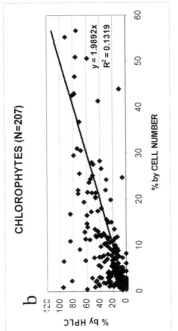

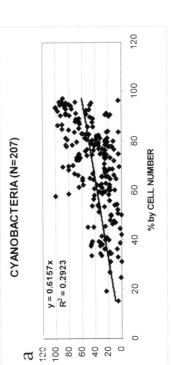

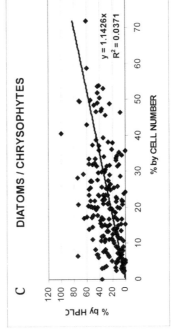

Figure 22. Comparison of HPLC and microscopically determined periphyton community structure for 207 Everglades periphyton samples: a) cyanobacteria, b) chlorophytes, c) diatoms (~chrysophytes).

[1.25(CHL*b*)] + **[1.09(FUCO)]** for the estimation of cyanobacteria, chlorophytes and diatoms, respectively. This might help in getting the slope closer to unity (i.e., m ~ 1.0) but would do nothing to alter the scatter and raise the correlation (R^2). Later in this section we forward other suggestions, each requiring additional research, to improve periphyton pigment-based chemotaxonomy.

Going on the resultant scatter of the overall Everglades data, it was decided to examine the individual areas separately. Therefore, these types of cross-plots (chemotaxonomy versus microscopy) were developed for samples from WCA-1, WCA-2/A-U, WCA-2/A-F, WCA-3A, WCA-3B, the Shark River and Taylor Sloughs. In these cases, we determined regressions with or without forcing the regression to the origin. There were certain apparent relationships but those relationships are certainly not straight forward. For example, Taylor Slough chlorophytes have Cell# = ~0.5 HPLC estimate (aka HPLC ~2 x Cell#) with R^2 ~0.5. However, some unexplainable results exist even in this relatively decent relationship. Namely, there are about a half dozen samples that microscopically showed 5–10% chlorophytes. These 6 or so samples yielded no CHL*b* or lutein. Since pigment analyses utilize many thousands more sample mass than does microscopy, these two pigments would not likely have been missed had chlorophytes actually been present. Inconsistencies such as this example are present in a great many other cases as well. Cryptophytes (~ 'hidden plants') were often not reported in the microscopic data when alloxanthin was present in the pigment analyses.

Conclusions again take us to the difficulty of analyzing samples containing a mix of living and dead microalgae and samples with highly variant internal light fields.

Additional parameters and study is certainly called for since the speed and turn-around times of pigment-based chemotaxonomy for utilization in Adaptive Feedback monitoring-management scenarios holds great promise.

Cell Number vs. 'unit' when Comparing HPLC Pigment-based Chemotaxonomy to Microscopy Data

Filamentous cyanobacteria and chlorophytes are often dominant forms in Everglades periphyton (Browder et al. 1994; McCormick et al. 1998; Hagerthey et al. 2011). Therefore, we asked for counts of the 207 sample set mentioned above to be counted both on the "unit" (e.g., 10 µm sections) and cell number bases. Such analyses of these samples gave surprisingly strange cross plots. *A priori*, one would assume that such measures internally would be much more similar (i.e., y ~ mx with $R^2 > 0.75$?). As we had this data, we included it but how it relates to the quantification of actual periphyton and then to pigment-based studies is entirely unknown and will not be speculated upon. Given the non-correlation and large scatter of unit vs. cell# comparisons, HPLC estimates were not attempted with 'unit' data.

Use of Biovolume for Comparison with HPLC Estimates

Finding that our current HPLC pigment-based chemotaxonomic SLE methodology did not work well when 'ground truthed' versus cell number, we next compared the same HPLC estimates versus biovolume for those samples from Water Conservation

Areas -1, -2A and -3A. Visually, overall relationships for appeared to be present but regressions were still generally poor. These results also direct additional study as to why such deviations occur. Looking at some of the plots, for example the diatoms in WCA-1, we also noted strange discrepancies. That is, by biovolume, a few samples were reported to be 40–50% diatoms yet extremely low amounts and even NO fucoxanthin was found during pigment analysis. This is highly unlikely if the diatoms were fresh viable cells. Also, a few samples were reported to have no chlorophytes by microscopy yet chlorophyll-*b* and lutein were present!

Next, we examined the amount of chlorophyll-*a* per cell. That is, in essentially all studies using pigment-based chemotaxonomy, CHL*a* is (unfortunately/by necessity) assumed to be equally distributed amongst the contributors and only pigment ratios are used to adjust estimates.

Chlorophyll-a per Cell as it Effects Cell Number Ground Truthing of Pigment-based Taxonomy

A variety of fresh water species have been cultured in the senior author's laboratory. From these cultures we choose a cyanobacterium (*Anacystis nidulans*), a desmid chlorophyte (*Scenedesmus* sp.) and a diatom (*Cyclotella meneghiniana*) to test.

These cultures were enumerated using a Coulter Cell Counter and following filtration and pigment analysis of a known amount, we found: *Anacystis nidulans* had 489,770 cells/mL and 2.299 µg CHL*a*/mL, giving 4.69 pg CHL*a*/cell; *Scenedesmus* sp. had 79,730 cells/mL and 0.06756 µg CHL*a*/mL, giving 847 fg pg CHL*a*/cell; *Cyclotella meneghiniana* had 667,454 cells/mL and 0.13523 µg CHL*a*/mL, giving 203 fg CHL*a*/cell.

We then used the cell concentration data to concoct five 'artificial communities' with cell number ratios for cyanobacteria/chlorophytes/diatoms of 1:1:1, 3:1:1, 1:3:1, 1:3:3, and 1:1:3 (Fig. 23).

Next we analyzed these 'artificial communities' and estimated their taxonomic composition using SLE equation #2. As can be noted from Fig. 23a, the % composition of these mixes as estimated by Eq. #2 (chemotaxonomy) highly underestimated the chlorophytes and diatoms. This was due to the higher concentration of CHL*a* in cyanobacteria relative to the other taxa. Plotting the actual amounts of CHL*a* in each taxon gave Fig. 23b.

For this mixture, in order to correctly calculate the community structure using CHL*a* as a proxy for cell numbers of cyanobacteria, chlorophytes and diatoms, respectively, the equation would need to be:

$$\Sigma CHLa = (1.0 \times ZEA) + (5.47 \times CHLb) + (22.7 \times FUCO) \qquad \text{Eq. \#8}$$

It must be noted that this is based on empirical data for these species, even though the internal ratios of CHL*a*/biomarker pigments are essentially what was derived for Eq.#2. Obviously, when equation #8 was applied to the pigment data from these analyses, the correct resultant mixtures (e.g., 1:1:1, 3:1:1, etc.) were obtained.

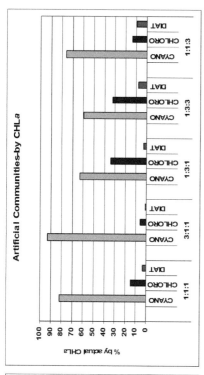

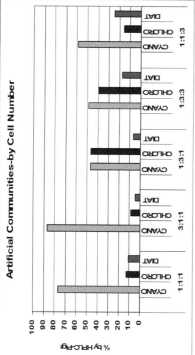

Figure 23. Comparison of the structures of artificial communities on the bases of (a) pigment-based chemotaxonomy (Eq. #2) and (b) the amount of CHL*a* actually present in each mixture.

Therefore, the mere use of internal ratios of CHL*a* to taxon-specific biomarker(s), or the inverse, as a proxy for each taxon's biomass or cell number needs to be normalized to the species actually being studied. In the case of Everglades periphyton, no such data exits to date.

Effects of Light/Shade

CHLa per cell

Many reports exist that attest to the fact that microalgal CHL*a* per cell increases with decreasing light (photon flux density) and conversely decreases in high light environments (e.g., Rosen and Lowe 1984; Delavega et al. 1993; Napolitano 1994; Steiger et al. 1999; Felip and Catalan 2000; Staehr et al. 2002; Bhandari and Sharma 2006; Soltani et al. 2006; Thomas et al. 2006; Greisberger and Teubner 2007; Grant and Louda 2010; *inter alia*).

In order to examine the effect of light on both CHL*a* per cell and internal pigment ratios, we performed a series of cultures under a wide variety of light conditions while holding other parameters (nutrients, T) constant (Grant and Louda 2010). That study investigated four cyanobacteria, two chlorophytes, two diatoms, a prymnesiophyte, and two dinoflagellates. Light conditions (photon flux density, PFD) were classified as low (30–44.5 µmol quanta·m^{-2}·s^{-1}), moderate (108–120 µmol quanta·m^{-2}·s^{-1}, high (300 µmol quanta·m^{-2}·s^{-1}), and intense (1,600–1,800 µmol quanta·m^{-2}·s^{-1}) and used full spectrum (daylight) fluorescent bulbs that were augmented with UVA bulbs during 'intense' light culturing.

Figure 24 contains the plots of CHL*a*/cell for two cyanobacteria (Fig. 24a) and two diatoms (Fig. 24b). The amount of CHL*a* per cell in *C. acerosum* (Fig. 24a solid) went from about > 400 to < 100 pg/cell, a nearly five-fold decrease of with increased light. Similarly, *C. turpinii* (Fig. 31a dashed) decreased by about one-half with increased light. The diatoms (Fig. 24b) also decreased but only by about one half. The anomalous increase at high light (300 µmol quanta·m^{-2}·s^{-1}) for *P. tricornutum* is unexplained. These analyses were run in triplicate and error bars (not shown) were of the order of $\pm 10\%$.

Data such as the above and that given in the references cited above point to a need to factor light conditions into any future studies on periphyton chemotaxonomy. That is, field light conditions need to be used to adjust CHL*a*/biomarker factors in an effort to fine-tune the methodology for such a diverse environment. Shading of periphyton by macrophytes, such as *Typha* (cat-tails) and *Cladium* (sawgrass) will decrease intensity, alter wavelengths in the remaining light and increase CHL*a*/cell in such periphyton communities (Delavega et al. 1993; Thomas et al. 2006). Likely, the strongest effect of light alteration as a driver of these effects is the rapid alteration of light with depth in the vertical profile of periphyton (cf. Losee and Wetzel 1983; Thomas et al. 2006; Hagerthey et al. 2011). Details of the rapid (µm to mm scale) alteration of downwelling irradiances (e.g., < 1% I_o at < 1.2 cm in an oligotrophic epipelic mat) in Everglades periphyton mats are given elsewhere (Hagerthey et al. 2011).

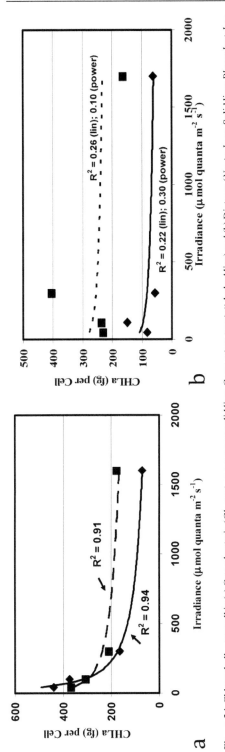

Figure 24. Chlorophyll-*a* per cell in (a) Cyanobacteria (*Closterium acerosum* solid line; *Cosmarium turpinii* dashed line) and (b) Diatoms (*Navicula* sp. Solid line; *Phaeodactylum tricornutum* dashed line). From Grant and Louda (2010).

CHLa-to-biomarker pigment ratios

To reiterate, the heart of pigment-based chemotaxonomy for the estimation of microalgal community structures rests first with discerning valid CHL*a* to taxon-specific biomarker ratios (CHL*a*/biomarker), or the inverse. There are also numerous reports on the effect of light and nutrients on internal pigment ratios (Gieskes and Kraay 1983; Wilhelm et al. 1991; Barlow et al. 1993; Grant and Louda 2010; and references in Wright and Jeffrey 2005 and Roy et al. 2011).

The master's (2006) and Ph.D. (2010) studies of co-author Cidya Grant examined the influences of photo-flux density (aka 'light') on pigment ratios and on a variety of biomass variables. Below are a few examples of pigment ratio relation to light intensity. These were taken from Grant and Louda (2010) with the permission of the publisher (Inter-Research Aquatic Biology).

Figure 25 contains the CHL*a*/zeaxanthin ratios (solid line trends) versus light field found for the cyanobacteria *Lyngbya* sp. (a) and *Anacystis nidulnas* (b). In both cases the ratio CHL*a*/ZEA decreases dramatically. This decrease is the combined effect of a large decrease in the amount of CHL*a* per cell plus larger amounts of ZEA per cell (Grant and Louda 2010). This is characteristic of photoprotectorant pigments (PPP). Figure 25a also has the trend for CHL*a*/echinenone ratios. While this ratio decreased a bit, both the large variability (error bars) and the fact that we have previously found CHL*a*/ECHIN ratios to be rather stable with increased light in *Anabaena flos-aquae* (Skoog 2003; Skoog and Louda 2002). This covariance with CHL*a* is characteristic of photosynthetic accessory pigments (PAP).

Figure 26a contains the CHL*a*/CHL*b* (lower trace, diamonds & squares) and CHL*a*/lutein (upper, circles & triangles) trends for the desmid chlorophytes *Closterium acerosum* (solid) and *Cosmarium turpinii* (dashed). It must be noted that these trends were found to be exactly the same for both species and, as such, were subjected to numerous verification checks to ensure that no data mix-ups had occurred. The trends are real. Again, CHL*a*/CHL*b* trends reveal the non-variance of a CHL*a* plus a PAP. Lutein, on the other hand, was shown to be a PPP and rather unfit for use as a chemotaxonomic marker unless light field is factored into the equation. The ratio CHL*a*/CHL*b* at about 2.5:1 fits well within the published values of 2:1 to 3:1 (see Halldal 1970; Strain et al. 1971; Meeks 1974).

Figure 26b contains the CHL*a*/fucoxanthin (solid) and CHL*a*/chlorophylls-c_1-c_2 (dotted) trends versus light intensity for the diatom *Phaeodactylum tricornutum*. Both trends reveal that these pigments co-vary with CHL*a* and are PAP pigments.

Figure 27 is the suggested (Grant and Louda 2010) overall CHL*a*/pigment trends for photosynthetic accessory pigments (PAP; solid & squares) and the photoprotectorant pigments (PPP; dashed & diamonds). The ratio values on the y-axis are totally hypothetical and serve only as order-of-magnitude indicators. From the above and as given in Fig. 27, it is easy to note that within small changes of irradiance the CHL*a*/PPP ratios can change 2–3 fold and up to an order of magnitude over large (e.g., 100$\rightarrow$ 500 µmol quanta m^{-2} s^{-1}). In the case of coccoidal cyanobacteria when only zeaxanthin is present to be used as a biomarker, light field will be one of the largest drivers in creating uncertainty in pigment-based chemotaxonomic estimates.

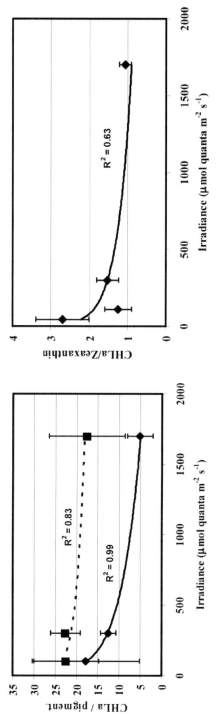

Figure 25. CHL*a*/biomarker ratios: (a) Zeaxanthin (solid) and Echinenone (dashed) in *Lyngbya* sp. (b) Zeaxanthin in *Anacystis nidulans* (from Grant and Louda 2010).

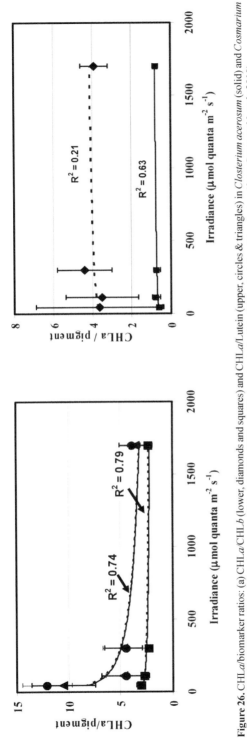

Figure 26. CHL*a*/biomarker ratios: (a) CHL*a*/CHL*b* (lower, diamonds and squares) and CHL*a*/Lutein (upper, circles & triangles) in *Closterium acerosum* (solid) and *Cosmarium turpinii* (dashed). (b) CHL*a*/fucoxanthin (squares, solid) and CHL*a*/CHL*s-c* (diamonds, dashed) in *Phaeodactylum tricornutum* (from Grant and Louda 2010).

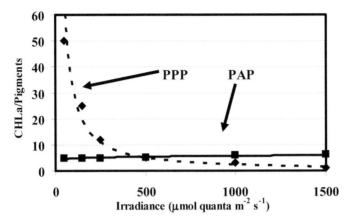

Figure 27. Generalized trends for the light-related trends of CHL*a* to photoprotectorant pigments (PPP) and photosynthetic accessory pigments (PAP) (from Grant and Louda 2010).

Effects of Degraded Algae

As detailed earlier in text, the presence of degraded chlorophyll-*a* species is a signal that senescence, death and/or predation has occurred within the algal matter under analysis (Currie 1962; Baker and Louda 1986; 2002; Carpenter et al. 1986; Hendry et al. 1987; Louda et al. 1998; 2000; 2002; 2008; 2011; Bale et al. 2011; references in Jeffrey et al. 1997; Roy et al. 2011).

The pheopigments are the Mg-free (pheo- aka phaeo- ≈'brown') derivatives of CHL*a* (e.g., pheophytin, pheophorbide and their pyro-, epimer- and allomer-forms: see Appendices A and B1). To assess the 'freshness' or viability of the periphyton samples that we were analyzing, we also routinely quantified all pigments present in the lipid extracts and determined the percentage pheopigment (%PHEO) complement as %PHEO = ([PHEOs]/[Σ PHEOs + CHLs-*a*])x100. "CHLs-*a*" are all pigments with the CHL*a* chromophore (CHL*a*, its epimer, allomers, pyro-forms plus those of chlorophyllide-*a*). Chlorophyllide(s)-*a* can indeed be a signal of senescence/ death (Louda et al. 1998; 2002; 2011) but these can also form during the handling and extraction of diatoms due to their high chlorophyllase activities (Jeffrey and Hallegraeff 1987; Louda et al. 1998; 2002; 2011). Therefore, the chlorophyllides-*a* (see Appendices A and B1) were not included in the breakdown products and only the PHEOs were considered.

Figure 28 is the plot of the number of periphyton samples versus their pheopigment contents. As shown, 206 out of 620 samples (33.2%) had > 10% pheopigments and 361 (58.2%) had > 5%! We consider > 10% pheopigments to be a minimum 'flag' for significant amounts of dead microalgae (see Baker and Louda 1986; 2002; Louda 2008; Louda et al. 1998; 2000; 2002; 2011). The ratio of pheophytin to CHL*a* has also been taken as "an indicator of community senescence" for benthic microalgae (Stevenson 1996).

In the future we will be reevaluating the present pigment-based chemotaxonomic estimation of Everglades periphyton using both a 5% and 10% PHEOs as cutoffs for non-inclusion of a sample in a data set. If it turns out that samples with > 5%

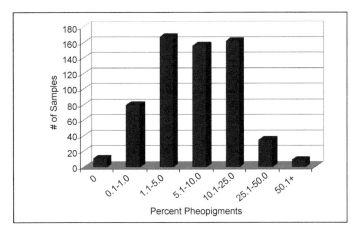

Figure 28. Histogram of the abundance of Everglades periphyton samples with various percentages of pheopigments in the total 'chlorophyll-*a* plus derivatives' pigment pool.

pheopigments are actually too degraded to yield valid pigment-based chemotaxonomy then more than half of the samples that we examined would fall into that category. As it is, one-third of these samples (Fig. 28) at the 10% pheopigment were already highly degraded.

Testing Known Mixtures with SLE, CHEMTAX and BCE Methods

The details of inverse simultaneous equations (SLE), CHEMTAX and the Bayesian Community Estimator (BCE) were covered earlier in text. Here we compare these approaches to the pigment-based chemotaxonomy of Everglades periphyton. Co-author J. Browne performed these calculations during her Master's thesis research (Browne 2010; Browne and Louda 2010) using the data generated during the six years of our overall study.

Prior to performing an inter method comparison; many iterations of the three methods were made to maximize the correspondence of each to the known mixtures. Additionally, all tests covered in this section utilize biovolume and not cell number, as was covered earlier in text. Such an initial test is exemplified here by the manipulation of the CHEMTAX program to include only the three divisions under study. That is, all parts of the algorithm which deal with dinoflagellates, cryptophytes, prymnesiophytes and other taxa were removed. Figure 29 illustrates the marked improvement in the community estimation when only taxa present are considered.

The first inter-method comparison was performed for the artificial mixes with but a single species per each of the three main divisions for in Everglades periphyton. That is, the mixes described earlier in text contained a cyanobacterium, a chlorophyte and a diatom. Figure 30 contains the plots of the optimized mathematical models versus the known fractions of each of the three taxa in the various mixtures (see Fig. 23).

Based on the Correlations (Pearson product-moments: not shown; see Browne 2010), it was concluded that for these artificial mixtures, both SLE and CHEMTAX performed equally and the best overall but BCE did a better job with the diatoms. It

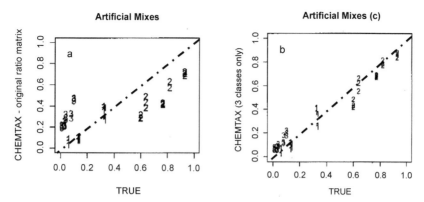

Figure 29. CHEMTAX estimation of the artificial microalgal community using (a) the original algorithm (Mackey et al. 1996; 1998) and (b) after constraining CHEMTAX to consider only chlorophytes (1), cyanobacteria (2) and diatoms (3). "TRUE" = microscopically determined cell numbers converted to biovolume.

needs to be noted that the CHL*a* data discussed earlier (Fig. 23 & associated text) was utilized here to generate taxon-specific biomarker multipliers. This is but the first step in modifying the procedure to mimic reality. However, adjusting these ratios and multipliers for field samples will be a daunting task indeed.

Next attention was turned to comparing the three 'optimized' methods using Everglades periphyton grab samples. This was performed on the 27 samples of Everglades periphyton for which biovolume data was provide at the time of evaluation (Browne 2010). Figure 31 contains plots for the community estimations of chlorophytes (1), cyanobacteria (2) and diatoms (3) performed using the revised SLE, CHEMTAX constrained to these three taxa and BCE. Correlations, using the CHEMTAX output, were in general only moderate, with the highest correlations (Pearson product-moments; aka Pearson's *r*) being for diatoms (0.47) and total cyanobacteria (0.57). Chlorophytes had the lowest *r* at 0.36 (see Browne 2010).

Pearson's *r* values for these tests are given in Table 1. Here CHEMTAX is a bit better than SLE for cyanobacteria, about equal for chlorophytes and SLE is a bit better for diatoms. Thus, for these Everglades periphyton grab samples, SLE and CHEMTAX performed about equally. The Bayesian model (BCE) was not as good compared to these methods. As with SLE and CHEMTAX, time is required for broader applications to be tested and the method modified.

Cross-plots of the revised SLE versus CHEMTAX estimations of periphytometer algal divisions in seven areas of the Everglades yielded excellent correlations (Pearson's $r = 0.787 \rightarrow 0.990$: see Browne 2010). However, without cell number and/or biovolume data on these samples, ground-truth testing could not be performed.

Given the arguments made earlier regarding the co-existence of senescent, dead and heterotrophically processed microalgal debris with living microalgae in the grab samples, it can reasonably be asserted that pigment-based chemotaxonomy 'should' work better on periphytometer samples.

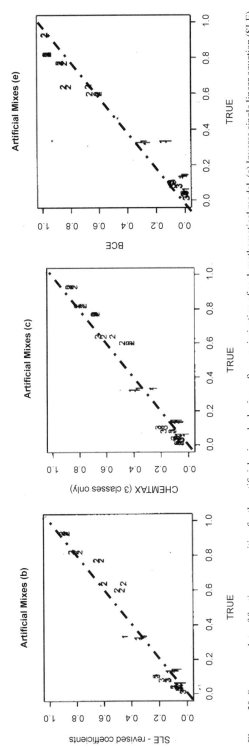

Figure 30. Summary plots of fraction composition for the artificial microalgal mixes after maximization of each mathematical model. (a) Inverse single linear equation (SLE). (b) CHEMTAX (see Fig. 29b). (c) Bayesian Community Estimator (BCE). *Code:* Chlorophyte (1), Cyanobacterium (2), Diatom (3). "TRUE" = known biovolume calculated from cell number data (Fig. 32).

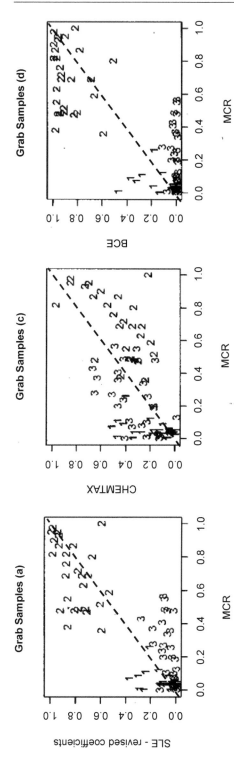

Figure 31. Taxon estimates for Everglades periphyton grab samples for which microscopically (MCR) determined biovolume data was available. *Code*: Chlorophyte (1), Cyanobacterium (2), Diatom (3).

Table 1. Pearson's *r* value correlations between microscopically determined biovolumes and pigment-based chemotaxonomic estimations of cyanobacteria, chlorophytes, and diatoms in Everglades periphyton grab samples using adjusted SLE, CHEMTAX and BCE methods (Adapted from Browne 2010).

BIOVOLUME vs	df	Total Cyanos (*r*)	Chloros (*r*)	Diats (*r*)
SLE-revised	25	0.445**	0.311*	0.475**
CHEMTAX (3 taxa)	25	0.568**	0.365**	0.448**
BCE	25	0.191	0.252	−0.003

p-value significance is indicated as (* $0.05 < p < 0.1$) and (** $p < 0.05$)

Bacterial Estimation-and Relation to Mercury Methylation

As given within the methods section earlier, we routinely integrate chromatograms at 440 nm (chlorophylls-*a*/-*b*/-$c_{1, 2, 3}$ & carotenoids), 410nm (pheopigments-*a*), 394 nm (internal standard), 360nm (bacteriochlorophylls-/bacteriopheopigments-*a*). Detection and quantification of bacteriochlorophyll-*a* gives an estimation of the amount of purple-sulfur bacteria present in a sample (cf. Cleckner et al. 1999). Using that data, one can then generate approximate protein biomass calculations.

Published 'generalized' chlorophyll to protein biomass conversions are as follows: For oxygenic photoautotrophs; CHLa = 1.45% DRY WT. OM; PROTEIN = 50% DRY WT. OM and therefore PROTEIN = 34.48 x CHLa (Meyers and Kratz 1955; APHA 1976; Hagmeir 1980). For anoxygenic photoautotrophs (purple and green sulfur bacteria); BCHLa/PROTEIN = 0.074; PROTEIN = 13.51 x BCHLa* (Adjusted by x 0.015 BCHLc if present) and GREEN &/or BROWN-S BACTERIA: BCHLc/ PROTEIN = 0.22 and therefore PROTEIN = 4.454 x BCHLc (Purple-S: vanGermerden 1980) (Green/Brown-S: Broch-Due et al. 1978; Blankenship et al. 1987; Gorlenko 1987; vanDorseen et al. 1987).

Using these values, we calculated the apparent oxygenic/anoxygenic protein biomass in a set of periphytometer samples collected in 2008. CHL*a* and bacteriochlorophyll-*a* (BCHL*a*) refers to their totals as measured during HPLC analyses. These estimates are given in Table 2.

It is noted here however that CHL*a*/biomass estimates for microalgae can vary up to five or ten fold relative to the methodolgies employed (Steeman-Nielsen 1975).

The importance of being able to detect the bacteriochlorophylls rests with their indication of microzones of anoxia and lowered redox (pE) conditions plus the presence

Table 2. Estimates of oxygenic and anoxygenic protein biomass is a series of Everglades periphytometer samples collected in 2008.

Sample	Cyano/Chloro/Diat	$\mu g/cm^2$ CHLa	Oxygenic Protein	% OXY	$\mu g/cm^2$ BCHLa	Anoxygenic Protein	%ANOXY
E-1	52/38/10	1.83	63.1	97.0%	0.143	1.9	3.0%
F-2	27/60/13	0.63	21.7	96.8%	0.054	0.07	3.2%
F-3	29/55/13	1.37	47.2	97.5%	0.089	1.2	2.5%
F-4	54/33/13	0.15	5.2	97.2%	0.011	0.1	2.8%
F-5	31/20/47	0.35	12.1	99.6%	0.004	0.1	0.4%
U-2	25/72/3	0.27	9.3	99.1%	0.006	0.1	0.9%
X-4	46/36/13	2.87	99	99.9%	0.110	0.1	0.1%
Z-2	32/53/7	0.19	6.6	97.6%	0.012	0.2	2.4%

of sulfate reducing bacteria (SRB). This then takes us to the identification of 'potential' sites of mercury metyhylation as discussed by Cleckner et al. (1999). In their paper, Cleckner et al. (1999) revealed a strong linear relationship ($r^2 = 0.63$) between the nmol BCHLa g^{-1} of periphyton and mercury methylation in WCA2A, the same area where the samples given here in Table 2 were collected. They had the highest mercury methylation when BCHLa (mw = 910 Da) was ≥ 100 nmol g^{-1} periphyton. In Table 2, the highest concentration of BCHLa, namely sample E-1 at 0.143 mg g^{-1}, is only 0.16 nmol. Though the samples in Table 2, at the time collected, were not highly likely to be harboring significant mercury methylation, the potential is shown. Additionally, it must be noted that these samples were periphytometers and, as such, did not have the bulk conducive for the establishment of anoxia as strong as possible in larger natural accumulations.

The methylation of mercury, forming the highly bioaccumlative lipophilic monomethylmercury (MMM), has been well studied by the United States Geologic Survey and associated scientists during the Aquatic Cycling of Mercury in the Everglades (ACME) project (Krabbenhoft 1996; Cleckner et al. 1998; Gilmour et al. 1998; Hurley et al. 1998; Krabbenhoft et al. 1998). The linkage of sulfur pollution from the Everglades Agricultural Area (EAA) to the elevated levels of sulfate in The Everglades and its support of sulfate reducing bacteria (SRB), generating hydrogen sulfide (H_2S) and lowering redox potentuals (pE), is also well documented (Bates et al. 2002; Ekstrom et al. 2003; Corrales et al. 2011).

Use of Pigment-based Chemotaxonomy in Relation to Water Quality Parameters and for Spatial, Temporal and Spatio-temporal Monitoring of Periphyton Communities

The *spatial* analysis of sites within WCA-1 (Fig. 32) reflect what is commonly considered the pattern of periphyton communities in this area. Here we find that both the dominance of cyanobacteria and increased total biomass (~productivity) reflect proximity to the L15/L39/L6 intersection. That is, sites Z2, Z3 and X1 are nearest these canals and, as such, likely experienced enhanced nutrient loading and a shift in the Redfield-Richards ratio to lower N:P values, favoring diazotrophic cyanobacteria as noted from the co-occurrence of aphanizophyll (see Fig. 35 below). Sites farther away or 'upstream' from this influence MESO/Y4 or X3/X4, respectively, were found to exhibit less biomass and to be dominated (> 80%) by chlorophytes. Increases in diatoms, though at levels of only about 10% of the total, were also detected in the near canal sites (Z2, Z3 and X1). We feel that data such as this, when eventually linked to water quality (nutrient, alkalinity, pH, etc.) and hydroperiod data, will be of great use in the timely monitoring of any changes accompanying the implementation of CERP changes in water distribution patterns.

Figure 33 is a *temporal* monitoring plot of the chemotaxonomically derived periphyton communities at the mesocosm site (WCA-1) in the Arthur R. Marshall National Wildlife Refuge (aka WCA-1). This plot is presented to illustrate the fact that cryptophytes can sometimes form a significant portion of the periphyton. In June of 2006, cryptophytes reached about 20% of the periphyton on CHLa-based biomass estimation. The presence of alloxanthin (Appendices A and B2), is found only in the

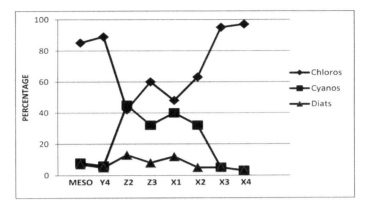

Figure 32. Pigment-based chemotaxonomic estimation of periphyton communities in WCA-1 in October 2005.

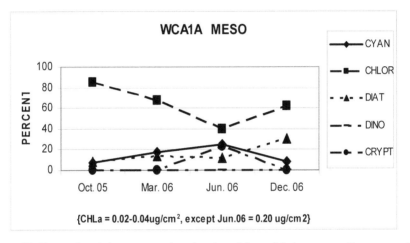

Figure 33. Pigment-based chemotaxonomic estimation of the periphyton communities recovered on periphytometers at the WCA-1 mesocosm site. Percentage is based on taxon-specific CHL*a* contributions to the ΣCHL*a*.

Cryptophyta, except for minor instances of its occurrence in marine dinoflagellates (Schnepf and Elbrachter 1999; Zapata et al. 2012), and, with the co-occurrence of α-carotene, verifies their presence here. It is also noted that the cryptophyte plus cyanobacteria surge in June 2006 accounted for a lot of the overall increased CHL*a* on the periphytometers. CHL*a*, in all cases here, equals the sum of taxon-specific calculated CHL*a*. However, in all cases the predicted and the actually measured (HPLC-PDA and total extract with spectrophotometric quantitation) were all in line with each other.

Figure 34 is a three year record of the periphyton communities at site T7-3 in the middle of Taylor Slough and about 17 km from the mangrove fringe of Florida Bay. Though sample spacing doesn't allow for a truly complete look at yearly cycles and water quality, especially salinity, was not available for all samplings, the potential for rapid feedback during long-term monitoring can easily be noted. That is, in these

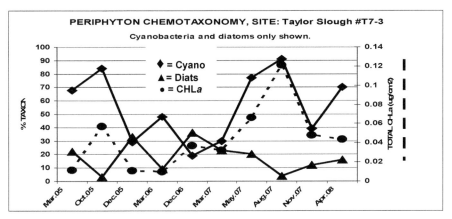

Figure 34. A three year record of total community CHLa and the cyanobacteria and diatoms in the periphytometer periphyton communities at Taylor Slough site T-1&-3.

cases the pigment-based chemotaxonomic estimation can have turnaround times on the order of days to a week at most. The low sum of cyanobacteria plus diatoms in the period December 2005 through April 2007 was due to a large increase in chlorophytes plus some cryptophytes. Then in June through September 2007, cyanobacteria became dominant and quite abundant. The driver for this productivity and selectivity is unknown. However, we suspect that this cyanobacterial increase was due to nitrogen fixing forms. The potential for such pigment based conclusions is detailed below.

Figures 33 and 34 above were simple examples of temporal monitoring using pigment data. Figure 35 below is an example of spatial monitoring along a known nutrient, notably phosphorus (P), trend along a transect in WCA-3A. Here is what we consider to be an excellent use of pigment-based chemotaxonomy. Sites E1 (Fig. 1) is close to the rim (Hillsboro) canal carrying high phosphorous loads from the EAA. As the transect goes into the center of WCA-2A, total phosphorus (TP) drops dramatically.

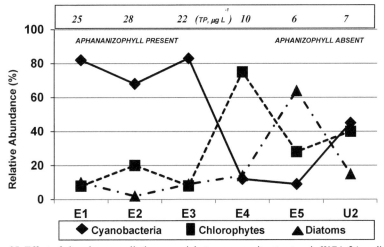

Figure 35. Effect of phosphorous pollution on periphyton community structure in WCA-2A as discerned by pigment-based chemotaxonomy.

Aphanizophyll (Appendices A and B2) is a well-known carotenoid biomarker for diazotrophic cyanobacteria (Hertzberg and Liaaen-Jensen 1966a-b; 1971; Leavitt et al. 2006; Patoine et al. 2006; Waters et al. 2009). We therefore conclude that the cyanobacterial dominance at E1 through E3 was due to diazotrophic (nitrogen-fixing) cyanobacteria. Chlorophytes became the major taxon at E4, diatoms at E5 and finally, in the interior at U2, non-nitrogen fixing cyanobacteria and diatoms co-dominated.

Cyanobacteria have evolved along a variety of lines, as represented by their variability in photosynthetic/photoprotective pigment arrays (e.g., Hertzberg and Liaaen-Jensen 1966a-b; O'Kelly 1993; Jeffrey and Vesk 1997; Graham and Wilcox 2000; Schagerl and Donabaum 2003; Jefferey et al. 2011). Aside from the water soluble phycobilins, the lipophilic pigments consist of chlorophyll-*a*, without any other chlorophyll accessory pigment, and taxon-specific carotenoid arrays. It is the varied carotenoid assemblages that are of interest as biomarkers with which to better dissect periphyton communities. Specifically, we are still investigating using the following pigments for their associated cyanobacterial types: Echinenone (filamentous marker), zeaxanthin (coccoidal forms, general marker), myxoxanthophyll (colonial), and aphanizophyll (diazotrophic {nitrogen fixing} forms). Thus, we hope to be able to tease apart cyanobacteria into 3 groups; unicellular, colonial and nitrogen fixing. This, at first, appears facile. However, all cyanobacteria and even chlorophytes, to a limited extent, produce zeaxanthin, often as a photoprotectorant in response to photon flux density (PFD: see Grant and Louda 2010). Many cyanobacteria also produce the keto-carotenoids echinenone and canthaxanthin. Thus, intertaxon corrections for overlapping contributions of zeaxanthin and echinenone are required. Authentic myxoxanthophyll (viz. myxol rhammnoside) and aphanizophyll on our RP-HPLC system have retention times of 22.9 and 18.8 minutes respectively and, as they have the same chromophore (i.e., myxol), give identical UV/Vis spectra (λ = (452), 476–8, 506–8 nm in eluant). Detailed 'ground-truthing', comparing pigment-based and microscopically derived taxonomies, will be especially useful in future trials.

Ultraviolet Photoprotective Pigments, the Scytonemins

A great many ultra-violet range (UVR) sunscreen pigments are known. The best known include the scytoneman series (see Appendix 2C) and mycosporine amino acids (MAAs: see Castenholz and Garcia-Pichel 2000; Groniger et al. 2000; Rastogi et al. 2010; Sinha and Hader 2008; Singh et al. 2010; inter alia). The MAAs are water soluble and were not analyzed during our studies. The scytonemins are lipid soluble and, as such, are extracted with the chlorophylls and carotenoids.

The scytonemin pigments are known from more than 300 cyanobacterial species that have yellow to brown sheaths (Edwards et al. 2000).

Reduced and oxidized scytonemin (Figs. 36–37) represent the REDOX pair of a very well-studied UVR sunscreen (see Singh et al. 2010) with a dimeric indole-phenol structure that has come to be termed the 'scytoneman' nucleus.

Throughout our studies on Everglades periphyton, the scytonemins have been very prevalent whenever cyanobacteria are abundant. Given the normal high light environment of southern Florida, this is to be expected. Cyanobacteria growing deep in the water column or in highly shaded areas had less to non-detectable scytonemins.

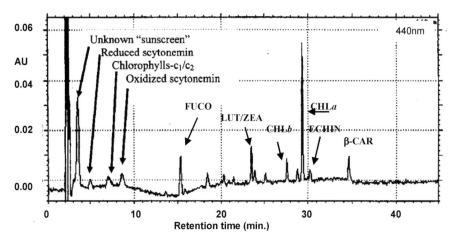

Figure 36. RP-HPLC chromatogram (λ = 440 nm) of the pigments extracted from periphyton recovered at site T-24 (FUCO = fucoxanthin, LUT/ZEA = lutein/zeaxanthin, ECHIN = echinenone, β-CAR = β-carotene. (NOTE: "unknown sunscreen" has been fully characterized as scytonmein-imine; Grant and Louda 2013).

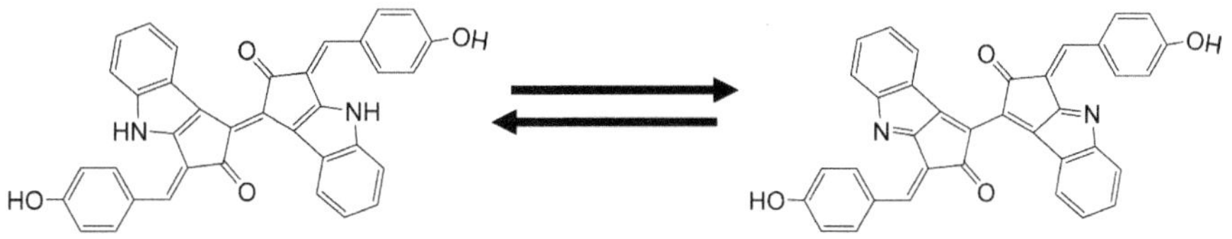

Figure 37. Structural relationship between the reduced and oxidized forms of scytonemin (see Garcia-Pichel and Castenholz 1991; Grant and Louda 2013).

When we began studying samples from the Shark River Slough low hydroperiod sites (Fig. 38a), a very polar pigment with a red-shifted 'scytonemin-like' spectrum (Fig. 38b) was observed. This we termed as the "unknown sunscreen", present above at 3.5 minutes (~1 min. after the solvent front) in Fig. 46.

Grant (2006), grew *Scytonema hofmannii*, as well as many other algae and cyanobacteria, under a wide range (30 to > 1,500 μmol photo• m^{-2}•s^{-1}), of light conditions. When cultured in moderate (~150 μmol quanta•m^{-2}•s^{-1}) to intense (~1,500 μmol quanta•m^{-2}•s^{-1}) light conditions (Figs. 7a and 7b, respectively), *S. hofmannii* synthesized this "unknown" in increasing amounts in parallel with increasing light (Grant and Louda 2010). Having observed this trend, we next undertook the isolation, purification and structural elucidation of this pigment from bulk periphyton as well as from *S. hofmannii*. The "unknown sunscreen" was subsequently identified as scytonemin-imine (Grant and Louda 2013: see Appendices A and B3).

Now that the high-light induction phenomenon and structure of scytonemin-imine are known, the next step is to ascertain its function. Given that it was synthesized by *S. hofmannii* in moderate to high light without added UV, we feel that its prime function is not that of a strict UVR photoprotectorant. Going on the assumption that evolution only generates and retains structural-functional materials for a reason, we examined its wavelength maxima. Many cyanobacteria, *S. hofmannii* included, evolved a variety

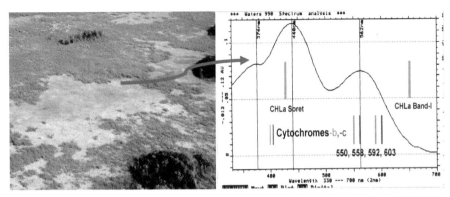

Figure 38. (a) An aerial view of a low hydroperiod periphyton mat in the Shark River Slough (SFWMD) (b) UV/Vis of scytonemin-imine isolated from Shark River Slough samples and cultures of *Scytonema hofmannii* (Grant and Louda 2013).

Color image of this figure appears in the color plate section at the end of the book.

of survival strategies to deal with such stressors as desiccation and photo-oxidation. Therefore, as a first speculative proposal we look towards the protection of both CHL*a* and the cytochromes by scytonemin-imine. That is, in Figure 38b we delineated the Soret and Band-I positions of CHL*a* as well as the Soret and α/β Bands of cytochromes –b and –c ($\alpha\lambda$ = 592, 603 and $\beta\lambda$ = 550, 558 nm respectively). The Soret or 'blue' absorbances of both CHL*a* and the cytochromes fall under a high absorbance region of scytonemin-imine. The broad absorption band of scytonemin-imine encompasses the α and β bands of the cytochromes very well. Protecting the cytochromes from high light during intense light periods could protect the cells from excess electron flow and generation of reactive oxygen species (ROS: Grant and Louda 2013).

Application to Monitoring/Adaptive Feedback and Future Research Directions

Statistical analysis strongly suggests that pigment-based chemotaxonomy holds promise for providing a relatively inexpensive and a speedy way to perform bulk periphyton community analysis. In marine water column studies, the work of Paerl et al. (2003 and references therein) refer to phytoplankton functional groups (PFP) when taxonomically classified using pigment data. The same acronym can be suggested here but interpreted as periphyton functional groups.

Referring to the large spatial and temporal studies that we conducted, Cohen and Lamsal (2008) stated, "... liquid chromatography permits the rapid and repeatable partitioning of pigments which are diagnostic of particular algal groups. If simple pigment extractions and relative importance information (i.e., fractional abundance of each pigment) co-varies in predictable ways with basic water quality variables, then pigment sampling can provide a viable environmental indicator."

Cohen and Lamsal (2008) go on to state "... there is strong evidence that emerges throughout our analysis that the relationship between pigment fractions and water quality measurements is contingent upon when and where a sample is collected.

When these categorical variables (hydrologic partition and season) are included in the analysis, the regression diagnostics improve markedly, in some cases sufficiently to provide viable prediction" (Cohen and Lamsal 2008).

Refinements regarding the pigment-based chemotaxonomy of grab samples must include detailed attention to the presence/absence, types and abundance of pheopigments. The fact that fucoxanthin (diatoms, other chrysophytes) and peridinin (peridinin-containing dinoflagellates) degrade more rapidly than the carotenols (zeaxanthin/cyanobacteria, alloxanthin/cryptophytes) must be remembered and factored into any and all conclusions. Additionally, light levels must be considered. For example, was the grab sample from open water or under the shade of emergent macrophytes? All of these and other factors are drivers for the exact CHL*a* concentration per cell and the internal ratios of taxon-specific marker pigments to CHL*a*.

Genetic variation (strain, clade) of a single species can yield variations in internal pigment ratios (Stolte et al. 2000). Examination of the CHL*a* contents in a single species (*Peridinium gatunese* Nygaard) within Lake Kinneret (aka the Sea of Galilee), Israel revealed that CHL*a* per cell established "correlation zones" within the lake based on the homogeneity/heterogeneity of the system (Lalikhman 1999). This also serves to reinforce the concept of 'patchiness' in microalgal growth patterns and physiology (cf. Steele 1976). These problems have been recognized by Mackey et al. (1998) in that they applied different sets of equations when studying phytoplankton at increasing depth in the western Pacific. It would therefore be important to reexamine Everglades periphyton with concurrent light field data including consideration of macrophyte shading, time of year (solar azimuth), and collection time of day.

Pigment based-chemotaxonomy does presently have two concrete applications within the scope of Everglades monitoring and assessment which deserve increased study and application. These are: (1) the rapid detection (viz. purple, green/brown -S bacteria) of anoxic sulfidic regions in microzones or whole samples wherein mercury methylation is likely; and (2) the use of periphytometers and epiphytometers (surrogate macrophytes) to assess the active short-term (1/2–2 mo.) recruitment and growth of major periphyton taxa. The use of periphytometers, especially if periphytometer growth is compared to coincident existing native periphyton, can aid in minimizing the variables mentioned above and provide rapid low cost evaluation of nutrient driven community alterations. Ground truthing of pigment-ratios and CHL*a* per cell to the species present should be performed at least in each of the four major seasons (i.e., sun angle/temperature regime). This would allow more valid multipliers (biomarker to CHL*a* related biomass) to be empirically determined and then either SLE or CHEMTAX estimations would apply equally well. Such rapid and potentially analyst-unbiased feedback could be a large asset in the strategy of adaptive management.

Acknowledgements

The senior author (JWL) acknowledges contract (DSR 93-313, PCP301640, ML061237, etc.) support between 1994 and 2009 from the South Florida Water Management District (SFWMD) and the Comprehensive Everglades Restoration Plan (CERP) through SFWMD as well as 2000–2003 grant support from the National Marine Fisheries Division of the National Oceanographic and Atmospheric Administration.

This funding supported pigment-based chemotaxonomic and other studies in the Lake Okeechobee-Everglades-Florida Bay portions of the K-O-E (Kissimmee-Okeechobee-Everglades) system from 1994–2009.

The Everglades studies reported herein were conducted from July 2003 through July 2009. Funding and samples were supplied by the South Florida Water Management District (SFWMD) under several contracts (#ML061237 and others) to the senior author (JWmL) of the Organic Geochemistry Group, Department of Chemistry and Biochemistry and the Environmental Sciences Program of Florida Atlantic University, Boca Raton, Florida. Oversight of the project was by co-author Hagerthey while a Senior Scientist in the Everglades Division of SFWMD in West Palm Beach, Florida.

Technical assistance from 2000–2009 by Ms. Panne (Pam) Mongkhonsri is greatly appreciated. *In vitro* studies on the effect of photon-flux density (*light*) on pigment ratios, allowing improvement of pigment-based linear equations, formed parts of both the M.S. and Ph.D. research of co-author Cidya Grant (Grant 2006; 2011). Comparisons of a variety of pigment-based mathematical methods (SLE, CHEMTAX, BCE, etc.) were performed by co-author Jaime Browne during her M.S. research (Browne 2010).

Inter-Research Science Center, Luhe, Germany is thanked for permission to reproduce figures from Grant and Louda (2010). These are present here as Figs. 32–35.

Continued support through the provision of facilities by Florida Atlantic University is noted and appreciated.

The diligent editorial suggestions and handling of this manuscript by co-editor Andy Gottlieb is noted and appreciated.

References

Agusti, S. 1991. Light environment within dense algal populations-cell size influences on self-shading. J. Plankton Res. 13: 863–871.

APHA. 1976. Standard Methods for the Examination of Water and Waste-water, 14th Edition. pp. 993–1057. American Public Health Association, Washington, D.C.

Atkins, W.R.G. and M. Parke. 1951. Seasonal changes in the phytoplankton as indicated by spectrophotometric chlorophyll estimations. J. Mar. Biol. Ass. U.K. 31: 495–508.

Baker, E.W. and J.W. Louda. 1986. Porphyrins in the geologic record. pp. 125–225. *In*: R.B. Johns (ed.). Biological Markers in the Sedimentary Record. Elsevier, Amsterdam.

Baker, E.W. and J.W. Louda. 2002. The Legacy of the Treibs' samples. pp. 3–128. *In*: A. Prashnowsky (ed.). Alfred Treibs Memorial Volume. Wurzburg.

Bates, A.L., W.H. Orem, J.W. Harvey and E.C. Spiker. 2002. Tracing sources of sulfur in the Florida Everglades. J. Environ. Qual. 31: 287–299.

Bautista, B. and F. Jimenez-Gomez. 1996. Ultraphytoplankton photoacclimation through flow cytometry and pigment analysis of Mediterranean coastal waters. Scientia Marina 60: 233–241.

Bellinger, B.J., S.E. Hagerthey, S. Newman and M.I. Cook. 2012. Detrital floc and surface soil microbial biomarker responses to active management of the nutrient impacted Florida Everglades. Microbiol. Aquat. Sys. 64: 893–908.

Bhandari, R. and P.K. Sharma. 2006. High-light-induced changes on photosynthesis, pigments, sugars, lipids and antioxidant enzymes in freshwater (*Nostoc spongiaeforme*) and marine (*Phormidium corium*) cyanobacteria. Photochem. Photobiol. 82: 702–710.

Biggs, B.J.F. and C. Kilroy. 2000. Stream Periphyton Monitoring Manual. New Zealand Ministry for the Environment (ISBN 0-478-09099-4). NIWA Publishers, Christchurch, NZ.

Bjørnland, T. and S. Liaaen-Jensen. 1989. Distibution patterns of carotenoids in relation to chromophyte phylogeny and systematics. pp. 37–60. *In:* J.C. Green, B.S.C. Leadbeater and W.L. Diver (eds.). The Chromophyte Algae: Patterns and Perspectives. Systematics Association Special Volume No. 38. Claredon Press, Oxford, U.K.

Bone, R.A. and J.T. Landrum. 1992. Distribution of macular pigment components, zeaxanthin and lutein, in human retina. pp. 360–366. *In*: L. Paker (ed.). Methods in Enzymology, Vol. 213: Carotenoids Part A. Chemistry, Separation, Quantitation, and Antioxidation. Academic Press, San Diego.

Boussiba, S., W. Bing, J.-P. Yuan, A. Zarka and F. Chen. 1999. Changes in pigments profile in the green alga *Haematococcus pluvialis* exposed to environmental stresses. Biotech. Letts. 21: 601–604.

Bouvier, F., A. D'harlingue, R.A. Backhaus, M.H. Kumagai and B. Camara. 2000. Identification of neoxanthin synthase as a carotenoid cyclase paralog. Eur. J. Biochem. 267: 6346–6352.

Breton, E., C. Brunet, B. Sautour and J.M. Brylinski. 2000. Annual variations of phytoplankton biomass in the Eastern English Channel: comparison by pigment signatures and microscopic counts. J. Plankton Res. 22: 1423–1440.

Broch-Due, M., J.G. Ormerod and B.S. Fjerdingen. 1978. Effect of light intensity on vesicle formation in *Chlorobium*. Arch. Microbiol. 116: 269–274.

Browder, J.A., S. Black, P. Schroeder, M. Brown, M. Newman, D. Cottrell, D. Black, R. Pope and P. Pope. 1981. Perspective on the Ecological Causes and Effects of the Variable Algal Composition Southern Everglades Periphyton. Everglades National Park, South Florida Research Center, Report T-643.

Browder, J.A., P.J. Gleason and D.R. Swift. 1994. Periphyton in the Everglades: spatial variation, environmental correlates and ecological implications. pp. 379–418. *In*: S.M. Davis and J.C. Ogden (eds.). Everglades: The Ecosystem and its Restoration. St. Lucie Press, Boca Raton.

Browne, J. 2010. Comparison of chemotaxonomic methods for determination of algal class composition in Florida Everglades periphyton. M.S., Environmental Sciences, Florida Atlantic University, Boca Raton, FL.

Browne, J. and J.W. Louda. 2010. Comparison of chemotaxonomic methods for determination of algal class composition in Florida Everglades periphyton. 74th Annual Meeting of the Florida Academy of Sciences. Ft. Pierce, FL. March 19. Abstr. ENV-24.

Bruland, G.L., T.Z. Osborne, K.R. Reddy, S. Grunwald, S. Newman and W.F. Debusk. 2007. Recent changes in soil total phosphorus in the Everglades: Water Conservation Area 3. Environ. Monit. Assess. 129: 379–395.

Carpenter, S.R., M.M. Elser and J.J. Elser. 1986. Chlorophyll production, degradation and sedimentation: Implications for paleolimnology. Limnol. Oceanogr. 31: 112–124.

Childers, D.L., R.D. Jones, J.C. Trexler, C. Buzzelli, S. Dailey, A.L. Edwards, E.E. Gaiser, K. Jayachandaran, A. Kenne, D. Lee, J.F. Meeder, J.H.K. Pechmann, A. Renshaw, J. Richards, M. Rugge, L.J. Scinto, P. Sterling and W. VanGelder. 2002. Quantifying the effects of low-level phosphorus additions on unenriched Everglades wetlands with *in situ* flumes and phosphorus dosing. pp. 127–152. *In*: J.W. Porter and K.G. Porter (eds.). The Everglades, Florida Bay and Coral Reefs of the Florida Keys: An Ecosystem Sourcebook. CRC Press, Boca Raton, FL.

Chimney, M.J. and G. Goforth. 2001. Environmental impacts to the Everglades ecosystem: a perspective and restoration strategies. Water Sci. Technol. 44: 93–100.

Cleckner, L.B., P.J. Gasrrison, J.P. Hurley, M.L. Olson and D.P. Krabbenhoft. 1998. Trophic transfer of methyl mercury in the northern Florida Everglades. Biogeochem. 40: 347–361.

Cleckner, L.B., C.C. Gilmour, J.P. Hurley and D.P. Krabbenhoft. 1999. Mercury methylation in periphyton of the Florida Everglades. Limnol. Oceanogr. 44: 1815–1825.

Cohen, M. and S. Lamsal. 2008. Statistical Analysis of Everglades Algal-Water Quality Relations. Draft Report, So. Fla. Water Manag. Dist. 31 pp.

Cooper, S.R., J. Huvane, P. Vaithiyananthan and C.J. Richardson. 1999. Calibration of diatoms along a nutrient gradient in Florida Everglades Water Conservation Area-2A, USA. J. Paleolimnol. 22: 413–437.

Corrales, J., G.M. Naja, C. Dziuba, R.G. Rivero and W. Orem. 2011. Sulfate threshold target to control methylmercury levels in wetland ecosystems. Sci. Total Environ. 409: 2156–2162.

Craft, N.E. 1992. Carotenoid reversed-phase high-performance liquid chromatography methods: reference compendium. pp. 185–205. *In*: L. Paker (ed.). Methods in Enzymology, Vol. 213: Carotenoids Part A. Chemistry, Separation, Quantitation, and Antioxidation. Academic Press, San Diego.

Currie, R.I. 1962. Pigment in zooplankton faeces. Nature 193: 956–957.

Dambeck, M., U. Eilers, J. Breitenbach, S. Stteiger, C. Buchel and G. Sandmann. 2012. Biosynthesis of fucoxanthin and diadinoxanthin and function of initial pathway genes in *Phaeodactylumtricornutum*. J. Exp. Botany, doi:10.1093/jxb/ers211.

Davies, B.H. 1965. Analysis of carotenoid pigments. pp. 489–532. *In*: T.W. Goodwin (ed.). Chemistry and Biochemistry of Plant Pigments. Academic Press, London.

del Pilar Sanchez Saavedra, M. and D. Voltolina. 1993. The chemical composition of *Chaetoceros* sp. (Bacillariophyceae) under different light conditions. Comp. Biochem. Physiol. 107B: 39–44.

Delavega, E.L., J.R. Cassani and H. Allaire. 1993. Seasonal relationship between southern naiad and associated periphyton. J. Aquat. Plant Manag. 31: 84–88.

Demmig-Adams, B. 1990. Carotenoids and photoprotection in plants: A role for the xanthophyll zeaxanthin. Biochim. Biophys. Acta 1020: 1–24.

Demmig-Adams, B. and W.W. Adams-III. 2006. Tansley Review: Photoprotection in an ecological context: the remarkable complexity of thermal energy dissipation. New Phytol. 172: 11–21.

Doi, M., T. Inage and Y. Shioi. 2001. Chlorophyll degradation in a *Chlamydomonas reinhardtii* mutant: An accumulation of pyropheophorbide *a* by anaerobiosis. Plant Cell. Phsiol. 42: 469–474.

Egeland, E.S., J.L. Garrido, L. Clementson, K. Andresen, C.S. Thomas, M. Zapata, R. Airs, C.A. Llewellyn, G.L. Newman, F. Rodriguez and S. Roy. 2011. Data sheets aiding identification of phytoplankton carotenoids and chlorophylls. pp. 665–822. *In*: S. Roy, C.A. Llewellyn, E.S. Egeland and G. Johnsen (eds.). Phytoplankton Pigments in Oceanography: Guidelines in Modern Methods. UNESCO, Paris.

Ekstrom, E.B., F.M.M. Morel and J.M. Benoit. 2003. Mercury methylation independent of the acetyl-coenzyme A pathway in sulfate-reducing bacteria. Appl. Environ. Microbiol. 69: 5414–5422.

Emmanouil, P., H.M. van Stokkum Ivo, F. Holger, C. Buchel and V.G. Rienk. 2005. Spectroscopic characterization of the excitation energy transfer in the fucoxanthin-chlorophyll protein of diatoms. Photosyn. Res. 86: 241–250.

Engel, N., C. Curty and A. Gossauer. 1996. Chlorophyll catabolism in *Chlorella protothecoides*. 8. Facts and artefacts. Plant Cell. Physiol. 34: 77–83.

Falkowski, P.G. and J.A. Raven. 2007. Aquatic Photosynthesis. Princeton University Press, Princeton, N.J.

Felip, M. and J. Catalan. 2000. The relationship between phytoplankton biovolume and chlorophyll in a deep oligotrophic lake: decoupling in their spatial and temporal maxima. J. Plankton Res. 22: 91–105.

Foppen, F.H. 1971. Tables for the identification of carotenoid pigments. Chromatogr. Revs. 14: 133–298.

Gaiser, E.E., D.L. Childers, R.D. Jones, J.H. Richards, L.J. Scinto and J.C. Trexler. 2006. Periphyton responses to eutrophication in the Florida Everglades: cross-system patterns of structural and compositional change. Limnol. Oceanogr. 51: 617–630.

Gaiser, E.E., P.V. McCormick, S.E. Hagerthey and A.D. Gottlieb. 2011. Landscape patterns of periphyton in the Florida Everglades. Crit. Revs. Envir. Sci. Technol. 41: 82–120.

Garcia-Pichel, F. and R.W. Castenholz. 1991. Characterization and biological implications of scytonemin, a cyanobacterial sheath pigment. J. Phycol. 27: 395–409.

Gilmour, C.C., G.S. Riedel, M.C. Ederington, J.T. Bell, J.M. Benoit, G.A. Gill and M.C. Stordal. 1998. Methylmercury concentrations and production rates across a tropic gradient in the northern Everglades. Biogeochem. 40: 327–345.

Gleason, P.J. and W. Spackman, Jr. 1974. Calcareous periphyton and water chemistry in the Everglades. pp. 146–181. *In*: P.J. Gleason (ed.). Environments of South Florida: Present and Past. Memoir No. 2. Miami Geological Society, Coral Gables, FL.

Glidenhoff, N., J. Herz, Gundermann, C. Buechel and J. Wachtveitl. 2010. The excitation energy transfer in the trimeric fucoxanthin-chlorophyll protein from *Cyclotella meneghiniana* analyzed by polarized transient absorption spectroscopy. Chem. Phys. 373: 104–109.

Gorlenko, V.M. 1987. Ecological niches of green sulfur and gliding bacteria. pp. 257–267. *In*: J.M. Olson, J.G. Omerod, J. Amesz, E. Stackenbrandt and H.G. Truper (eds.). Photosynthetic Green Bacteria. Plenum, New York.

Gossauer, A. and N. Engel. 1996. Chlorophyll catabolism—Structures, mechanisms, conversions. J. Photochem. Photobiol.-B. Biology 32: 141–151.

Gottlieb, A.D., J.H. Richards and E.E. Gaiser. 2005. Effects of desiccation duration on the community structure and nutrient retention of short and long-hydroperiod Everglades periphyton mats. Aquatic Bot. 82: 99–112.

Gottlieb, A.D., J.H. Richards and E.E. Gaiser. 2006. Comparative study of periphyton community structure in long and short-hydroperiod marshes. Hydrobiol. 569: 195–207.

Graham, J.E. and D.A. Bryant. 2009. The biosynthetic pathway for myxol-2'-fucoside (myxoxanthophyll) in the cyanobacterium *Synechococcus* sp. Strain PCC 7002. J. Bacteriol. 191: 3292–3300.

Graham, L.E. and L.W. Wilcox. 2000. Algae. Prentice Hall, Upper Saddle River, New Jersey.

Grant, C. 2006. Effect of light on microalgal pigment ratios. M.S. Chemistry and Biochemistry, Florida Atlantic University, Boca Raton, FL.

Grant, C. 2011. Effect of photic flux on microalgal pigment ratios, carbon and protein biomass in microalgae. Ph.D. Chemistry and Biochemistry, Florida Atlantic University, Boca Raton, FL.

Grant, C.S. and J.W. Louda. 2010. Microalgal pigment ratios in relation to light intensity—Implications for chemotaxonomy. Aquatic Biol. 11: 127–138.

Grant, C.S. and J.W. Louda. 2013. Scytonemin-imine, a mahogany-colored UV/VIS sunscreen of cyanobacteria exposed to intense solar radiation. Org. Geochem. 65: 29–36.

Greisberger, S. and K. Teubner. 2007. Does pigment composition reflect phytoplankton community structure in differing temperature and light conditions in a deep alpine lake? An approach using HPLC and delayed fluorescence techniques. J. Phycol. 43: 1108–1119.

Grimm, B., R.J. Porra, W. Rudiger and H. Scheer (eds.). 2006. Chlorophylls and Bacteriochlorophylls. Biochemistry, Biophysics, Functions and Applications. Vol. 25, Advances in Photosynthesis and Respiration. Govindgee (Series Ed.). Springer, Dordrecht, Netherlands.

Hager, A. 1980. The reversible, light induced conversions of xanthophylls in the chloroplast. pp. 57–79. *In:* F.-C. Czygan (ed.). Pigments in Plants. Gustav Fischer Verlag, Stuttgart.

Hagerthey, S.E., M. Jacoby, J.W. Louda and P. Monghkronsri. 2003. Development of a High Performance Liquid Chromatography (HPLC) Protocol for Monitoring Periphyton in the Florida Everglades. Joint Conference on the Science and Restoration of the Greater Everglades and Florida Bay Ecosystem. Palm Harbor, FL. April 13–18, 2003. GEER Abstracts, pp. 235–236.

Hagerthey, S.E., J.W. Louda and P. Mongkronsri. 2006. Evaluation of pigment extraction methods and a recommended protocol for periphyton chlorophyll *a* determination and chemotaxonomic assessment. J. Phycology 42: 1125–1136.

Hagerthey, S.E., B.J. Bellinger, K. Wheeler, M. Gantar and E.E. Gaiser. 2011. Everglades Periphyton: A Biogeochemical Perspective. Critical Revs. Envir. Sci. Technol. 41: 309–343.

Hagerthey, S.E., S. Newman and S. Xue. 2012. Periphyton-based transfer functions to assess ecological imbalance and management of a subtropical ombrotrophic peatland. Freshwater Biol., doi:10.1111/j.1365-2427.2012.024848.x.

Hagmeir, E. 1960. Untersuchungen uber die Menge und die Zusammensetzung von Seston und Plankton in Wasser Proben von Reisen in die Nordsee und nach Island. Ph.D. Diss. Kiel. 166 pp.

Halldal, P. 1970. Photosynthetic apparatus of microalgae. pp. 17–55. *In:* P. Halldal (ed.). Microbiology of Microorganisms. Wiley Interscience, New York.

Hambrook-Berkman, J.A. and M.G. Canova. 2007. Algal biomass indicators (ver. 1.0): U.S. Geological Survey Techniques of Water-Resources Investigations, book 9, chap. A7, section 7.4, August.

Hanrahan, G. 2009. Environmental Chemometrics: Principles and Modern Applications. CRC Press, Boca Raton, 292 pp.

Harvey, H.W. 1934. Amount of phytoplankton population. J. Mar. Biol. Assoc. U.K. 19: 761–773.

Havens, K.E., A.D. Steinman, H.J. Carrick, J.W. Louda and E.W. Baker. 1999. A comparative analysis of periphyton communities in a subtropical lake using HPLC pigment analysis and microscopic cell counts. Aquatic Sci. 61: 307–322.

Havskum, H., L. Schluter, R. Scharek, E. Berdalet and S. Jacquet. 2004. Routine quantification of phytoplankton groups – microscopy or pigment analyses? Mar. Ecol. Prog Ser. 273: 31–42.

Hertzberg, S. and S. Liaaen-Jensen. 1966a. The carotenoids of blue-green algae-I. The carotenoids of *Oscillatoria rubescens* and an *Athrospira* sp. Phytochem. 5: 557–563.

Hertzberg, S. and S. Liaaen-Jensen. 1966b. The carotenoids of blue-green algae-II. The carotenoids of *Aphanizomenon flos-aquae*. Phytochem. 5: 565–570.

Hertzberg, S. and S. Liaaen-Jensen. 1971. The constitution of aphanizophyll. Phytochem. 10: 3251–3252.

Higgins, H.W., S.W. Wright and L. Schulter. 2011. Quantitative interpretation of chemotaxonomic pigment data. pp. 257–313. *In:* S. Roy, C.A. Llewellyn, E.S. Egeland and G. Johnsen (eds.). Phytoplankton Pigments: Characterization, Chemotaxonomy and Applications in Oceanography. Cambridge University Press, Cambridge.

Holden, M. 1976. Chlorophylls. pp. 2–37. *In:* T.W. Goodwin (ed.). Chemistry and Biochemistry of Plant Pigments, 2nd Ed. Academic Press, London.

Hurley, J.P. and D.E. Armstrong. 1991. Pigment preservation in lake sediments: A comparison of sedimentary environments in Trout Lake, Wisconsin. Can. J. Fish. Aquat. Sci. 48: 472–486.

Hurley, J.P., D.P. Krabbenhoft, L.B. Cleckner, M.L. Olson, G.R. Aiken and P.S. Rawlik, Jr. 1998. System controls on the aqueous distribution of mercury in the northern Florida Everglades. Biogeochem. 40: 293–310.

Iwaniec, D.M., D.L. Childers, D. Rondeau, C.J. Madden and C. Saunders. 2006. Effects of hydrologic and water quality drivers on periphyton dynamics in the southern Everglades. Hydrobiol. 569: 223–235.

Jeffrey, S.W. and S.W. Wright. 1987. A new spectrally distinct component in preparations of chlorophyll *c* from the micro-alga *Emiliania huxleyi* (Prymnesiophyceae). Biochim. Biophys. Acta 894: 180–188.

Jeffrey, S.W. and G.M. Hallegraeff. 1987. Chlorophyllase distribution in ten classes of phytoplankton: a problem for chlorophyll analysis. Mar. Ecol. Prog. Ser. 35: 293–304.

Jeffrey, S.W. and M. Vesk. 1997. Introduction to marine phytoplankton and their pigment signatures. pp. 37–84. *In*: S.W. Jeffrey, R.F.C. Mantoura and S.W. Wright (eds.). Phytoplankton Pigments in Oceanography: Guidelines in Modern Methods. UNESCO, Paris.

Jeffrey, S.W., R.F.C. Mantoura and S.W. Wright (eds.). 1997. Phytoplankton Pigments in Oceanography: Guidelines in Modern Methods. UNESCO, Paris.

Jeffrey, S.W., S.W. Wright and M. Zapata. 2011. Microalgal classes and their signature pigments. pp. 3–77. *In*: S.W. Jeffrey, R.F.C. Mantoura and S.W. Wright (eds.). Phytoplankton Pigments in Oceanography: Guidelines in Modern Methods. UNESCO, Paris.

Kalikhman, I. 1999. Distribution fields for aquatic ecosystem components: determination of significance of correlation zones. Hydrobiologia 400: 1–11.

Krabbenhoft, D.P. 1996. Mercury Studies in the Florida Everglades. U.S. D.O.I.-U.S.G.S. Fact Sheet FS-166-96.

Krabbenhoft, D.P., J.P. Hurley, M.L. Olson and L.B. Cleckner. 1998. Diel variability of mercury phase and species distribution in the Florida Everglades. Biogeochem. 40: 311–325.

Kreps, E. and N. Verbinskaya. 1930. Seasonal changes in the Barents Sea. J. Cons. Intern. Explor. Mer. V: 329–345.

Larsen, L.G., J.W. Harvey, G.B. Noe and D.J. Nowacki. 2008. Transport dynamics of floc in ridge and slough vegetation communities: A laboratory flume experiment and numerical study. Greater Everglades Ecosystem Restoration Conference, Naples, FL. July 28–Aug. 1.

Larsen, L.G., J.W. Harvey, G.B. Noe and J.P. Crimaldi. 2009. Predicting organic floc transport dynamics in shallow aquatic ecosystems: Insights from the field, the laboratory, and numerical modeling. Water Resources Res. 45: W01411, doi:10.1029/2008WR007221.

Leavitt, P.R. 1993. A review of factors that regulate carotenoid and chlorophyll deposition and fossil pigment abundance. J. Palolimneol. 9: 109–127.

Leavitt, P.R., C.S. Brock, C. Ebel and A. Pantoine. 2006. Landscape-scale effects of urban nitrogen on a chain of freshwater lakes in central North America. Limnol. Oceanogr. 51: 2262–2277.

Li, Y. and D.C. Walton. 1990. Violaxanthin is an abscisic acid precursor in water-stressed dark-grown bean leaves. Plant Physiol. 92: 551–559.

Lohr, M. and C. Wilhelm. 1999. Algae displaying the diadinoxanthin cycle also possess the violaxanthin cycle. Natl. Acad. Sci. USA 96: 8784–8789.

Lohr, M. and C. Wilhlm. 2001. Xanthophyll synthesis in diatoms: quantification of putative intermediates and comparison of pigment conversion kinetics with rate constants derived from a model. Planta 212: 382–391.

Louda, J.W. and E.W. Baker. 1986. The Biogeochemistry of Chlorophyll. pp. 107–126. *In*: M. Sohn (ed.). Organic Marine Geochemistry. ACS Symposium Series #305, American Chemical Society, Washington, D.C.

Louda, J.W. and P. Monghkonsri. 2004. Comparison of spectrophotometric estimates of chlorophylls-*a*, -*b*, -*c* and 'pheopigments' in Florida Bay seston with that obtained by high performance liquid chromatography-photodiode array analyses. Fla. Sci. 67: 281–292.

Louda, J.W., J. Li, L. Liu, M.N. Winfree and E.W. Baker. 1998. Chlorophyll degradation during senescence and death. Org. Geochem. 29: 1233–1251.

Louda, J.W., J.W. Loitz, D.T. Rudnick and E.W. Baker. 2000. Early diagenetic alteration of chlorophyll-*a* and bacteriochlorophyll-*a* in a contemporaneous marl ecosystem. Org. Geochem. 31(12): 1561–1580.

Louda, J.W., L. Liu and E.W. Baker. 2002. Senescence- and death-related alteration of chlorophylls and carotenoids in marine phytoplankton. Org. Geochem. 33: 1635–1653.

Louda, J.W., S.E. Hagerthey and P. Monghkronsri. 2003. Pigment-based chemotaxonomic studies of Everglades periphyton. 67th Annual Meeting of the Florida Academy of Sciences. Orlando, Fl., Abstract # BIO-14. March 21–22, 2003.

Louda, J.W., J.W. Loitz, A. Melisiotis and W.H. Orem. 2004. Potential sources of hydrogel stabilization of Florida Bay lime mud sediments and implications for organic matter preservation. J. Coastal Res. 20: 448–463.

Louda, J.W., S.E. Hagerthey and P. Mongkhronsri. 2006. Pigment-Based Chemotaxonomic Evaluation of Everglades Periphyton Communities. Greater Everglades Ecosystem Restoration Conference. Lake Buena Vista, Florida. June 5–9.

Louda, J.W., R.R. Neto, A.R.M. Magalhaes and V.F. Schneider. 2008. Pigment alterations in the brown mussel *Perna perna*. Comp. Biochem. Physiol. –B 150: 385–394.

Louda, J.W., P. Mongkhonsri and E.W. Baker. 2011. Chlorophyll degradation during senescence and death-III: Three to ten year experiments, implications for ETIO-series generation. Org. Geochem. 42: 688–699.

Mackey, M.D., D.J. Mackey, H.W. Higgins and S.W. Wright. 1996. CHEMTAX—a program for estimating class abundances from chemical markers: application to HPLC measurements of phytoplankton. Mar. Ecol. Prog. Ser. 144: 265–83.

Mackey, M.D., H.W. Higgins, D.J. Mackey and S.W. Wright. 1997. CHEMTAX user's manual: a program for estimating class abundances from chemical markers – application to HPLC measurements of phytoplankton pigments. CSIRO Marine Laboratories Report 229, Hobart, Australia.

Mackey, D.J., H.W. Higgins, M.D. Mackey and D. Holdsworth. 1998. Algal class abundances in the western equatorial Pacific: estimation from HPLC measurements of chloroplast pigments using CHEMTAX. Deep-Sea Res. I 45: 1441–68.

Mantoura, R.F.C. and C.A. Llewellyn. 1983. The Rapid determination of algal chlorophyll and carotenoid pigments and their breakdown products in natural waters by reverse-phase high-performance liquid chromatography. Anal. Chim. Acta 151: 297–314.

Mantoura, R.F.C. and D.J. Repeta. 1997. Calibration methods for HPLC. pp. 407–428. *In*: S.W. Jeffrey, R.F.C. Mantoura and S.W. Wright (eds.). Phytoplankton Pigments in Oceanography: Guidelines in Modern Methods. UNESCO, Paris.

Matile, P., S. Hortensteiner, H. Thomas and B. Krautler. 1996. Chlorophyll breakdown in senescent leaves. Plant Physiol. 112: 1403–1409.

McCormick, P.V., P.S. Rawlik, K. Lurding, E.P. Smith and F.H. Sklar. 1996. Periphyton-water quality relationships along a nutrient gradient in the northern Florida Everglades. J. N. Am. Benthol. Soc. 15: 433–449.

McCune, B. and J.B. Grace. 2002. Analysis of Ecological Communities. MjM Software Design, Glendale Beach, OR.

Meeks, J.C. 1974. Chlorophylls. pp. 161–175. *In*: W.D.P. Stewart (ed.). Algal Physiology and Biochemistry. Blackwell Scientific. Oxford.

Meyers, J. and W.A. Kratz. 1955. Relations between pigment content and photosynthetic characteristics in a blue-green alga. J. Gen. Physiol. 39: 11–22.

Millie, D.F., H.W. Paerl and J.P. Hurley. 1993. Microalgal pigment assessments using high-performance liquid chromatography: A Synopsis of organismal and ecological applications. Can. J. Fish. Aquat. Sci. 50: 2513–2527.

Mineeva, N.M. 2011. Pigments as indicators of Phytoplankton Biomass (Review). Int. J. Algae. 13: 330–340.

Mochimaru, M., H. Masukawa, T. Maoka, H. Mohamed, W. Vermaas and S. Takaichi. 2008. Substrate specificities and availability of fucosyltransferases and β–carotene hydroxylase for myxol-2'-fucoside synthesis in *Anabaena* sp. strain PCC 7120 compared with *Synechococcus* sp. Strain PCC 6803. J. Bacteriol. 190: 6726–6733.

Mohamed, H., A. van de Meene, R. Roberson and W. Vermaas. 2005. Myxoxanthophyll is required for normal cell wall structure and thylakoid organization in the cyanobacterium *Synechocystis* sp. strain PCC 6803. J. Bacteriol. 187: 6883–6892.

Monette, D. and S.H. Markwith. 2012. Hydrochory in the Florida Everglades: temporal and spatial variation in seed dispersal phenology, hydrology, and restoration of wetland structure. Ecol. Restor. 30: 180–191.

Mongkhronsri, P., S.E. Hagerthey and J.W. Louda. 2007. Utilization of pigment-based chemotaxonomy for rapid spatial-temporal assessment of Everglades periphyton. 71st Annual Meeting of the Florida Academy of Sciences. St. Petersburg, FL. March 16–17. Abstr. ENV.

Morden, C.W and A.R. Sherwood. 2002. Continued evolutionary surprises among dinoflagellates. Proc. Natl. Acad. Sci. 99: 11558–11560.

Muller, P., X.-P. Li and K.K. Niyogi. 2001. Non-photochemical quenching: a response to excess light energy. Plant Physiol. 125: 1558–1566.

Nair, A., S. Sathyyendranath, T. Platt, J. Morales, V. Stuart, M.-H. Forget, E. Devred and H. Bouman. 2008. Remote sensing of phytoplankton functional types. Remote Sensing Environ. 112: 3366–3375.

Napolitano, G.E. 1994. The relationship of lipids with light and chlorophyll measurements in freshwater algae and periphyton. J. Phycol. 30: 943–950.

Neto, R.R., R.N. Mead, J.W. Louda and R. Jaffe. 2006. Organic biogeochemistry of detrital flocculent material (floc) in a subtropical, coastal, wetland. Biogeochem. 77: 283–304.

O'Kelly, C.J. 1993. Relationships of eukaryotic algae to other protists. pp. 269–294. *In*: T. Berner (ed.). Ultrastructure of Microalgae. CRC Press Inc., Boca Raton.

Owens, T.G., P.G. Falkowski and T.E. Whitledge. 1980. Diel periodicity in cellular chlorophyll content in marine diatoms. Mar. Biol. 59: 71–77.

Paerl, H.W., L.M. Valdes, J.L. Pinckey, M.F. Piehler, J. Dyble and P.H. Moisander. 2003. Phytoplankton photopigments as indicators of estuarine and coastal eutrophication. Biosci. 53: 953–964.

Pandolfini, E., L. Thys, B. Leporeq and J.-P. Descy. 2000. Grazing experiments with two freshwater zooplankters: fate of chlorophyll and carotenoid pigments. J. Plankton Res. 22: 305–319.

Patoine, A., M.D. Graham and P.R. Leavitt. 2006. Spatial variation of nitrogen fixation in lakes of the northern Great Plains. Limnol. Oceanogr. 51: 1665–1677.

Picnkney, J.L., D.F. Millie, B.T. Vinyard and H.W. Paerl. 1997. Environmental controls of phytoplankton bloom dynamics in the Neuse River Estuary, North Carolina, U.S.A. Can. J. Fish. Aquat. Sci. 54: 2491–2501.

Pisani, O., J.W. Louda and R. Jaffe. 2013. Biomarker assessment of spatial and temporal changes in the composition of flocculent material (floc) in a subtropical wetland. Environ. Chem. 10: 424–436.

R Development Core Team. 2009. R: a language and environment for statistical computing. Vienna: R Foundation for Statistical Computing. http://www.R-project.org.

Repeta, D.J. 1989. Carotenoid diagenesis in recent marine sediments: II. Degradation of fucoxanthin to lolioide. Geochim. Cosmochim. Acta 53: 699–707.

Repeta, D.J. and R.B. Gagosian. 1982. Carotenoid transformations in coastal marine waters. Nature 295: 51–54.

Repeta, D.J. and R.B. Gagosian. 1987. Carotenoid diagenesis in recent marine sediments-I. The Peru continental shelf (15°S, 75°W). Geochim. Cosmochim. Acta 51: 1001–1009.

Reuss, N. 2005. Sediment pigments as biomarkers of environmental change. Ph.D. Thesis. National Environmental Research Institute, Roskilde, Denmark.

Richards, F.A. and T.F. Thompson. 1952. The estimation and characterization of plankton populations by pigment analyses. II. A spectrophotometric method for the estimation of plankton pigments. J. Mar. Res. 11: 156–172.

Rosen, B. and R.L. Lowe. 1984. Physiological and ultrastructural responses of *Cyclotella meneghiniana* (Bacillariophyta) to light intensity and nutrient limitation. J. Phycol. 20: 173–183.

Rowan, K.S. 1989. Photosynthetic Pigments of Algae. Cambridge University Press, Cambridge.

Roy, S., C.A. Llewellyn, E.S. Egeland and G. Johnsen (eds.). 2011. Phytoplankton Pigments: Characterization, Chemotaxonomy and Applications in Oceanography. Cambridge University Press, Cambridge.

Schagerl, M. and C. Pichler. 2000. Pigment composition of freshwater charophyceae. Aquatic Biol. 67: 17–129.

Schagerl, M. and K. Donabaum. 2003. Patterns of major photosynthetic pigments in freshwater algae, 1. Cyanoprokaryota, Rhodophyta, and Cryptophyta. Ann. Limnol.– Int. J. Limnol. 39: 49–62.

Schagerl, M., C. Pichler and K. Donabaum. 2003. Patterns of major photosynthetic pigments in freshwater algae, 2. Dinophyta, Euglenophyta, Chlorophyceae and Charales. Ann. Limnol.– Int. J. Limnol. 39: 49–62.

Scheer, H. 1991. Structure and occurrence of chlorophylls. pp. 3–30. *In*: H. Scheer (ed.). Chlorophylls, CRC Press, Boca Raton, FL.

Schnepf, E. and M. Elbrachter. 1999. Dinophyte chloroplasts and phylogeny—A review. Grana 38: 81–97.

Singh, S.P., D.-P. Hader and R.P. Sinha. 2010. Cyanobacteria and Ultraviolet radiation (UVR) stress: Mitigation strategies. Ageing Res. Revs. 9: 79–90.

Sinha, R.P. and D.-P. Hader. 2008. UV-protectants in cyanobacteria. Plant Sci. 174: 278–289.

Skoog, K.O. 2003. Pigment-Based Chemotaxonomy of Phytoplankton Communities in Lake Okeechobee, Florida. M.S. Environmental Sciences, Florida Atlantic University, Boca Raton, FL.

Skoog, K.O. and J.W. Louda. 2002. Effects of Light Field upon the Chemotaxonomic Estimation of Cyanobacteria in Lake Okeechobee, Florida. 66th. Annual Meeting, Florida Academy of Sciences, Miami, Florida. March 7–9.

Soltani, N., R.A. Kahvari-Nejad, M.T. Yazdi, S. Shokravi and E. Fernadez-Valient. 2006. Variation of nitrogenase activity, photosynthesis and pigmentation of the cyanobacterium *Fischerella ambigua* strain FS18 under different irradiance and pH values. World J. Microbiol. Biotech. 22: 571–576.

Staehr, P.A., P. Henriksen and S. Markager. 2002. Photoacclimation of four marine phytoplankton species to irradiance and nutrient availability. Mar. Ecol. Prog. Ser. 238: 47–59.

Stauber, J.L. and S.W. Jeffrey. 1988. Photosynthetic pigments from fifty-one species of marine diatoms. J. Phycol. 24: 158–172.

Steele, J.H. 1976. Patchiness. pp. 98–115. *In:* D.H. Cushing and J.J. Walsh (eds.). The Ecology of the Seas. Blackwell Scientific Publ., London.

Steiger, S., L. Schäfer and G. Sandmann. 1999. High-light-dependant upregulation of carotenoids and their antioxidative properties in the cyanobacterium *Synechocystis* PCC 6803. J. Photochem. Photobiol. 52: 14–18.

Steinman, A., K.E. Havens, J.W. Louda, N.M. Winfree and E.W. Baker. 1998. Characterization of the Photoautotrophic Bacterial Communities in a Sub-Tropical Lake (Lake Okeechobee, Florida). Can. J. Fish. Aquat. Sci. 55: 206–219.

Stevenson, R.J. 1996. 1. Algal Ecology in Freshwater Benthic Habitats. pp. 3–30. *In:* R.J. Stevenson, M.L. Bothwell and R.L. Lowe (eds.). Algal Ecology: Freshwater Benthic Ecosystems. Academic Press, San Diego.

Stevenson, R.J. and L.L. Bahls. 1999. Periphyton protocols. *In:* M.T. Barbour, J. Gerritsen, B.D. Snyder and J.B. Stribling (eds.). Rapid Bioassessment Protocols for Use in Streams and Wadeable Rivers: Periphyton, Benthic Macroinvertebrates and Fish, Second Edition. EPA 841-B-99-002. U.S. Environmental Protection Agency; Office of Water; Washington, D.C.

Stolte, W., G.W. Kraay, A.A.M. Noordeloos and R. Riegman. 2000. Genetic and physiological variation on pigment composition of *Emiliania huxleyi* (Prymnesophyceae) and the potential use of its pigment ratios as a quantitative physiological marker. J. Physiol. 36: 529–539.

Stoneman-Douglas, M. 1947. The Everglades: River of Grass. Rinehart and Company, New York.

Strain, H.H., B.T. Cope and W.A. Svec. 1971. Procedures for isolation, identification, estimation, and investigation of the chlorophylls. Methods Enzymol. 13: 452–476.

Strickland, J.D.H. and T.R. Parsons. 1968. A Practical Handbook of Seawater Analysis. Fish. Res. Brd. Canada Bull. No. 167, 293 pp.

Swift, D.R. 1984. Periphyton and water quality relationships in the Everglades water conservation areas. pp. 97–117. *In:* P.J. Gleason (ed.). Environments of South Florida: Present and Past-II. (2nd Ed.). Miami Geological Society, Coral Gables, FL.

Takaichi, S. 2011. Carotenoids in algae: Distributions, biosyntheses and functions. Mar. Drugs 9: 1101–1118.

Takaichi, S. and M. Mimuro. 1998. Distribution and geometric isomerism of neoxanthin in oxygenic phototrophs: 9'-cis, a sole molecular form. Plant Cell Physiol. 39: 968–977.

Takaichi, S., T. Maoka and K. Masamoto. 2001. Myxoxanthophyll in *Synechocystis* sp. PCC 6803 is Myxol-2'-dimethyl-fucoside (3R, 2'S) – myxol-2'-(2,4-di-O-Methyl-α-L-fucoside), not rhamnoside. Plant Cell Physiol. 42: 756–762.

Takaichi, S., T. Maoka, K. Takasaki and S. Hanada. 2010. Carotenoids of *Gemmatimonas aurantiaca* (Gemmatimonadetes): identification of a novel carotenoid, deoxyscillol-2-rhamnoside, and proposed biosynthetic pathway of oscillol-2,2'-dirhamnoside. Microbiol. 156: 757–763.

Taylor, F.J.R., M. Hoppenrath and J.F. Saldarriaga. 2008. Dinoflagellate diversity and distribution. Biodiv. Cons. 17: 407–418.

Thomas, S., E.E. Gaiser, M. Gantar, A. Pinowska, L.J. Scinto and R.D. Jones. 2002. Growth of calcareous epilithic mats in the margin of natural and polluted hydrosystems: phosphorus removal implications in the C-111 basin, Florida Everglades, USA. Lake Resevoir. Manag. 18: 324–330.

Thomas, S., E.E. Gaiser and F.A. Tobias. 2006. Effects of shading on calcareous benthic periphyton in a short-hydroperiod oligotrophic wetland (Everglades, FL, USA). Hydrobiol. 569: 209–221.

Trexler, J., E. Gaiser and D. Childers. 2006. Interaction of hydrology and nutrients in controlling ecosystem function in oligotrophic coastal environments of South Florida. Hydrobiol. 569: 1–2.

Van den Meersche, K. and K. Soetaert. 2009. BCE: Bayesian composition estimator: estimating sample (taxonomic) composition from biomarker data. R package version 1.4. http://CRAN.R project.org/package=BCE.

Van den Meersche, K., K. Soetaert and J.J. Middleburg. 2008. A Bayesian compositional estimator for microbial taxonomy based on biomarkers. Limnol. Oceanogr. Methods 6: 190–99.

Van den Meersche, K., K. Soetaert and D. VanOevelen. 2009. xsample(): an R function for sampling linear inverse problems. J. Stat. Software 30 (Code Snippet 1): 1–15.

Van Germerden, H. 1980. Survival of *Chromatium vinosum* at low light intensities. Arch. Hydrobiol. 125: 115–121.

Van Heukelem, L. and S.B. Hooker. 2011. The importance of a quality assurance plan or method validation and minimizing uncertainties in the HPLC analysis of phytoplankton pigments. pp. 195–256. *In*: S.W. Jeffrey, R.F.C. Mantoura and S.W. Wright (eds.). Phytoplankton Pigments in Oceanography: Guidelines in Modern Methods. UNESCO, Paris.

Veldhuis, M.J.W. and G.W. Kraay. 2004. Phytoplankton in the subtropical Atlantic Ocean: towards a better assessment of biomass and composition. Deep-Sea Res.-I. 51: 507–530.

Vymazal, J. and C.J. Richardson. 1995. Species composition, biomass, and nutrient content of periphyton in the Florida Everglades. J. Phycology 31: 343–354.

Waters, M.N., C.L. Schelske, W.F. Kenney and A.D. Chapman. 2005. The use of sedimentary pigments to infer historic algal communities in Lake Apopka, Florida. J. Paleolimnol. 33: 53–71.

Waters, M.N., M.F. Piehler, A.B. Rodriguez, J.M. Smoak and T.S. Bianchi. 2009. Shallow lake trophic status linked to late Holocene climate and human impacts. J. Paleolimnol. 42: 51–64.

Watson, S., E. McCauley and J.A. Downing. 1992. Sigmoid relationships between phosphorus, algal biomass, and algal community structure. Can. J. Fish. Aquat. Sci. 49: 2605–2610.

Watson, S., E. McCauley and J.A. Downing. 1997. Patterns in phytoplankton taxonomic composition across temperate lakes of differing nutrient status. Limnol. Oceanogr. 42: 487–495.

Winfree, N.M., J.W. Louda, E.W. Baker, A. Steinman and K.E. Havens. 1997. Application of Chlorophyll and Carotenoid Pigments for the Chemotaxonomic Assessment of the Waters and Surficial Sediments of Lake Okeechobee, Florida. pp. 77–91. *In*: R. Eganhouse (ed.). Application of Molecular Markers in Environmental Geochemistry. ACS Symposium Series #671, American Chemical Society, Washington, D.C.

Woitke, P., K. Hesse and J.-G. Kohl. 1997. Changes in lipophilic photosynthetic pigment contents of different *Microcystis aeruginosa* strains in response to growth irradiance. Photosynthetica 33: 443–453.

Wright, S.W. and S.W. Jeffrey. 2005. Pigment markers for phytoplankton production. pp. 71–104. *In*: J.K. Volkman (ed.). Handbook of Environmental Chemistry, Vol. 2N: Marine Organic Matter, Biomarkers, Isotopes and DNA. Springer-Verlag, Berlin.

Wright, S.W., S.W. Jeffrey, R.F.C. Mantoura, C.A. Llewellyn, T. Bjørnland, D. Repeta and N. Welschmeyer. 1991. Improved HPLC method for the analysis of chlorophylls and carotenoids from marine phytoplankton. Mar. Ecol. Prog. Ser. 77: 183–96.

Wright, S.W., R.L. van den Enden, I. Pearce, A.T. Davidson, F.J. Scott and K.J. Westwood. 2010. Phytoplankton community structure and stocks in the Southern ocean (30-80-0-E) determined by CHEMTAX analysis of HPLC pigment signatures. Deep-Sea Res.-II 57: 758–778.

Yacobi, I. and Y.Z. Ostrovsky. 2012. Sedimentation of phytoplankton: role of ambient conditions and life strategies of algae. Hydrobiologa, DOI 10.1007./s10750-012-1215-9.

Zapata, M., J.L. Garrido and S.W. Jeffrey. 2006. Chlorophyll *c* pigments: current status. pp. 39–53. *In*: B. Grimm, R.J. Porra, W. Rudiger and H. Scheer (eds.). Chlorophylls and Bacteriochlorophylls. Biochemistry, Biophysics, Functions and Applications, Vol. 25, Advances in Photosynthesis and Respiration. Govindgee (Series Ed.). Springer, Dordrecht, Netherlands.

Zapata, M., S. Fraga, F. Rodriguez and J.L. Garrido. 2012. Pigment-based chloroplast types in dinoflagellates. Mar. Ecol. Prog. Ser. 465: 33–52.

Zonneveld, C. 1998. A cell-based model for chlorophyll *a* to carbon ratio in phytoplankton. Ecol. Model. 113: 55–70.

APPENDICES*

APPENDIX A: Plant/algal pigments, selected terminology and pigment implications to Chemotaxonomy and CERP-RECOVER/MAP activities with Everglades periphyton communities.

Appendix B: Pigment Structures

Appendix B1: Structures of chlorophylls and their derivatives

Appendix B2: Structures of carotenoids

Appendix B3: Structures of scytonemin and derivatives.

---

*Appendices can be found on pages 455 to 468

14

Detecting Calcareous Periphyton Mats in the Greater Everglades Using Passive Remote Sensing Methods

Daniel Gann,[1,2,*] *Jennifer Richards,*[2,3] *Sylvia Lee*[2,3,a] and *Evelyn Gaiser*[2,3,b]

Introduction

Use of remotely sensed data for environmental and ecological assessment has recently become more widespread in wetland research and management, and advantages and limitations of this approach have been addressed (Ozesmi and Bauer 2002). Applications of remote sensing (RS) methods vary in spatial and temporal extent and resolution, in the types of data acquired, and in digital image processing and pattern recognition algorithms used. Remote sensors can acquire spatially extensive regions at high temporal frequencies using either passive or active sensors. Passive sensors include panchromatic, multi-spectral or hyper-spectral detectors; active sensors, such as RADAR (radio detection and ranging) and LiDAR (light detection and ranging),

---

[1] Geographic Information Center & Remote Sensing Center, Florida International University, Miami, Florida 33199.
   Email: gannd@fiu.edu
[2] Department of Biological Sciences, Florida International University, Miami, FL 33199.
   Email: richards@fiu.edu
[3] Southeast Environmental Research Center, Florida International University, Miami, FL 33199.
   [a]Email: slee017@fiu.edu
   [b]Email: gaisere@fiu.edu
* Corresponding author

emit signals and record the amount of returned energy after signal interaction with the earth's surface (Jensen 2005). Sensors can be mounted on different acquisition platforms, such as unmanned aerial systems (UAS or drones), airplanes or satellites.

Algal mats (periphyton) are a dominant and widespread feature of the Everglades wetlands, are major contributors to ecosystem structure and function, and are highly sensitive to changes in the environment (McCormick and Stevenson 1998; Gaiser et al. 2011). Although recent applications of RS to detect wetland vegetation in the Everglades have developed fine-scale vegetation classification systems (Madden et al. 1999; Rutchey et al. 2006; Gann et al. 2012), RS approaches for vegetation detection have yet to fully incorporate the detection of periphyton presence and estimation of its abundance. Detection and assessment of periphyton quantity, quality, and composition using RS methods would be advantageous because periphyton is a useful indicator of nutrient and hydrologic conditions in wetlands (Gaiser 2009; Lee et al. 2013). When evaluated with computer processing algorithms, remotely sensed data provides a powerful tool that can capture system-wide variability and change in periphyton patterns in a consistent and reproducible fashion. Reliable detection of periphyton presence is a crucial first step in establishing correlations between RS data and ecologically meaningful periphyton biophysical characteristics such as biomass, which in combination with periphyton composition, could serve as assessment tools for ecosystem integrity (Gaiser et al. 2009).

Although there is an apparent disconnection between the "micro" of microbiology and the scale of most remotely sensed data, Everglades periphyton forms spatially extensive mats that are detectable with RS methods. Remote sensing has been used to detect microorganisms in other environments, e.g., microbial crusts in deserts (O'Neill 1994; Schmidt and Karnieli 2000; Weber et al. 2008; Ustin et al. 2009; Kidron et al. 2012), coastal algal blooms (Richardson 1996; Tang et al. 2004; Stumpf and Tomlinson 2005; Carvalho et al. 2010; Klemas 2011; Palanisamy 2011), algal communities in oceans or lakes (Matthews et al. 2010; 2012), and cyanobacteria blooms (Kutser et al. 2006; 2008; Reinart and Kutser 2006). Environmental settings in lakes and ocean waters allow spectral signatures of algal blooms to be differentiated from the relatively stable, open water signature. Vast floating or suspended bacterial or algal mats show distinct photosynthetic pigment signatures, which separate them from signatures for deep water or sediment-loaded water (Klemas 2011; Palanisamy 2011). Detection of algal and bacterial blooms in shallow coastal waters, however, may need to address the presence of floating vegetation (Matthews et al. 2010; 2012) and reflectance of bottom substrates and submerged aquatic vegetation (Cannizzaro and Carder 2006).

Detection of seasonal algal communities (periphyton) in a wetland environment is even more challenging because the biotic and environmental conditions under which periphyton develops are more variable, and the matrix is spatially more heterogeneous. Periphyton signatures are mixed with spectral responses of diverse vegetation, as well as submerged or exposed surface materials such as peat, marl, and limestone bedrock. Our ability to use RS to detect periphyton presence and temporal dynamics in a wetland such as the Everglades depends on three components that affect the spectral properties of periphyton and its detectability. The first component concerns the complexity of the periphyton mat matrix, which can be considered its own micro-ecosystem with biotic constituents and abiotic precipitates (mainly calcium carbonate); this micro-ecosystem is spatially and temporally dynamic, varying with environmental conditions (Gaiser et al. 2009). The second component is the range of habitats where periphyton is encountered across the Everglades landscape and the phenological cycle of periphyton development in those habitats during the course of the wet and dry seasons. The third component involves detectability and separability of spectral characteristics of remotely sensed data derived from periphyton environments. For each of these three components, we consider below aspects that aid or inhibit the ability to distinguish and detect different Everglades periphyton types using remotely sensed data. We consider only the use of passive remote sensors mounted on satellites and are, therefore, interested in the reflection properties of electromagnetic radiation in the range of the spectral wavelengths emitted by the Earth's sun.

Effects of Periphyton Composition and Growth Forms on Spectral Properties

Periphyton morphological growth forms differentiated by Hagerthey et al. (2011) include cohesive, laminated mats; thin, sheet-like, desmid-rich communities that are loosely attached to substrates; amorphous clouds of filamentous green algae suspended in the water column; and epipelic crusts, epiphyton and metaphyton (Hagerthey et al. 2011). Thus, periphyton occurs in different growth forms and each form is composed of multiple constituents that affect the optical properties of absorption, transmittance and reflectance of electromagnetic radiation for different ranges along the electromagnetic spectrum. Below, we describe first the effects of periphyton composition on spectral properties, then the effects of periphyton growth form.

Periphyton Composition Spectral Effects

The most abundant biological components of Everglades periphyton can be categorized as species of (a) cyanobacteria, (b) green algae, and (c) diatoms. The periphyton-associated abiotic component with highest variability in spectral effect is precipitated calcium carbonate.

Cyanobacteria

Cyanobacteria dominate Everglades periphyton, making up about 90 percent of the total volume, and provide mat structure and specialized microhabitats for other microbes in the mat (Donar et al. 2004; Thomas et al. 2006). The photosynthetic apparatus of the prokaryotic cyanobacteria has chlorophyll *a* (absorption peaks in blue wavelengths at 435 nm and in red at 670–680 nm) and phycobilins (absorption peak at 618 nm); data are for pigments *in vivo* (Rabinowitch and Govindjee 1969). Spectral reflectance for chlorophyll *a*, therefore, is low below 500 nm, increases to a peak around 550 nm or the green portion of the spectrum, decreases around 670, but has a sharp peak in the near infra-red around 730 nm. In cyanobacterial cells, these pigments combine with phycobilin spectral reflectance in the 530–550 nm or cyan-green range of the visible spectrum (Hunter et al. 2008) (Fig. 1). Additionally, in the Everglades and similar wetlands, cyanobacteria often produce a superficial golden-brown or rusty-red layer

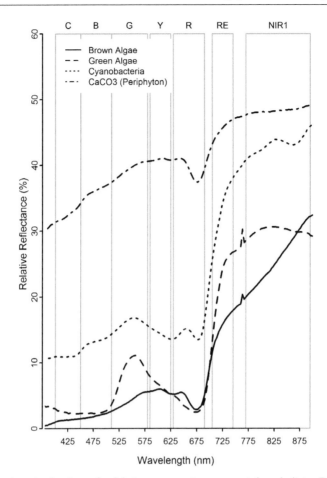

Figure 1. Spectrometer signatures of periphyton components or representative substitutes. Diatom pigment fucoxanthin is represented by the brown alga *Sargassum fluitans*, green algae signature of *Spirogyra* sp. CaCO$_3$ signature was recorded from dried, ground periphyton precipitate, and cyanobacteria signature was recorded from the undersurface of an Everglades cyanobacterial mat where the blue-green layer without scytonemin was exposed. The light gray bars in the background show sensor bandwidths for WorldView-2 satellite. C = coastal (violet) band; B = blue; G = green; Y = yellow; R = red; RE = red edge; NIR1 = near infrared 1; see Table 1 for additional data.

of the photo-protective pigment scytonemin on the surface of the periphyton mat (Sirová et al. 2006; Hagerthey et al. 2011); scytonemin absorbs strongly below 500 nm (Garcia-Pichel and Castenholz 1991) and produces an orange/rusty-red reflectance when combined with chlorophyll *a* and phycobilins.

Scytonemin is abundant in the extracellular polysaccharide matrix of the common Everglades cyanobacterial species *Scytonema hoffmannii* and *Schizothrix calcicola*, especially under high light conditions (Smith 2009). However, the typical cyan-green color of cyanobacteria is preserved in the interior of attached periphyton mats, as well as the underside of floating mats, which may become exposed when mats are disturbed.

Green Algae species

Filamentous green algae such as *Mougeotia*, *Spirogyra*, and *Bulbochaete* species are abundant in the Everglades in loosely attached, floating periphyton (Gottlieb et al. 2006; Hagerthey et al. 2011). Desmids are more common in deep soft-water environments, such as Loxahatchee National Wildlife Refuge and sloughs in Big Cypress National Park, where a deep peat layer impedes direct interaction between the limestone bedrock and water column. Filamentous species are also common in deep environments but are indicators of phosphorus enrichment. Green algae, like plants, have chlorophyll *a*, chlorophyll *b* and carotenoids as photosynthetic pigments. Thus, green algae resemble plants in absorbing light mainly in the blue (435 nm peak for chlorophyll *a*, 480 nm peak for chlorophyll *b*, 440–515 nm for carotenoids) and red (670–680 nm for chlorophyll *a*, 650 nm for chlorophyll *b*) (data are for pigments *in vivo*, (Rabinowitch and Govindjee 1969). Spectral reflectance for green algae, therefore, is as described for cyanobacteria but lacking strong reflectance in the yellow-red portion of the visible spectrum (570–675) nm range (Lee et al. 2011) (Fig. 1).

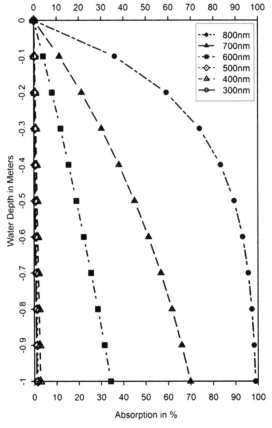

Figure 2. Attenuation for different wavelengths of light (300–800 nm) in water from 0 to 1 meter depth. Graph was generated using Lambert-Beers Law under the assumption of clear shallow water with solar and sensor zenith incidence angles.

Diatom species

The diatom component of Everglades periphyton has received much attention because diatoms have morphologically complex, siliceous cell walls that make them readily identifiable, and their community composition responds rapidly to environmental change, making them useful as environmental indicators (Gaiser et al. 2006; Gottlieb et al. 2006; Gaiser 2009). The environmental stresses imposed on Everglades diatoms, such as the seasonally fluctuating hydrology and very low nutrient levels, maintain a relatively less-diverse assemblage compared to more temperate systems (Gaiser et al. 2011). However, a characteristic assemblage of diatoms has been identified for the Everglades and similar wetlands in the Caribbean, which may indicate endemism and evolution of biological dependence on the periphyton mat habitat (Slate and Stevenson 2007; Gaiser et al. 2011; La Hée and Gaiser 2012; Lee et al. 2014). Diatoms indicative of unenriched conditions in the Everglades include *Mastogloia calcarea*, *Encyonema evergladianum*, and *Fragilaria synegrotesca*. In contrast, more diverse assemblages that are presumably not endemic to the region occur in enriched environments, where the cohesive, native periphyton mats are not dominant (Gaiser et al. 2005; Gaiser et al. 2011). Diatoms indicative of enriched conditions include *Eunotia incisa*, *Nitzschia amphibia*, and *Gomphonema parvulum*. Diatoms have the photosynthetic pigments chlorophyll *a* and *c* (absorption peaks at 445 nm and 645 nm (Rabinowitch and Govindjee 1969), as well as fucoxanthin (absorption peaks at 425, 450 and 475 in hexane, 550 nm *in vivo* (Rabinowitch and Govindjee 1969); these three pigments account for approximately 70 percent of diatom photosynthetic pigments (Brown 1988). Chlorophyll *a* is the most abundant pigment in diatoms, with ratios of pigment to chlorophyll *a* being 0.3–0.4 for fucoxanthin and 0.09–0.1 for chlorophyll *c* (Méléder et al. 2003). The combined absorption peaks of these pigments, as well as other factors, produce spectral reflectances for diatoms at 570–610 nm or the yellow portion of the visible spectrum; this reflectance typically has a broad major peak around 570–600 nm; there is a secondary peak, however, around 650 nm (Hunter et al. 2008; Lee et al. 2011) (Fig. 1, where we have used brown algal reflectance as a proxy for diatom reflectance).

Calcareous Inorganic Matter

Calcium carbonate is a prominent component of Everglades periphyton mats, as calcium carbonate can accumulate on the surfaces of cyanobacterial filaments by abiotic precipitation or biotic uptake and reprecipitation (Gleason and Spackman 1974; Merz 1992; Browder et al. 1994; Hagerthey et al. 2011). The reflectance of these mats becomes especially prominent in the dry season, when the mat dries out to form a white crust, since the uppermost layers of periphyton mats contain the highest density of calcium carbonate (Donar et al. 2004). This calcium carbonate contributes to formation of marl soil found in some parts of the Everglades (Gleason and Spackman 1974; Gottlieb et al. 2006). Exposed limestone bedrock can also produce a spectral signature composed primarily of calcium carbonate reflectance. Pure calcium carbonate reflects strongly in the visible spectrum, with a gradual increase in reflectance from

400 to 700 nm, followed by a gradual decline (Figs. 1 and 3). Similar spectra have been found for relatively homogeneous calcium carbonate-derived ocean sediments (Hochberg 2003; Louchard et al. 2003).

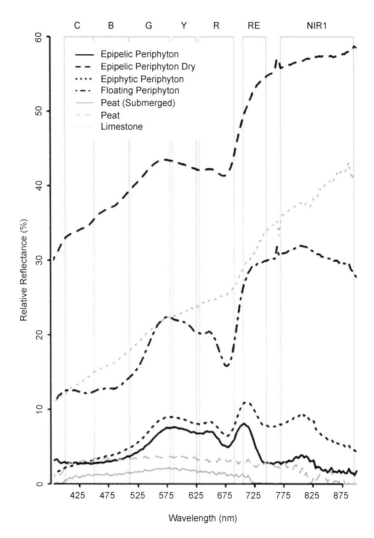

Figure 3. Field spectroradiometer spectra of Everglades periphyton types and common substrates sampled in northeast Shark River Slough, Everglades National Park. Dry periphyton signature from epipelic periphyton after 21 days of drying; limestone signature from an exposed limestone rock; peat from exposed but hydrated sample; all other media sampled *in situ*. Background shading and band abbreviations as in Fig. 1.

Periphyton Growth Form Spectral Effects

Biotic microbial constituents and abiotic precipitates of a periphyton mat matrix are spatially and temporally dynamic and vary in association with their environmental conditions. In unenriched environments of the Everglades, low nutrient availability, high abundance of carbonate ions from dissolution of limestone, and frequent desiccation encourages the growth of periphyton mats that are dominated by cyanobacteria (Gaiser et al. 2011; Hagerthey et al. 2011), particularly by *Scytonema hoffmannii* and *Schizothrix calcicola* (Browder et al. 1994). The extracellular polymeric substances (EPS) that coat filaments or encase coccoid forms of cyanobacteria contribute to mat cohesion and provide the main structure of thick, sometimes laminated, periphyton mats (Stal 1995), although some EPS-secreting diatoms also contribute to the mat matrix (Gaiser et al. 2010). Calcium carbonate can accumulate on the surfaces of cyanobacterial filaments by abiotic precipitation mediated by the EPS or biotic reprecipitation after inorganic carbon acquisition from the water column (Merz 1992; Browder et al. 1994; Hagerthey et al. 2011). In enriched Everglades environments, higher phosphorus concentrations promote disintegration of the mat structure and an assemblage dominated by solitary, non-colony-forming diatoms and green algae, in particular *Spirogyra* and *Mougeotia* (Gaiser et al. 2006; Gottlieb et al. 2006). This type of periphyton can be loosely attached to substrates or form green, gelatinous clouds that are more typical of less alkaline, mesotrophic or eutrophic aquatic systems (Hagerthey et al. 2011); these algae also appear in oligotrophic environments at sites of localized enrichment or nutrient pulses.

Cohesive, calcium-rich mats can be spatially continuous and are more likely to be detectable by RS methods than loose mats. Because the composition of photosynthetic pigments differs among autotrophs, differences in the relative abundances of cyanobacteria, diatoms, and green algae in these two periphyton types are expected to produce differences in spectral signatures in RS imagery. Spectral signatures of pigments or compounds are also modified by the context in which they are organized in living organisms (e.g., differences in pigment absorption *in vitro* vs. *in vivo*) and by the consortia in which the organisms occur. Additionally, environmental conditions can influence pigment abundance, e.g., high light vs. low light responses (Méléder et al. 2003), and species composition can change in response to environmental changes (Gaiser et al. 2005; Lee et al. 2013). The periphyton spectral signature seen by RS sensors reflects these spectral properties in a cumulative, integrative manner.

Effects of Environmental and Phenological Cycles on Periphyton Spectral Reflectance Properties

Detection of periphyton assumes that there is a distinct periphyton spectral signature when compared to vegetation and substrate reflectance. However, periphyton occurs in different habitats, where it is interspersed or attached to a variety of plant species (epiphytic) and substrates (epipelic). It frequently forms below the water column, attached to submerged aquatic or emergent vegetation such as *Eleocharis* spp. or *Cladium jamaicense*, or it can be attached to peat or limestone substrates. Epiphytic periphyton also occurs attached to and/or interspersed with floating vegetation such as *Utricularia* spp. and *Nymphaea odorata* (Gaiser et al. 2011). These different habitats affect the spectral properties of the periphyton in different ways. Epipelic and epiphytic periphyton, for instance, occur along a gradient of water depths, and because the light attenuation by water differs among wavelengths (Fig. 2), the spectral reflectance intensities are altered differently by depth for the various wavelength intervals (bandwidths) observed by a remote sensor. Likewise, floating mats, which are loose aggregations that are richer in green algae than epipelic mats and commonly occur with floating vegetation, are expected to have more plant-like spectral signatures than the cohesive, epipelic mats in shallower and more desiccation-prone regions.

Wetting and desiccation of periphyton during flooding and drying episodes of a short-hydroperiod marsh alter periphyton composition and growth form (Gottlieb et al. 2006) and provoke responses in pigment quality and quantity, which in turn lead to a change in reflectance properties. The onset of the dry season initially promotes high growth rates of desiccation-adapted cyanobacteria (Komárek et al. 2014, this volume), until density-dependent competition for nutrients likely becomes a limiting factor in the growth of late successional benthic mats (Stevenson et al. 1991). Then, in the late dry season, desiccation of the periphyton matrix exposes the accumulated calcium carbonate and other abiotic materials as a white layer on the surface of the dried periphyton crust. Complete drying and decomposition often occurs in very shallow water or when the water table falls below the sediment surface. The change in reflective properties of epipelic and epiphytic periphyton will, therefore, be a combination of decreasing reflectance from photosynthetic pigments and organic tissue and increasing exposure of calcium carbonate. As cyanobacteria dry, they produce the photo-protective pigment scytonemin, which changes the visual reflectance properties from a light blue-green or olive green appearance to a golden brown/rusty red (Hagerthey et al. 2011). In conjunction with decreasing water depth and exposure of the mats above the water column, this development is expected to increase reflectance across the full spectrum, with the change in reflectance increasing from the shorter to longer wavelengths as the water level decreases. Completely exposed calcium carbonate on the surfaces of desiccated periphyton mats is expected to become the dominant component in the periphyton signature toward the end of the dry season, leading to significant increase in reflectance across the full visible spectrum (bright white), and especially in the near-infrared portion of the spectrum.

A similar spectral change can be observed in vegetation. As vegetation undergoes water stress and wilting, the spectral reflectance will also increase throughout the full spectrum and especially in the near-infrared. The phenological cycle of periphyton therefore needs to be considered in the broader context of phenological cycles of wetland plants, and detection and classification algorithms using RS data must be able to distinguish spectral signatures of locations where the periphyton types are present from locations where periphyton is absent. Consequently, when acquiring and interpreting RS data used to detect and classify periphyton, it is important to consider the phenological cycles of periphyton and associated vegetation, as well as the hydrological cycles of the system.

Another aspect of periphyton cycles affecting spectral reflectance is vertical movement of periphyton mats. Cohesive cyanobacteria-dominated mats can start out as submerged epipelic mats attached to substrate (peat or marl); gas production within the mat can increase buoyancy to a point where patches of periphyton detach from the benthic layer and rise to the water surface, where they continue to grow until reaching maximum biomass in the mid-dry season (Komárek et al. 2014, this volume). Well-developed periphyton mats can also exhibit vertical movement in the water column as gas builds up during the day and is lost at night (Iwaniec et al. 2006), or when heavy precipitation pushes the mats below the water surface (Komárek et al. 2014, this volume). In contrast, floating mats in long-hydroperiod marshes that do not dry down or dry down infrequently persist from year-to-year, although composition in these mats varies seasonally, and the mats undergo a winter reduction in biomass (Gottlieb 2003). Composition of benthic and floating mats differs significantly, although the difference is primarily one of abundance rather than presence/absence (Gottlieb 2003; Gottlieb et al. 2006). Differentiating floating mats that arise as floating vs. epipelic mats is only possible if they have a spectrally-significant difference due to compositional differences or different amounts of precipitated calcium carbonate, and if the RS sensor is capable of recording at radiometric and spectral resolutions that capture these differences.

The interpretation of spectral signatures of epiphytic periphyton is even more complicated, since water depth varies along epiphytic sweaters from the bottom of the plants to just below the water surface and, in some cases, extending above the water surface. The relative surface exposure in different depths of the water column affects the spectral reflectance properties, and the signature is, therefore, expected to resemble a mix of floating and benthic mats. In addition, epiphytic periphyton is attached to emergent vegetation (e.g., graminoids and, to a lesser degree, broadleaved plants), which confounds the pure periphyton signature by adding a significant component of the plant signature to which it is attached. Finally, similar to epipelic periphyton transforming into floating mats, epiphytic periphyton sweaters can turn into carpets of floating cylinders when the host plant dies and detaches from the substrate (Thomas et al. 2002).

This narrative of growth forms and phenological cycles suggests there are two periphyton spectral gradients with associated confusion and detection probabilities. The first gradient ranges from shallow to deep submerged epipelic and epiphytic periphyton and will most likely invite confusion at the shallow end of this gradient, where submerged or partially exposed epiphytic periphyton might be misinterpreted as calcareous floating mats.

The second periphyton spectral gradient ranges from wet, exposed, floating mats to dry decomposed matter, where the most anticipated confusion is expected to occur between exposed rock substrate (limestone) and dry decomposed periphyton mats, and between floating mats and exposed live mats (former floating or epipelic during the dry season's receding water level). Some of this confusion can be eliminated by analyzing the data as a multi-seasonal record that allows interpretation of the dry form as a result of its origin in the wet season. Using wet and dry sets of data to discriminate among periphyton types, however, is not expected to completely eliminate confusion—some residual confusion will always remain if the floating mats move or are disturbed between data acquisition dates.

The presence or absence of water and the biological responses of periphyton to water provide an inherently strong RS signal for the different periphyton types. Differentiating periphyton types and eliminating confusion of submerged periphyton from floating mats during the wet season is expected to be possible due to the absorption of a large percentage of the near-infrared (NIR) radiance by the water column above the submerged periphyton, leading to lower NIR reflectance when compared to a floating mat of equal biotic and calcareous composition. If the water column is not too deep, this might be the single best indicator to separate floating from epipelic or epiphytic periphyton. As the water gets deeper, the water column will eventually attenuate the radiance across the full electromagnetic spectrum to the point at which the benthic layers are not detectable; benthic mats, however, are unlikely to occur in deeper water (> 1 m) because of light limitation. The difference between wet floating and dry exposed periphyton will partly be dependent on the degree of moisture and stage of decomposition of the periphyton organic matter, as well as the amount of calcification.

Effects of Remote Sensor Specifications and Limitations on Periphyton Detection

Using remote sensing methods to detect and monitor periphyton relies on the resolving power of the sensor to capture the spectral variability of the different periphyton types in contrast to other vegetation and abiotic land-cover classes (e.g., limestone, peat). In choosing adequate sensors, it is important to understand how spatial, spectral, radiometric and temporal resolution of remotely sensed data relates to the spectral properties of periphyton types in their respective environments and during their phenological cycle. Detecting periphyton and monitoring its changes with RS methods needs to be evaluated with regard to resolution characteristics of existing remote sensors and the data they produce.

Spectral and Radiometric Resolution

The biotic component of Everglades periphyton that differentiates it from other vegetation is the presence of cyanobacteria with their photo-protective pigment scytonemin, of diatoms, and of calcium carbonate precipitates; each of these has characteristic reflectance properties, as described above. To detect periphyton that has an abundance of diatoms, or that produces scytonemin, a sensor with a spectral band in the yellow wavelength range will be advantageous. Since attenuation in water is reduced for shorter wavelengths, shorter wavelength spectral bands that penetrate deeper into the water column allow for differentiation of submerged materials. Thus, sensors with a spectral resolution that includes bands in the shorter blue wavelengths are also advantageous.

Radiometric resolution reflects the information content in a given pixel. High radiometric resolution allows for differentiation of subtle differences in the reflective properties of a landscape. For example, an image in only black and white conveys less information than an image with 256 gray tones. The higher the radiometric resolution, however, the larger the data volume, which translates into more data intensive, and hence more costly, data management in terms of data processing time and storage capacity.

The minimal spectral and radiometric resolutions required of a sensor will be those that differentiate signatures of materials of interest. These resolutions, in turn, are determined by the magnitude of difference in spectral reflectance properties and by the portion of the spectrum where these differences are maximal.

Spatial and Temporal Resolution

The spatial resolution of a sensor system defines two aspects of RS data: the grain size equivalent to the smallest objects that are to be detected (i.e., smallest periphyton patches to be detected) and the extent of the study area (i.e., entire wetland or small subareas). The instantaneous field of view (IFOV) of a sensor determines the spatial resolving power of the sensor, and the relative abundance of macrophytes, algae, open water and periphyton within the IFOV will determine the spectral signature mix. For a landscape with a highly heterogeneous land cover, the sensor must have greater spatial resolution to be able to resolve relatively homogenous patches. The Landsat TM satellite IFOV produces 30 x 30 m pixels; the SPOT 5 satellite produces 10 x 10 m pixels; the WorldView-2 satellite produces 2 x 2 m pixels. The RS data from these different satellites will record very different spectral reflectance mixes per pixel for the same site. A sensor with a higher spectral resolution whose range reaches into the infra-red generally has a coarser spatial resolution, but if signatures of interest are most distinct in the longer wavelengths, a trade-off between spectral and spatial resolution might be advantageous.

Based on these considerations, the optimal choice of sensor for periphyton detection and monitoring is a sensor that can deliver a spectral resolution that allows differentiation of cyanobacteria or green-algae-dominated periphyton from plants and from each other and that can penetrate water to capture reflective characteristics of submerged periphyton. In addition, the spatial resolution of the sensor needs to be able to resolve small but abundant patches of periphyton in order to avoid underestimating periphyton presence and extent. In summary, detection of presence and abundance of periphyton will depend on (1) the identification of spectral reflectance properties that differentiate periphyton types; (2) differentiation of periphyton spectral properties from spectral reflectance of other vegetative or abiotic land-cover types; and (3) the remote sensor's capacity (radiometric, spectral, spatial and temporal resolutions) to resolve these spectral reflectance differences.

All three components apply to detection at a single point in time and to monitoring through time. For monitoring change over time, temporal metrics (i.e., difference of seasonal or intra-annual vs. inter-annual variability) need to be defined for ecologically meaningful intervals. If the goal is to analyze changes in spatial coverage of periphyton throughout a season, temporal resolution of the RS data (i.e., frequency of data acquisition) becomes more critical, since seasonal phenology (e.g., growth, senescence), water depth and nutrient changes, but also the movement of floating periphyton mats, e.g., by wind, surface flow, or fauna, need to be considered. Understanding these variations will allow us to correctly interpret images taken at any time of the year, and phenological and other natural variations will not be misinterpreted as increase or decrease in abundance.

Below we present results of a preliminary study on Everglades periphyton detection using RS methods from WorldView-2 satellite (WV2) data sets. WV2 satellites have a multi-spectral sensor (8 spectral bands) that includes a short wavelength blue (coastal) spectral band, as well as a yellow band (Table 1); WV2 spatial resolution is 2 x 2 m.

In our study, we first gathered and analyzed spectral reflectance signatures acquired by a portable spectrometer (< 10 nm bandwidth). We acquired and evaluated signatures

Table 1. Spectral and spatial characteristics of the 8 reflective WorldView 2 bands (Updike and Comp 2010).

Band Name	Center Wavelength (nm)	50% Band Pass (nm)	5% Band Pass (nm)	Spatial Resolution (m)
(1) coastal	427	401–453	396–458	2
(2) blue	478	448–508	442–515	2
(3) green	546	511–581	506–586	2
(4) yellow	608	589–627	584–632	2
(5) red	659	629–689	624–694	2
(6) red edge	724	704–744	699–749	2
(7) near-infrared 1	831	772–890	765–901	2
(8) near-infrared 2	908	862–954	856–1,043	2

of periphyton types, associated vegetation and abiotic factors in their environments. Using these reflectances as background, we then acquired and qualitatively interpreted RS data (i.e., RS spectral reflectances) for floating periphyton and associated vegetation and abiotic factors from a ridge and slough landscape represented by a site in the southern portion of Water Conservation Area 3 (WCA-3) and epipelic mats and associated vegetation and abiotic factors from a wet prairie landscape represented by the northeast Shark River Slough (NESRS) in the northeastern corner of Everglades National Park (ENP) (Fig. 4).

We also compared spectral signatures acquired with the spectrometer to signatures derived from the WV2 data. Finally we mapped periphyton presence and evaluated map accuracies within a 3 km² study area in the ridge and slough landscape of WCA3A. The questions we addressed in this study were

1. How do spectral signatures of submerged and dry epipelic periphyton mats and floating periphyton differ, and how do they compare to vegetation and bare substrate?
2. How do the signatures acquired by a spectrometer compare to spectral signatures acquired by a satellite-mounted multi-spectral sensor?
3. Can periphyton types be distinguished among themselves and from vegetation and abiotic components of the environment using WV2 data?
4. Using WV2 data, with what accuracies can we detect periphyton presence and differentiate it from vegetation and abiotic classes in an Everglades ridge and slough landscape?

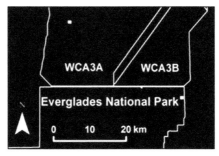

Figure 4. Map of northern Everglades National Park (ENP) and southern Water Conservation Areas (WCA) 3A and 3B showing locations of ridge and slough (WCA3A) and wet prairie (ENP) where periphyton and other signatures (Fig. 5) were extracted from WV2 data (white rectangles).

We discuss lessons-learned and ways to move forward with using RS data to estimate areal coverage and spatial and temporal variability in monitoring periphyton in the larger Everglades ecosystem.

Materials and Methods

We determined spectral reflectance properties of Everglades periphyton types and common substrates using a UniSpec-DC spectrometer (PP Systems International, Inc., Amesbury, MA); with a spectral resolution of < 10 nm, for a wavelength range of 310–1,100 nm. Spectral reflectance measurements of epipelic, epiphytic and floating periphyton of epipelic and epiphytic origin were collected in the northeast Shark River Slough, Everglades National Park in September–December 2013. A periphyton sample was returned to the lab and allowed to dry, with spectral signatures recorded throughout dehydration. The sample was completely dry after 21 days. In the field we also measured spectral reflectances of common substrates, including peat (both in air but hydrated and submerged in 50 cm water) and limestone, sampled from an exposed rock. Data are reported as averages of three consecutively acquired spectra, with each acquisition averaging 10 spectral scans.

We evaluated the detection and separability of periphyton types from atmospherically corrected satellite data acquired by the multispectral sensor of the WorldView-2 (WV2) satellite (Digital Globe, Inc., Longmont, CO). For a bi-seasonal signature evaluation satellite data were acquired on 6 November 2010 for the wet season and on 5 May 2011 for the dry season. Using this data, we examined periphyton in a ridge and slough landscape in Water Conservation Area 3A (WCA3A) and in a wet prairie landscape in northeast Shark River Slough (NESRS) (Fig. 4). Periphyton types, associated vegetation and environmental conditions were taken into consideration when developing a class scheme. The classes we considered in this analysis were epiphytic floating periphyton, *Nymphaea odorata* in slough, submerged epipelic periphyton, and submerged peat in wet prairie.

Signatures for the wet and dry season comparison were extracted from the wet season 2010 and dry season 2011 images for representative samples; sample selection was guided by visual interpretation of aerial photography from 2009 and 2011 and by field data samples collected during multiple campaigns in NESRS and WCA 3A between 2010 and 2012 (Gann et al. 2012).

When mapping periphyton presence in the ridge and slough landscape, we differentiated periphyton dominant from periphyton present but secondary, *Nymphaea*, open water with submerged aquatic vegetation (SAV), vs. all other ridge/tree island vegetation types. For training sample selection and design-based accuracy assessment we utilized very high resolution aerial photography (~3 cm) acquired by an unmanned aerial system (UAS; aka drone) developed and operated by the UF/IFAS UAS Research group (http://uav.ifas.ufl.edu/index.shtml). Photographs were acquired in August 2012 in WCA 3A (Gann and Richards 2013). In order to match the reference data source (geo-referenced high-resolution photography mosaic), we atmospherically and geographically corrected a WV2 image acquired on 20 October 2012. The study site covered a 3 km^2 area for which we had two 1 km^2 high resolution mosaics of geo-referenced aerial photography on the east and west. We used a random forest

(Liaw and Wiener 2002) classifier to predict periphyton presence using the percent spectral reflectance values of the atmospherically corrected WV2 data. Random forest classifiers are based on the recursive partitioning and random forest principles pioneered by Breiman (Breiman 2001) and have been successfully used in Everglades wetland vegetation mapping applications (Gann et al. 2009; 2012). We set the number of decision trees to be constructed to 10,000 with 3 variables randomly selected at each decision node of a tree. The 2 x 2 m classified map was aggregated to a minimum mapping unit of 20 m^2 (5 contiguous cells) before the accuracy assessment was performed. We applied a post-classification stratified-random sample design to evaluate overall and class-specific map accuracies. Based on a multinomial distribution for 5 classes we determined that a total number of 575 samples was adequate to evaluate a map accuracy with a 95 percent confidence and a precision of 5 percent (Congalton and Green 1999). Samples were evaluated from the 3 cm high-resolution mosaic.

Results

1. How do spectral signatures of submerged and dry epipelic periphyton mats and floating periphyton differ, and how do they compare to vegetation and bare substrate?

   Spectral reflectance properties of different Everglades periphyton types and common substrates differ in relative intensity overall and for different parts of the spectrum (Fig. 3). Peat has the lowest reflectance, which is relatively uniform across the visible spectrum; water (submergence) reduces the peat's overall reflectance and greatly decreases it above 725 nm. Exposed limestone has a very high reflectance that increases relatively linearly from 375 to 875 nm. All of the periphyton types have high reflectance between 570 and 650 nm, a decrease in reflectance around 675 nm, and an increase that peaks or begins to level at 720–730 nm. The signature above 730 nm flattens (floating and dry periphyton) or decreases (epipelic and epiphytic submerged periphyton) (Fig. 3). The decrease above 675 nm for epipelic and epiphytic periphyton parallels the decrease in peat reflectance when it is submerged and provides a signal for submergence. The broad range of reflectance from 570 to 650 nm, as well as the secondary peak at 670 nm, for the different forms of periphyton is found in cyanobacteria and brown alga (and diatom) spectral signatures (Fig. 1). The intensity of the secondary peak varies among the periphyton samples, being strongest in the floating periphyton (Fig. 3). The sharp decrease at 675 nm, then rapid increase around 730 nm, reproduces the expected chlorophyll signature. The dried periphyton, which has the highest overall relative reflectance, has a spectral signature that integrates the floating periphyton signature with the calcium carbonate signature (Fig. 3).

2. How do the spectral signatures acquired by a spectrometer compare to spectral signatures acquired by a satellite-mounted multi-spectral sensor?

   The WV2 spectral reflectances integrate reflectances across band widths that range from 400 to 700 nm in the visible to red-edge spectrum, with even larger bandwidths in the infrared (Fig. 5A), reducing the fine-scale detail seen in the spectrometer data (Fig. 3 vs. 5A). Additionally, because the WV2 sensor samples a

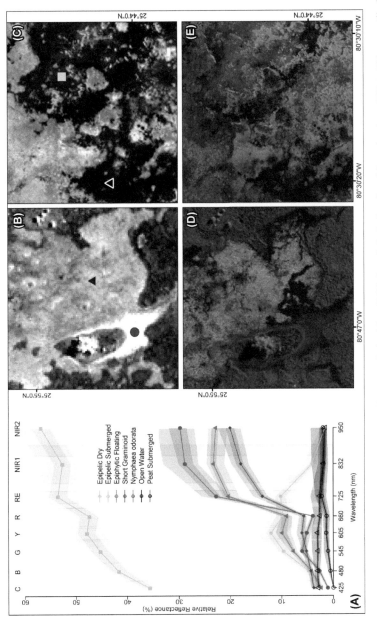

Figure 5. (A) Selected spectral signatures of slough and wet prairie classes derived from WV2 data; dark shades of confidence bands are 25th to 75th percentile; lighter shades are below 5th and above 95th percentiles. Background shading and band abbreviations as in Fig. 1. (B, C) WV2 imagery showing extraction locations of signatures for epiphytic floating periphyton (triangle) vs. *Nymphaea odorata* (circle) in slough within WCA3A (B) and submerged epipelic periphyton (square) vs. submerged peat (triangle) in wet prairie within ENP (C). (D, E) Corresponding aerial photography for B (D) and C (E). Sample locations given in Fig. 4.

Color image of this figure appears in the color plate section at the end of the book.

much larger area (2 x 2 m pixels) than the spectroradiometer, the WV2 signatures are more variable, reflecting environmental and biological heterogeneity (confidence bands, Fig. 5A). The periphyton secondary peak at 670 nm is lost in the WV2 reflectances, as is the sharp decrease in reflectance at 675 nm and rapid increase at 730 nm that comes from chlorophyll (periphyton signatures in Fig. 3 vs. Fig. 5A). In the WV2 data, these peaks contribute to a general peak in the yellow, decreased reflectance in the red, and greatly increased reflectance in the red-edge (Fig. 5A). The decrease in reflectance above 730 seen in epipelic periphyton spectrometer signature, as compared to the leveling out or continued increase above 730 nm seen in floating periphyton (Fig. 3), becomes a distinct difference between these two types of periphyton in the WV2 red-edge band (high for epiphytic floating, lower for epipelic submerged, Fig. 5A). The difference in WV2 spectral reflectance for floating and epipelic periphyton becomes especially distinct when the value for the red-edge band is compared to the red band (Fig. 5A); this difference could form the basis of an index to distinguish these two types in the WV2 data.

3. Can periphyton types be distinguished among themselves and from vegetation and abiotic components of the environment using WV2 data?

The WV2 data accurately reflect landscape features seen in the visible spectrum (Fig. 5, B vs. D and C. vs. E). Both wet and dry season WV2 satellite imagery delineate areas of epipelic and floating periphyton identifiable from the visible wavelengths in aerial photography. In the WV2 data, the abiotic components of the environment (open water and submerged peat) are easily distinguished from the biotic components by their low relative reflectance and lack of distinct peaks in any spectral band (Fig. 5A). The floating aquatic plant *Nymphaea odorata* has a typical plant spectral signature that is low in the coastal (violet) and blue bands, has a peak in the green band that falls to a low in the red band, then has a steep increase in reflectance in the red-edge and increases even more in the near-infrared (Fig. 5A). The three periphyton classes (epipelic dry, epipelic submerged and epiphytic floating) differ from the *N. odorata* signature in having high reflectance in the yellow band, as compared to green (Fig. 5A). These three types of periphyton have very different spectral signatures (Fig. 5A). Epipelic submerged has highest reflectance in the yellow band, with lower reflectance in the red-edge and very low reflectance in the near-infrared. This drop in the near-infrared, which results from absorption by water, differentiates it from the floating and epipelic dry periphyton. These latter two both have high relative reflectance in the red-edge and highest in the near-infrared. The epipelic dry signature differs from the floating periphyton in the much higher reflectance in every spectral band. The spectral signature for short graminoids is intermediate between the *N. odorata* signature and that of floating epiphytic periphyton (Fig. 5A). Its reflectance in the green, yellow and red bands is more similar to the periphyton signatures, perhaps reflecting some periphyton associated with these graminoids, while its reflectance in the red-edge and near-infrared is more similar to *N. odorata*.

4. Using WV2 data, with what accuracies can we detect periphyton presence and differentiate it from vegetation and abiotic classes in an Everglades ridge and slough landscape?

The WV2 map of periphyton and other classes captured the distribution and spatial heterogeneity of periphyton in this ridge and slough landscape (Fig. 6). Overall and class-specific accuracies for floating periphyton and associated vegetation classes in a ridge and slough landscape was very high (Table 2). Design-based overall accuracy of the map was 93.7 percent. The class with the highest accuracy was floating periphyton (95.7 percent), while the lowest accuracy was recorded for the *N. odorata*-floating periphyton mix class (90.7 percent, Table 2). This mixed class was confused only with the two dominant classes *N. odorata* (5.1 percent) and periphyton (4.2 percent) that it was composed of (Table 2). The ridge class was most frequently confused with open water (4.9 percent), with a small amount of confusion with *N. odorata* (< 1 percent, Table 2).

Slough and ridge classes were mapped at fairly balanced proportions (56 percent slough classes vs. 44 percent ridge). The mixed periphyton-*N. odorata* was 38 percent of the combined slough classes, followed by dominant periphyton with 31 percent, and dominant *N. odorata* with 25 percent (Table 2). The periphyton dominant class was encountered mainly in the eastern portion of the study area, whereas the sloughs of the western area were dominated by a periphyton-*N. odorata* mix class (Fig. 6).

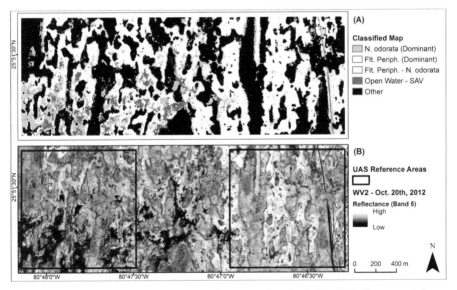

Figure 6. (A) Classified map of 3 km² study area in WCA3A (location in Fig. 4); the five mapped classes are water lily (*N. odorata* (Dominant)), floating periphyton (Flt. Periph. (Dominant)), mixed water lily-floating periphyton (Flt. Periph. – *N. odorata*), open water that may have submerged aquatic vegetation (e.g., *Utricularia* spp.) (Open Water – SAV), and ridges and other vegetation (Other). (B) WV2 near-infrared percent reflectance for area mapped in (A).

Table 2. Contingency matrix of design-based post-classification map accuracy estimates (n = 575). Accuracies are reported as column percentages. Diagonal elements (grey) are class-specific accuracies, and off-diagonal elements are exclusion errors for the class in that column.

Class Name	*N. odorata* (Dom.)	Flt. Periphyton (Dom.)	Flt. Periphyton + *N. odorata*	Open Water + SAV	Ridge Classes	Pct.	Pct. of Slough
N. odorata (Dom.)	92.1	0.0	5.1	2.8	0.8	14.3	25.5
Flt Periphyton (Dom.)	0.0	95.7	4.2	0.0	0.0	17.1	30.5
Flt Periphyton + *N. odorata*	1.8	4.3	90.7	0.9	0.0	21.2	37.8
Open Water + SAV	6.1	0.0	0.0	96.2	4.9	3.5	6.26
Ridge Classes	0.0	0.0	0.0	0.0	94.3	43.8	

Discussion and Conclusions

Monitoring periphyton in the greater Everglades presents major challenges because of the high temporal and spatial variability in periphyton presence and abundance in all its growth forms and composition variations, the natural intra-annual and inter-annual temporal dynamics, and the large spatial extent of the Everglades wetland. We have presented theoretical rationale, as well as preliminary reflectance analyses, that suggest that it is possible to successfully detect periphyton types at different stages in their phenological cycles throughout the wet and dry season. This will allow researchers to ask new questions, such as are periphyton location and extent stable on a landscape scale inter-annually, and what is the connectivity of periphyton mats across the landscape?

On-going hydrologic restoration is expected to change existing and historic patterns of periphyton type presence and abundance (Gaiser 2009; Gaiser et al. 2011), which requires a relatively flexible, rapid and affordable method to monitor these patterns. Remote sensing, when supported by reliable field measurements, offers data acquisition and analysis techniques that allow for spatially explicit modelling and monitoring of periphyton dynamics across entire landscapes. Remote sensing can generate information over large spatial extents at high spatial and temporal resolution without disturbing the environment. The amount of surrogate data measurements that are captured by remote sensors are unparalleled, and if a strong correlation exists between the remotely sensed data and the actual variables of interest, the results have a higher accuracy and greater confidence. In addition, the spatially explicit nature of the information may allow for new insights, such as what particular environments or habitat types are co-variates, or what landscape-level spatial patterns occur.

The first step in using RS techniques is acquiring the data. Although some satellites capture images on a regular basis (e.g., Landsat images taken every 16 days), others require tasking, i.e., an image is taken and stored only if requested. An advantage of RS, however, is that once acquired, the data obtained from RS imagery provides a permanent record that can be re-used; thus, the data can be analyzed repeatedly using different algorithms and having different purposes. This aspect provides both reproducibility and flexibility in data analysis. This, combined with the repeatability of data capture, makes RS especially attractive for analyzing trends.

Remotely sensed data sets can also help us to understand the effects of major environmental drivers and stressors, such as fires and hurricanes, on seasonal and inter-annual periphyton variability. These drivers and stressors affect large spatial extents, and RS data is arguably the only way to capture their landscape effects. Of particular interest in the Everglades is the potential to track changes in the Everglades coastal ecotone in relation to sea level rise. Periphyton is abundant in the freshwater-marine transition zone, producing a bright RS signature, but is scarce in the mangroves, which have a typical vegetation signature. Thus, RS methods can provide large extent data on where and how this ecotone is changing, expediting calculations of the rate of biological change and potentially providing insight into what accelerates or delays the transition.

Requirements for Biomass and Biophysical Parameter Estimation Using Remote Sensing Techniques

Periphyton type, presence or absence is only one characteristic of interest in periphyton research. A second characteristic is estimation of the biophysical parameter biomass. Most of our current data on periphyton presence, abundance, and composition comes from field data collected from ≤ 1 m^2; this data may include species composition, biomass, biovolume, and/or TN, TP, and organic content. Sampling extent and distribution varies with each particular study. A major challenge in using RS techniques is to connect this detailed plot scale of biological information (e.g., biomass) to the landscape scale of the remote sensing data (Mutanga et al. 2012). To relate periphyton biophysical data acquired and derived from 1 m^2 field samples to spectral reflectances for each pixel in a remotely sensed data set requires that the field data be scaled appropriately to the RS data pixel size and that the field sample location has sufficient spatial accuracy to be matched to the RS data location. Remotely sensed data is generally geo-referenced with an accuracy of 0.5 pixels (or ~1 m for WV2), while commonly-used field GPS devices are accurate to $\geq \pm 3$ m. Thus, the spatial uncertainty of the field data is too great to accurately align the two data sets, which poses a major limiting factor of co-registering field sampling data with RS data at this time. Ways to overcome this limitation are to use high accuracy ($\leq \pm 0.5$ m) GPS to record locations or to collect periphyton data from the center of spatially homogeneous periphyton patches whose diameters are > 2 times the accuracy of the GPS measurements.

The temporal difference between periphyton sample data collection and image acquisition is a second limiting factor in matching field and RS data. The importance of this factor depends on the phenological cycle of the landscape under consideration. For a system like the Everglades, that has a strong bi-annual wet/dry cycle, image acquisition within the season sampled may be adequate to correlate field data to RS data.

When estimating biomass or deriving other biophysical parameters from the remotely sensed data knowing their spatial and temporal variances is essential. Understanding the requirements for correlating laboratory, field, and RS data will allow us to develop this knowledge. Collaborative efforts from all three approaches will then support understanding Everglades landscape and community structure at scales that have not previously been possible.

References

Breiman, L. 2001. Random forests. Mach. Learn. 45: 5–32. Springer, The Netherlands.

Browder, J.A., P.J. Gleason and D.R. Swift. 1994. Periphyton in the Everglades: spatial variation, environmental correlates, and ecological implications. pp. 379–418. *In*: S.M. Davis and J.C. Ogden (eds.). Everglades: the Ecosystem and its Restoration. St. Lucie Press, Boca Raton.

Brown, J.S. 1988. Photosynthetic pigment organization in diatoms (*Bacillariophyceae*). J. Phycol. 24: 96–102.

Cannizzaro, J.P. and K.L. Carder. 2006. Estimating chlorophyll *a* concentrations from remote-sensing reflectance in optically shallow waters. Remote Sens. Environ. 101: 13–24.

Carvalho, G.A., P.J. Minnett, L.E. Fleming, V.F. Banzon and W. Baringer. 2010. Satellite remote sensing of harmful algal blooms: a new multi-algorithm method for detecting the Florida red tide (Karenia brevis). Harmful Algae 9: 440–448.

Congalton, R.G. and K. Green. 1999. Assessing the Accuracy of Remotely Sensed Data: Principles and Practices. Lewis Publishers, Boca Raton, Florida.

Donar, C.M., K.W. Condon, M. Gantar and E.E. Gaiser. 2004. A new technique for examining the physical structure of Everglades floating periphyton mat. Nov. Hedwigia 78: 107–119.

Gaiser, E. 2009. Periphyton as an indicator of restoration in the Florida Everglades. Ecol. Indic. 9: S37–S45.

Gaiser, E.E., D.L. Childers, R.D. Jones, J.H. Richards, L.J. Scinto, J.C. Trexler, D. Lee, A.L. Edwards, K. Jayachandran and G.B. Noe. 2005. Cascading ecological effects of low-level phosphorus enrichment in the Florida Everglades. J. Environ. Qual. 34: 717–723.

Gaiser, E.E., D.L. Childers, R.D. Jones, J.H. Richards, L.J. Scinto and J.C. Trexler. 2006. Periphyton responses to eutrophication in the Florida Everglades: cross-system patterns of structural and compositional change. Limnol. Oceanogr. 51: 617–630.

Gaiser, E.E., L.J. Scinto, J. Trexler, D. Johnson and F. Tobias. 2009. Developing ecosystem response indicators to hydrologic and nutrient modifications in Northeast Shark River Slough, Everglades National Park. Contractor Report to Everglades National Park. 128 pp.

Gaiser, E.E., J. La Hée, F.A.C. Tobias and A.H. Wachnicka. 2010. *Mastogloia smithii var lacustris* Grun.: a structural engineer of calcareous mats in karstic subtropical wetlands. Proc. Natl. Acad. Sci. Philadelphia 160: 99–112.

Gaiser, E.E., P.V. McCormick, S.E. Hagerthey and A.D. Gottlieb. 2011. Landscape patterns of periphyton in the Florida Everglades. Crit. Rev. Environ. Sci. Technol. 41: 92–120.

Gann, D. and J.H. Richards. 2009. Determine the effectiveness of plant communities classification from satellite imagery for the Greater Everglades freshwater wetlands & community abundance, distribution and hydroperiod analysis for WCA 2A. Contractor Report to the South Florida Water Management District, West Palm Beach, FL, 278 pp.

Gann, D. and J.H. Richards. 2013. Evaluating high-resolution aerial photography acquired by unmanned aerial systems for use in mapping Everglades wetland plant associations. Subcontractor Report to the University of Florida, Gainesville, FL, and US Army Corps of Engineers, Jacksonville, FL, 37 pp.

Gann, D., J.H. Richards and H. Biswas. 2012. Determine the effectiveness of vegetation classification using WorldView 2 satellite data for the Greater Everglades. Contractor Report to the South Florida Water Management District, West Palm Beach, FL, 62 pp.

Garcia-Pichel, F. and R.W. Castenholz. 1991. Characterization and biological implications of scytonemin, a cyanobacterial sheath pigment. J. Phycol. 27: 395–409.

Gleason, P.J. and J. Spackman. 1974. Calcareous periphyton and water chemistry in the Everglades. pp. 146–181. *In*: P.G. Gleason (ed.). Environments of South Florida: Past and Present. Memoir No. 2. Miami Geological Society.

Gottlieb, A.D. 2003. Seasonal variation in short and long hydroperiod Everglades periphyton mat community structure and function. pp. 55–95. *In:* Short and long hydroperiod Everglades periophyton mats: Community characterization and experimental hydroperiod manipution. Ph.D. Dissertation. Florida International University, Miami, FL.

Gottlieb, A.D., J.H. Richards and E.E. Gaiser. 2006. Comparative study of periphyton community structure in long and short-hydroperiod Everglades marshes. Hydrobiologia 569: 195–207.

Hagerthey, S.E., B.J. Bellinger, K. Wheeler, M. Gantar and E. Gaiser. 2011. Everglades periphyton: a biogeochemical perspective. Crit. Rev. Environ. Sci. Technol. 41: 309–343.

Hochberg, E. 2003. Spectral reflectance of coral reef bottom-types worldwide and implications for coral reef remote sensing. Remote Sens. Environ. 85: 159–173.

Hunter, P.D., A.N. Tyler, M. Presing, A.W. Kovacs and T. Preston. 2008. Spectral discrimination of phytoplankton colour groups: the effect of suspended particulate matter and sensor spectral resolution. Remote Sens. Environ. 112: 1527–1544.

Iwaniec, D.M., D.L. Childers, D. Rondeau, C.J. Madden and C. Saunders. 2006. Effects of hydrologic and water quality drivers on periphyton dynamics in the southern Everglades. Hydrobiologia 569: 223–235.

Jensen, J.R. 2005. Introductory Digital Image Processing: a Remote Sensing Perspective. 3rd ed. Prentice Hall, Upper Saddle River, NJ.

Kidron, G.J., S. Barinova and A. Vonshak. 2012. The effects of heavy winter rains and rare summer rains on biological soil crusts in the negev desert. Catena 95: 6–11.

Klemas, V. 2011. Remote sensing of algal blooms: an overview with case studies. J. Coast. Res. 34–43.

Kutser, T., L. Metsamaa, N. Strömbeck and E. Vahtmäe. 2006. Monitoring cyanobacterial blooms by satellite remote sensing. Estuar. Coast. Shelf Sci. 67: 303–312.

Kutser, T., L. Metsamaa and A.G. Dekker. 2008. Influence of the vertical distribution of cyanobacteria in the water column on the remote sensing signal. Estuar. Coast. Shelf Sci. 78: 649–654.

La Hée, J.M. and E.E. Gaiser. 2012. Benthic diatom assemblages as indicators of water quality in the Everglades and three tropical karstic wetlands. Freshw. Sci. 31: 205–221.

Lee, B.S., K.C. McGwire and C.H. Fritsen. 2011. Identification and quantification of aquatic vegetation with hyperspectral remote sensing in western Nevada rivers, USA. Int. J. Remote Sens. 32: 9093–9117.

Lee, S.S., E.E. Gaiser and J.C. Trexler. 2013. Diatom-based models for inferring hydrology and periphyton abundance in a subtropical karstic wetland: implications for ecosystem-scale bioassessment. Wetlands. 33: 157–173.

Lee, S.S., E.E. Gaiser, B. Van de Vijver, M.B. Edlund and S.A. Spaulding. 2014. Morphology and typification of *Mastogloia smithii* and *M. lacustris*, with descriptions of two new species from the Florida Everglades and the Caribbean region. Diatom Res., doi:10.1080/0269249X.2014.889038.

Liaw, A. and M. Wiener. 2002. Classification and regression by random forest. R. News 2: 18–22.

Louchard, E.M., P.R. Reid and C.F. Stephens. 2003. Optical remote sensing of benthic habitats and bathymetry in coastal environments at Lee Stocking Island, Bahamas: a comparative spectral classification approach. Limnol. Oceanogr. 48: 511–521.

Madden, M., D. Jones and L. Vilchek. 1999. Photointerpretation key for the Everglades Vegetation Classification System. Photogramm. Eng. Remote Sensing 65: 171–177.

Matthews, M.W., S. Bernard and K. Winter. 2010. Remote sensing of cyanobacteria-dominant algal blooms and water quality parameters in Zeekoevlei, a small hypertrophic lake, using MERIS. Remote Sens. Environ. 114: 2070–2087.

Matthews, M.W., S. Bernard and L. Robertson. 2012. An algorithm for detecting trophic status (chlorophyll-*a*), cyanobacterial-dominance, surface scums and floating vegetation in inland and coastal waters. Remote Sens. Environ. 124: 637–652.

McCormick, P.V. and R.J. Stevenson. 1998. Periphyton as a tool for ecological assessment and management in the Florida Everglades. J. Phycol. 34: 726–733.

Méléder, V., L. Barillé, P. Launeau, V. Carrère and Y. Rincé. 2003. Spectrometric constraint in analysis of benthic diatom biomass using monospecific cultures. Remote Sens. Environ. 88: 386–400.

Merz, M.U.E. 1992. The biology of carbonate precipitation by cyanobacteria. Facies 26: 81–101.

Mutanga, O., E. Adam and M.A. Cho. 2012. High density biomass estimation for wetland vegetation using WorldView-2 imagery and random forest regression algorithm. Int. J. Appl. Earth Obs. Geoinf. 18: 399–406.

O'Neill, A.L. 1994. Reflectance spectra of microphytic soil crusts in semi-arid Australia. Int. J. Remote Sens. 15: 675–681.

Ozesmi, S.L. and M.E. Bauer. 2002. Satellite remote sensing of wetlands. Wetl. Ecol. Manag. 10: 381–402. Kluwer Academic Publishers.

Rabinowitch, E. and Govindjee. 1969. Photosynthesis. John Wiley and Sons, Inc., New York.

Reinart, A. and T. Kutser. 2006. Comparison of different satellite sensors in detecting cyanobacterial bloom events in the Baltic Sea. Remote Sens. Environ. 102: 74–85.

Richardson, L.L. 1996. Remote sensing of algal bloom dynamics. Bioscience 46: 492–501.

Rutchey, K., T.N. Schall, R.F. Doren, A. Atkinson, M.S. Ross, D.T. Jones, M. Madden, L. Vilchek, K.A. Bradley, J.R. Snyder, J.N. Burch, T. Pernas, B. Witcher, M. Pyne, R. White, T.J. Smith, III, J. Sadle, C.S. Smith, M.E. Patterson and G.D. Gann. 2006. Vegetation Classification for South Florida Natural Areas. Saint Petersburg, FL, United States Geological Survey, Open-File Report 2006-1240. 142 pp.

Schmidt, H. and A. Karnieli. 2000. Remote sensing of the seasonal variability of vegetation in a semi-arid environment. J. Arid Environ. 45: 43–59.

Shanmugam, P. 2011. A new bio-optical algorithm for the remote sensing of algal blooms in complex ocean waters. J. Geophys. Res. Ocean. 116, (C04016): 1–12.

Sirová, D., J. Vrba and E. Rejmánková. 2006. Extracellular enzyme activities in benthic cyanobacterial mats: comparison between nutrient-enriched and control sites in marshes of northern Belize. 44: 11–20.

Slate, J.E. and R.J. Stevenson. 2007. The diatom flora of phosphorus-enriched and unenriched sites in an Everglades marsh. Diatom Res. 22: 355–386.

Smith, T.E. 2009. Spatial and temporal response of scytonemin and photosynthetic pigments in calcareous mats from southern Florida (USA). Int. J. Algae 11: 199–210.

Stal, L.J. 1995. Physiological ecology of cyanobacteria in microbial mats and other communities. New Phytol. 131: 1–32.

Stevenson, R.J., C.G. Peterson, D.B. Kirschtel, C.C. King and N.C. Tuchman. 1991. Density-dependent growth, ecological strategies, and effects of nutrients and shading on benthic diatom succession in streams. J. Phycol. 27: 59–69.

Stumpf, R. and M. Tomlinson. 2005. Remote sensing of harmful algal blooms. pp. 277–296. *In*: R. Miller, C. Del Castillo and B. McKee (eds.). Remote Sensing of Coastal Aquatic Environments. Springer, The Netherlands.

Tang, D.L., H. Kawamura, H. Doan-Nhu and W. Takahashi. 2004. Remote sensing oceanography of a harmful algal bloom off the coast of southeastern Vietnam. J. Geophys. Res. Ocean. 109: C03014.

Thomas, S., E.E. Gaiser, M. Gantar, A. Pinowska, L.J. Scinto and R.D. Jones. 2002. Growth of calcareous epilithic mats in the margin of natural and polluted hydrosystems: phosphorus removal implications in the C-111 Basin, Florida Everglades, USA. Lake Reserv. Manag. 18: 324–330.

Thomas, S., E.E. Gaiser, M. Gantar and L. Scinto. 2006. Quantifying the responses of calcareous periphyton crusts to rehydration: a microcosm study (Florida Everglades). Aquat. Bot. 84: 317–323.

Updike, T. and C. Comp. 2010. Radiometric Use of WorldView-2 Imagery. Longmont, Colorado, USA: DigitalGlobe.

Ustin, S.L., P.G. Valko, S.C. Kefauver, M.J. Santos, J.F. Zimpfer and S.D. Smith. 2009. Remote sensing of biological soil crust under simulated climate change manipulations in the Mojave Desert. Remote Sens. Environ. 113: 317–328.

Weber, B., C. Olehowski, T. Knerr, J. Hill, K. Deutschewitz, D.C.J. Wessels, B. Eitel and B. Büdel. 2008. A new approach for mapping of biological soil crusts in semidesert areas with hyperspectral imagery. Remote Sens. Environ. 112: 2187–2201.

SECTION III

Microbiology of the Everglades

15

The Microbial Ecology of Mercury Methylation and Demethylation in the Florida Everglades

Christopher Weidow[a] and *Andrew Ogram*[b,*]

Introduction

The Florida Everglades is a large, ecologically-sensitive wetland located at the southernmost tip of the US state of Florida. The Everglades is home to diverse flora and fauna, many of which are endemic, in addition to providing habitat for many species of overwintering birds and newly-introduced, invasive plant and animal species. The Everglades also serves as an important freshwater source for large human populations in South Florida, such that the sustainability and quality of water in the Everglades is of critical importance. As a result of the importance of the Everglades to South Florida, it is the subject of the largest environmental restoration project in history. Among the most important issues facing the restoration of the Everglades are the transformation and fate of the various forms of mercury, particularly methylmercury (CH_3Hg^+). Methylmercury is a strong neurotoxin that binds to and destabilizes proteins essential to cell division and repair (Clarkson 2002). It is especially toxic to developing fetuses, neonates, and young children as these periods of growth involve rapid rates of cellular differentiation and division that, if exposed to biologically-significant concentrations of methylmercury, fail to develop normally and can lead to spontaneous abortion as well as neurological and physical abnormalities that vary in extent and mortality. Because of its toxicity and potential for both human and wildlife health impacts, the factors controlling both methylmercury formation and degradation are of importance.

Soil and Water Science Department, University of Florida Gainesville, FL 32611-0290.
[a] Email: cweidow@ufl.edu
[b] Email: aogram@ufl.edu
* Corresponding author

The primary route for human exposure to methymercury is via consumption of seafood; fish and shellfish may accumulate significant amounts of methylmercury due to the relatively high bioaccumulation factors of methyl mercury. For more information on human health effects of methyl mercury, the interested reader is directed to the comprehensive review by Clarkson (2002).

Most mercury in the Everglades is introduced via atmospheric deposition, which ranges from 19 to 25 μg Hg m^{-2} yr^{-1} in South Florida (Liu et al. 2008). Mercury is introduced into the atmosphere by both natural and anthropogenic processes, and is deposited in the Everglades as the inorganic, divalent cation (Hg^{2+}). Certain areas of the Everglades contain sufficiently high concentrations of mercury that the Florida Department of Environmental Protection issued fish consumption warnings or outright bans on fish consumption (Vaithiyanathan et al. 1996).

As stated previously, the organic forms of mercury pose the greatest risk to animals and humans, and the most toxic form is methylmercury (CH_3Hg^+). Conversion of Hg^{2+} to methylmercury is exclusively microbial, and the biogeochemical controls on methylmercury cycling, including formation and demethylation, are complex topics and are the subjects of this review. Recent research into methylmercury cycling has revealed new insight about how this toxic compound is cycled. Recent discoveries in both the microbiology and geochemistry of methylmercury formation have provided scientists and environmental managers with new insights and a greater understanding of how best to counter what remains a growing concern for South Florida.

To date, it is thought that methylation and demethylation of mercury in the environment is mediated by either facultative or obligately-anaerobic bacteria, especially the sulfate reducing bacteria (SRB) and iron reducing bacteria (IRB) (Compeau and Bartha 1985; 1987; King et al. 2000; Fleming and Nelson 2006; Kerin et al. 2006), and to a lesser extent, the strictly anaerobic Archaeal methanogens (ex., *Methanosarcinales*) (Oremland et al. 1991; 1995; Marvin-Dipasquale and Oremland 1998; Hamelin et al. 2011) and Firmicutes (Gilmour et al. 2013).

Mercury Methylation

The ability to methylate mercury methylation appears to be distributed across broad phylogenetic boundaries; however, the most efficient methylators described to date for the Everglades are particular genera within the SRB and IRB (Gilmour et al. 2013). The role of SRB in mercury cycling has been recognized for many years (Compeau and Bartha 1985), and has been studied extensively in geographically diverse environments (Compeau and Bartha 1985; 1987; Oremland et al. 1991; Vaithiyanathan et al. 1996; Marvin-Dipasquale and Oremland 1998; Gilmour et al. 1998; Pak and Bartha 1998a;b; King et al. 2000; Marvin Dipasquale et al. 2000). SRB are a phylogenetically and metabolically diverse group of microorganisms whose principal method of energy production is anaerobic respiration using sulfate (SO_4^{2-}) as the terminal electron acceptor. They are widely distributed throughout the Everglades (Castro et al. 2002; 2005).

Sulfate reduction is an important process in certain areas of the Everglades, particularly in the northern areas that are subject to runoff from the Everglades Agricultural Area (EAA). Elemental sulfur (S^0) is added to the muck soils of the EAA to reduce the pH of the typically alkaline soils; sulfur oxidizing bacteria convert the S^0 to H_2SO_4, leaving $SO_4^=$ to runoff into the northern Everglades through runoff from rain and irrigation. In addition, increased primary productivity in the Everglades resulted from phosphorus runoff, increasing the availability of electron donors for SRB, methanogens, and IRB, thereby increasing the activities of these groups. Processes other than sulfate reduction, such as iron reduction or methanogenesis, may be important in patches.

Preliminary work to define the role of SRB in mercury methylation in the Everglades was conducted by Gilmour and colleagues (1998). Selective inhibitors and terminal electron acceptors were employed to demonstrate the relative importance of SRB in mercury methylation in various locations of the Everglades. Addition of molybdate, an inhibitor of sulfate reduction, greatly reduced mercury methylation potentials, while bromoethanesulfonate, an inhibitor of methanogens, had no significant effect on methylation rates, or may have been stimulatory (Gilmour et al. 1998). Of the remaining compounds added to soil cores, only nitrate and sulfide had an inhibitory effect on methylation rates, while sulfate, ammonium, phosphate, and Fe (III) did not significantly affect mercury methylation. At least in the areas of the Everglades studied by Gilmour et al. (1998), SRB appeared to be the dominant mercury methylators, with little contribution by either IRB or methanogens. It should be noted that methylation rates are highly variable in the Everglades, and do not always correlate with sulfate reduction rates.

The exact biochemical mechanism by which inorganic mercury is methylated, and the relationship of this mechanism to metabolism is not completely understood at this time; however, Parks et al. (2013) recently reported on the involvement of the genes *hgcAB* as required for methylation. These genes are widely distributed across broad phylogenetic boundaries (Gilmour et al. 2013), although the detaproteobacteria (including SRB and IRB) appear to be the most efficient at gratuitous mercury methylation. Methylmercury formation does not produce energy for growing cells and does not typically convey resistance to inorganic mercury toxicity at high levels; however, there is some evidence that at lower concentrations of Hg^{2+}, methylation serves as a detoxification mechanism by increasing volatilization of the molecule out of the cell (Marvin-Dipasquale and Oremland 2000; Schaefer et al. 2004). Considerable data indicate that methylation is a gratuitous process associated with the reductive acetyl-CoA, or Wood-Ljungdahl, pathway for CO_2 fixation in at least some strains, although methylation may not be exclusively associated with that pathway (Ekstrom et al. 2003). Among those SRB with the reductive acetyl-coA, a critical step is transference of a methyl group to acetyl-coA synthase via a corrinoid protein. The gene encoding the putative corrinoid protein that likely acts as methyl carrier for mercury methylation has been cloned and sequenced. This gene is present in all mercury methylating bacteria and archaea for which genome sequences are available, and do not appear to be present in those that do not methylate mercury for which genome sequences are available (Parks et al. 2013). It is not known at this time, however, why many prokaryotes that express the reductive acetyl-CoA pathway do not carry these particular genes, and therefore do not methylate mercury.

Carbon metabolism in SRB may be complex and highly variable, which likely impacts the efficiencies with which SRB methylate mercury. SRB may be broadly classified as complete oxidizers (for their ability to completely oxidize substrates through acetate, or the number 2 carbon in acetyl-CoA, to CO_2) or incomplete oxidizers (which cannot completely oxidize acetate) (Castro et al. 2000). Complete oxidizers typically carry the reductive acetyl-CoA pathway (apparently important to mercury methylation), while incomplete oxidizers do not.

The relative environmental distribution of these two different metabolisms in the Everglades, and consequently the functions required for mercury methylation, is likely controlled by environmental factors. Factors related to nutrient enrichment in WCA-2A of the Everglades (Castro et al. 2005), in which complete oxidizers dominate in the nutrient impacted regions while incomplete oxidizers dominate in the lower nutrient areas. It is not known at this time how, or if, this selection for SRB metabolisms as a function of nutrient status might impact mercury methylation in WCA-2A.

Of potential importance to mercury methylation in the Everglades is the importance of SRB as secondary fermenters in low sulfate environments, such as in freshwater marshes like the Everglades. Secondary fermenters (also known as syntrophs due to their requirement for a H_2-consuming partner) are distributed throughout the Everglades (Chauhan et al. 2004), and at least some SRB are capable of mercury methylation during fermentative growth (King et al. 2000). As with most SRB, the phylogenetic distribution of secondary fermenters in the Everglades is dependent on nutrient status in the Everglades. It is not known at this time if syntrophic growth is an important process for mercury methylation in the Everglades, however.

The potential role that methanogens play in mercury methylation was recently recognized (Parks et al. 2013; Gilmour et al. 2013), although it is not known if methanogens make a significant contribution to mercury methylation in the Everglades. Using stable isotopes and specific metabolic inhibitors, Hamelin et al. (2011) found that methanogens were the principal mercury methylators in a shallow fluvial lake (Lake St. Pierre) along the St. Lawrence River, in Quebec, Canada. Mercury methylation and demethylation rates in periphyton were determined by stable isotope probing using stable isotopes of mercury as either ^{199}HgO or methyl-^{200}Hg (Me^{200}Hg). Conversion of ^{199}HgO to Me^{199}Hg was used to determine methylation rates while degradation of Me^{200}Hg to ionic ^{200}Hg^{2+} was used to determine demethylation rates. Incubation with molybdate confirmed that SRB were not involved in either process.

As with the SRB, the distribution of specific groups of methanogens in the Everglades is controlled by interaction between physiology and environmental factors. This will be discussed in greater detail below, in the context of methylmercury demethylation.

Methylmercury Demethylation

Three main mechanisms have been proposed by which methylmercury is demethylated: 1) oxidative demethylation (OD) to CO_2 and Hg^{2+}; 2) *merB*-mediated reductive demethylation (RD) to CH_4 and Hg^{2+} via organomercurial-lyase; 3) *mer*-independent RD via an unstable dimethylmercury-sulfide intermediate (Oremland et al. 1991; Marvin-Dipasquale and Oremland 1998; Pak and Bartha 1998b; Marvin-Dipasquale et al. 2000). *merA*-mediated reduction of Hg^{2+} to Hg^0 appears to be unassociated with either OD or RD in SRB, where the final products are either CO_2 and Hg^{2+} or

CH_4 and Hg^{2+} (Marvin-Dipasquale and Oremland 1998). Reduction of mercuric ions to volatile elemental mercury (Hg^0) via *mer*A-mediated mercuric reductase may be carried out by separate groups of microbes, possibly aerobes (Schaefer et al. 2004; Oregaard and Sørensen 2007).

Processes controlling demethylation of methylmercury in the Everglades may differ significantly from those effecting methylation. In a study of mercury demethylation in the Everglades, Marvin-Dipasquale and Oremland (1998) showed that both SRB and methanogens were responsible for demethylation; however, methanogens were the dominant demethylators and increased in relative importance in an apparent inverse relationship with nutrient status.

Numerous studies have indicated the significance in methylmercury degradation (Oremland et al. 1991; Pak and Bartha 1998a;b; Marvin-Dipasquale and Oremland 1998; 2000). Demethylation by methanogens has only been described to date via the oxidative demethylation of mercury in both sediments and pure culture (Marvin-Dipasquale and Oremland 1998; 2000; Pak and Bartha 1998a). Marvin-Dipasquale and Oremland (1998) hypothesize that the mechanism by which methanogens oxidatively demethylate mercury proceeds via a reaction that is stoichiometrically similar to methanogen-based degradation of monomethylamine (Marvin-Dipasquale and Oremland 1998):

$$4CH_3NH_3^+ + 2H_2O \rightarrow 3CH_4 + CO_2 + 4NH_4^+;$$

such that

$$4CH_3Hg^+ + 2H_2O + 4H^+ \rightarrow 3CH_4 + CO_2 + 4Hg_2^+ + 4H_2$$

Monomethylamine degradation, as well as the use of other methyl-group containing compounds, are not widely distributed functions among methanogens and are unique to methanogens within the order Methanosarcinales (Liu and Whitman 2008). The order contains two families: the Methanosarcinaceae, of which the genus *Methanosarcina* is representative; and the Methanosaetaceae, which contains only one genus, *Methanosaeta*. In addition to metabolism of methyl-group containing compounds, the Methanosarcinales are capable of utilizing H_2/CO_2 and acetate as substrates for methanogenesis, and therefore have the widest substrate range among currently known methanogens (Liu and Whitman 2008). If methylmercury degradation by methanogens proceeds via a mechanism similar to that of other methyl-group containing compounds (Marvin-Dipasquale and Oremland 1998), it is likely that methanogens belonging to the order Methanosarcinales are the most likely agents of methanogen-driven mercury demethylation in the Florida Everglades. As with SRB, the phylogenetic distribution of methanogens in the Everglades is controlled by the interaction of physiology with environment. This may, in turn, impact the potentials for demethylation of methylmercury.

There are two classically recognized pathways methanogenesis, as well as one which has only come into the research field in the last thirty years (Liu and Whitman 2008). The distribution of these pathways differ with nutrient status in the Everglades (Chauhan et al. 2004; Castro et al. 2005; Holmes et al. 2014). The classic pathways are acetotrophic methanogenesis and hydrogenotrophic methanogenesis. Acetotrophic methanogens oxidize acetate with the production of CH_4, while hydrogenotrophic methanogens use exogenous H_2 to reduce CO_2 to methane (Liu and Whitman 2008). The third group of methanogens, the methylotrophs, metabolize methyl-group containing compounds to CO_2 and CH_4. The acetotrophs dominate in the lower nutrient areas of the Everglades, while the hydrogenotrophs dominate in the higher nutrient areas of WCA-2A (Castro et al. 2004; Chauhan et al. 2004). It is not known at this time if the selection of acetotrophs in the nutrient impacted areas is related to the higher rates of demethylation observed by Marvin-Dipasquale and Oremland (1998).

Contrary to the major demethylation pathway proposed for methanogens, the OD pathway may be more important in SRB, and may proceed via a reaction similar to the oxidation of acetate (Marvin-Dipasquale and Oremland 1998):

$$SO_4^{2-} + CH_3COO^- + 3H^+ \rightarrow H_2S + 2CO_2 + 2H_2O$$

such that

$$SO_4^{2-} + CH_3Hg^+ + 3H^+ \rightarrow H_2S + CO_2 + Hg^{2+} + 2H_2O.$$

Oremland et al. (1991) studied methylmercury demethylation in sediments taken from three different aquatic systems: an estuarine salt marsh; a freshwater lake; and an alkaline-hypersaline lake. As expected, SRB competed with methanogens for electron donors in both estuarine and freshwater environments; however, they were only involved in mercury demethylation in freshwater. Both SRB and methanogens may be responsible for methylmercury demethylation in freshwater; however, the demethylation activities of SRB are likely to be more important estuaries (Oremland et al. 1991).

Methylation vs. Demethylation

If SRB are the dominant mercury methylators, and methanogens are the dominant demethylators, in the Everglades, the distribution and relative activities of specific physiologies of SRB and methanogens may play a significant role in determining the relative rates of methylation vs. demethylation, and hence the concentrations of methylmercury potentially available for bioaccumulation. Despite the greater energy conserved by sulfate reduction, several studies have shown that methanogens will outcompete SRB for acetate if sulfate is limiting (Compeau and Bartha 1985; 1987; Oremland et al. 1991; Pak and Bartha 1998a;b; Liu and Whitman 2008). In anaerobic soils, only when sufficient concentrations of sulfate are present will SRB prevail over methanogens as the dominant group. Different groups have suggested that the coexistence of SRBs and methanogens Everglades soils most likely results from non-competitive substrate utilization or excess electron donor (carbon) relative to sulfate (Marvin-Dipasquale and Oremland 1998; Castro et al. 2005).

The microbial ecology of methylmercury degradation in the Everglades is a major gap in knowledge at this time. Perhaps of significance to mercury cycling, the relative proportion of hydrogenotrophic to acetotrophic methanogenesis is significantly higher than predicted in the nutrient impacted areas of the Everglades (Holmes et al. 2014), and the relative proportions of complete oxidizing SRB to incomplete oxidizers are significantly higher in nutrient impacted areas (Castro et al. 2005). As discussed previously, the relationships between these different physiologies and mercury cycling are not completely understood at this time; however, coupled with the geochemical factors discussed below, this specific selection of SRB and methanogen physiologies may have implications for determining which biogeochemical factors and microbial consortia will have the most influence on methylmercury degradation in Everglades' soils.

Geochemical Factors Affecting Bioavailability

Regardless of the dominant microbial physiology controlling methylation and demethylation, it may be that the most important factor controlling the fate of mercury is the biological availability of Hg^{2+}. The availability of Hg^{2+} for uptake and methylation by prokaryotes is controlled by the complex geochemistry of the Everglades. Within the Everglades, mercury and its various organic and inorganic forms are found in the following environmental compartments: surface water; periphyton; floc (recently deposited, partially-decomposed organic matter); and peat (Liu et al. 2009). Both inorganic and organic forms of mercury are likely to be more concentrated in one (or a few) of these compartments than to be evenly distributed throughout all of them. The distribution depends on the predominating biogeochemical factors of any given site (Liu et al. 2009).

The transformation and translocation of Hg^{2+} is largely controlled by biogeochemical factors related to the ability of Hg^{2+} (or Hg^+) to form complexes with other organic and inorganic materials. Gilmour et al. (1998) reported that mercury methylation was inhibited by relatively high rates of sulfate reduction to sulfide; sulfide forms insoluble complexes with Hg^{2+}, thereby eliminating its availability for biological uptake ("sulfide inhibition"). Methylmercury concentrations (and production rates) are therefore much lower in WCA-2A than in some areas of the Everglades National Park, likely because SRB produce so much sulfide that ionic mercury is precipitated as insoluble mercuric sulfide, HgS. Colloquially known as cinnabar, HgS is a dark red or purplish mineral with a solubility in water that is near zero. Between pH 4 and 9 (and the absence of any dissolved mineral or organic matter) HgS only dissolves to a concentration of 11×10^{-17} ppb.

When considering the various chemical and biological reactions of mercury on scales relevant to animals and humans, the formation of HgS would appear to be an effectively-permanent removal of mercury from environmental waters and soils, with few exceptions. One of these exceptions is the possibility of increased dissolution of cinnabar in the presence of increased concentrations of dissolved organic matter (DOM). Ravichandran et al. (2001) found that the concentration of aqueous mercury derived from HgS increased from undetectable to 1.7 µmol/mg of dissolved carbon in a 20 mg DOC/L solution, such that DOM in surface waters might increase the amount of mercury available to microorganisms for methylation from insoluble minerals. This solublization is complicated by concentrations of divalent cations, however; calcium at concentrations as low as 2.5×10^{-4} M, HgS solubility in water with DOC was reduced up to 85%. This is significant because of the relatively high concentrations of Ca^{++} in much of the Everglades.

Organic matter that is consolidated in floc or soil has been shown to reduce the bioavailability of mercury by strong binding to reduced thiol ligands (Drexel et al.

2002); mercury binds with different affinities to organic matter depending on which functional groups it is associated with. Characterized as either strong or weak bonds ($DOC_{S/W}$, for DOC and $Peat_{S/W}$ for solid peat) mercury binds very strongly with thiol and other sulfur-containing functional groups while binding weakly to carboxylic acids and phenols. Environmental concentrations of inorganic mercury are lower than the total number of binding sites in DOC, such that mercury will readily bind to the strong binding sites, or thiol groups, in DOC in the absence of significant concentrations of Ca^{2+}. When mercury concentrations are higher than the number of binding sites available in DOC, mercury partitions to the strong binding sites in peat. It is only when mercury concentrations exceed the total number of strong binding sites available in both DOC and peat will mercury proceed to bind to the weak binding sites for both DOC and peat. Drexel et al. (2002) also concluded that there is competition between DOM/peat and dissolved sulfides for binding mercury.

A more recent study by Gu et al. (2011) concluded that DOC has both mercury-complexing groups (thiols, etc.) and mercury-reducing moieties, such as semiquinones. Whether or not DOC complexes with (and immobilizes) mercury or reduces it to gaseous form appears to depend on the relative availability of mercury-complexing or mercury-reducing active sites. An artificially reduced standard humic acid reduced as much as 70% of a ~10 nM solution of Hg^{2+} to gaseous mercury at low DOC concentrations (~0.2 mg HA/L). However, when DOC concentrations were increased (and mercury concentrations kept constant), a significantly lower percentage of total mercury was volatilized (~40% Hg^0 at 1 mg HA/L and ~10% Hg^0 at 2 mg HA/L), indicating that reduced DOC reduces Hg^{2+} at lower DOC concentrations because the reducing moieties outcompete the complexing groups for mercury interactions. With increasing DOC concentrations, both mercury-complexing and mercury-reducing moieties increase; however, interactions between mercury and the complexing groups become dominant at DOC concentrations as low as 1 mg/L (where only 40% of inorganic mercury is volatilized and the remainder is believed to be complexed).

It is clear that multiple organic and inorganic geochemical properties simultaneously influence mercury availability. The ultimate disposition of mercury in the environment will depend on the predominating geochemical factors of the specific location. In general, DOM concentrations in the Everglades are much higher in the north where primary productivity and nutrient contamination is high and lower in the south, where nutrient contamination from the EAA is reduced. Conversely, dissolved inorganic minerals (DIM) (calcium, magnesium, chloride) are higher in the southern Everglades than in the north. How mercury will behave at any given location in the Everglades likely depends on these properties.

Liu et al. (2009) provide a focused evaluation on the geochemical factors that influence mercury and methylmercury partitioning to different environmental compartments in the Everglades. The environmental factors that predominantly govern the partitioning of inorganic mercury into the various compartments are surface water DOC (DOC_{SW}), surface water chloride (Cl_{SW}), surface water pH (pH_{SW}), soil and floc organic matter, or ash-free dry weight ($AFDW_{SD/FC}$), soil mineral content (MC_{SD}), and floc chlorophyll A concentrations ($CHLA_{FC}$).

Liu et al. (2009) showed that the concentration of inorganic mercury in soil and floc is inversely-related to pH_{SW}, DOC_{SW}, and Cl_{SW}. Conversely, mercury concentrations in soil and floc were directly related to MC_{SD} and $AFDW_{SD/FC}$ concentrations. As discussed above, increased surface water DOC concentrations may result in increased binding of ionic mercury with DOC functional groups. When the general acidity of surface water increases the concentration of aqueous H^+ increases.

Soil and floc organic matter contents influence mercury distribution in much the same way that DOC_{SW} does (Ravichandran et al. 2001; Drexel et al. 2002). In the Everglades, recently deposited floc material tends to be higher in organic matter than the underlying soil and total mercury could be expected to be at higher concentrations in the floc than in the soil. Soil also has additional mineral constituents (MC_{SD}) such as Fe-, Al-, Ca-, Mg-, and Mn-containing oxyhydroxides that are not always present in floc. Because addition of calcium ions to a solution containing DOC and HgS reduces the solubility of HgS. However, in the absence of sulfide and other polyvalent cations with which ionic mercury might bind, the presence of polyvalent metal cations usually increases the concentrations of ionic mercury in solution by displacing mercury ions from organic and inorganic complexes and into the aqueous phase.

Polyvalent cations drive mercury into solution from peat/DOC when sulfide is absent by means of cation exchange from the adsorbed to the aqueous phase. If sulfide is present, mercury will bind to it and form insoluble HgS, which will remain insoluble even in the presence of DOC as long as there are sufficient cations to prevent DOC-mercury interactions. Mineral contents do not typically play a role in mercury compartmentalization to either floc or periphyton where such aggregations are few, if present at all. In general, partitioning of methylmercury to various environmental compartments of the Everglades is governed by the same factors as inorganic mercury (Liu et al. 2008; 2009), but its partitioning to periphyton is of greater concern as periphyton serves as the primary food source of the guppy-sized mosquitofish *Gambusia holbrooki*, an essential organism at the lower end of the Everglades food web and, therefore, an important factor in the bioaccumulation of methylmercury in higher trophic organisms.

Conclusions

The cycle of methylation and demethylation of mercury is complex, and much of the fundamental mechanisms governing those processes is poorly understood in the Everglades. Recent advances in knowledge of the biochemistry and genetics of mercury methylation (Parks et al. 2013; Gilmour et al. 2013) are likely to lead to studies that will significantly improve our understanding of the environmental controls on the distribution of methylating strains in the Everglades. Of particular interest will be studies that elucidate the regulation of those genes in the environment. Little attention has been paid to the biochemistry of mercury demethylation or its potential role in controlling ambient methyl mercury concentrations in the Everglades, although it may be a significant process in some environmental compartments.

Regardless of the microbial ecology of mercury methylation and demethylation, it may be that the complex geochemistry of the Everglades provides the ultimate control over the concentrations of methyl mercury in this complex environment. Concentrations of sulfide, DOC, and calcium, and pH, play important roles in the availability of mercury to uptake and transformation by microorganisms. Geochemistry and microbial ecology are highly variable in different regions and environmental compartments of the Everglades, such that conclusions made in one area are not likely to hold across the entire Everglades ecosystem. Much more work must be done to describe both the detailed microbial ecology of mercury cycling and interaction of mercury with geochemical factors before the factors controlling the distribution of mercury in the Everglades are understood.

References

Castro, H., K.R. Reddy and A. Ogram. 2002. Composition and function of sulfate-reducing prokaryotes in eutrophic and pristine areas of the Florida Everglades. Appl. Environ. Microbiol. 68: 6129–6137.

Castro, H., S. Newman, K.R. Reddy and A. Ogram. 2005. Distribution and stability of sulfate-reducing prokaryotic and hydrogenotrophic methanogenic assemblages in nutrient-impacted Regions of the Florida Everglades. Appl. Environ. Microbiol. 71: 2695–2704.

Castro, H.F. and A.V. Ogram. 2000. Phylogeny of sulfate reducing bacteria. FEMS Microbiol. Ecol. 31: 1–9.

Chauhan, A., A. Ogram and K.R. Reddy. 2004. Syntrophic-methanogenic associations along a nutrient gradient in the Florida Everglades. Appl. Environ. Microbiol. 70: 3475–3484.

Clarkson, T.W. 2002. The three modern faces of mercury. Env. Health Perspect. Supplements 110: 11–23.

Compeau, G.C. and R. Bartha. 1985. Sulfate-reducing bacteria: Principal methylators of mercury in anoxic estuarine sediment. Appl. Environ. Microbiol. 50: 498–502.

Compeau, G.C. and R. Bartha. 1987. Effect of salinity on mercury-methylating activity of sulfate-reducing bacteria in estuarine sediments. Appl. Environ. Microbiol. 53: 261–265.

Conrad, R. 1999. Contribution of hydrogen to methane production and control of hydrogen concentrations in methanogenic soils and sediments. FEMS Microbiol. Ecol. 28: 193–202.

Drexel, R.T., M. Haitzer, J.N. Ryan, G.R. Aiken and K.L. Nagy. 2002. Mercury(II) sorption to two Florida Everglades peats: Evidence for strong and weak binding and competition by dissolved organic matter released from the peat. Environ. Sci. Technol. 36: 4058–4064.

Ekstrom, E.B., F.M.M. Morel and J.M. Benoit. 2003. Mercury methylation independent of the acetyl-coenzyme A pathway in sulfate-reducing bacteria. Appl. Environ. Microbiol. 69: 5414–5422.

Fleming, E.J., E.E. Mack, P.G. Green and D.C. Nelson. 2006. Mercury methylation from unexpected sources: Molybdate-inhibited freshwater sediments and an iron-reducing bacterium. Appl. Environ. Microbiol. 72: 457–464.

Gilmour, C., M. Podar, A.L. Bullock, A.M. Graham, S.D. Brown, A.C. Somenahally, A. Johs, R.A. Hurt, Jr., K.L. Bailey and D.A. Elias. 2013. Mercury methylation by novel microorganisms from new environments. Environ. Sci. Technol. 47: 11810–11820.

Gilmour, C.C., G.S. Riedel, M.C. Ederington, J.T. Bell, J.M. Benoit, G.A. Gill and M.C. Stordall. 1998. Methylmercury concentrations and production rates across a trophic gradient in the northern Everglades. Biogeochemistry 40: 327–345.

Gu, B., Y. Bian, C.L. Miller, W. Dong, X. Jiang and L. Liang. 2011. Mercury reduction and complexation by natural organic matter in anoxic environments. Proc. Natl. Acad. Sci. USA 108: 1479–1483.

Hamelin, S., M. Amyot, T. Barkay, Y. Wang and D. Planas. 2011. Methanogens: principal methylators of mercury in lake periphyton. Environ. Sci. Technol. 45: 7693–7700.

Holmes, M.E., J.P. Chanton, H.-S. Bae and A. Ogram. 2014. Effect of nutrient enrichment on δ13CH4 and the methane production pathway in the Florida Everglades, J. Geophys. Res. Biogeosci. 119: 1267–1280.

Kerin, E.J., C.C. Gilmour, E. Roden, M.T. Suzuki, J.D. Coates and R.P. Mason. 2006. Mercury methylation by dissimilatory iron-reducing bacteria. Appl. Environ. Microbiol. 72: 7919–7921.

Liu, G., Y. Cai, Y. Mao, T. Philippi, P. Kalla, D. Scheidt, J. Richards, L.J. Scinto and C. Appleby. 2008. Distribution of total and methylmercury in different ecosystem compartments in the Everglades: implications for mercury bioaccumulation. Environ. Poll. 153: 257–265.

Liu, G., Y. Cai, Y. Mao, D. Scheidt, P. Kalla, J. Richards, L.J. Scinto, G. Tachiev, D. Roelant and C. Appleby. 2009. Spatial variability in mercury cycling and relevant biogeochemical controls in the Florida Everglades. Environ. Sci. Technol. 43: 4361–4366.

Liu, Y. and W.B. Whitman. 2008. Metabolic, phylogenetic, and ecological diversity of the methanogenic archaea. Ann. N.Y. Acad. Sci. 1125: 171–189.

Marvin-Dipasquale, M.C. and R.S. Oremland. 1998. Bacterial methylmercury degradation in Florida Everglades peat sediment. Environ. Sci. Technol. 32: 2556–2563.

Marvin-Dipasquale, M.C., J. Agee, C. McGowan, R.S. Oremland, M. Thomas, D. Krabbenhoft and C. Gilmour. 2000. Methyl-mercury degradation pathways: a comparison among three mercury-impacted ecosystems. Environ. Sci. Technol. 34: 4908–4916.

Øregaard, G. and S. Sørensen. 2007. High diversity of bacterial mercuric reductase genes from surface and sub-surface floodplain soil (Oak Ridge, USA). ISME J. 1: 453–467.

Oremland, R.S., C.W. Culbertson and M.R. Winfrey. 1991. Methylmercury decomposition in sediments and bacterial cultures: involvement of methanogens and sulfate reducers in oxidative demethylation. Appl. Environ. Microbiol. 57: 130–137.

Pak, K.R. and R. Bartha. 1998a. Mercury methylation and demethylation in anoxic lake sediments and by strictly anaerobic bacteria. Appl. Environ. Microbiol. 64: 1013–1017.

Pak, K.R. and R. Bartha. 1998b. Mercury methylation by interspecies hydrogen and acetate transfer between sulfidogens and methanogens. Appl. Environ. Microbiol. 64: 1987–1990.

Parks, J.M., A. Johs, M. Podar, R. Bridou, R.A. Hurt, Jr., S.D. Smith, S.J. Tomanicek, Y. Qian, S.D. Brown, C.C. Brandt, A.V. Palumbo, J.C. Smith, J.D. Wall, D.A. Elias and L. Liang. 2013. The genetic basis for bacterial mercury methylation. 339: 1332–1335.

Schaefer, J.K., J. Yagi, J.R. Reinfelder, T. Cardona, K.M. Ellickson, S. Tel-Or and T. Barkay. 2004. Role of the bacterial organomercury lyase (merB) in controlling methylmercury accumulation in mercury-contaminated natural waters. Environ. Sci. Technol. 38: 4304–4311.

Vaithiyanathan, P., C.J. Richardson, R.G. Kavanaugh, C.B. Craft and T. Barkay. 1996. Relationships of eutrophication to the distribution of mercury and to the potential for methylmercury production in the peat soils of the Everglades. Environ. Sci. Technol. 30: 2591–2597.

16

Methanogens within the Sawgrass Communities of the Everglades and Biscayne Bay Watersheds

Priyanka Kushwaha,[1] *Jacqueline Zayas,*[2,a] *Yanie Oliva,*[2,b]
Maria Mendoza[2,3] and *DeEtta Mills*[2,3,*]

Introduction

Over 10^9 bacteria cells can be detected in a single gram of soil (Travers et al. 1987). This phenomenal abundance and biodiversity presents challenges to understanding soil microbial community structure and function, as the majority of environmental microbes cannot be cultured in the laboratory (Torsvik et al. 1990). Microbial metagenomics, the isolation of whole genomic DNA and subsequent clone library screening (Handelsman et al. 1998), and next-generation sequencing technologies (Sundquist et al. 2007) are current methods used by ecologists to establish differences between microbial communities (Daniel 2005; Kakirde et al. 2010) in a culture-independent manner.

---

[1] Department of Chemistry and Biochemistry, CP-304, 11200 SW 8th Street, Florida International University, Miami, FL 33199.
   Email: pkush003@fiu.edu
[2] Department of Biological Sciences, OE 167, 11200 SW 8th Street, Florida International University, Miami, FL 33199.
[a] Email: jzaya001@fiu.edu
[b] Email: yoliv001@fiu.edu
[3] International Forensic Research Institute, Department of Chemistry and Biochemistry, OE 116, 11200 SW 8th Street, Florida International University, Miami, FL 33199.
   Email: Maria.Mendoza@ventura.org
* Corresponding author: millsd@fiu.edu

When microbial metagenomic analyses are applied to a soil sample, it can produce a unique and total genomic fingerprint that can be used to assess diversity as well as discern community dynamics and ecological interrelationships within the environment (Osborn et al. 2000; Horswell et al. 2002).

Function versus Structural Biodiversity of Microbial Communities

Although prokaryotic diversity limits are still unknown, microbial ecologists struggle with how to even define a prokaryotic species. Most prokaryotes cannot be cultured or studied as individual species. Currently, culture-independent methods define species-level taxa as operational taxonomic units (OTUs) based on full-length 16S rRNA gene sequences ($\approx$1,500 bp). If OTUs of a queried sample share $\geq$ 97% sequence homology with known species, based on public databases such as GenBank, the two are considered to be the same species (Stackebrandt and Goebel 1994; Stackebrandt and Ebers 2006). Herein lies the weakness of 16S classification schemes. The ability to classify an unknown phylotype as a particular species based on 16S gene sequences is only as informative as the known species in the database. In other words, novel uncultured microbial species (bacteria or archaea) with few or no related sequences archived in the database are often misclassified using only 16S rRNA sequences (Fox et al. 1992). It is also possible that 16S rRNA gene identification does not adequately reflect the inherent functional phenotype found in bacteria or archaea classified as the same species (Robinson et al. 2010). For example, Robinson et al. noted that *Escherichia coli* strains collected from various different environments were taxonomically identical (using the 16S rRNA, 99% similarity) but had very different functional capabilities, ranging from pathogenic to commensal. Although it is important to characterize soil microbial community assemblages, given the variation within and among species, it is ecologically important to understand the functional complexity of species and communities to sustain soil ecosystems and the ecosystem services they provide.

Carbon Cycling

Microbes are important decomposers in the global carbon cycle. The breakdown of plant material and other detritus (Millard and Singh 2010) by microbes is the first step in supplying both belowground and aboveground communities with the necessary nutrients for growth (e.g., C, N, P, K, S). Aboveground vegetation is the major carbon source available and has been found to be one of the forces responsible for belowground microbial diversity (Marschner et al. 2001; Wieland et al. 2001). Cellulose degradation at the beginning of the decomposition cycle and methanogenesis at the end are important biochemical processes in aerobic and anaerobic carbon decomposition, respectively. Important to the anoxic or anaerobic soil ecosystems are methanogens (methane producing archaea) that can reduce the final products of the carbon cycle such as acetate, formate, CO_2, methylamines, and methanol to methane. Understanding functional diversity (guilds) provides important information on the critical roles microbes play in biogeochemical cycles and nutrient partitioning within a system that cannot always be ascertained from 16S rRNA data (Torsvik and

Øvreås 2002). An understanding of functional diversity of methanogens, therefore, would provide a better understanding of carbon cycle dynamics and methane release to the atmosphere. Subsequently, disruptions of the carbon cycle via anthropogenic perturbations and the global impact of such changes can be assessed.

Biochemistry of Methane Formation

There are several carbon substrates and pathways that ultimately result in methane production and several of the genes in the terminal steps of the pathways are highly conserved across all methanogens (Bapteste et al. 2005). Methane production from several different substrates is shown below (see Blaut 1994 for an in-depth review of methane biochemistry).

(1) $CO_2 + 4H_2 \rightarrow CH_4 + 2H_2O$

(2) $4\ HCOO^- + 4H^+ \rightarrow 3CO_2 + CH_4 + 2H_2O$

(3) $4CH_3OH \rightarrow 3CH_4 + CO_2 + 2H_2O$

(4) $4(CH_3)_3NH^+ + 6H_2O \rightarrow 9CH_4 + 3CO_2 + 4NH_4^+$

(5) $CH_3COO^- + H^+ \rightarrow CH_4 + CO_2$

As methanogens cannot break down complex organic molecules for methane production, they rely on presence of other anaerobes in their habitat to breakdown organic molecules into simple sugars or fatty acids followed by fermentation by syntrophs to produce formate, acetate, hydrogen and carbon dioxide—the major substrates for methanogenesis. In addition, acetogens (acetate-producing bacteria) play an important role in this syntrophic association of methanogens, and thus assimilate hydrogen and formate effectively (Nazaries et al. 2013).

The pathway that is most widely distributed across all orders of methanogens is the seven-step hydrogenotropic pathway. In this pathway, carbon dioxide is used as the substrate that is reduced by a hydrogen molecule that acts as an electron donor (Reeve et al. 1997). Formate can be converted into carbon dioxide and, thus, utilizes the same pathway for methane production. Two other pathways that are involved in methanogenesis are the aceticlastic pathway and the methylotrophic pathway. Acetate is the substrate for the aceticlastic pathway, whereas, methanol and methyl-amines are employed as substrates for methylotropic pathway. Acetate is broken down into methyl and carbon monoxide (CO) in the aceticlastic pathway. The methyl group thus produced is linked to methanopterin before it is reduced to methane in two steps of enzymatic reactions, reactions that are homologous to the last two steps of the hydrogenotropic pathway. In the methylotropic pathway, the C-1 compounds can be converted to three molecules of carbon dioxide using the reverse hydrogenotropic pathway in order to follow the forward pathway to release three molecules of methane. Alternatively, C-1 compounds can be directly reduced to methane following the last step of hydrogenotropic pathway.

These three pathways converge at the last step where methyl-coenzyme M reductase catalyzes the conversion of methyl-CoM to methane (Bapteste et al. 2005). One enzyme, methyl-coenzyme M reductase, universally conserved, catalyzes the last step in the methanogenic cycle. Therefore, the gene, *mcrA,* that encodes for the alpha subunit is often used to taxonomically classify methanogens (Reeve et al. 1997). This gene is ubiquitous to all methanogens regardless of the carbon substrates utilized earlier in methanogenesis pathways. However, it is not uniform in sequence across taxa (Blaut 1994).

The Everglades and Biscayne Bay Watersheds

The greater Florida Everglades Watershed is a unique freshwater marsh ecosystem that begins with the northern border at the Kissimmee River where the flow of water moves south toward Florida Bay and the Gulf of Mexico (Galloway et al. 1999). It is the largest freshwater marsh ecosystem in the North America. This flow-dependent, low-nutrient ecosystem has been greatly impacted by agricultural nutrient inputs, principally phosphorus, and extensive water management that has disrupted the natural water flow (Davis et al. 1994; Sklar et al. 2005; McVoy et al. 2011). In addition, rainfall during Florida's wet season acts as the main source for freshwater input into the Everglades ecosystem where many soil microhabitats remain saturated or soil moisture remains high year round. This chronic, elevated soil moisture maintains an anoxic habitat for functional guilds such as the methanogenic archaea. The ability to convert carbon compounds from, for example, hydrogen and CO_2, to methane are contained within the unique enzyme (methyl-coenzyme M reductase) found exclusively, to date, in methanogens (Ermler et al. 1997).

Because of the immediate and long-term effects humans have on this unique ecosystem, the largest restoration effort in US history is underway (WRDA 2000) to try to mitigate the problems associated with excess nutrient inputs that phosphorus and drainage canals have had on the ecosystem. Although the visible shift from sawgrass-dominated prairies to cattail plant communities is easily noted at impacted sites (Hagerthy et al. 2008), less is known about the belowground impact on the microbes responsible for carbon cycling in the system.

Effects of Phosphorus on Methanogenesis

There have been several studies that looked at the impact that excess nutrients have on the Everglades microbial communities (Pennanen et al. 1998; Bell et al. 2009) and specifically, the methanogens. Castro and colleagues (2004) studied the agricultural phosphorus runoff into the Everglades and the impact it had on the microbial community structure and function. The above ground plant biomass shifted from *Cladium* sawgrass dominating the wetland plains to dense cattail plant communities in the eutrophic areas of Water Conservation Area 2A (WCA-2A). This, in turn, changed the amount and type of carbon input into the system via detritus and root exudates and increased the organic content of the soils. These trophic shifts caused a dramatic change in the belowground biogeochemical cycling. Overall microbial activity increased as well as sulfate reduction rates and methanogenesis. By looking not only at the 16S phylogenetic marker but also the methyl-coenzyme M reductase, alpha subunit gene (*mcrA*)—the functional marker—they could assess the nutrient impact on the terminal step of carbon cycling. Compared to oligotrophic control sites, the methane production in the eutrophic zones increased dramatically. There was also an increase in the biodiversity of methanogens in the eutrophic zone compared to the oligotrophic sites. They also found the 16S clone libraries were more diverse than that of the functional libraries, not surprisingly so, since methanogenesis is limited to a few specialized archaeal clades. Looking at the *mcrA* clone libraries, they observed an obvious shift in the dominant clades and some minor clades both across the nutrient gradients and to a lesser extent, during seasons. This suggested that the nutrient pollution was affecting both the structure and the activity in the methanogenic community. Eutrophic soil was dominated by clusters related to the Methanomicrobiales (clusters MRC-5, 6, 7) and a summer appearance (rainy season) of clusters aligned with Methanobacteriales, Methanosaeta and one unknown cluster that did not align with any archived sequences. There was also a change in the dominant phylotypes as well as the appearance and disappearance of other minor clades. Shannon's diversity indices showed a decrease in diversity between the eutrophic and the oligotrophic site but little change within the sites based on season (Castro et al. 2004).

Effects of Cellulose Degradation on Methanogenesis

In another study at the same Everglades sites as the Castro 2004 study, Uz and Ogram (2006) looked at the cellulose degrading communities, specifically *Clostridium* spp. and set up soil microcosms from the eutrophic, transitional and oligotrophic sites. Sulfate reducers and methanogens depend on fermentation products and those populations were greater in number in the eutrophic and transitional zones versus the oligotrophic sites. They found the functional guilds of sulfate reducers and hydrogen-scavenging methanogens were significantly affected by the nutrient status of the soils versus the plant type or residue. Plant type did not seem to have an effect on the cellulolytic clostridia populations but did seem to influence the fermentation guilds and syntrophs responsible for the C-1 substrates needed for methanogenesis (Uz and Ogram 2006).

Therefore, enhanced knowledge of the carbon cycle and the functional guilds responsible for the carbon cycling can be very useful for understanding anthropogenic impact on the Everglades ecosystem. In addition to understanding impacts after the fact, important base line data on the functional guilds should also be gathered so as to better understand the consequences of any remediation strategy to return the impacted lands back to their "natural" state (Lovley 2003).

Restoration Effects on Methanogen Communities

Phosphorus reductions and restored hydrology efforts in the Everglades are ongoing and will have important implications for soil microbial community structure and processing of carbon. It is therefore important to monitor the changes to the belowground communities that are so integral to the overall recovery, resilience and health of this ecosystem. As hydrology is restored, it becomes even more critical to monitor the baseline functional capabilities of methanogens in both restored and impacted areas of the Everglades. In addition, physical impoundment by canal levies has created anaerobic habitats that are now flooded year-round. Removal of these levies will dramatically change both aboveground plant community and belowground microbial populations. With reduction in phosphorous, hydrological changes, and removal of levies, nutrient flow through the system can impact plant communities and thus, the carbon input into the system. This in turn can affect the rates of microbial metabolism and the pathways used, their ability (or lack of ability) to adapt to the disturbance and the changing pool of carbon available in the restored Everglades.

Shifts in plant type and the resulting detritus (Millard and Singh 2010) will drive changes in microbial community functional diversity, available electron acceptors, carbon biotransformations and ultimately, shifts in the metabolic products and substrates available to the methanogenic communities. Shifts along a nutrient gradient in the northern Everglades (Chauhan and Ogram 2006) were dominated by the aceticlastic methanogens in the eutrophic habitat and changed to a more hydrogenotrophic dominated communities in the oligotrophic zones. Therefore, the substrates used for methane production changed with the plant and nutrient levels. Smith et al. (2007) followed the recovery (and subsequent decline of some methanogen functional groups) of Hole-in-the-Donut site within the Everglades after eradication efforts removed invasive plants. Using culture-independent gene analyses of the methyl coenzyme M reductase genes (*mcrA*) and clone libraries, they showed a dominance of the hydrogenotrophic methanogens of the orders Methanobacteriales and Methanococcales and decline in the relative abundance of Methanobacteriales *mcrA* genes that were correlated with the recovery time of the site. These limited studies indicate a growing need to more clearly understand the intrinsic functional ecology of the Everglades microbes driving essential biogeochemical cycles (Smith et al. 2007).

Current Ongoing Study

Functional ecology studies of the Miami-Dade County soils in the Everglades and Biscayne watersheds—soils that will be impacted by changes in hydrology from the restoration effort—are ongoing within our research group. The diversity of methanogens is of interest to our group as part of a larger soil ecological study. Based on the hypothesis that soil type structures the microbial community that occupies a soil (Bossio et al. 1998; Fierer and Jackson 2006), the question can be asked, does it also drive the functional diversity within the soil or is function redundant within structure?

In order to assess this question, two different soil types (with similar above ground habitats), Lauderhill Dania-Pahokee (listed as soil type 2, transect, KNT in this study) and Perrine -Biscayne-Pennsuco (soil type 4, transect, CS), in Miami-Dade County, Florida (Fig. 1), were compared using the *mcrA* clone libraries and sequencing.

The objective of this particular study was to assess the methyl-coenzyme M reductase (*mcrA*) gene diversity in two different parent soil types with similar habitats—saturated and dominated by marsh grasses. As the terminal step in methanogenesis is catalyzed by the highly conserved methyl-coenzyme M reductase, the null hypothesis would be that no differences in the *mcrA* gene diversity and its taxonomic associations would be seen, regardless of soil type.

BLAST analyses and subsequent sequence alignment and phylogenetic tree construction were performed (Fig. 2). The neighbor joining consensus tree was derived from *mcrA* DNA sequences aligned using MUSCLE within Mega 5.0 software (http://www.megasoftware.net/) and bootstrap values from 1000 iterations are shown.

■ **Soil Type 2:** Lauderhill Dania-Pahokee
  Soil Type 4: Perrine-Biscayne-Pennsuco

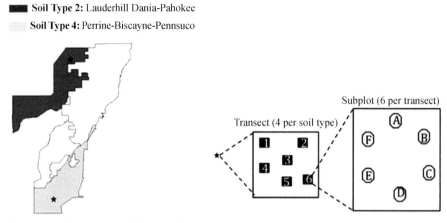

Figure 1. A schematic map of the two soil types in Miami-Dade, County, FL. Stars indicate the sampling sites and the blown up schematic, the sampling scheme.

The tree used aligned sequences from this current study, two other previously discussed Florida Everglades studies (Castro et al. 2004; Smith et al. 2007; shown in the tree as the DQ and AY sequences), other uncultured methanogens and sequences of six reference methanogens from archival databases.

Dominance of Unknown *mcrA* Sequences in the Florida Everglades Soils

Methanogens are classified into five orders: Methanococcales, Methanopyrales, Methanobacteriales, Methanomicrobiales, and Methanosarcinales. All the five orders are identified as having hydrogenotropic pathway. However, aceticlastic and methylotropic pathways are restricted to Methanosarcinales. On the basis of Bergey's taxonomy, methanogens are grouped into two classes: 1) Methanopyrales, Methanobacteriales, and Methanococcales; and 2) Methanosarcinales, and Methanomicrobiales (Garitty 2001). To date few studies (Castro et al. 2004; Castro et al. 2005; Smith et al. 2007) have assessed functional genes in methanogens in the Everglades, perhaps due to limitations in growing the archaeal species using traditional culturing methods. In the current study, there were two major clades that associated with known reference sequences. Some of the sequences from the other Florida studies (labeled with AY or DQ in the tree) more closely aligned with *Methanothermobacter thermautotrophicus* but yet many were not associated with any known reference samples. The KNT and CS labeled sequences from our study associated almost exclusively with either uncultured archaea or with the uncultured euryarchaeote from oligotrophic soils in the northern Everglades. While some of the clones from the two soil types examined in Miami-Dade County grouped together with a particular soil type, there was overlap in the clones from the different soil types.

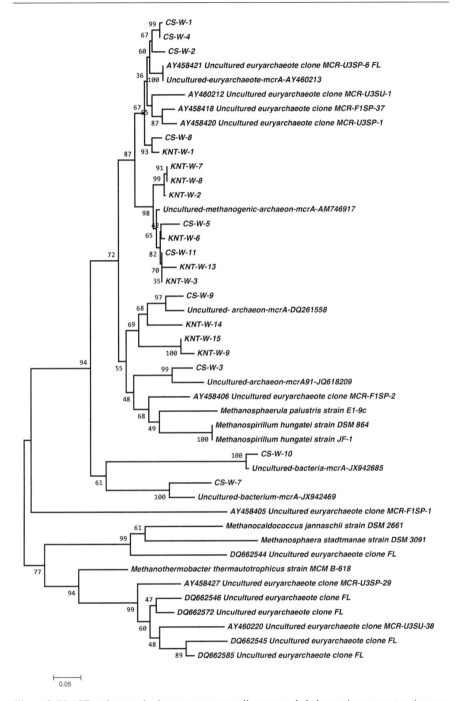

Figure 2. BLAST analyses and subsequent sequence alignment and phylogenetic tree construction were derived from *mcrA* DNA sequences and were then aligned with MUSCLE. The Neighbor Joining, linearized tree used aligned sequences from the current study, two other Florida Everglades studies (DQ and AY sequences), other uncultured methanogens and sequences of six reference methanogens.

Discussion and Conclusions

There are substantial amounts of data supporting the claim that microbial community structure is driven by soil type (Bossio et al. 1998; Dunbar et al. 2000; Marschner et al. 2001). However, evidence is lacking as to whether methanogenesis is also driven by soil type alone or is influenced more by habitat-selective factors. Many of the sequences derived from this study and others (Smith et al. 2007) are from unknown phylotypes and demonstrated the unknown biodiversity of *mcrA* genes in the Everglades and Biscayne watersheds. Results from these studies do not clearly define soil as the only determinant of phenotypic diversity. Many of the *mcrA* sequences did cluster more closely with one soil type, but not exclusively. There is a clear lack of knowledge about these critical guilds and their functions in any wetland soils, as seen by the many novel gene sequences not associated with any known *mcrA* sequences in the database. In the case of methanogenensis, the dominant anaerobic habitat conditions may more heavily influence the functional guilds than the soil's chemical and abiotic drivers.

In the studies conducted by Castro et al. (2004) and Uz and Orgam (2006), nutrient pollution in form of phosphorous coupled with hydrological alteration, influenced the *mcrA* gene diversity with shifts in the dominant sequences. In the presence of excess nutrients, nutrient limitations within microbial communities are lifted, and competitive exclusion is no longer a driver. What these changes mean in terms of ecosystem function is unknown. Will restored hydrology alone 'restore' the functional integrity of soil methanogens in the Everglades soil? Are the phosphorus legacies going to continue to affect the carbon cycling in these soils? Does functional gene diversity infer different levels of enzymatic efficiency? Or is the diversity representative of a long evolutionary history of this ancient process? It is essential to find answers to any long term effect that the phosphorous pollution and subsequent change in hydrology may have on carbon cycling in Everglades soil and devise a method to use methanogen diversity to monitor these changes.

These cumulative results are indicative of how little is still known about the evolutionary affiliation of microbes driving such critical biochemical functions as methanogenesis. The limitation of assessing only the 16S rDNA composition of the community is that metabolic capacity and function are often inferred from known phylotypes and yet, many of the *mcrA* gene sequences from this study and others had no known phylogenetic affiliation in the databases. To date, a handful of genomes from culturable methanogens have been sequenced (Liu and Whitman 2008) but as studies are finding out, these genomes do not capture the overall functional gene diversity (*mcrA*) of the methanogens in soils (Grosskopf et al. 1998).

Knowledge about the functional capabilities and diversity of these critical methanogen guilds could provide vital information in regards to the recovery effort of the Everglades. Therefore, it may be more informative to screen for functional guilds and genes rather than structural genes in order to understand the true restoration impact on ecosystem services. Understanding the baseline functional ecology of currently undisturbed sites within the Everglades restoration areas will allow for better assessment of the overall impact, recovery and resilience of the belowground microbial communities—communities that are so critical to the health of the ecosystem (i.e., the interdependency of the syntrophic bacteria that supply the hydrogen substrates for methanogenesis) (Walker et al. 2012).

The global carbon cycle is solely dependent on microbes. Microbial communities participate in the carbon cycle by either fixing carbon from the atmosphere or supplementing plant growth, or degrading organic materials in the environment. Responsibility of releasing carbon as a greenhouse gas solely lies in the functional capacity of microbes and the establishment a balanced ecosystem. Clean water, healthy soils and suppression of diseases are all benefits of healthy soil systems.

The microbiology of the Everglades should be at the core of the Everglades health assessment and more studies are needed to better protect and understand this vast 'river of grass' and its contribution to South Florida's sustainability into the future. To better understand the sustainability, resilience and future impact of climate change, rising sea levels and most importantly, ecosystem services conferred by these microbes, infers functional gene ecology has to become an important facet in microbial community studies. Perhaps it may even be more important than which microbes are present as genetic function links directly to carbon sequestration, methane and other green house emissions and can quickly signal how natural and anthropogenic disturbances impact these crucial ecosystem services (Nazaries et al. 2013). The Everglades is one of the largest wetlands in the North America and deserves protection for generations to come.

Acknowledgements

The study was supported in part by the NGA HM 1582-09-1-00 II to DKM and supported MM. JZ was supported by the MARC-USTAR program (NIH/NIGMS T34 GM083688) at FIU. YO was supported in part by the MBRS-RISE program (NIH/NIGMS R25 GM061347). The content is solely the responsibility of the authors and does not necessarily represent the official views of the National Institutes of Health or the National Geospatial Intelligence Agency.

References

Bapteste, E., C. Brochier and Y. Boucher. 2005. Higher-level classification of the Archaea: evolution of methanogenesis and methanogens. Archaea 1: 353–363.

Bell, C., V. Acosta-Martinez, N.E. McIntyre, S. Cox, D.T. Tissue and J.C. Zak. 2009. Linking microbial community structure and function to seasonal differences in soil moisture and temperature in a Chihuahuan desert grassland. Microbial Ecol. 58: 827–842.

Blaut, M. 1994. Metabolism of methanogens. A. van Leeuw. 66: 187–208.

Bossio, D.A., K.M. Scow, N. Gunapala and K.J. Graham. 1998. Determinants of soil microbial communities: effects of agricultural management, season, and soil type on phospholipid fatty acid profiles. Microbial Ecol. 36: 1–12.

Castro, H., A. Ogram and K.R. Reddy. 2004. Phylogenetic characterization of methanogenic assemblages in eutrophic and oligotrophic areas of the Florida Everglades. Appl. Environ. Microbiol. 70: 6559–6568.

Castro, H., S. Newman., K.R. Reddy and A. Ogram. 2005. Distribution and stability of sulfate-reducing prokaryotic and hydrogenotrophic methanogenic assemblages in nutrient-impacted regions of the Florida Everglades. Appl. Environ. Microbiol. 71: 2695–2704.

Chauhan, A. and A. Ogram. 2006. Phylogeny of acetate-utilizing microorganisms in soils along a nutrient gradient in the Florida Everglades. Appl. Environ. Microbiol. 72: 6837–6840.

Daniel, R. 2005. The metagenomics of soil. Nat. Rev. Microbiol. 3: 470–478.

Davis, S.M., L.H. Gunderson, W.A. Park, J.R. Richardson and J.E. Mattson. 1994. Landscape dimension, composition, and function in a changing Everglades ecosystem. pp. 419–444. *In*: S.M. Davis and

J.C. Ogden (eds.). Everglades: The Ecosystem and its Restoration. St. Lucie Press, St. Lucie Press, Delray Beach.

Dunbar, J., L.O. Ticknor and C.R. Kuske. 2000. Assessment of microbial diversity in four southwestern United States soils by 16S rRNA gene terminal restriction fragment analysis. Appl. Environ. Microbiol. 66: 2943–2950.

Ermler, U., W. Grabarse, S. Shima, M. Goubeaud and R.K. Thauer. 1997. Crystal structure of methyl-coenzyme M reductase: the key enzyme of biological methane formation. Science 278: 1457–1462.

Fierer, N. and R.B. Jackson. 2006. The diversity and biogeography of soil bacterial communities. Proc. Natl. Acad. Sci. USA 103: 626–631.

Fox, G.E., J.D. Wisotzkey and P. Jurtshuk, Jr. 1992. How close is close: 16S rRNA sequence identity may not be sufficient to guarantee species identity. Int. J. Syst. Bacteriol. 42: 166–170.

Galloway, D., D.R. Jones and S.E. Ingebritsen. 1999. Land subsidence in the United States. USGS 1182: 95–106.

Garrity, G. 2001. Bergey's Manual of Systematic Bacteriology. Springer-Verlag, New York, 721 p.

Grosskopf, R., P.H. Janssen and W. Liesack. 1998. Diversity and structure of the methanogenic community in anoxic rice paddy soil microcosms as examined by cultivation and direct 16S rRNA gene sequence retrieval. Appl. Environ. Microbiol. 64: 960–969.

Hagerthy, S.E., S. Newman, K. Rutchey, E.P. Smith and J. Godin. 2008. Multiple regime shifts in a subtropical peatland: Community specific thresholds to eutrophication. Ecol. Monographs 78: 547–565.

Handelsman, J., M.R. Rondon, S.F. Brady, J. Clardyand and R.M. Goodman. 1998. Molecular biological access to the chemistry of unknown soil microbes: a new frontier for natural products. Chem. Biol. 5: R245–R249.

Horswell, J., S.J. Cordiner, E.W. Maas, T.M. Martin, K.B.W. Sutherland, T. Speir, B. Nogales and A.M. Osborn. 2002. Forensic comparison of soils by bacterial community DNA profiling. J. Forensic Sci. 47: 350–353.

Kakirde, K.S., L.C. Parsley and M.R. Liles. 2010. Size does matter: Application-driven approaches for soil metagenomics. Soil Biol. Biochem. 42: 1911–1923.

Liu, Y. and W.B. Whitman. 2008. Metabolic, phylogenetic, and ecological diversity of the methanogenic archaea. Ann. NY Acad. Sci. 1125: 171–189.

Lovley, D.R. 2003. Cleaning up with genomics: applying molecular biology to bioremediation. Nat. Rev. Microbiol. 1: 35–44.

Marschner, P., C.H. Yang, R. Lieberei and D.E. Crowley. 2001. Soil and plant specific effects on bacterial community composition in the rhizosphere. Soil Biol. Biochem. 33: 1437–1445.

McVoy, C.W., W.P. Said, J. Obeysekera, J.A. Van Arman and T.W. Dreschel. 2011. Landscapes and hydrology of the predrainage Everglades. University Press of Florida.

Millard, P. and B. Singh. 2010. Does grassland vegetation drive soil microbial diversity? Nutr. Cycl. Agroecosys. 88: 147–158.

Nazaries, L., Y. Pan, L. Bodrossy, E.M. Baggs, P. Millard, J.C. Murrell and B.K. Singh. 2013. Evidence of microbial regulation of biogeochemical cycles from a study on methane flux and land use change. Appl. Environ. Microbiol. 79: 4031–4040.

NRCS, U. 2009. Soil Survey of Dade County Area, Florida.

Osborn, A.M., E.R.B. Moore and K.N. Timmis. 2000. An evaluation of terminal-restriction fragment length polymorphism (T-RFLP) analysis for the study of microbial community structure and dynamics. Environ. Microbiol. 2: 39–50.

Pennanen, T., H. Fritze, P. Vanhala, O. Kiikkila, S. Neuvonen and E. Baath. 1998. Structure of a microbial community in soil after prolonged addition of low levels of simulated acid rain. Appl. Environ. Microbiol. 64: 2173–2180.

Reeve, J.N., J. Nölling, R.M. Morgan and D.R. Smith. 1997. Methanogenesis: genes, genomes, and who's on first? J. Bacteriol. 179: 5975–5986.

Robinson, C.J., B.J. Bohannan and V.B. Young. 2010. From structure to function: the ecology of host-associated microbial communities. Microbiol. Mol. Biol. Rev. 74(3): 453–476.

Sklar, F.H., M.J. Chimney, S. Newman, P. McCormick, D. Gawlik, S. Miao, C. McVoy, W. Said, J. Newman, C. Coronado, G. Crozier, M. Corvela and K. Rutchey. 2005. The ecological–societal underpinnings of Everglades restoration. Front. Ecol. Environ. 3: 161–169.

Smith, J.M., H. Castro and A. Ogram. 2007. Structure and function of methanogens along a short-term restoration chronosequence in the Florida Everglades. Appl. Environ. Microbiol. 73: 4135–4141.

Stackebrandt, E. and B.M. Goebel. 1994. Taxonomic note: a place for DNA-DNA reassociation and 16S rRNA sequence analysis in the present species definition in bacteriology. Int. J. Syst. Bacteriol. 44: 846–849.

Stackebrandt, E. and J. Ebers. 2006. Taxonomic parameters revisited: tarnished gold standards. Microbiol. Today 33: 152–155.

Sundquist, A., S. Bigdeli, R. Jalili, M.L. Druzin, S. Waller, K.M. Pullen, Y.Y. El-Sayed, M.M. Taslimi, S. Batzoglou and M. Ronaghi. 2007. Bacterial flora-typing with targeted, chip-based Pyrosequencing. BMC Microbiol. 7: 108–118.

Torsvik, V., J. Goksøyr and F.L. Daae. 1990. High diversity in DNA of soil bacteria. Appl. Environ. Microbiol. 56(3): 782–787.

Torsvik, V. and L. Øvreås. 2002. Microbial diversity and function in soil: from genes to ecosystems. Curr. Opin. Microbiol. 5: 240–245.

Travers, R.S., P.A. Martin and C.F. Reichelderfer. 1987. Selective process for efficient isolation of soil *Bacillus* spp. Appl. Environ. Microbiol. 53: 1263–1266.

Uz, I. and A.V. Ogram. 2006. Cellulolytic and fermentative guilds in eutrophic soils of the Florida Everglades. FEMS Microbiol. Ecol. 57: 396–408.

Walker, C.B., A.M. Redding-Johanson, E.E. Baidoo, L. Rajeev, Z. He, E.L. Hendrickson, M.P. Joachimiak, S. Stolyar, A.P. Arkin, J.A. Leigh, J. Zhou, J.D. Keasling, A. Mukhopadhyay and D.A. Stahl. 2012. Functional responses of methanogenic archaea to syntrophic growth. ISME J. 6: 2045–2055.

Water Resources Development Act. 2000. Pub. L. No. 106–541, 114 Stat. 2711.

Wieland, G., R. Neumann and H. Backhaus. 2001. Variation of microbial communities in soil, rhizosphere, and rhizoplane in response to crop species, soil type, and crop development. Appl. Environ. Microbiol. 67: 5849–5854.

Ecological Perspective on the Associations of Syntrophic Bacteria, Methanogens and Methanotrophs in the Florida Everglades WCA-2A Soils

Ashvini Chauhan,[1,*] *Ashish Pathak*[1] and *Andrew Ogram*[2]

Introduction

Historically, phosphorus (P) runoff from the Everglades Agricultural Area (EAA) has caused significant alterations in both vegetation type and density that have been shown to correlate with changes in the composition and activities of native soil microbial communities (Castro et al. 2004; 2005; Chauhan and Ogram 2006a;b). This chapter highlights findings from studies of methanogenic consortia conducted along the P gradient in Water Conservation Area 2A (WCA2A) of the northern Florida Everglades. Phosphorus runoff from the EAA into WCA2A produced a gradient in P-concentrations in soil and water, providing an excellent system for studying the

[1] School of the Environment, Environmental Biotechnology Lab, 1515 S. MLK Blvd., Suite 305B, Building FSHSRC, Florida A&M University, Tallahassee, FL-32307, USA
  Email: ashishpathak72@gmail.com
[2] Soil and Water Science Department, University of Florida, PO Box 110290, Gainesville, FL 32611, USA.
  Email: aogram@ufl.edu
* Corresponding author: ashvini.chauhan@famu.edu

response of the Everglades marshland microbial communities to nutrient enrichment. The study sites include three stations along the gradient, including the eutrophic regions (station F1), contrasted with the transition region (station F4) and the oligotrophic region (station U3).

Fatty Acid-oxidizing Hydrogen-producing Syntrophic Bacteria and the Methane-producing, Methanogens in the Everglades Marshland

As is typical of any wetland, detritus and other organic matter is decomposed in the Everglades marshland by a network of functional groups of microorganisms that may respond rapidly to changes in aboveground environmental conditions. The general scheme of organic matter decomposition in a wetland involves the primary fermentative bacteria that convert the monomeric carbohydrates to small alcohols and fatty acids, such as propionate and butyrate. The fatty acids are then taken up by secondary fermentative bacteria, also known as fatty acid- oxidizing, hydrogen-producing syntrophic bacteria (Fig. 1).

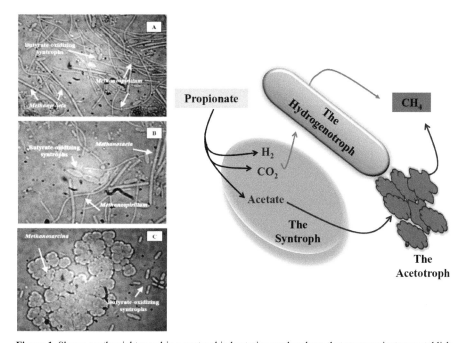

Figure 1. Shown on the right panel is a syntrophic bacterium and archaea that communicate to establish a syntrophic (cross-feeding) relationship of mutual benefit. The bacterium ferments propionate to acetate, CO_2, and H_2, a highly endergonic or energy unfavorable conversion. The reaction is driven by the hydrogen being siphoned off by the archaeon for methanogenesis. Propionate fermentation can proceed as long as the partial pressure of H_2 is kept low by the archaeon. The left panel shows images obtained from differential interference contrast microscopy of consortia enriched from F1 (eutrophic), F4 (transition), and U3 (oligotrophic) microcosms spiked with VFAs. A, consortium from F1 (eutrophic) microcosm. *Methanosaeta* characteristic of long filaments with blunt ends, juxtaposed with butyrate-oxidizing syntrophic bacteria; B, consortium from F4 (transition) microcosm; C, consortium from U3 (oligotrophic) microcosm. *Methanosarcina* in clumps with very few presumed fatty-acid oxidizing bacteria.

Syntrophs occur and function in close proximity to the methanogens to provide them with substrates used for methane production: H_2 and acetate. Energetic constraints require that syntrophs be exposed to low H_2 concentrations. Hydrogenotrophic methanogens typically live adjacent to syntrophs, thereby providing the methanogen with H_2 for reduction of CO_2 to methane, while maintaining the necessary low H_2 concentrations required for secondary fermentation. Acetate is consumed by a separate physiological group, the acetotrophs (also known as aceticlasts), which ferment acetate to methane, as outlined in Fig. 1.

The composition of microorganisms participating in fermentation, syntrophy and methanogenesis is governed by a suite of environmental factors, including the quality and quantity of electron donors and acceptors, and concentrations of nutrients. The physiological activities of these microorganisms, in turn, determine the pathway through which biogeochemical cycles proceed in the marshland soils. Therefore, shifts in the structure and compositional makeup of these consortia may be indicative of changes in the environment and may result in changes in specific function(s) of the consortia. Obtaining a comprehensive understanding of the microbial groups responsible for these critical marshland productivity processes will potentially serve in the development of sensitive and precise indicators of nutrient status in the Everglades, and additionally, characterization of the carbon flow through these food-webs will provide needed insights into the underlying mechanisms through which nutrient-loading impacts ecosystem level processes and health in the Everglades marshland.

Of major interest is the methanogenesis process because methane can be a potent greenhouse gas and it can also serve as an indicator of ecosystem productivity. Specifically, studies performed across a variety of wetlands, including the Everglades, have shown that ecosystem function and methane emissions are strongly correlated such that higher the productivity of microorganisms that form methane, higher is the rate and concentration of methane produced (Whiting and Chanton 1993; 1992; Bellasario et al. 1999; Chanton et al. 1995; Chasar et al. 2000). Enhanced methane formation is typically observed in nutrient-loaded wetlands, and natural wetlands account for over 20% of global methane budget. CH_4 is approximately 20-fold more potent as a greenhouse gas than CO_2, and therefore, a complete understanding of the methane producing pathways and how these pathways are influenced by eutrophication will facilitate refined models of CH_4 production and emission from the Everglades. Such studies will also likely be the basis for estimating productivity and environmental inputs of CH_4 from other eutrophic marshlands and peatlands.

Primary fermentative bacteria of the *Clostridiaceae* family ferment the monomers into volatile fatty acids (VFAs) such as propionate and butyrate in the Everglades marshland soils (Uz et al. 2007). As shown in Fig. 1, the VFAs are then oxidized by the syntrophic secondary fermentative bacteria, resulting in the formation of acetate, CO_2, and H_2. To investigate the pathways through which carbon is processed by syntrophs in the Florida Everglades, three sites along the nutrient gradient in WCA2A were studied. In addition to differences in activities and numbers, syntrophic consortia in enrichments from eutrophic regions of WCA-2A differed in composition from those from oligotrophic regions, indicative of different selective forces that are likely present in the nutrient-loaded regions. To this end, studies on other systems have shown that fatty acids were utilized more rapidly in tricultures of *Methanospirillum hungatei* (hydrogenotroph), *Methanothrix soehngenii* (acetate utilizer) and MPOB (mesophilic propionate oxidizing bacteria) than in biculture (*Methanospirillum* with MPOB), indicating that low acetate and H_2 concentrations were favorable for syntrophy in those systems (Voolapalli and Stuckey 1999; Imachi et al. 2001). Butyrate utilization has been shown to depend on acetate levels, such that accumulation of acetate thermodynamically controls the extent of substrate degradation (Warikoo 1996; Jackson and McInerney 2002). In line with these observations, we also found tripartite member consortia in F1 and F4 enrichments, but not in U3 enrichments (Fig. 1A–C), which could affect the methane production pathways (Hines et al. 2001; 2008) and rates as have been reported from the nutrient-enriched regions of the Florida Everglades (Castro et al. 2004; 2005; Chauhan et al. 2004; Ogram et al. unpublished). Our findings provide a strong foundation to indicate that the syntrophic bacteria varied as a function of the soil nutrient status, and the specific distribution of these guilds provides critical information regarding the effects of P-enrichment on the flow of carbon cycling pathways in the Everglades.

Comparisons using predicted versus observed methane production from propionate, butyrate and acetate amended microcosms (Chauhan et al. 2004; Chauhan and Ogram 2006) were also drawn. For this calculation, we used previous estimates showing that complete oxidation of 1 mol of propionate generates 1.75 mol of CH_4, 1 mol of butyrate generates 2.5 mol of CH_4 and 1 mol of acetate utilization generates 1 mol of CH_4, respectively (Mormile et al. 1996; Schink 1997). We found that methanogenesis rates from amended microcosms containing F1 soils spiked with propionate, yielded 65% of the predicted CH_4; F4, 69%, and U3, 44%. When butyrate was used as the substrate, F1 microcosms yielded 67% of the predicted CH_4; F4, 81%, and U3, 44% and finally when acetate was the provided substrate, F1 microcosms produced 78% of the predicted CH_4; F4, 89%, and U3, 66%, respectively. This clearly showed that U3 microcosms produced much less methane than was theoretically expected, compared to other sites, suggesting existence of distinctly different microbial activities or pathways in eutrophic than in oligotrophic soils of the Everglades marshland.

To further link the structure to the function of syntrophic bacteria, additional studies were conducted using stable isotope probing (SIP). Briefly, SIP was conducted by spiking ^{13}C-propionate and 13C-butyrate to a concentration of 10 mM in anaerobic soil slurry microcosms, and the assimilation of ^{13}C by the syntrophs were followed (Chauhan and Ogram 2006a). As shown in Fig. 2, for propionate-amended microcosms, 16S rRNA gene clone libraries from P-enriched and transition sites were dominated by sequences related to previously described propionate oxidizers, such as *Pelotomaculum* spp. and *Syntrophobacter* spp.

Of major interest was the significant representation of 16S rDNA sequences clustering with *Smithella propionica*, an organism that is known to dismutate propionate to butyrate, i.e., propionate is first dismutated to acetate and butyrate and then degraded via β-oxidation (de Bok et al. 2001; Liu et al. 1999). Therefore, it is possible that a multiple trophic interaction between propionate- and butyrate-oxidizing syntrophic bacteria occurred during the degradation of propionate in the nutrient-enriched soils of the Florida Everglades. In addition, sequences associated with known sulfate reducing prokaryotes (SRP) constituted up to 20% of the total

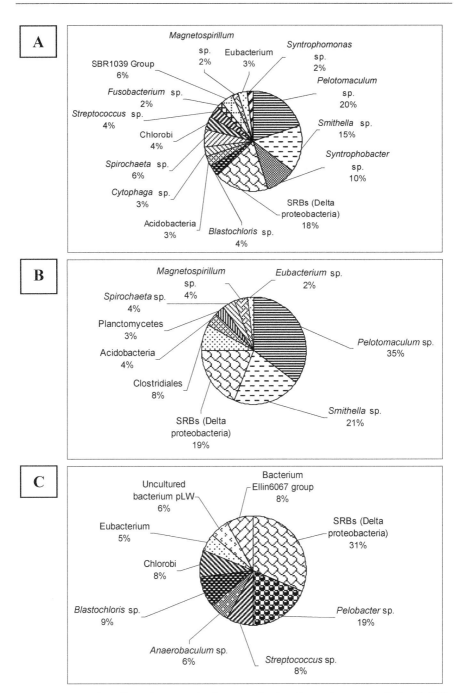

Figure 2. Distribution of bacterial clones in [¹³C]-DNA fractions from microcosms spiked with [¹³C]-propionate. A, F1 (eutrophic); B, F4 (transition); and C, U3 (oligotrophic) microcosms. Adapted from Chauhan and Ogram (2006a).

relative phylotype abundance within the impacted sites. Unlike the nutrient enriched sites, sequences of dominant phylotypes from the oligotrophic samples did not cluster with known syntrophic bacteria but were taxonomically related to various sulfate-reducing prokaryotes and *Pelobacter* spp.

In the ^{13}C-butyrate-amended microcosms, sequences clustering with delta-proteobacteria or the low-G+C gram-positive groups were found, many associated with known syntrophs such as *Syntrophospora* spp. and *Syntrophomonas* spp. (Chauhan et al. 2006a). Sequences related to *Pelospora* spp. and SRPs dominated clone libraries from oligotrophic microcosms. As with the propionate library, the dominant cluster in the U3 [13C]-butyrate library was associated with the SRPs. Moreover, in contrast to the U3 propionate library, sequences clustering with *Clostridium* sp. comprised a significant portion (26%) in the U3 [13C]-butyrate library (Chauhan et al. 2006a). *Clostridium* species are typically primary fermenters, but some may have been misclassified and hence can potentially function as syntrophs (Zhao et al. 1990).

When we performed UniFrac PCA analyses on the sequences from the unlabeled and labeled soil microcosm experiments, we found that bacterial (Fig. 3A) and archaeal (Fig. 3B) communities from the eutrophic soils clustered away and separately to the transition and oligotrophic sites respectively.

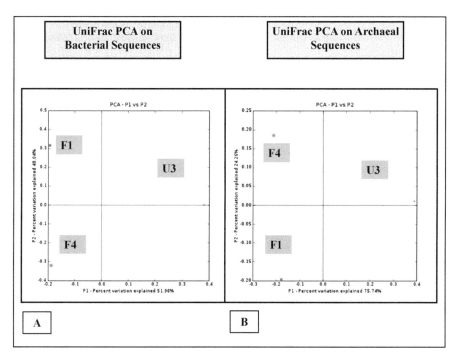

Figure 3. First two coordinates from a principal coordinate analysis (PCA) of sequences belonging to A. bacteria and B. archaeal spp., originating from soils collected from F1 (eutrophic), F4 (transition) and U3 (oligotrophic). Microcosms were established by spiking either [13C]-propionate, [13C]-butyrate or [13C]-acetate, respectively. Percentages represented in the axis labels are percentages of variation that are explained by the principal coordinates.

Additionally, significant differences (P-values < 0.001) were found to exist between eutrophic, transition and oligotrophic carbon cycling bacteria and archaea in the Florida Everglades.

Because the carbon is transferred from the syntrophic bacteria onto the methanogens, we also probed the ^{13}C-DNA fractions to investigate the methane producing partners. These analyses showed that the archaeal sequences from the eutrophic ^{13}C microcosms belonged to *Methanomicrobiaceae*, *Methanospirillaceae*, and *Methanosaetaceae* families of methanogens, respectively. Conversely, the low P-microcosms were dominated by acetotrophs, including sequences taxonomically related to *Methanosarcina*, suggesting easier availability of acetate in these microcosms.

In addition, stable isotope studies using ^{13}C-acetate showed distinct differences between the acetate-utilizing microbial guilds in eutrophic and oligotrophic libraries. Specifically, in F1 and F4 clone libraries, 35 to 40% of the sequences clustered as a distinct clade within the *Syntrophus* spp. (Chauhan and Ogram 2006b), which are known to syntrophically oxidize a variety of compounds but not acetate. The predominant sequences in the U3 library clustered mainly with *Geobacter* spp. and *Clostridium* spp. which are well known to utilize acetate (Coates et al. 1998; Cervantes et al. 2000). In addition, sequences representing SRPs formed significant proportions of all libraries. Overall, ^{13}C-acetate studies showed that it is very likely that most acetate is consumed by syntrophs in nutrient impacted soils, and by methanogens and SRPs in the oligotrophic soils of the Florida Everglades (Chauhan and Ogram 2006a;b).

When carbon is microbially-cycled in methanogenic environments, approximately 70% of the methane produced is from acetate via the activity of acetotrophic methanogenesis. The remaining 30% of CH_4 is produced from the reduction CO_2 by H_2 by the hydrogenotrophic methanogenesis (Conrad 1999). Several exceptions to this stoichiometry of methanogenesis have been demonstrated where the majority of methane is produced exclusively by the hydrogenotrophic route, including the Everglades marshland (Castro et al. 2004; 2005; Chauhan et al. 2004; Ogram et al. unpublished). Significantly higher methane formation from the eutrophic Everglades soils can be due to the differences in the composition and activities of methanogenic assemblages due to P-loading. Specifically, methanogens from P-enriched and transition regions were shown to be dominated by hydrogenotrophic methanogens, with significantly greater most probable numbers of hydrogenotrophs than of acetotrophs (Chauhan et al. 2004; Castro et al. 2004; 2005). This general trend has been supported by recent work (Bae and Ogram, unpublished data) by qPCR with group specific primers. The reason for the relatively high proportion of hydrogenotrophs to acetotrophs is not known at this time.

Methane-oxidizing Methanotrophs in the Everglades Marshland

Numerous studies in various environments have demonstrated that relatively little of the methane produced in the marshland soils is released into the environment (Westerman 1993; Boon and Lee 1997). This is largely attributed to the activities of methane oxidizing bacteria (MOB), or methanotrophs. MOB consume methane as their source of carbon and energy; the first step in the biochemical pathway is initiated by the conversion of methane to methanol by methane monooxygenase (MMO). MMO is found as a membrane-associated form (pMMO) and as a soluble, cytoplasmic form (sMMO). Methanotrophs are a diverse group at the metabolic and phylogenetic levels (Hanson and Hanson 1996), but are broadly divided into either Type I, affiliated with *Gammaproteobacteria,* and Type II, affiliated with *Alphaproteobacteria.* Among this division, the Type II MOB are metabolically more efficient in oxidizing methane at lower O_2 concentrations (Amaral and Knowles 1995).

King (1994) previously showed significant methane oxidation in the Everglades peat soils, consuming as much as 91% of the potential methane diffusive flux; however, little is known of the types of MOB that may be metabolically active in these environments. Potential methane oxidation rates in eutrophic (F1) soils were at least 2-fold higher compared with U3 soils. Additionally, across all soils, highest methane oxidation were observed from the top 0–2 cm soil fraction and declined rapidly in the 8–10 cm depths (Chauhan et al. 2012). In accordance with methane oxidation, MPNs showed that methanotroph cell numbers in the F1 soils were approximately 4-log higher (10^{11} MPN/g) than the U3 soils (10^9 MPNs/g). Transition (F4) soil methanotrophic bacterial numbers mirrored those observed in the eutrophic soils.

Additionally, as shown in table 1, Pearson and Spearman R square values analyzed on methane oxidation rates indicated a slightly better correlation of methane oxidation rates between F4 and U3 than those measured between F1 and U3; R values that fall between 0–1.0 indicates that the measured variables correlate well such that they tend to increase or decrease together. In other words, this analysis suggests that if methane oxidation rates increase in F1 and F4 soils, the rates will increase in U3 soils as well; if rates decrease in these sites, then U3 will also show reduced methane oxidation rates. Moreover, ANOVA and Chi-square p-values (< 0.0001) also showed significant differences in cell numbers and methane oxidation rates along the nutrient gradient.

Therefore, it appears that the 0–10 cm Everglades soils existing in close vicinity of the predominant plant communities have the propensity to oxidize methane produced and released by the lower anoxic soils. These findings are similar to those shown for the composition and activities of microbial communities that drive respiration (Reddy et al. 1999), syntrophy (Chauhan et al. 2004; 2006a;b), methanogenesis (Chauhan et al. 2004), and sulfate reduction (Castro et al. 2002; 2005), which are enhanced in the nutrient enriched areas of the Everglades.

Similar to the studies on syntrophy, the ability to link the structure of metabolically active MOB to their function along the nutrient gradient in the Florida Everglades was provided by stable isotope probing analysis (SIP) using $^{13}CH_4$ followed by PCR-cloning of the alpha subunit of particulate methane monooxygenase (*pmo*A) gene. This showed that *pmo*A sequences related to type I methanotrophs were present in both eutrophic and oligotrophic soil microcosms suggesting that majority of methane is oxidized by type I methanotrophs in these environments (Chauhan et al. 2012) (Fig. 4).

Relative abundance and taxonomic analysis of *pmo*A gene sequences revealed that the most dominant sequences (63%) retrieved from the eutrophic (F1) SIP microcosms affiliated approximately between 86 and 90% homology with *Methylobacter pmoA* sequences, followed by *Methylomonas* spp. (31%). In the transition (F4) microcosms, *Methylococcus* spp. dominated at 44% followed by *Methylobacter* (27%) and *Methylomonas* spp. (21%). As shown in Fig. 4, the *Methylobacter*-like operational taxonomic units (OTUs) identified from this study grouped into two clades (groups I

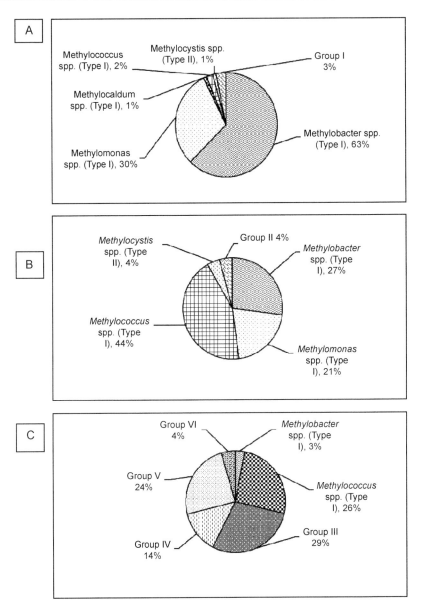

Figure 4. Relative abundance of functionally active methane-oxidizing bacteria (MOB) in $^{13}CH_4$ microcosms established with soils collected from F1 (eutrophic), F4 (transition) and U3 (oligotrophic) soils, based on frequencies of *pmoA* genes in clone libraries constructed from the [^{13}C]-DNA fractions. Adapted from Chauhan et al. (2012).

and II) (Chauhan et al. 2012). Group I sequences from F1 soils clustered with *pmoA* sequence of *Methylobacter* sp. SV96, and the F-4 soil group II was related to *pmoA* from *Methylomonas methanica*. A minor fraction of the F1 clone library also contained *pmoA* gene sequences from *Methylococcus* (2%) and *Methylocaldum* (1%). *Methylocystis* spp., were also identified from F1 soils (1%) and F-4 soils (4%), respectively. In the oligotrophic (U3) SIP microcosms, four additional *pmo*A gene clades were identified that clustered between 86–90% with *Methylocaldum szegediense* (Chauhan et al. 2012). Additionally, *Methylococcus* (26%) and *Methylobacter* sp. (3%) were also identified in U3 SIP microcosms. It remains unclear at this time if the group I–VI *pmoA* sequences represent novel lineages specific to the Florida Everglades but it is interesting to speculate that these unique *pmoA* gene sequences might belong to specialized methane-oxidizing bacteria that are ecologically adapted to the Florida Everglades habitat.

We found that *pmo*A sequence diversity was statistically different between F1, F4 and U3 soils (Table 1). Specifically, both Shannon's species diversity (H), evenness (E) declined in F1 and F4 soils when compared with U3 soils. Other statistical analyses performed included Simpson's diversity index, Chao 1 and 2 richness estimates at log-linear 95% confidence intervals, abundance and incidence-based coverage estimator of species richness, first and second-order Jackknife richness estimator, bootstrap richness estimator, Michaelis-Menten richness estimator, and Fisher's alpha diversity index; these analyses confirmed that methanotroph diversity declined in the nutrient enriched soils of the Florida Everglades (Table 1).

Table 1. Statistical analyses on the methanotroph *pmo*A gene sequence diversity in the ^{13}C-microcosms constructed with soils from F1, F4 and U3.

Site	Shannon's Diversity (H)	Shannon's Evenness (E)	Simpson's Diversity	Chao1 Estimate	Chao2 Estimate	α-Diversity Index
F1	0.39	0.5	3.7	6.4 (13; 5.8)*	14.8 (48; 7.7)	1.3
F4	0.3	0.44	5.6	9.4 (15; 8.8)	19.4 (66; 10)	1.8
U3	0.63	0.81	6.3	11 (11; 110	23.2 (79; 13)	2.2

*Upper and lower confidence limits are shown in parentheses.

Ecological Perspective and Future Directions

The Everglades Comprehensive Restoration Plan (1999) includes removal of P from water and altering current water flow patterns to approximate historic flows. The Florida Everglades is unique with variable landscape providing habitats and specific roles played by microbial communities in the decomposition of organic matter coupled with the biogeochemical cycling of phosphorus, sulfur, and mercury. Characterization of the compositions and activities of microbial communities in the Everglades therefore provides important information regarding fundamental processes impacted by human activities, as well as information required for development of restoration strategies. With regard to the flow of carbon through the trophic groups of syntrophic bacteria, methanogens, sulfate reducers and the methanotrophs, the nutrient-enriched marshland soils differ significantly from their counterpart oligotrophic soils. This conceptual model of carbon flow is depicted in Fig. 5.

Specifically, syntrophic bacteria, hydrogenotrophic methanogens and Types I and II methanotrophs are the principal assemblages through which carbon is mineralized to methane and CO_2 in soils from the P-enriched regions. Conversely, in the oligotrophic

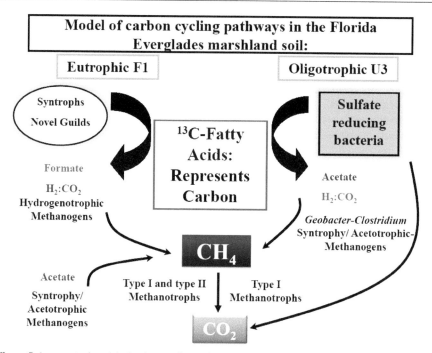

Figure 5. A conceptual model of carbon cycling pathways in the P-impacted and non-impacted marshland soils of the Florida Everglades. Distinct differences were observed with P-enrichment such that carbon was primarily recycled via strong linkages between the syntrophs, the hydrogenotrophic methanogens and the type I and II methanotrophs. Conversely, in the non-impacted sites, carbon was recycled mainly by the sulfate reducing prokaryotes (SRPs).

soils, this route of carbon is likely through the sulfate reducing prokaryotes (SRP), which generate CO_2, respectively.

It is generally accepted that a reduction of species richness occurs as a direct consequence of pollution, whereby, unperturbed species-rich assemblages are typically evenly distributed. Subsequent to perturbation, species-rich guilds are replaced by species-poor ones exhibiting high dominance (Magurran and Phillip 2001). Correspondingly, the diversity and evenness analyses of the Everglades soil microorganisms indicated that competition resulted in domination by certain species that are able to survive in the highly productive, nutrient-rich soils; conversely, oligotrophic soils consisted of a more even distribution of bacteria with less degree of competition between species.

Despite the understanding these studies have made available on the impacts of nutrient-loading to marshland soil microbiota, very little is currently known of the potential roles of these guilds during the restoration trajectory occurring in the Florida Everglades. In particular, can some of the soil fungi, bacteria and archaea serve as pioneer species or early colonizers and set the stage for reducing the course of restoration in the Florida Everglades, as depicted in Fig. 6.

Conversely, it could also be that soil microbiota merely colonize the niches made available during the course of restoration and do not play a major role during the

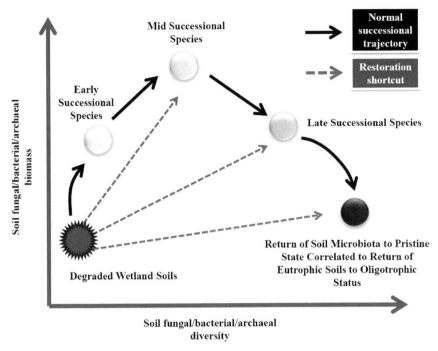

Figure 6. Potential roles of soil microorganisms during the restoration course of degraded marshland soil environments. Because soil microorganisms rapidly change as a function of their environment, of major interest is the possibility of manipulating some of the shifts observed in the P-impacted soils such that quick reversals of the biogeochemical cycles in the impacted marshland soils can be accomplished.

restoration process. This remains a debatable area in restoration ecology (Harris 2009) and clearly more research is required before we can begin to address these knowledge gaps especially in context to the Florida Everglades.

Conclusions

The Florida Everglades represents a unique freshwater marshland, consisting of a myriad of interspersed ecosystems, including ridges, sloughs, marshes, and tree islands. The Florida Everglades is also the site of the largest restoration project ever undertaken. The restoration attempts are part of a comprehensive plan to first understand the different ecological components of the Everglades and how these entities interact to drive the predominant biogeochemical cycles. Such extensive studies will potentially facilitate the reversal of disturbances caused by drainage and nutrient loading to this marshland.

Decades of study on the Florida Everglades has documented that the northern locations of this marshland have become nutrient loaded, specifically in phosphorus and sulfate, and this enrichment correlates strongly to shifts in anaerobic carbon cycling processes (Chauhan et al. 2004; Chauhan and Ogram 2006a; Castro et al. 2004; 2005). Most restoration efforts have focused on water quality, with little if any, emphasis given to the marshland microorganisms that are the chief drivers of nutrient cycling in wetlands. Changes in animal or plant communities have been studied as indicators of ecosystem level changes following anthropogenic disturbances. A growing body of evidence now suggests that microorganisms are more sensitive to environmental change than higher organisms, including plants and animals (Tokeshi 1993; Magurran and Phillip 2001). In fact, relationships between diversity and recovery of microbial functions induced by different types of stresses have been characterized, including changes in land use, soil organic carbon (Griffiths et al. 2001; 2008; Lewis et al. 2010) and other anthropogenic impacts (Coke et al. 1987; Chauhan et al. 2004; 2006; Cleveland et al. 2007; Castro et al. 2010). These studies indicate that microbial diversity shifts and resilience is relative to the functions being studied, viable species, their interactions in a particular environment being studied and physicochemical structure which directly shapes microbial community composition and associated biogeochemical functions (Griffiths et al. 2008). Whether soil microbiota serve as pioneer species and facilitate recovery processes of degraded environments or merely colonize the niches being made available during the reestablishment of higher life forms remains to be rigorously tested and shown (Harris 2009). Regardless, a better understanding of wetland soil microbial communities and how they respond to anthropogenic inputs will facilitate development of bioindicators of human stressors and also provide a better understanding of the impacts of nutrient loading to microbial communities and the biogeochemical cycles they control.

References

Amaral, J.A. and R. Knowles. 1995. Growth of methanotrophs in methane and oxygen counter gradients. FEMS Microbiol. Lett. 126: 215–220.

Bellisario, L.M., T.R. Moore, J.L. Bubier and J.P. Chanton. 1999. Controls on methane emission from a northern peatland. Global Biogeochemical Cycles 13: 81–91.

Boon, P. and K. Lee. 1997. Methane oxidation in sediments of a floodplain wetland in south-eastern Australia. Lett. Appl. Microbiol. 25: 138–142.

Castro, H., K.R. Reddy and A. Ogram. 2002. Composition and function of sulfate reducing prokaryotes in eutrophic and pristine areas of the Florida Everglades. Appl. Environ. Microbiol. 68: 6129–6137.

Castro, H., S. Newman, K.R. Reddy and A. Ogram. 2005. Distribution and stability of sulfate reducing prokaryotic and hydrogenotrophic methanogenic assemblages in nutrient impacted regions of the Florida Everglades. Appl. Environ. Microbiol. 71: 2695–2704.

Castro, H.F., A. Ogram and K.R. Reddy. 2004. Phylogenetic characterization of methanogenic assemblages in eutrophic and oligotrophic areas of the Florida Everglades. Appl. Environ. Microbiol. 70: 6559–6568.

Castro, H.F., A.T. Classen, E.E. Austin, R.J. Norby and C.W. Schadt. 2010. Soil microbial community responses to multiple experimental climate change drivers. Appl. Environ. Microbiol. 76: 999–1007.

Cervantes, F.J., S. van der Velde, G. Lettinga and J.A. Field. 2000. Competition between methanogenesis and quinone respiration for ecologically important substrates in anaerobic consortia. FEMS Microbiol. Ecol. 34: 161–171.

Chanton, J., J. Bauer, P. Glaser, D. Siegel, E. Ramonowitz, S. Tyler, C. Kelley and A. Lazrus. 1995. Radiocarbon evidence for the substrates supporting methane formation within northern Minnesota peatlands. Geochim. Cosmochim. Acta 59: 3663–3668.

Chasar, L., J. Chanton, P. Glaser, D. Siegel and J. Rivers. 2000. Radiocarbon and stable carbon isotopic evidence for transport and transformation of dissolved organic carbon, dissolved inorganic carbon and CH_4 in a northern Minnesota peatland. Global Biogeochemical Cycles 14: 1095–1108.

Chauhan, A. and A. Ogram. 2006a. Stable isotope probing of fatty acid oxidizing guilds in the Florida Everglades. Appl. Environ. Microbiol. 72: 2400–2406.

Chauhan, A. and A. Ogram. 2006b. Phylogeny of acetate utilizing microorganisms in soils along a nutrient gradient in the Florida Everglades. Appl. Environ. Microbiol. 72: 6837–6840.

Chauhan, A., A. Ogram and K.R. Reddy. 2004. Syntrophic-methanogenic associations along a nutrient gradient in the Florida Everglades. Appl. Environ. Microbiol. 70: 3475–3484.

Chauhan, A., A. Ogram and K.R. Reddy. 2006. Syntrophic archaeal associations differ with nutrient impact in a freshwater marsh. J. Appl. Microbiol. 100: 73–84.

Chauhan, A., A. Pathak and A. Ogram. 2012. Composition of methane-oxidizing bacterial communities as a function of nutrient loading in the Florida everglades. Microb. Ecol. 64: 750–9.

Cleveland, C.C., D.R. Nemergut, S.K. Schmidt and A.R. Townsend. 2007. Increases in soil respiration following labile carbon additions linked to rapid shifts in soil microbial community composition. Biogeochemistry 82: 229–240.

Coates, J.D., D.J. Ellis, E. Roden, K. Gaw, E.L. Blunt-Harris and D.R. Lovley. 1998. Recovery of humics-reducing bacteria from a diversity of sedimentary environments. Appl. Environ. Microbiol. 64: 1504–1509.

Coke, L.B., C.C. Weir and V.G. Hill. 1987. Environmental impact of bauxite mining and processing in Jamaica. Soc. Econ. Stud. 36: 289–333.

Conrad, R. 1999. Contribution of hydrogen to methane production and control of hydrogen concentrations in methanogenic soils and sediments. FEMS Microbiol. Ecol. 28: 193–202.

de Bok, F.A., A.J. Stams, C. Dijkema and D.R. Boone. 2001. Pathway of propionate oxidation by a syntrophic culture of *Smithella propionica* and *Methanospirillum hungatei*. Appl. Environ. Microbiol. 67: 1800–1804.

Ewe, S.M.L., E.E. Gaiser, D.L. Childers, V.H. Rivera-Monroy, D. Iwaniec, J. Fourquerean and R.R. Twilley. 2006. Spatial and temporal patterns of above ground net primary productivity (ANPP) in the Florida Coastal Everglades LTER Hydrobiologia. 569: 459–474.

Griffiths, B.S., M. Bonkowski, J. Roy and K. Ritz. 2001. Functional stability, substrate utilization and biological indicators of soils following environmental impacts. Appl. Soil Ecol. 16: 49–61.

Griffiths, B.S., P.D. Hallett, H.L. Kuan, A.S. Gregory, C.W. Watts and A.P. Whitmore. 2008. Functional resilience of soil microbial communities depends on both soil structure and microbial community composition. Biol. Fertil. Soil. 44: 745–754.

Hanson, R.S. and T.E. Hanson. 1996. Methanotrophic bacteria. Microbiol. Rev. 60: 439–771.

Harris, J. 2009. Soil microbial communities and restoration ecology: facilitators or followers. Science 325: 573–574.

Hines, M.E., K.N. Duddleston, J. Rooney-Varga, D. Fields and J.P. Chanton. 2008. Uncoupling of acetate degradation from methane formation in Alaskan wetlands: Connections to vegetation distribution. Global Biogeochemical Cycles 22: GB2017, doi: 10.1029/2006GB002903.

Imachi, H., Y. Sekiguchi, Y. Kamagata, A. Ohashi and H. Harada. 2000. Cultivation and *in situ* detection of a thermophilic bacterium capable of oxidizing propionate in syntrophic association with hydrogenotrophic methanogens in a thermophilic methanogenic granular sludge. Appl. Environ. Microbiol. 66: 3608–3615.

Jackson, B.E. and M.J. McInerney. 2002. Anaerobic microbial metabolism can proceed close to thermodynamic limits. Nature 415: 454–456.

King, G.M. 1994. Associations of methanotrophs with the roots and rhizomes of aquatic vegetation. Appl. Environ. Microbiol. 60: 3220–3227.

Lewis, D., J.R. White, D. Wafula, R. Athar, H.N. William and A. Chauhan. 2010. Soil Functional Diversity analysis of a bauxite mined restoration chronosequence. Microb. Ecol. 59: 710–723.

Liu, Y., D.L. Balkwill, H.C. Aldrich, G.R. Drake and D.R. Boone. 1999. Characterization of the anaerobic propionate-degrading syntrophs *Smithella propionica* gen. nov., sp. nov. and *Syntrophobacter wolinii*. Int. J. Syst. Bacteriol. 49: 545–556.

Magurran, A.E. and D.A.T. Phillip. 2001. Implications of species loss in freshwater fish assemblages. Ecography 24: 645–650.

Mormile, M.R., K.R. Gurijala, J.A. Robinson, M.J. McInerney and J.M. Suflita. 1996. The importance of hydrogen in landfill fermentations. Appl. Environ. Microbiol. 62: 1583–1588.

Reddy, K.R., J.R. White, A. Wright and T. Chua. 1999. Influence of phosphorus loading on microbial processes in the soil and water column of wetlands. pp. 249–273. *In*: K.R. Reddy, G.A. O'Connor and C.L. Schelske (eds.). Phosphorus Biogeochemistry in Subtropical Ecosystems. Lewis Publishers, New York, N.Y.

Schink, B. 1997. Energetics of syntrophic cooperation in methanogenic degradation. Microbiol. Mol. Biol. Rev. 61: 262–280.

Tokeshi, M. 1993. Species abundance patterns and community structure. Adv. Ecol. Res. 24: 112–186.

Uz, I., A. Chauhan and A. Ogram. 2007. Cellulolytic, fermentative, and methanogenic potential in benthic periphyton mats from the Florida Everglades. FEMS Microbiol. Ecol. 61: 337–347.

Voolapalli, R.K. and D.C. Stuckey. 1999. Relative importance of trophic group concentrations during anaerobic degradation of volatile fatty acids. Appl. Environ. Microbiol. 65: 5009–5016.

Warikoo, V., M.J. McInerney, J.A. Robinson and J.M. Suflita. 1996. Interspecies acetate transfer influences the extent of anaerobic benzoate degradation by syntrophic consortia. Appl. Environ. Microbiol. 62: 26–32.

Westerman, P. 1993. Wetland and swamp microbiology. pp. 215–238. *In*: T.E. Ford (ed.). Aquatic Microbiology. Blackwell Scientific, Cambridge, MA.

Whiting, G. and J. Chanton. 1993. Primary production control of methane emission from wetlands. Nature 364: 794–795.

Zhao, H.X., D.C. Yang, C.R. Woese and M.P. Bryant. 1990. Assignment of *Clostridium bryantii* to *Syntrophospora bryantii* gen. nov., comb. nov. on the basis of a 16S rRNA sequence analysis of its crotonate-grown pure culture. Int. J. Syst. Bacteriol. 40: 40–44.

18

Potential for Biological Control of Invasive Plants in the Everglades Ecosystem using Native Microorganisms

Kateel G. Shetty[a,*] and *Krish Jayachandran*[b]

Introduction

Biological invasion are considered as one of the most serious threats to ecosystems around the world, the spread of introduced, non-native species in the United States are causing major environmental damages and losses adding up to $120–137 billion per year (Pimentel et al. 2000; Pimentel et al. 2005). The cost associated with management of invasive species, which includes prevention, monitoring, mitigation, control and restoration, are huge (Westbrooks 1998). Majority of the native US species, about 42% on the Threatened or Endangered species lists are at risk primarily because of alien-invasive species (Pimentel et al. 2005). Within the United States the impact threat from invasive species is not uniform for all geographic locations, Hawaii and Florida being more prone to higher species rate of arrival compared to the other states with 15 species per year on average (CSIR-UCR 2014).

The National Invasive Species Council (NISC) was established by Executive Order (EO) 13112 to ensure that Federal programs and activities to prevent and control invasive species are coordinated, effective and efficient. An alien species (nonindigenous, exotic, non-native) in a particular ecosystem is defined as any species "that is not native to that ecosystem" (Executive Order 13112, 1999). An invasive

Earth and Environment Department, Florida International University, Miami, FL 33199.

[a] Email: shettyk@fiu.edu
[b] Email: jayachan@fiu.edu
* Corresponding author

species is defined as "an alien species (plant, animal, insect, bacteria, and fungi) whose introduction does or is likely to cause economic or environmental harm or harm to human health" (Executive Order 13112, 1999). Naturalized exotic is an exotic that sustains itself outside cultivation, if it is still exotic—it has not become native and an exotic that not only has naturalized but is expanding on its own in native plant communities is called invasive exotic.

Nonindigenous invasive plant species damage both natural and agricultural environments. As a consequence of establishment and spread of invasive species economic interests associated with agricultural environment are adversely affected. While within a natural environment, the vital multilayered and interconnected components such as native biodiversity, natural ecosystem structure and ecosystem functions gets drastically altered and downgraded, ultimately these negative impacts lead to significant losses in ecosystem integrity (Richardson and van Wilgen 2004).

According to the Enemy Release hypothesis (ER) plants escape from their coevolved enemies when introduced to a new environment, which in turn helps them to become invasive pests (Keane and Crawley 2002). The introduced exotic plant in its new location is also likely to encounter novel, non-coevolved herbivores and pathogens, and this is also an important component of ER as it pertains to invasion success (Keane and Crawley 2002; Colautti et al. 2004; Agrawal et al. 2005). Successful establishment and spread of an exotic species may be prevented at the invaded area by the action of native enemies (which may also include competing native plant species), this is known as the Biotic Resistance hypothesis (BR) (Darwin 1859; Naeem et al. 2000; Maron and Vila 2001) and the BR may also be responsible for low number of species becoming invasive from the population of introduced species (Williamson 2006).

For long-term prediction, prevention and management of invasive species, a clear understanding of fundamental ecological components responsible for invasive species success is critical. Understanding natural areas ecosystem is challenging, it is a highly complex system with diverse community interactions and ecosystem-level processes involving many species. However, understanding potential evolutionary changes among invaders and soil ecosystems will help us to accurately predict the long-term effects of biological invasions.

On the basis of field observations both ER and BR appears to be true. ER is true (positive feedback), but BR (negative feedback) is also true. However, the invasion successes can mostly be attributed to the dominance of both known and unknown factors. The factors often involve positive feedbacks between invasive plant species and the soil community (Thorpe and Callaway 2006), these are caused by the absence of pathogens and by the effects of invasive plants on the soil biota that drive nutrient cycles or on the abiotic components of the nutrient cycle themselves. In addition it is also possible that invasive species possess tolerance and phenotypic plasticity to overcome the pathogen induced damages (Gilbert and Parker 2006; Alexander 2010).

The survey and observations indicate that factors associated with BR do exist in the background although at a lower threshold. The benefits of ER to the invasive species within the native ecosystems are of limited benefit if pathogens evolve and accumulate over time leading to negative feedback interactions restricting invasive's dominance (Klironomos 2002; Mitchell et al. 2010). It has been observed that invasive species does not always experience enemy release, and also better plant performance will not be always follow with enemy release (Chun et al. 2010). In their introduced habitats many invasive plants, including grasses, forbs, trees and ferns were found to be infected by pathogens suggesting that pathogen accumulation on invasive species is widespread and potentially important for regulating invasive plant populations (Gilbert 2002; Flory and Clay 2013). These pathogen accumulations in general are more likely occur in habitats containing closely related species compared to habitats with more distantly related species (Agrawal and Kotanen 2003; Parker and Gilbert 2007).

The invasive plant species although free of co-evolved pathogens can still acquire new ones in its introduced region. This is possible because in both plant and pathogen the corresponding traits related to infection and virulence are subject to evolutionary change, rapid pathogen evolution can generate the ability to infect novel plant hosts (Gilbert and Parker 2006; 2010). New-encounter diseases develop when a plant is confronted with a pathogen with which it has not had an evolutionary history (Parker and Gilbert 2004). There are many examples of devastating new encounter disease on tropical perennial crops (Ploetz 2007). A new pathogen can emerge by evolutionary adaptation to a new host following a host shift—involving a new host but genetically related and host jump—new host that is genetically distant from the original host.

The theoretical and empirical proofs predict more negative effects of pathogen build-up on high-density species than less common species (Clay et al. 2008; Mordecai 2011). Although this is a possibility there are no documented cases of natural pathogen accumulation leading to invasive species decline (Flory and Clay 2013). Our studies indicate that while the BR factors are present, the effects observed are not at the population level, even when the invasive plant forms monospecific stands. These BR effects from native pathogens are found to be isolated, in patches and discontinuous.

Invasive Plant Species in Everglades's Ecosystem

In South Florida, the invasive plants such as Old World climbing fern (*Lygodium microphyllum* (Cav.) R. Br.), melaleuca (*Melaleuca quinquenervia* (Cav.) S.F. Blake), and Brazilian pepper (*Schinus terebinthifolius* Raddi) are considered priorities among the sixty-nine species of nonindigenous plants, while in the Kissimmee Basin and Lake Okeechobee aquatic plants such as hydrilla (*Hydrilla verticillata* (L.f.) Royle), water hyacinth (*Eichhornia crassipes* (Mart.) Solms), and tropical American water grass (*Luziola subintegra* Swallen) are considered priorities (Rodgers et al. 2012). The South Florida water management district (SFWMD), is one of the agencies involved in widespread efforts towards management of invasives. The agency's successful management of melaleuca and its continued maintenance is widely recognized as an interagency program model.

Located in South Florida over an area of 1.3 million acres the Everglades National Park (ENP) is designated as the only subtropical wilderness in the continental United States. The principal ecosystem types within the park include shallow-water marine habitats, saltwater wetland forests and marshes, freshwater marshes and prairies, and upland complexes of pine and hardwood forests. The Everglades National Park native species and the ecosystem are under significant threat due to the presence of more than 220 non-native species in the park. According to the Florida Exotic Pest Plant Council (http://www.fleppc.org/list/list.htm), the non-natives are grouped under two categories, the 67 exotic plant species that are causing ecological damage to native plant communities are grouped under Category-I list and the 71 exotic plant species that are spreading and increasing in range, but have yet to cause actual ecological damage are grouped under Category-II list.

Focused efforts towards containment and eradication of invasive species is usually restricted to certain number of species, as full scale removal of all the exotic species and restoration is not possible due to the huge area that needs to be covered and funding limitation. Brazilian pepper, melaleuca, Australian pine (*Casuarina equisetifolia* L.), seaside mahoe (*Thespesia populnea* (L.) Sol. ex Corrêa), lather leaf (*Colubrina asiatica* (L.) Brongn.), Old World climbing fern are among the most commonly targeted exotic plant species.

Biological Control of Invasive Plant Species

Long-term use of herbicides to control invasive plant species often leads to negative effects on non-target flora and fauna, water quality, and other environmental health components (Pimentel 2005; Weidenhamer and Callaway 2010). Exposure to herbicides may lead to the problem of evolution of resistance in target plant species, so increased amounts of herbicides may have to be used for long term suppression. Another major issue is cost; in natural areas use of herbicides over large areas can become prohibitively expensive. In view of these problems, the natural area managers are increasingly promoting the use of classical biological control as a part of comprehensive integrated management of exotic invasive plant species.

Biological control is defined as the use of living organisms to suppress the population density or impact of a specific pest organism, making it less abundant or less damaging than it would otherwise be (Eilenberg et al. 2001). Biological control (biocontrol) methods are based on natural systems, and natural ecosystems represent the epitome of efficiency and conservation of energy (Shrum 1982). In situations where only one dominant weed species needs to be managed without reducing the species richness of the flora and fauna, the ideal choice would be biological control, because of its high level of selectivity.

The most common and successful biological control strategy is the classical biological control approach, which involves "intentional introduction of an exotic (non-native), usually coevolved, biological control agent for permanent establishment and long-term pest control" (Eilenberg et al. 2001). The exotic agents (natural enemies) are obtained from the target plant's native range. Another approach is augmentation or inundation; it is based on increasing the abundance and impact of biocontrol agents through inundative release and/or distribution of either exotic or native agents to achieve biological control of exotic weeds.

Biological control of weeds by using plant pathogens has gained acceptance as a practical, safe, environmentally beneficial, weed management method applicable to agroecosystems (Charudattan 2001). This approach has achieved considerable success in the use of plant pathogens as agents for biological control of certain weeds. Plant pathogens typically have a fairly narrow host range, making them relatively safe to use as biological weed control agents in terms of possible non-target plant effects.

The classical strategy using plant pathogens is also termed the "inoculative" method, as it usually involves a small "dose" of an exotic biological control organism, applied once or only occasionally, to a host population of weeds that is usually of foreign origin. The inundative method involves single or multiple applications of sufficiently high levels of inoculum to the weed population, under conditions that favor disease onset, to quickly create a disease epidemic among plants (Templeton et al. 1979; Charudattan 1991). The term mycoherbicide usually refers to an indigenous fungus applied in an inundative manner to control native weeds (TeBeest and Templeton 1985). Following application, mycoherbicides will generally persist locally at an elevated level, and then return to endemic levels when the targeted weed population is depressed (Charudattan 1988).

The Case for Native Biological Control Agents

The use of biological control agents as an environmentally compatible alternative for pest management is more common and is recognized for its effectiveness, low cost and relatively safe environmental impact. However, one of the serious ecological drawbacks of biological control is the introduction of more exotic species into new ranges. The classical biocontrol agents like their target host (exotic invasive plant) are themselves non-native introductions and in their new habitat they will be free of their co-evolved natural enemies, the long term ecological implications of these additional introductions of exotics are difficult to predict.

The potential impact of introduced biological control agents on the native biota is large. Non-native biocontrol agents may attack non-target species (Messing and Wright 2006); these may be mediated through direct host-shifting and also through indirect interactions. The introduced agents may spread to other areas and become invasive there (Simberloff et al. 2005). Introduced enemies can switch hosts, and in field situations they may have a stronger impact on indigenous plants than on their original hosts (Callaway et al. 1999).

However, recent examination of *Centaurea maculosa* biocontrol has established that even highly host-specific biocontrol agents can have rather significant nontarget effects, indicating a need to better understand these side effects to guard against them (Pearson and Callaway 2003; 2005). Additionally specialist enemies may enhance the aggressiveness of an exotic plant species in some situations (Marler et al. 1999), in terms of enhancing the concentrations of defense and allelochemicals (Zangerl and Berenbaum 2005).

Laboratory and quarantine studies on introduced agents have inherent limitations and are not expected to replicate the varied conditions, and numerous complex ecological interrelationships that exist in the natural world. As a living organism the biological control agent is bound to move and spread, the associated ecological risk gets further compounded when multiple species are introduced (Cock 1986; Muller 1990). Another concern is potential for negative consequences from introduction of exotic pathogens and parasites along with biological control agents (Hawkins and Marino 1997; Guy et al. 1998). This problem gets further complicated if the novel pathogen vectored by the introduced insect is difficult to detect using conventional methods and poses serious threat to commercial agricultural crops (Thompson et al. 2013).

The classical biocontrol requires a long period of agent screening prior to introduction and release. In addition, the cost involved in discovery and screening in the foreign country and quarantine is very high. Therefore, more ecologists are now recommending classical bicontrol only as a last resort (Louda and Stiling 2004; Simberloff et al. 2005; ShiLi et al. 2012).

Alternatively, native enemies for exotic, invasive species may provide a viable control strategy. If natural enemies cause more damage to exotic, invasive species than to native, non-invasive species, this strategy would be advantageous. Native insects were found to be effective in controlling exotic invasive weeds, *Cassida rubiginosa* (shield beetle) on creeping thistle plant (*Cirsium arvense*) in Europe (Bacher and Schwab 2000) and weevil *Euhrychiopsis lecontei* on Eurasian watermilfoil (*Myriophyllum spicatum*), in North America (Creed 1998; Newman 2004). Significant population reduction of exotic invasive plant species mile-a-minute (*Mikania micrantha*) has been reported using native parasitic dodders (*Cuscuta* spp.) in China (Yu et al. 2011).

Most of the potential hazards of classical biocontrol agents (introduced from other regions/countries) can be avoided with microbial herbicides by selecting pathogens that are already endemic in the area where they are to be used. During their long coexistence with the endemic pathogen, the native plant species have evolved to cope with the genetic variability of the pathogen. It is preferable to consider indigenous fungi as candidate species for mycoherbicide development. The advantage to this approach is that native pathogens are subject to natural controls and normally do not persist in the environment at greater than endemic levels (Wall et al. 1992).

Native pathogens are locally available and locally adapted, and to some extent local agricultural crops/native species have been successfully screened against their pathogenic capabilities (Shrum 1982; Yang and TeBeest 1993). Displacement of non-target organisms by the use of a mycoherbicide is not generally considered to be a significant risk, as the effects of local population inundation by indigenous fungi are temporary and limited in a scale compared with potential displacement of non-target organisms by exotic biological control pathogen release (Charudattan 1991).

It is not always true that exotic species will be more successful in its new location, where it is free from its natural enemies. Saltcedar (*Tamarix* spp.) a Mediterranean plant taken to Argentina where within two years the plantings were wiped out by *Botryosphaeria tamaricis* (Worsham 1982). Successful controls have been reported with native pathogens in control of weeds (Julien and Griffiths 1998). Three native plant pathogens have been reported to be potential biocontrol agents against Kudzu, an invasive exotic species, *Pseudomonas syringae* pv. *phaseolicola* (Zidak and Backman 1996), *Myrothecium verrucaria* (Boyette 2002) and *Colletotrichum gloeosporiodes* (Farris and Reilly 2000).

There are reports of biocontrol projects using pathogens in Hawaii, Kahili ginger (*Hedychium gardnerianum*) host-specific strain of the common plant pathogenic bacterium *Pseudomonas solanacearum* (Anderson and Gardner 1997), combination of a rust fungus and an eriophyid mite for introduced fuschia in Hawaii Volcanoes National Park (Gardner 1987) and the attack of ivy gourd (*Coccinia grandis*), an aggressive introduced vine invading lower-elevation forests, by a powdery mildew fungus, *Oidium* sp. (Gardner 1994).

Survey and Discovery of Native Pathogenic Fungi

Brazilian Peppertree

Brazilian peppertree *Schinus terebinthifolius* Raddi (Sapindales: Anacardiaceae) is an aggressive exotic invasive species, it now covers hundreds of thousands of hectares in south and central Florida, as well as many of the islands on the east and west coasts of the state. It is an evergreen, hardwood tree species native to Argentina, Brazil, and Paraguay (Mytinger and Williamson 1987) and was introduced to the United States in the mid-1800s as an ornamental. Brazilian peppertree is a pioneer of disturbed sites, but is also successful in undisturbed natural environments (tropical hardwood forests, pine rocklands, sawgrass marshes, and mangrove swamps) in Florida (Jones and Doren 1997).

The "Hole-in-the-Donut" (HID), is perhaps the largest and most infamous of the Everglades National Park ENP Brazilian peppertree infestation, it involves an area of over 4,000 hectares of abandoned agricultural lands in the midst of natural subtropical ecosystems. Since the whole of ENP has over 100,000 acres that are affected by Brazilian peppertree, the infestation within the HID is only part of a much larger issue. Herbicide application and mechanical removal methods are being used for removal and management of Brazilian peppertree (Dalrymple et al. 2003; Cuda et al. 2006), these are labor intensive and expensive because of large area that needs to be covered (Manrique et al. 2009). These operations within the natural areas raise concerns regarding pollution and the effects on non-target species (Doren and Jones 1997). Use of classical insect biological control agents are some of alternative possibilities that are being considered (Cuda et al. 2005; Manrique et al. 2009).

Field surveys in the Florida natural areas done by our group have shown the prevalence of native phytopathogen–induced diseases on Brazilian peppertree which included incidences of various types of severe foliar diseases, die-back, blight, seedling and sapling death, witches broom, tumors and galls, and inflorescence blight symptoms caused by native pathogens. Detached leaf assay was employed for initial selection of pathogenic isolates. It was also observed that several unidentified endophytic fungi associated are with Brazilian peppertree with *Pestalotiopsis* sp. being the most frequent one. They may have a role in the tree's success as an invasive species in its new habitat.

Brazilian peppertree seed profusely, it is a contributing factor for its local reproduction as well as its potential for dispersal (Tobe et al. 1998). Seed mortality can potentially limit plant invasiveness (Fenner 1992). A naturalized chalcid wasp *Megastigmus transvaalensis*, discovered in Florida was reported to damage the Brazilian peppertree drupe and make it incapable of germination (Habeck et al. 1989). Seed predators are frequently employed as biological control agents (Kremer 2000), but role of microorganisms in the seed pathology were not reported. Pathogens have been tested for management of weeds through seed attack (Medd and Cambell 2005). The efficacy of fungal pathogen seed infection can be augmented by combining it with selective seed-attacking insects (Kremer 2000).

Laboratory testing of Brazilian peppertree seed batches from different years showed significant differences in percent fungal infection; more than 50% of the infected seeds did not germinate (Shetty et al. 2011). These seed-borne fungal infections lead to either germination failure or fungal attack of seedlings after germination. The Brazilian peppertree seed germination failure due to seed rot fungi and fungal seedling attack confirms the involvement of native pathogens in the life cycle of this exotic invasive species. Field observations on poor establishment of Brazilian peppertree seeds on certain sites dominated by native ragweed (McMullen 2003; Shetty et al. 2007) support the need for additional studies on changes in soil biogeochemistry and rhizosphere microflora near competing native plant species.

An endophytic fungi *Neofusicum batangarum* isolated from Brazilian peppertree seed was found to be a virulent pathogen (Shetty et al. 2011). In greenhouse inoculation studies *N. batangarum* was capable of killing a year old Brazilian peppertree saplings. Field inoculation of Brazilian peppertree branches with *N. batangarum* resulted in dieback symptoms. Host range studies on one related native species (winged sumac, *Rhus copallinum*) and one non-native species (mango, *Mangifera indica*) showed that neither was affected by girdle inoculation of stems. The fungus was able to colonize and damage vascular tissue as indicated by extensive dark discoloration and necrosis. It was also observed that inoculation of Brazilian peppertree leaves, inflorescence and drupes in the field with *N. batangarum* mycelial inoculum resulted in necrosis and blight symptoms. These results demonstrate possibilities for future efforts towards integration of compatible biological agents for effective reduction of weed seed viability prior to entry into the seedbank.

Regarding the origin of *N. batangarum*, it may have been introduced into Florida with plant materials from African origin or it is also likely that the fungus has a wider geographic range than is currently known. The pathogenic interaction between a South American host, Brazilian peppertree, with a native or naturalized fungus, *N. batangarum*, can be considered as a new association (Hokkanen and Pimentel 1984).

Old World Climbing Fern

The fragile natural ecosystems in the South Florida are being threatened by an invasive old World climbing fern, *Lygodium microphyllum* (Cav.) R. Br. (Lygodiaceae, Pteridophyta) (Pemberton and Ferriter 1998). The Old World climbing fern, is native to tropical and subtropical areas of Africa, southeastern Asia, northern and eastern Australia, and the Pacific islands (Pemberton 1998). It was first found to be naturalized in Florida 1965; however, its rapid spread is now a serious concern because of its dominance over native vegetation. *Lygodium* is classified as a Category I invasive species by the Florida Exotic Pest Plant Council (Langeland and Craddock Burks 1998) and is spreading at an alarming rate and already over 100,000 acres have been found to be infested by this aggressive weed. Old World climbing fern occurs in a variety of forested and non-forested sites in southern Florida, including hardwood hammocks, cypress swamps, savannas, woodlands, marshes, and wet prairies (Wunderlin 1982; 1998; Pemberton and Ferriter 1998). The state and nation are incurring huge economic cost to manage these invasive plant species.

Lygodium control method involves prescribed burning, biological controls, mechanical removal, and herbicides all part of an integrated management system. Currently integrated approach combining biocontrols with other control methods (herbicides, etc.) is considered the best long-term management practice by the Florida

Exotic Pest Plant Council *Lygodium* Task Force (http://fleppc.org). The South Florida water management district (SFWMD) and the USDA-Agricultural Research Service (ARS) began cooperative co-funded biological control research on Old World climbing fern in 1998. This biocontrol research has been predominantly oriented towards re-association of co-evolved natural enemies by introduction of selected insect pests of *Lygodium* collected from different countries as biological control agents. After eight years of surveys (1997–2005), potentially useful insects and a mite have been found. Three insect agents an Australian moth (*Austromusotima camptonozale*), an eriophyid gall mite (*Floracarus perrepare*), and another pyralid moth (*Neomusotima conspurcatalis*) were evaluated for release and in addition additional insect agents are also under investigation. In February 2005, a moth (*Austromusotima camptonozale*) was released in southeast Florida as the first biocontrol agent. To assess the effects of insect biocontrol agents on *Lygodium* in the field may take many years. According to the *Lygodium* Task Force management plan (2006) additional searches need to be conducted for insects, fungi, and microbes as potential control agents.

Because of the paucity of scientific studies, it is assumed that ferns in general are immune from pathogenic disease problems (Hoshizaki 1979; Swain 1980). However, subsequent studies have shown that it is not the case. Bracken ferns within the United Kingdom forests were found to be susceptible to fungal diseases such as pinnule die-back (*Certaobasidium anceps*) and curl-tip disease (*Ascochyta pteridis* and *Phoma aquiline*) (Green 2003). The Leather ferns in Hawaii are affected by disease caused fungi *Calonectria theae*, *Cylindrocladium pteridis*, *Rhizoctonia* species, *Cercospora* species, and *Pythium* species. In Florida, *Cylindrocladium floridanum, C. heptaseptatum, C. pteridis*, and *C. scoparium, Rhizoctonia* spp. have been reported as pathogens of leatherleaf fern (www.ctahr.hawaii.edu/oc/freepubs).

Exotic invasive plant species including *Lygodium* are susceptible to native microbial pathogens. Localized epidemics or plants with severe symptoms are good source of potential biocontrol agents. There are reports on fungal pathogens of *Lygodium*, Jones et al. (2002) have identified a pathogenic fungus *Colletotrichum gloeosporioides* from *L. microphyllum* grown in shade house. The rust fungus *Puccinia lygodii* which is native to tropical South America has been collected from *L. japonicum* in Louisiana and the fungus was found to be pathogenic to *L. microphyllum* (Rayamajhi et al. 2005). Another pathogenic fungus *Bipolaris sacchari* was isolated from *L. japonicum* in shade house was found to cause leaf spot on *L. japonicum* and *L. microphyllum* in Florida (Elliott and Rayamajhi 2005).

A mycoherbicide *Myrothecium verrucaria* has been successfully used in the control of invasive plants such as leafy spurge (Yang and Jong 1995) and kudzu (Boyette et al. 2002). *M. verrucaria* is native to the soils of the southern United States; it is a facultative parasite that can be found in many plant species (Domsch et al. 1980). In greenhouse studies *Lygodium* was found to be susceptible to *Myrothecium verrucaria* infection, spray inoculation with *M. verrucaria* resulted in successful disease development with leaf necrosis symptoms and growth suppression (Clarke et al. 2007). A disease index close to 3 on a scale of 0 to 4 was observed at 1×10^8 ml^{-1} conidial concentration.

Host range screening of selected native plant species for susceptibility to *M. verrucaria* showed low disease indices after repeated spray inoculations; the highest index attained was 0.4 by Slash pine (*Pinus elliottii* L.). Additional growth chamber studies on *Lygodium* studies were conducted using *Myrothecium* inoculation in combination with surfactant Silwet L-77 (Loveland Industries, Greeley, CO). The results from the study showed significantly faster rate of infection and higher mean severity rating at high temperature/high humidity combination than at low temperature/ low humidity combination (Table 1). The combination of high temperature and high humidity condition required for successful infection process naturally exists during summer months in Florida. Since there is no selective media available for isolation of *Myrothecium*, collection of large number of native strains and screening for desirable biocontrol features is not feasible at the present moment.

Table 1. Comparison of disease indices of *L. microphyllum* plants under varied temperature and humidity parameters in the growth chamber.

Humidity	Day	Temperature–Disease index	
		22°C/70°F	30°C/80°F
40%	3	0.5	0.69
	7	1.02	1.28
	10	1.61	1.76
90%	3	0.78	2.87
	7	1.52	3.28
	10	2.07	3.61

The disease survey done in the *Lygodium* infested sites within Florida natural areas showed the occurrence of severe disease symptoms on *Lygodium* (Plate 1 and 2), demonstrating the susceptibility of non-native invasive *Lygodium microphyllum*

Plate 1. Severe outbreak of leaf spot on *Lygodium microphyllum*, background shows dead plants with disease symptoms.
Color image of this figure appears in the color plate section at the end of the book.

Plate 2. Severe outbreak of leaf spot and blight on leaves and fertile fronds of *Lygodium microphyllum*.
Color image of this figure appears in the color plate section at the end of the book.

to native phytopathogens. At some sites large sections of Lygodium were seen with severe outbreak of disease symptoms, including leaf spots, chlorosis, blight and defoliation, dead and dying *Lygodium* sporlings with necrotic lesions were also observed, proving that native phytopathogen/s can be highly virulent and are indeed capable of killing *Lygodium*.

Among the 78 fungal isolates tested 12 isolates caused extensive necrosis of detached leaflets and 9 of the isolates were able cause leaflet tissue necrosis in less than 72 hours after inoculation. Majority of the fungal isolates are unidentified and the known isolates belonged to *Alternaria* sp., *Curvularia* sp., and *Cylindrocladium* sp. Inoculation experiment using selected 11 fungal isolates were conducted under greenhouse conditions. Three of the fungal isolates caused more than 50% disease incidence on *Lygodium* plants and disease severity that was based on a 0 to 4 scale ranged from 2 to 3.3 (Fig. 1). *Cylindrocladium* sp. (N-1) has been reported to be pathogenic to ferns, and is also known to infect underground rhizome.

Host range study using native ferns species—*Anemia adiantifolia* (Pine fern), *Nephrolepsis exaltata* (Boston fern), *Polypodium polypodioides* (Resurrection fern), *Adiantum tenerum* (Mainden hair fern), *Pteris bahamensis* (Bahama ladder brake) and *Thelypteris* cf. *kunthii* (Southern shield fern)—showed that at 14 days after inoculation none of native species except Resurrection fern showed any disease symptoms. The symptoms seen in Resurrection fern leaflet braches inoculated with isolate J-5 showed curling, upon close inspection of curled leaflets showed no necrosis or chlorosis. The native phytopathogens do have the potential to impact the growth and spread of *Lygodium* through their extensive damaging effects on the vegetative (leaflet) and reproductive tissues (fertile frond) of the plant.

Minor indigenous insect pests could play an important synergistic role in disease vectoring and reducing plant vigor and resistance. A severe infestation of a native brown soft scale insect (possibly a *Coccus* or *Pulvinaria* family Coccidae) on *Lygodium* occurred during our greenhouse experiments (Plate 3). It is likely that the insect came with the sporeling plant and soil material from the natural area as the pots were kept in a separate enclosed chambers and rest of the greenhouse was periodically sprayed as a prophylactic measure. The plants exhibited severe leaf spots with chlorosis, and necrosis of rachis. The scales were able to transfer pathogens deeper into hardy rachis tissue and into underground rhizomes causing severe disease symptoms and plant death.

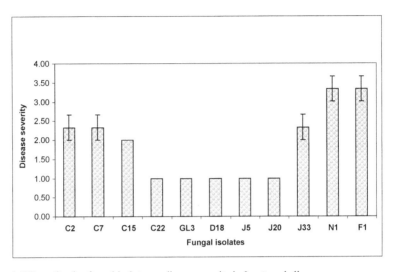

Figure 1. Effect of native fungal isolates on disease severity in *L. microphyllum.*

Plate 3. Close-up of *Lygodium microphyllum* rachis showing brown soft scale insect and tissue necrosis. *Color image of this figure appears in the color plate section at the end of the book.*

Synthesis and Research Needs

Plant diseases play a very important role in the natural ecosystem; plant pathogens can bring about changes in host plant mortality and fecundity, population dynamics, and affect the structure and composition of natural plant communities (Gilbert 2002). There is increasing interest in the role played by density-dependent mortality from natural enemies (negative density dependence, NDD), particularly plant pathogens, in promoting the coexistence and diversity of tropical trees. The Janzen–Connell hypothesis (Janzen 1970; Connell 1971) suggests that specialized natural enemies such as insect herbivores and fungal pathogens maintain high diversity by elevating mortality when plant species occur at high density. Recent reports support the role of native fungi through NDD in structuring tropical plant communities and in maintaining diversity (Bagchi et al. 2014).

Potential for successful application of BR hypothesis is possible in managing spread of invasive species if it is to encounter novel enemies in its new environment (Creed 2000; Siemann et al. 2006). Therefore among the available options, initiating concerted efforts towards discovering and developing natural enemies from invaded ecosystem is more sustainable than depending solely on introduced biocontrol agents.

Design and execution of project to specifically discover and develop novel native host specific pathogen of invasive plant species will be complex and challenging. Population of generalist herbivores or pathogens with broad host range can not be augmented for the purpose of biocontrol. More lethal new encounter disease association between invasive plant and native pathogen may be occurring in an isolated site due to spatiotemporal heterogeneity within the large natural area, and the pathogen may have little chance for infecting new host and propagating due to spatial separation and absence of a suitable vector. Therefore the novel pathogen has to be discovered before the plant dies out and becomes colonized by saprophytes. Area wide deliberate set-up for short term exposure of container grown (with native soil) weakened invasive plants in the natural areas for trapping novel pathogens is another possibility. A generalist native herbivore could be deployed for wider transmission if it could vector a host specific native pathogen. Another possibility is the use of highly competitive native plant species capable of supporting native suppressive pathogens antagonistic to invasive species to fill in open or susceptible sites within the natural area.

These challenging approaches to develop potential alternative to biological control can be efficiently met only through the collaborations of multidisciplinary scientists in different organizations and institutions, natural areas managers, invasive species working groups, private land owners and interested stake holders.

Taxonomic identification, molecular characterization, special growth conditions, inoculum production and formulation, identification of host plant species, epidemiological modeling will require collaborations among plant pathologists, mycologists, ecologists, taxonomists and private industries, and the use of morphological, microbiological and advanced molecular techniques.

Conclusions

Given the serious impacts of invasive species on nation's fragile natural ecosystems, there is an urgent need for increase in our fundamental understanding of potential control agents and their interaction with hosts across diverse ecosystems. Also considering that the new arrival and establishment of invasive species is still continuing, and the number of invasive plant species that are in the natural areas, continued introduction of multiple exotic biological control agents is unsustainable in the long run. Effective alternative strategies needs to be developed without further delay, as it takes considerable amount of time for a product or strategy to be ready for successful field application.

A number of endemic pathogens are found to attack exotic invasive plant species in South Florida natural areas. Several of these possess attributes that indicate these pathogens have potential for development into biological agents for control of their specific natural hosts. Further research into development of native pathogens could have a significant impact on availability of potential tools to manage the spread of invasive such as Brazilian peppertree and *L. microphyllum* in South Florida.

Acknowledgments

We thank the National Park Service (Everglades National Park, SFNRC) and the South Florida Water Management District for funding support. We thank Tainya Clarke, Jose Pacheco, and Binod Pande for technical help.

References

Agrawal, A.A. and P.M. Kotanen. 2003. Herbivores and the success of exotic plants: a phylogenetically controlled experiment. Ecol. Lett. 6: 712–715.

Agrawal, A.A., P.M. Kotanen, C.E. Mitchell, A.G. Power, W. Godsoe and J. Klironomos. 2005. Enemy release? An experiment with congeneric plant pairs and diverse above- and belowground enemies. Ecology 86: 2979–2989.

Alexander, H.M. 2010. Disease in natural plant populations, communities, and ecosystems: insights into ecological and evolutionary processes. Plant Dis. 94: 492–503.

Anderson, R.C. and D.E. Gardner. 1997. Biological control of the alien weed kahili ginger (*Hedychium gardnerianum*) in Hawaii Volcanoes National Park. Abstracts of the 1997 Hawaii Conservation Conference, Maui. 11.

Bacher, S. and F. Schwab. 2000. Effects of herbivore density, timing of attack and plant community on performance of creeping thistle *Cirsium arvense* (L.) Scop. (Asteraceae). Biocontrol Sci. Techn. 10: 343–352.

Bagchi, R., R.E. Gallery, S. Gripenberg, S.J. Gurr, L. Narayan, C.E. Addis, R.P. Freckleton and O.T. Lewis. 2014. Pathogens and insect herbivores drive rainforest plant diversity and composition. Nature 506: 85–88.

Boyette, C.D., H.L. Walker and H.K. Abbas. 2002. Biological control of kudzu (*Pueraria lobata*) with an isolate of *Myrothecium verrucaria*. Biocontrol Sci. Techn. 12: 75–82.

Callaway, R.M., T. DeLuca and W.M. Belliveau. 1999. Herbivores used for biological control may increase the competitive ability of the noxious weed *Centaurea maculosa*. Ecology 80: 1196–1201.

Center for Invasive Species Research – University of California, Riverside. 2014. Invasive species FAQ's. Retrieved from https://cisr.ucr.edu/invasive_species_faqs.html, Last Accessed: 5/09/2014.

Charudattan, R. 1988. Inundative control of weeds with indigenous fungal pathogens. pp. 86–110. *In*: M.N. Burge (ed.). Fungi in Biological Control Systems. Manchester University Press, New York.

Charudattan, R. 1991. The mycoherbicide approach with plant pathogens. pp. 24–57. *In*: D.O. TeBeest (ed.). Microbial Control of Weeds. Chapman & Hall, New York.

Charudattan, R. 2001. Biological control of weeds by means of plant pathogens: significance for integrated weed management in modern agro-ecology. BioControl 46: 229–260.

Chun, Y.J., M. van Kleunen and W. Dawson. 2010. The role of enemy release, tolerance and resistance in plant invasions: Linking damage to performance. Ecol. Lett. 13: 937–946.

Clarke, T.C., K.G. Shetty, K. Jayachandran and M.R. Norland. 2007. *Myrothecium verrucaria*—a potential biological control agent for the invasive 'old world climbing fern' (*Lygodium microphyllum*). Biocontrol 52: 399–411.

Clay, K., K.O. Reinhart, J.A. Rudgers, T. Tintjer, J.M. Koslow and S.L. Flory. 2008. Red queen communities. pp. 145–178. *In*: R.S. Ostfeld, F. Keesing and V.T. Eviner (eds.). Infectious Disease Ecology: Effects of Ecosystems on Disease and of Disease on Ecosystems. Princeton University Press, Princeton, NJ.

Cock, M.J.W. 1986. Requirements for biological control: an ecological perspective. Biocontrol News and Information 7: 7–16.

Colautti, R.I., A. Ricciardi, I.A. Grigorovich and H.J. MacIsaac. 2004. Is invasion success explained by the enemy release hypothesis? Ecol. Lett. 7: 721–733.

Connell, J.H. 1971. On the role of natural enemies in preventing competitive exclusion in some marine animals and in rain forest trees. pp. 298–312. *In*: P.J. Boer and G.R. Graadwell (eds.). Dynamics of Numbers in Populations. Proc. Adv. Stud. Inst., Osterbeek 1970. Center Agr. Publ. Doc., Wageningen.

Creed, R.P. 1998. A biogeographic perspective on Eurasian water milfoil declines: additional evidence for the role of herbivorous weevils in promoting declines. J. Aquat. Plant Manage. 36: 16–22.

Creed, R.P. 2000. Is there a new keystone species in North American lakes and rivers? Oikos 91: 405–408.

Cuda, J.P., J.C. Medal, M.D. Vitorino and D.H. Habeck. 2005. Supplementary host specificity testing of the sawfly *Heteroperryia hubrichi*, a candidate for classical biological control of Brazilian peppertree, *Schinus terebinthifolius*, in the USA. BioControl 50: 195–201.

Cuda, J.P., A.P. Ferriter, V. Manrique and J.C. Medal. 2006. Florida's Brazilian Peppertree Management Plan Recommendations from the Brazilian peppertree task force Florida Exotic Pest Plant Council. http://www.fleppc.org/Manage Plans/2006BP managePlan5.pdf.

Dalrymple, G.H., R.F. Doren, N.K. O'Hare, M.R. Norland and T.V. Armentano. 2003. Plant colonization after complete and partial removal of disturbed soils for wetland restoration of former agricultural fields in Everglades National Park. Wetlands 22: 1015–1029.

Darwin, C. 1859. The Origin of Species. John Murray, London, UK, 502 pp.

Domsch, K.H., W. Gams and T.H. Anderson. 1980. *Myrothecium. In*: Compendium of soil Fungi, Vol. 1. Academic Press, New York. pp. 481–487.

Doren, R.F. and D.T. Jones. 1997. Management in Everglades National Park. pp. 275–286. *In*: D. Simberloff, D.C. Schmitz and T.C. Brown (eds.). Strangers in Paradise: Impact and Management of Nonindigenous Species in Florida. Island Press, Washington, D.C.

Eilenberg, J., A. Hajek and C. Lomer. 2001. Suggestions for unifying the terminology in biological control. BioControl 46: 387–400.

Elliott, M.L. and M.B. Rayamajhi. 2005. First report of *Bipolaris sacchari* causing leaf spot on *Lygodium japonicum* and *L. microphyllum* in Florida. Plant Dis. 89: 1244.

Farris, J. and C.C. Reilly. 2000. The biological control of kudzu (*Pueraria lobata*). Proceeding of the American Association for the Advancement of Science, Washington, D.C. 116: AB2.

Fenner, M. 1992. Seeds: The Ecology of Regeneration in Plant Communities. CAB International, Wallingford, UK.

Flory, L.S. and K. Clay. 2013. Pathogen accumulation and long-term dynamics of plant invasions. J. Ecol. 101: 607–613.

Gardner, D.E. 1987. Biocontrol of fuchsia. Newsletter of the Hawaiian Botanical Society 26: 19.

Gardner, D.E. 1994. Biocontrol of ivy gourd (*Coccinia grandis*) on Windward O`ahu. Newsletter of the Hawaiian Botanical Society 33: 13–16.

Gilbert, G.S. and I.M. Parker. 2006. Invasions and the regulation of plant populations by pathogens. pp. 289–305. *In*: M.W. Cadotte (ed.). Conceptual Ecology and Invasion biology. Springer, Dordrecht.

Gilbert, G.S. and I.M. Parker. 2010. Rapid evolution in a plant-pathogen interaction and the consequences for introduced host species. Evol. Appl. 3: 144–156.

Gilbert, S.G. 2002. Evolutionary ecology of plant disease in natural ecosystems. Ann. Rev. Phytopathol. 40: 13–43.

Green, S. 2003. A review of the potential for the use of bioherbicides to control forest weeds in the UK. Forestry 76: 285–298.

Guy, P.L., D.E. Webster, L. Davis and R.L.S. Forster. 1998. Pests of non-indigenous organisms, hidden costs of introduction. Trends Ecol. Evol. 13: 111.

Habeck, D.H., F.D. Bennett and E.F. Grissell. 1989. First record of a phytophagous seed chalcid from Brazilian peppertree in Florida. Fla. Entomol. 72: 378–379.

Hawkins, B.A. and P.C. Marino. 1997. The colonization of native phytophagous insects in North America by exotic parasitoids. Oecologia 112: 566–571.

Hokkanen, H.M.T. and D. Pimentel. 1984. New approach for selecting biological control agents. Can. Entomol. 116: 1109–1121.

Hoshizaki, B.J. 1979. Fern Growers Manual. Alfred A. Knopf, New York.

Janzen, D.H. 1970. Herbivores and the number of tree species in tropical forests. Am. Nat. 104: 501–528.

Jones, D.T. and R.F. Doren. 1997. The distribution, biology and control of *Schinus terebinthifolius* in Southern Florida, with special reference to Everglades National Park. pp. 81–93. *In*: J.H. Brock, M. Wade, P. Pysek and D. Green (eds.). Plant Invasions: Studies from North America and Europe. Backhuys Publishers, Leiden, Netherlands.

Jones, K.A., M.B. Rayamajhi, P.D. Pratt and T.K. Van. 2003. First report of the pathogenicity of *Colletotrichum gloeosporioides* on invasive ferns, *Lygodium microphyllum* and *L. japonicum*, in Florida. Plant Dis. 87: 101.

Julien, M.H. and M.W. Griffiths. 1998. Biological Control of Weeds: A World Catalogue of Agents and their Target Weeds. CAB International North America, Wallingford, UK, 223 pp.

Keane, R.M. and M.J. Crawley. 2002. Exotic plant invasions and the enemy release hypothesis. Trees 17: 164–170.

Klironomos, J. 2002. Feedback with soil biota contributes to plant rarity and invasiveness in communities. Nature 417: 67–70.

Kremer, R.J. 2000. Combinations of microbial and insect biocontrol agents for management of weed seeds. pp. 799–806. *In*: N.R. Spencer (ed.). Proceedings of the X International Symposium on Biological Control of Weeds. 4–14 July 1999, Montana State University, Bozeman, Montana, USA.

Langeland, K.A. and K. Craddock Burks. 1998. Identification and Biology of Non-Native Plants in Florida's Natural Areas. University of Florida, Gainesville, USA.

Louda, S.M. and P. Stiling. 2004. The double-edged sword of biological control in conservation and restoration. Conserv. Biol. 18: 50–53.

Manrique, V., J.P. Cuda, W.A. Overholt and S.M.L. Ewe. 2009. Synergistic effect of insect herbivory and plant parasitism on the performance of the invasive tree *Schinus terebinthifolius*. Entomol. Exp. Appl. 132: 118–125.

Marler, M.J., C.A. Zabinski and R.M. Callaway. 1999. Mycorrhizae indirectly enhance competitive effects of an invasive forb on a native bunchgrass. Ecology 80: 1180–1186.

Maron, J.L. and M. Vila. 2001. When do herbivores affect plant invasion? Evidence for the natural enemies and biotic resistance hypotheses. Oikos 95: 361–373.

Mordecai, E.A. 2011. Pathogen impacts on plant communities: unifying theory, concepts, and empirical work. Ecol. Monogr. 81: 429–441.

McMullen, R.T. 2003. An investigation of soil suppression of Brazilian pepper (*Schinus terebenthifolius* Raddi) on spoil mounds in the Hole-in-the Donut restoration program of Everglades National Park. M.S. Dissertation. Florida International University, Florida.

Medd, R.W. and M.A. Campbell. 2005. Grass seed infection following inundation with *Pyrenophora semeniperda*. Biocontrol Sci. Technol. 15: 21–36.

Messing, R.H. and M.G. Wright. 2006. Biological control of invasive species: solution or pollution. Front. Ecol. Environ. 4: 132–140.

Mitchell, C.E., D. Blumenthal, V. Jarosik, E.E. Puckett and P. Pysek. 2010. Controls on pathogen species richness in plants' introduced and native ranges: roles of residence time, range size and host traits. Ecol. Lett. 13: 1525–1535.

Muller, H. and R.D. Goeden. 1990. Parasitoids acquired by *Coleophora parthenica* (Lepidoptera: Coleophoridae) ten years after its introduction into southern California for the biological control of Russian thistle. Entomophaga 35: 257–268.

Mytinger, L. and G.B. Williamson. 1987. The invasion of *Schinus* into saline communities of Everglades National Park. Fl. Sci. 50: 7–12.

Naeem, S., J.M.H. Knops, D. Tilman, K.M. Howe, T. Kennedy and S. Gale. 2000. Plant diversity increases resistance to invasion in the absence of covarying extrinsic factors. Oikos 91: 97–108.

Newman, R.M. 2004. Invited review—biological control of Eurasian watermilfoil by aquatic insects: basic insights from an applied problem. Arch. Hydrobiol. 159: 145–184.

Parker, I.M. and G.S. Gilbert. 2004. The evolutionary ecology of novel plant-pathogen interactions. Ann. Rev. Ecol. Evol. S. 35: 675–700.

Parker, I.M. and G.S. Gilbert. 2007. When there is no escape: the effects of natural enemies on native, invasive, and noninvasive plants. Ecology 88: 1210–1224.

Pearson, D.E. and R.M. Callaway. 2003. Indirect effects of host-specific biological control agents. Trends Ecol. Evol. 18: 456–461.

Pearson, D.E. and R.M. Callaway. 2005. Indirect nontarget effects of host-specific biological control agents: implications for biological control. Biol. Control 35: 288–298.

Pemberton, R.W. 1998. The potential of biological control to manage Old World climbing fern (*Lygodium microphyllum*), an invasive weed in Florida. Am. Fern J. 88: 176–182.

Pemberton, R.W. and A.P. Ferriter. 1998. Old World climbing fern (*Lygodium microphyllum*), a dangerous invasive weed in Florida. Am. Fern J. 88: 165–175.

Pimentel, D. 2005. Environmental and economic costs of the application of pesticides, primarily in the United States. Environ. Dev. Sustainability 7: 229–52.

Pimentel, D., L. Lach, R. Zuniga and D. Morrison. 2000. Environmental and economic costs of nonindigenous species in the United States. BioScience 50: 53–65.

Pimentel, D., R. Zuniga and D. Morrison. 2005. Update on the environmental and economic costs associated with alien-invasive species in the United States. Ecol. Econ. 52: 273–288.

Ploetz, R.C. 2007. Diseases of tropical perennial crops: challenging problems in diverse environments. Plant Dis. 91: 644–663.

Rayamajhi, M.B., R.W. Pemberton, T.K. Van and P.D. Pratt. 2005. First report of infection of *Lygodium microphyllum* by *Puccinia lygodii*, a potential biocontrol agent of an invasive fern in Florida. Plant Dis. 89: 110.

Richardson, D.M. and B.W. van Wilgen. 2004. Invasive alien plants in South Africa: how well do we understand the ecological impacts? S. Afr. J. Sci. 100: 45–52.

Rogers, L., M. Bodle, D. Black and F. Laroche. 2012. Status of nonindigenous species. pp. 7-1–7-35. *In*: 2012 South Florida Environmental Report, Vol. I—The South Florida Environment. South Florida Water Management District, West Palm Beach, FL. http://www.sfwmd.gov/portal/page/portal/pg_grp_sfwmd_sfer/portlet_prevreport/2012_sfer/v1/chapters/v1_ch7.pdf.

Shetty, K.G., K. Jayachandran, K. Quinones, K.E. O'Shea, T.A. Bollar and M.R. Norland. 2007. Allelopathic effects of ragweed compound thiarubrine-A on Brazilian pepper. Allelopathy J. 20: 371–378.

Shetty, K.G., D. Minnis, A.Y. Rossman and K. Jayachandran. 2011. The Brazilian peppertree seed-borne pathogen, *Neofusicoccum batangarum*, a potential biocontrol agent. Biol. Control 56: 91–97.

ShiLi, M., Y. Li, Q. Guo, H. Yu, J. Ding, F. Yu, J. Liu, X. Zhang and M. Dong. 2012. Potential alternatives to classical biocontrol: using native agents in invaded habitats and genetically engineered sterile cultivars for invasive plant management. Tree Forestry Sci. Biotechnol. 6: 17–21.

Shrum, R.D. 1982. Creating epiphytotics. pp. 113–138. *In*: R. Charudattan and H.L. Walker (eds.). Biological Control of Weeds with Plant Pathogens. John Wiley and Sons, Inc. New York.

Siemann, E., W.E. Rogers and S.J. DeWalt. 2006. Rapid adaptation of insect herbivores to an invasive plant. Proc. Royal Society, B, Biological Sciences 273: 2763–2769.

Simberloff, D., I.M. Parker and P.N. Windle. 2005. Introduced species policy, management and future research needs. Front. Ecol. Environ. 3: 12–20.

Swain, T. 1980. The importance of flavonoids and related compounds in fern taxonomy and ecology: an overview of the symposium. B. Torrey Bot. Club 107: 113–115.

TeBeest, D.O. and G.E. Templeton. 1985. Mycoherbicides: progress in the biological control of weeds. Plant Dis. 69: 6–10.

Templeton, G.E., D.O. TeBeest and R.J. Smith, Jr. 1979. Biological weed control with mycoherbicides. Annu. Rev. Phytopathol. 17: 301–310.

Thompson, S., J.D. Fletcher, H. Ziebell, S. Beard, P. Panda, N. Jorgensen, S.V. Fowler, L.W. Liefting, N. Berry and A.R. Pitman. 2013. First report of '*Candidatus* Liberibacter europaeus' associated with psyllid infested Scotch broom. New Disease Reports 27: 6.

Thorpe, A.S. and R.M. Callaway. 2006. Interactions between invasive plants and soil ecosystems: will feedbacks lead to stability or meltdown? pp. 323–342. *In*: M.W. Cadotte, S.M. McMahon and

T. Fukami (eds.). Conceptual Ecology and Invasions Biology: Reciprocal Approaches to Nature. Springer, The Netherlands.

Tobe, J.D., K.C. Burks and R.W. Cantrell. 1998. Florida Wetland Plants: An Identification Manual. Florida Department of Environmental Protection, Tallahassee, FL.

Wall, R.E., R. Prasad and S.F. Shamoun. 1992. The development and potential role of mycoherbicides for forestry. For. Chron. 68: 736–741.

Weidenhamer, J.D. and R.M. Callaway. 2010. Direct and indirect effects of invasive plants on soil chemistry and ecosystem function. J. Chem. Ecol. 36: 59–69.

Westbrooks, R.G. 1998. Invasive Plants, Changing the Landscape of America: Fact Book. Federal Interagency Committee for the Management of Noxious Weeds, Washington, D.C.

Williamson, M. 2006. Explaining and predicting the success of invading species at different stages of invasion. Biol. Invasions 8: 1561–1568.

Worsham, A.D. 1982. Discussion of topics. pp. 219–236. *In*: R. Charudattan and H.L. Walker (eds.). Biological Control of Weeds with Plant Pathogens. John Wiley and Sons, Inc., New York.

Wunderlin, R.P. 1982. Guide to the Vascular Plants of Central Florida. University Presses of Florida, Tampa, FL, 472 p.

Wunderlin, R.P. 1998. Guide to the Vascular Plants of Florida. University Press of Florida, Gainesville, FL, 806 p.

Yang, S.M. and S.C. Jong. 1995a. Host range determination of *Myrothecium verrucaria* isolated from leafy spurge. Plant Dis. 79: 994–997.

Yang, X.B. and D.O. TeBeest. 1993. Epidemiological mechanisms of mycoherbicide effectiveness. Phytopathology 83: 891–893.

Yu, H., J. Liu, W.M. He, S.L. Miao and M. Dong. 2011. *Cuscuta australis* restrains three exotic invasive plants and benefits native species. Biol. Invasions 13: 747–756.

Zangerl, A.R. and M.R. Berenbaum. 2005. Increase in toxicity of an invasive weed after reassociation with its coevolved herbivore. Proc. Natl. Acad. Sci. USA 102: 15529–15532.

Zidak, N.K. and P.A. Backman. 1996. Biological control of kudzu (*Pueraria lobata*) with the plant pathogen *Pseudomonas syringae* pv. *phaseolicola*. Weed Sci. 44: 645–649.

19

Algal Toxin Degradation by Indigenous Bacterial Communities in the Everglades Region

Krish Jayachandran[a,*] and *Kateel G. Shetty*[b]

Introduction

Algal blooms have been a persistent problem for Florida's fresh and marine waters often due to the indiscriminate agricultural use of nitrogen and phosphorus rich fertilizers, poor water and land management practices and inadequate waste water treatment methods (Philips et al. 2005). The profuse growth of algae on the surface of waters not only increases turbidity, particulate matter and the production of taste and odor causing compounds in water, but blocks sunlight and kills submerged vegetation and fish which also reduces species diversity within the water body (Fleming and Stephan 2001). When the algal cells finally begin to lyse, some species release their toxic cell metabolites directly into the water body (Metcalf and Codd 2004). These metabolites, referred to as cyanotoxins, are potentially dangerous to many organisms, including humans (Codd et al. 1997; WHO 1999).

Microcystins are the most commonly encountered cyanotoxins in freshwaters, and microcystin LR (MC-LR) is the most extensively studied (Fleming et al. 2002). Figure 1 shows the chemical structure of microcystin LR. The toxin was first isolated from the cyanobacteria *Microcystis aeruginosa*, but later from species in the genera

Earth and Environment Department, Florida International University, Miami, FL 33199.
[a] Email: jayachan@fiu.edu
[b] Email: shettyk@fiu.edu
* Corresponding author

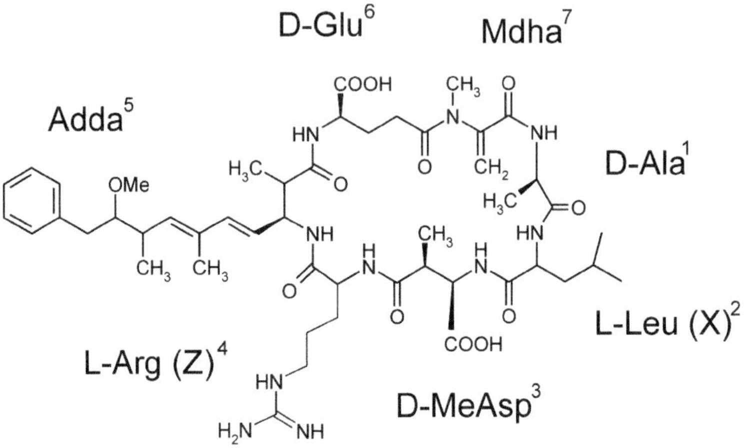

Figure 1. Structure of Microcystin-LR.

Anabaena, Oscillatoria, Nostoc and *Anabaenopsis* (Burns 2007). Microcystin is toxic to mammals, fish, plants and invertebrates (Codd et al. 1997; Pflugmacher et al. 1999) and is reported to bioaccumulate in several crop plants (irrigated with contaminated water), mussels, crayfish, and fish (grown in contaminated estuarine water) commonly consumed by humans (Figueiredo et al. 2004). Studies have also shown the possible correlation between consumption of surface water contaminated with microcystins and the occurrence of HCC (Hepato Cellular Carcinoma), the most prevalent type of liver cancer in humans (Falconer 1991; Fleming et al. 2001). Microcystin toxicity poses a great concern for public health safety, and thus The World Health Organization recommends a level of less than 1 µg/L of the toxin in drinking water to be safe for humans (WHO 1998).

Chemically, microcystins are a group of monocyclic heptapeptides having a general structure composed of five D-amino acids and two variable L-amino acids (which contributes to the two-letter suffix). Adda (3-amino-9 methoxy-2,6,8 trimethyl-10-phenyldeca-4,6-dienoic acid) is the characteristic β-amino acid responsible for acute toxicity (Harada et al. 1990; 2004). The Lethal Dose (LD_{50}) value for microcystins in mice ranges from 50 to 300 µg/kg body weight (bw), with MC-LR having an LD_{50} of 50 µg/kg bw comparable to that of chemical organophosphate nerve agents (Dawson 1998). Liver is the main target organ for microcystin toxicity (Robinson et al. 1991; Fischer et al. 2000). The major exposure route for humans is the consumption of contaminated water (WHO 1998). Minor routes include recreational use of lakes and rivers (dermal), contaminated food and inhalation (if microcystin is present as aqueous aerosols in air) (Codd et al. 1997).

Microcystin is recalcitrant to conventional water treatment (Lahiti and Hiisverta 1989) and is often detected even after chemical treatment of waters. Physical methods of microcystin removal have been tried but are constantly hampered by the presence of natural organic matter (Newcombe et al. 2003). Chemicals used to kill the algae are potentially toxic and may remain in the water to pose greater danger when consumed

(Kenefick et al. 1993). These chemicals also kill the algal cells (releasing more toxins into the water body) and the beneficial bacteria that can degrade the toxin (Jones and Orr 1994). Continued and increased use of such chemicals can induce resistance in organisms, making the management practices much more difficult. Biological methods of removal are being widely researched and are gaining more popularity (Cousins et al. 1996; Ho et al. 2007; Edwards and Lawton 2009). Several microcystin degrading bacteria have been reported so far (Valeria et al. 2006; Tsuji et al. 2006; Ho et al. 2007; Lemes et al. 2007; Pathmalal 2009).

The first study of biological removal of microcystin was carried out by Jones et al. (1994) in Murrumbidgee River, New South Wales, Australia. The MC-LR concentration of 1 mgL^{-1} was degraded within 5d using native bacteria previously exposed to the toxin. Cousins et al. (1996) also identified the possibility of using an indigenous mixed bacterial population to degrade low levels of MC-LR in reservoir waters that have been previously exposed to *Microcystis* blooms. Ho et al. (2007) reported that during MC-LR and MC-LA biodegradation with *Sphingopyxis* sp. LH21, no cytotoxic by-products were detected, thus demonstrating the possibility of using such bacterial species for biodegradation in drinking or recreational water supplies. While several studies have been carried out around the world, no research to our knowledge regarding isolation of microcystin degrading bacteria and microcystin biodegradation process has been published in United States particularly from South Florida's Lake Okeechobee region.

Lake Okeechobee (LO) is the largest freshwater lake of Florida providing for majority of drinking water treatment facilities. It is also an ecologically and recreationally important water body (Gray et al. 2005) and a potential Aquifer Storage and Recovery zone (ASR), influencing several other lakes, rivers and canals including Caloosahatchee River, St. Lucie Canal and the Kissimmee River in Florida (Steidinger 1999). Phosphorus levels in LO rose from 40 parts per-billion (ppb) in the 1970s to more than 130 ppb by the 1990s due to runoff from nearby agricultural operations, improved pasture, and dairy operations. Elevated levels of phosphorus in Lake Okeechobee have increased the occurrence of algal blooms and changed the species composition of the algal community (Gray et al. 2005). Species of *Anabaena, Microcystis,* and *Cylindrospermopsis* which produce toxins, are dominant in LO, making the water unpalatable and therefore a poor base for the aquatic food chain (Fleming 2002). Chlorophyll *a* and microcystin toxin concentrations were at elevated levels during August and October of 2005 with the highest concentrations observed in the Harney Pond, Indian Prairie Canal and Fish Eating Bay area (SFWMD). Sites such as Clewiston, Pahokee and Taylor slough also show raised levels of microcystin very frequently. Williams et al. (2007) studied the cyanobacterial toxins in Florida's freshwater systems during 2005 and reported up to 95 μg/L of microcystin in some zones of Lake Okeechobee. Being frequently exposed to such algal bloom scenarios, these locations would have native bacteria adapted to high toxin levels and that could exhibit toxin degradation abilities. The objective of this study was to isolate and characterize microcystin biodegrading bacteria or bacterial consortium from Florida's Lake Okeechobee.

Materials and Methods

Water Sample Collection

Four sites were selected in Lake Okeechobee area that represented high bloom area (HB), low bloom area (LB) and medium bloom area (MB), based on historic microcystin levels in these sites as reported by the South Florida Water Management District (SFWMD 2008). The sampling sites and their GPS co-ordinates are given in Table 1. Samples were collected in pre-cleaned 125 mL Nalgene bottles. Sub-surface water samples were collected at each site and transferred to sample bottles. The bottles were rinsed with the water sample before collection. Three samples were collected per site (but were later pooled together for experiments and analyses). Samples were collected at two different locations on the Indian Prairie Canal (sites D and E), since floating algal biomass was seen at two different points of Site D. The bottles were labeled with the site name, placed in an insulated cooler, transported to the lab, and stored at 4°C.

Table 1. Description of Lake Okeechobee sampling sites.

Site	Site Name	Bloom frequency	GPS co-ordinates	Number of samples collected
A	Clewiston	LB	N 26°45.617' W 80°55.054'	3 (A1, A2, A3)
B	Fish Eating Creek Canal	HB	N 26°57.752' W 81°07.276'	3 (B1, B2, B3)
C	Harney Pond Canal	MB	N 26°59.158' W 81°03.955'	3 (C1, C2, C3)
D	Indian Prairie Canal	HB	N 27°02.578' W 80°56.940'	3 (D1, D2, D3)
E	Indian Prairie Canal	HB	N 27°03.134' W 80°57.664'	3 (E1, E2, E3)

Microcystin and Enrichment Media

Microcystin-LR (MC-LR) (from ARCH Core Facility at Florida International University, Miami) stock solution was prepared and stored at 4°C (1 mL stock = 50 µg MC-LR). The initial acclimatization (enrichment) and later experiments was carried out in Minimal Salts Media (MSM) suggested by Valeria et al. (2006). The composition of the media per liter includes 112 mg $MgSO_4 \cdot H_2O$, 5 mg $ZnSO_4 \cdot H_2O$, 2.5 mg $Na_2MoO_4 \cdot 2H_2O$, 340 mg KH_2PO_4, 670 mg $Na_2HPO_4 \cdot 7H_2O$, 14 mg $CaCl_2$ and 0.13 mg $FeCl_3$. Twenty-five milliliters of MSM (with 0.25 µg mL^{-1} MC-LR) was dispensed into sterilized Erlenmeyer flasks under sterile condition.

Enrichment Study

Water samples collected from each site served as inoculum for enrichment studies. One mL of pooled sample from each site was added to 25 mL of sterile MSM with 0.25 µg mL^{-1} of MC-LR. Duplicate samples were maintained. A control, consisting of media without MC-LR and inoculated with sample was also maintained under similar conditions. Similar procedure was repeated for other sites. The inoculated flasks, controls and duplicates were incubated at 26°C in a rotary shaker incubator for 20 days after which they were sub-cultured with fresh toxin and continued to incubate for another 20 days. The process was repeated one more time at the end of 40 days. After each 20 day sub-culture, the bacterial populations in the samples were checked using dilution plate method on dilute nutrient agar (10%) and the plates were incubated for 3 days at 26°C. Colonies were represented as colony forming units (CFU) per mL of inoculum.

Microcystin Biodegradation Studies

After three cycles of enrichment studies (60 days period), one mL of inoculum from each of the incubated flasks was transferred to fresh 25 mL MSM amended with 0.25 μg mL^{-1} of MC-LR and incubated at 26°C in a rotary shaker incubator for a series of 10, 20 and 30 days to observe the pattern of microcystin degradation by the isolates of each site. Separate sets of control flasks without inoculum and without MC-LR were incubated under similar conditions as experimental flasks. Colony counts were carried out at the end of 10, 20 and 30 days and plated on dilute nutrient agar (10%) plates. Colonies from the nutrient agar plate were then sub-cultured and maintained.

Microcystin Analyses

At the end of incubation period the samples were analyzed for microcystin using High Performance Liquid Chromatography (HPLC). A series of steps were carried out before the samples were injected into the HPLC, to remove bacterial cells suspended in the sample that cause turbidity, which would otherwise interfere with the chromatographic process.

Each sample was sonicated to lyse the bacterial cells and release the cell-bound toxins for complete analysis. This was done by inserting the sonicating probe into the sample solution for one minute. The sonicating probe was sprayed with 70% alcohol after each sample and wiped clean to prevent contamination. The sonicated samples were then centrifuged at 8°C for 10 minutes at 10000 rpm. The supernatant was filtered using a 0.22 μm sterile membrane filter. The filtered samples were then allowed to pass through a C18 Solid Phase Extraction (SPE) cartridge (Extract Clean ™ C18-HC 1000 mg from GRACE). The cartridge was first activated with 10 mL of 100% methanol followed by 10 mL of distilled water. To the activated cartridge, the sample was slowly added and collected separately. Ten mL of methanol was finally added to the cartridge to elute the adsorbed microcystin and the eluent was collected separately. The eluent was then evaporated completely by purging nitrogen gas into the sample for 6–8 hours. To the evaporated sample 1 mL methanol was added. If samples showed turbidity, they were centrifuged again at 10000 rpm for 4 minutes. The supernatant was then transferred to clean HPLC vials and sealed. HPLC was run on a Hewlett-Packard system equipped with a UV-VIS spectrophotometer, using a 4.6 × 25 cm SUPELCO-Discovery C18 HPLC column (5 μm), with acetonitrile: 0.05% and ammonium acetate (20 mM; pH5) as mobile phase, flow rate: 0.1 mL min^{-1}, column temperature: 20°C and UV detection at 238 nm. Standards of MC-LR (4, 6 and

8 μg) were prepared and analyzed for determining the calibration curve. Trend line and R^2 value was generated using Microsoft Excel ™ software (Version 1997–2003). The HPLC peak area was used in the trend-line equation to calculate the amount of MC-LR recovered from the sample and controls.

Selection of Isolates and Optimal Condition Studies

From the results of HPLC analyses, the sample showing maximum degradation over the 30-day incubation period was identified. Controls with no inoculum were also analyzed for MC-LR content to account for photo-degradation. The amount of toxin degradation was calculated using the difference between initial and final toxin levels divided by the initial toxin concentration expressed as a percentage value.

A consortium of two bacterial isolates DC7 and DC8 from site D responsible for maximum toxin degradation was used for further experiments. The bacterial isolates DC7 and DC8 were grown separately in 250 mL Erlenmeyer flasks containing 50 mL of MSM+ medium (MSM amended with glucose and nitrogen to increase cell yield) on a rotary shaker (200 rpm) at 26°C. Bacterial cells were harvested during exponential growth phase by centrifugation (10,000 × g, 10 min, 4°C), washed once with sterile MSM solution and resuspended in MSM medium and used immediately. Each treatment flask received 5 × 10⁵ CFU mL⁻¹ of DC7 and 5 × 10⁵ CFU mL⁻¹ of DC8 separately, adding up to an initial consortium inoculum of 1 × 10⁶ CFU mL⁻¹. The bacterial consortium was grown in MSM media amended with 0.3 μg mL⁻¹ of MC-LR to study their toxin tolerance level at 26°C for 15 and 20 days. The effect of incubation temperature was determined by carrying out the biodegradation at 24, 26 and 28°C. Twenty-five mL of MSM (with 0.25 μg mL⁻¹ MC-LR) was inoculated with the bacterial consortium and incubated for 20 days. Since high phosphate levels are quite common in Lake Okeechobee the possible interference of phosphate in the biodegradation process was studied by using media rich in phosphate which consists of 112 mg $MgSO_4 \cdot H_2O$, 5 mg $ZnSO_4 \cdot H_2O$, 2.5 mg $Na_2MoO_4 \cdot 2H_2O$, 340 mg KH_2PO_4, 1340 mg $Na_2HPO_4 \cdot 7H_2O$, 14 mg $CaCl_2$ and 0.13 mg $FeCl_3$ per 1 L of deionized water. The bacterial consortium was incubated with high phosphate MSM media along with 0.25 μg mL⁻¹ MC-LR at 26°C for 20 days. Control with no MC-LR and duplicates were maintained for all experiments.

DC7 and DC8 Isolate 16S rRNA Sequence Analysis

The site D sample bacterial isolates DC7 and DC8 that showed maximum microcystin degradation were first characterized using standard bacteriological methods. The 16S rRNA gene was PCR amplified from genomic DNA isolated from pure bacterial colonies. Primers used are universal 16S primers that correspond to positions 0005F and 0531R for a 500 bp sequence, and 0005F and 1513R for the 1500 bp sequence. Amplification products were purified from excess primers and dNTPs and checked for quality and quantity by running a portion of the products on an agarose gel. Cycle sequencing of the 16S rRNA amplification products was carried out using DNA polymerase and dye terminator chemistry. Excess dye-labeled terminators were then removed from the sequencing reactions. The samples were electrophoresed on ABI 3130 Genetic Analyzer. Sequence analysis was performed using Sherlock® DNA microbial analysis software and database.

Results and Discussion

Bacterial Population During Enrichment

Heterotrophic colony count studies carried out every 20 days (for three cycles) during initial enrichment phase revealed that sample from site D showed high turbidity and constantly higher number of colonies on the dilute nutrient agar plates when compared to samples from other sites (Fig. 2). During first cycle G-1 (at the end of 20 days), number of colonies from site D was as high as 290×10^6 CFUs mL^{-1} and with third cycle G-3 (at the end of 60 days) the colonies were still as high as 160×10^6 CFUs mL^{-1}. Even though samples from other sites showed higher colony counts during first cycle, the numbers started to dwindle rapidly over the course of 60 days. The constantly high number of colonies surviving from sample D showed the prospective presence of microcystin assimilating organisms present in the sample.

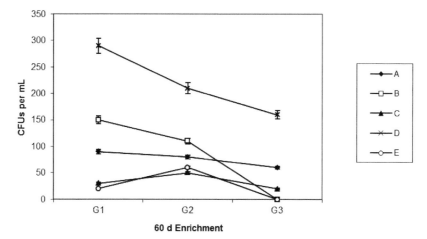

Figure 2. Changes in heterotrophic bacterial population (CFU $\times$ 10^6 mL^{-1}) in microcystin enrichment cultures derived from Lake Okeechobee sampling sites. Interval between each cycle (G1, G2, and G3) was 20 days. A, B, C, D and E are sampling sites.

Microcystin Degradation Studies

During the microcystin degradation study, sample from site D showed highest degradation rate (Table 2). At the end of 10 days of incubation, the recovery of microcystin from controls averaged to 3.14 µg in 25 mL of medium (initial concentration was 6.25 µg in 25 mL of medium). Loss of MC-LR in controls could be attributed to photo degradation during incubation. Except sample D, all other samples showed lower degradation than the controls. Sample D showed about 6% more degradation than the controls and this pattern continued during 20 and 30 day

Table 2. Microcystin degradation (%) in enrichment cultures after 10, 20 and 30 day incubation.

Incubation period and sample site	Average toxin recovered (µg)	Degradation (%)
Day 10		
A	3.78 (± 0.42)	39.52
B	3.88 (± 0.05)	37.92
C	3.75 (± 0.47)	40
D	2.80 (± 0.09)	55.2
E	3.64 (± 0.37)	41.84
Control	3.14 (± 0.09)	49.76
Day 20		
A	2.05 (± 0.05)	67.2
B	3.26 (± 0.55)	47.44
C	2.46 (± 0.31)	60.72
D	1.59 (± 0.16)	74.56
E	2.93 (± 0.66)	53.20
Control	3.79 (± 0.06)	39.36
Day 30		
A	1.72 (± 0.55)	72.48
B	2.09 (± 0.37)	66.56
C	2.17 (± 0.04)	65.36
D	1.00 (± 0.45)	84
E	2.29 (± 0.13)	63.28
Control	3.41 (± 0.17)	45.44

Values in parenthesis are SE.

incubation too. The initial degradation rate predominantly depends on the amount of microcystin available as well as the population of organisms present to make use of the toxin. Once the population establishes and starts to degrade the toxin, the rate of degradation increases and occurs at a faster rate. Cousins et al. (1996) report that during the first 12 day of microcystin degradation, the rate was high only when they incubated the sample in raw reservoir waters, while the rate did not change rapidly when similar studies were carried out in deionized water.

At the end of 20 day incubation, the degradation was much faster with sample D. While 3.79 µg MC per 25 mL of MSM were recovered from the controls, only about 1.59 µg MC were recovered from sample D. There was a corresponding increase in bacterial cell count along with microcystin degradation (Fig. 3). Bacterial cell count studies during day 0, 11, 21 and 31 confirms that the increasing cell numbers were indeed responsible for subsequent microcystin degradation (Table 3). Controls (with no microcystin) did not show any colonies on dilute nutrient agar plates. At the end of 30 day incubation the degradation rate began to get slower, which could be due to dwindling microbial count or the inability of the microbes to breakdown the toxin any further. Analysis of metabolites arising from this degradation would give a clear idea about further progress in degradation.

The bacterial population from the site D enrichment was found to be consisting of two bacterial colony types. The two isolates DC7 and DC8, when tested individually

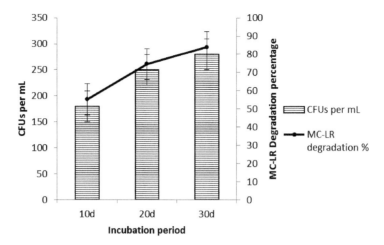

Figure 3. Microcystin-LR biodegradation (%) by site D bacterial consortium and changes bacterial cell count (CFU × 10⁶ mL⁻¹) during incubation period.

Table 3. Changes in heterotrophic bacterial population (CFU × 10⁶ mL⁻¹) during microcystin enrichment culture.

Sample site	Day 0	Day 11	Day 21	Day 31
		CFU* mL⁻¹		
A	40	110	150	180
B	40	100	130	200
C	40	170	200	180
D	40	180	250	280
E	40	170	190	210

*CFU = colony forming units, mean three replications.

in a replicated study showed no significant growth or biodegradation of microcystin (data not shown), therefore the two isolates were used together as a consortium for further studies. With the HPLC results, it was confirmed that bacterial consortium from site D had maximum MC-LR degradation capacity. The maximum amount of microcystin was degraded during the first 20 days of incubation with sample D. Hence this consortium was chosen for further characterization studies. When microcystin concentration was increased from 0.25 to 0.3 μg mL⁻¹, the site D consortium was still able to degrade 51% more toxin compared to the controls in 15 days and 45% more than the control in 20 days (Fig. 4). Temperature seems to have some effect on microcystin degradation (Fig. 5). While at 28°C, the degradation was poor and at 24 and 26°C, the degradation remained more or less the same. Phosphate level in the media seems to have no effect on microcystin degrading bacteria (Fig. 6). Even at higher phosphate levels the bacteria was still able to degrade the toxin.

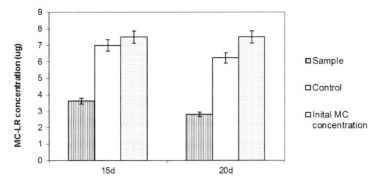

Figure 4. Microcystin-LR recovery from site D consortium culture after 20 day incubation. Initial microcystin content in each flask was 7.5 μg for both sample and control, with no bacterial inoculum in control.

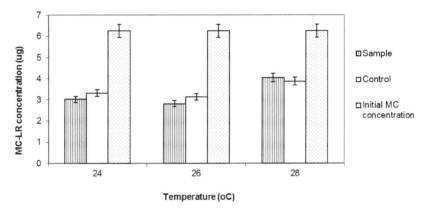

Figure 5. Microcystin-LR recovery from site D consortium culture after 20 day incubation at different temperatures. Initial microcystin content in each flask was 6.25 μg for both sample and control, with no bacterial inoculum in control.

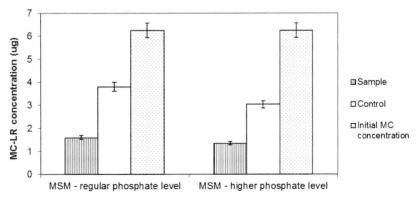

Figure 6. Effect of high phosphate concentration on microcystin-LR recovery after 20 day incubation of site D consortium culture. Initial microcystin content in each flask was 6.25 μg for both sample and control, with no bacterial inoculum in control.

Identification of Microcystin Degrading Bacteria

The site D consortium consisted of two colony types. DC8: Large, more or less circular, pale yellow, glistening colonies with slightly uneven margins, Gram positive rod, non-motile.

DC7: Small, circular, bright white, mucoid colonies with smooth edges, Gram negative bacilli, actively motile.

The partial sequence of the 16S rRNA gene for the bacterial isolates DC8 and DC7 was used for database search and construction of phylogenetic tree using neighbor joining method. Based on the results, the isolate DC8 belongs phylogenetically to the genus *Microbacterium*, while DC7 has its closest match with *Ochrobactrum anthropi*. The bacterial isolate DC8 has in fact 0% difference with two sp. of *Microbacterium* (*Microbacterium liquefaciens* and *Microbacterium maritypicum/oxydans*). The bacterial strains presenting 16S rRNA gene similarities lower than 99.5% and higher than 97% may belong to different species; a full 16S rDNA sequence analysis and other tests are needed for definitive taxonomic identity (Tindall et al. 2010). Definitive identification of isolate DC8 may require additional biochemical tests or sequencing of additional gene loci. The second bacterial isolate DC7 has sequence matches with multiple genera, with *Ochrobactrum anthropi* as the closest match (2.13% difference). However, based on the online homology BLAST search in the openly available Genbank data, the DC7 isolate was found to be 100% matching with *Rhizobium gallicum* (AY972457). Definitive identification of isolate DC7 may require additional biochemical tests or sequencing of additional gene loci. There are problems in identifying groups of bacteria for which 16S rRNA gene sequences are not discriminative enough; in addition the use of unverified public databases such as GenBank can be problematic in that errors in sequence data are not uncommon (Woo et al. 2009).

Microbacterium species are usually Gram-positive, aerobic or facultatively anaerobic, non-motile, non-spore-forming rods. It grows between 15 and 37°C. The pH range for growth is 5–8. *Ochrobactrum anthropi* is a gram negative aerobic, peritrichously flagellated, nonfermentative, nonfastidious motile rods that are positive for oxidase, catalase and urease production. The organism is ubiquitous in nature but most frequently encountered in aquatic bodies. Laura et al. (1996) report the aerobic degradation of atrazine using this microbe, isolated from an activated sludge. *Rhizobium* species are Gram negative, aerobic, motile non-spore forming rods. *Rhizobium* sp. are common soil bacteria that form plant nodules and have the enzyme nitrogenase for fixing nitrogen. Biodegradation of sulfonated Phenylazonaphthol Dye (Ruiz-Arias et al. 2010) and phenol (Wei et al. 2008) was found to be associated with *Rhizobium* sp. While *Microbacterium* sp. has never been reported so far in biodegradation or bioremediation studies, it could still play an important role in this particular study. Further research is needed to understand the role of each of these bacteria in degrading the toxin, whether they act together or separately, which organism precedes the other, and determining the optimal concentration (bacterial count) needed for effective degradation. This could be an important next step in commercial applications to remove microcystins from drinking water supply.

The results from the present study showed that a bacterial consortium consisting of *Microbacterium* sp. and *Rhizobium gallicum* (AY972457) isolated from a high frequency algal bloom site at Lake Okeechobee (Indian Prairie canal), was effective in degrading up to 74% of microcystin within 20 days at 26°C. The time frame observed for such an effective degradation seems to be similar to other microcystin biodegradation studies. Mattheinsen et al. (2000) report 70% of Microcystin-LR degradation within 10–15 days using samples from Patos Lagoon, Brazil (degradation rate 0.03–0.06 mg MC–LR mL^{-1} d^{-1}). The time period for MC-LR removal showed that the degradation was highest during the initial 20 d and then decreased gradually. While many authors (Jones et al. 1994; Lam et al. 1995) report an initial lag phase during biodegradation, Christofferson (2002) did not observe a lag phase. These differences in the presence and absence of lag phase as well as its duration could be the result of different bacterial community composition and its degree of adaptation (Valeria et al. 2006). But most authors agree that native microbial community can be utilized to perform biodegradation of these cyanotoxins quite rapidly. The degradation also seemed to have been influenced by the bacterial cell density, with higher the bacterial cell density yielding higher degradation rates. Even though samples from other sites showed some amount of degradation, isolates from site D showed the highest. Temperature had an effect on MC-LR degradation. Optimal degradation was achieved at 24 and 26°C and any further increase resulted in decreased degradation rate. Similar results have been reported by Park et al. (2001) during their biodegradations studies using Y2 strain isolated from a hypertrophic lake in Lake Suwa, Japan. They report that maximum biodegradation occurred between 20 to 25°C. It is interesting to note that higher phosphate levels in the media did not hamper the biodegradation process in our studies.

With Florida's growing population, subtropical environment, climate change, and growing agricultural operations all over the state, it is quite natural to expect increasing bloom scenarios (and subsequent toxin release) in its lakes, rivers and estuaries. Adopting suitable water treatment technologies, preferably a combination of physical and biological methods, would effectively remove such toxins from drinking and recreational water supply and will help safe-guard Floridian's health and protect the lives of aquatic species.

References

Burns, J. 2007. Toxic cyanobacteria in Florida waters. pp. 117–126. *In*: K. Hudnell (ed.). Proceedings of the Interagency International Symposium on Cyanobacterial Harmful Algal Blooms. Advances in Experimental Medicine and Biology. U.S. EPA, Research Triangle Park, NC.

Chorus, I. and J. Bartram. 1999. Cyanobacterial Toxins. Toxic Cyanobacteria in Water: A guide to their public health consequences, monitoring and management. World Health Organization. ISBN 0-419-23930-8. 400 pp. Accessed on June 23, 2014. http://www.who.int/water_sanitation_health/resourcesquality/toxcyanbegin.pdf.

Christofferson, K., S. Lyck and A. Winding. 2002. Microbial activity and bacterial community structure during degradation of microcystins. Aquatic Microb. Ecol. 27: 125–136.

Codd, G.A., C.J. Ward and S.G. Bell. 1997. Cyanobacterial toxins: occurrence, modes of action, health effects and exposure routes. Arch. Toxicol. Suppl. 19: 399–410.

Cousins, I.T., D.J. Bealing, H.A. James and A. Sutton. 1996. Biodegradation of MC-LR by indigenous mixed bacterial populations. Wat. Res. 30: 481–485.

Dawson, R.M. 1998. Toxicology of microcystins. Toxicon 36: 953–962.

Edwards, C. and L.A. Lawton. 2009. Bioremediation of cyanotoxins. pp. 109–129. *In*: A.I. Laskin, G.M. Gadd and S. Sariaslani (eds.). Advances in Applied Microbiology. Academic Press, New York, USA.

Falconer, I.R. 1991. Tumor promotion and liver injury caused by oral consumption of cyanobacteria. Environ. Toxicol. Wat. Qual. 6: 177–184.

Figueiredo, D.R., U.M. Azeiteiro, S.M. Esteves, F.J.M. Goncalves and M.J. Pereira. 2004. Microcystin producing blooms—A serious global public health issue. Ecotoxicol. Environ. Saf. 59: 151–163.

Fischer, W.J., B.C. Hitzfeld, F. Tencalla, J.E. Eriksson, A. Mikhailov and D.R. Dietrich. 2000. Microcystin-LR toxicodynamics, induced pathology, and immunohistochemical localization in livers of blue–green algae exposed rainbow trout (*Oncorhynchus mykiss*). Toxicol. Sci. 54: 365–373.

Fleming, L.E. and W. Stephan. 2001. Blue Green Algae, Their Toxins and Public Health Issues. Report to the Florida Harmful Algal Bloom Taskforce. NIEHS Marine and Freshwater Biomedical Sciences Center, University of Miami, Miami, FL.

Fleming, L.E., C. Rivero, J. Burns, C. Williams, J.A. Bean, K.A. Shea and J. Shin. 2002. Blue green algal (Cyanobacterial) toxins, surface drinking water and liver cancer in Florida. Harmful Algae 1: 157–168.

Gray, P.N., C.J. Farrell, M.L. Kraus and A.H. Gromnicki. 2005. Lake Okeechobee: A synthesis of information and recommendations for its restoration. Audubon of Florida, Miami. http://fl.audubon.org/sites/default/files/documents/audubon_stateofthelake_2006.pdf. Accessed on June 23, 2014.

Harada, K.I., K. Matsuura, M. Suzuki, M.F. Watanabe, S. Oishi, A.M. Dahlem, V.R. Beasley and W.W. Carmichael. 1990. Isolation and characterization of the minor components associated with microcystins LR and RR in the cyanobacterium (blue-green algae). Toxicon 28: 55–64.

Harada, K.I., S. Imanishi, H. Kato, M. Masayoshi, E. Ito and K. Tsuji. 2004. Isolation of Adda from microcystin-LR by microbial degradation. Toxicon 44: 107–109.

Ho, L., A.L. Gaudieux, S. Fanok, G. Newcombe and A.R. Humpage. 2007. Bacterial degradation of microcystin toxins in drinking water eliminates their toxicity. Toxicon 50: 438–441.

Jones, G.J. and P.T. Orr. 1994. Release and degradation of Microcystin following algicide treatment of a *Microcystis aeruginosa* bloom in a recreational lake, as determined by HPLC and protein phosphatase inhibition assay. Water Res. 28: 871–876.

Jones, G.J., D.G. Bourne, R.L. Blakeley and H. Doelle. 1994. Degradation of the cyanobacterial hepatotoxin microcystin by aquatic bacteria. Nat. Toxins 2: 228–235.

Kenefick, S.L., S.E. Hrudey, H.G. Peterson and E.E. Prepas. 1993. Toxin release from *Microcystis aeruginosa* after chemical treatment. Water Sci. Technol. 27: 433–440.

Lahiti, K. and L. Hiisverta. 1989. Removal of Cyanobacterial toxins in water treatment processes—review of studies conducted in Finland. Water Supply 7: 149–154.

Lam, A.K.Y., P.M. Fedorak and E.E. Prepas. 1995. Biotransformation of the cyanobacterial hepatotoxin microcystin LR, as determined by HPLC and protein phosphatase bioassay. Environ. Sci. Technol. 29: 242–246.

Laura, D., G. De Socio, R. Frassanito and D. Rotilio. 1996. Effects of atrazine on *Ochrobactrum anthropi* membrane fatty acids. Appl. Environ. Microbiol. 62: 2644–2646.

Lemes, G.A.F., R. Kersanach, L.S. Pinto, O.A. Dellagostin, J.S. Yunes and A. Mattheinsen. 2007. Biodegradation of microcystins by aquatic *Burkholderia* sp. from a south Brazilian coastal lagoon. Ecotoxicol. Environ. Saf. 69: 358–65.

Matthiensen, A., K.A. Beattie, J.S. Yunes, K. Kaya and G.A. Codd. 2000. [DLeu] microcystin-LR, from the cyanobacterium *Microcystis* RST 9501 and from a *Microcystis* bloom in the Patos lagoon estuary, Brazil. Phytochem. 55: 383–387.

Metcalf, J.S. and G.A. Codd. 2004. Cyanobacterial toxins in the water environment. A review of current knowledge. Foundation for Water Research. University of Dundee. Accessed on June 23, 2014. http://www.jlakes.org/web/cyanobacterial-toxins-in-the-water-environment-ROCKS2004.pdf.

Mikhailov, A., A.S. Harmala-Brasken, J. Hellman, J. Meriluoto and J.E. Eriksson. 2003. Identification of ATP-synthetase as a novel intracellular target for microcystin-LR. Chem. Biol. Interact. 142: 223–237.

Newcombe, G., D. Cook, S. Brooke, L. Ho and N. Slyman. 2003. Treatment options for microcystin toxins: Similarities and differences between variants. Environ. Toxicol. 24: 299–308.

Park, H.D., Y. Sasaki, T. Maruyama, E. Yanagisawa, A. Hiraishi and K. Kato. 2001. Degradation of cyanobacterial hepatotoxin microcystin by a new bacterium isolated from a hypertrophic lake. Environ. Toxicol. 16: 337–343.

Pathmalal, M.M., C. Edwards, B.K. Singh and L.A. Lawton. 2009. Isolation and identification of novel microcystin-degrading bacteria. Appl. Environ. Microbiol. 75: 6924–6928.

Pflugmacher, S., G.A. Codd and C.E.W. Steinberg. 1999. Effects of the cyanobacterial toxin microcystin-LR on detoxication enzymes in aquatic plants. Environ. Toxicol. 14: 111–115.

Philips, E.J., E. Bledsoe, M. Cichra S. Badylak and J. Frost. 2005. The Distribution of potentially toxigenic cyanobacteria in Florida. Proceedings of Health Effects of Exposure to Cyanobacteria Toxins, State of Science, Florida Department of Health, The Center for Disease Control, Mote Marine Laboratory and University of South Florida.

Robinson, N.A., J. Pace, C.F. Matson, G.A. Miura and W.A. Lawrence. 1991. Characterization of chemically titrated microcystin-LR and its distribution in mice. Toxicol. *in vitro* 5: 341–345.

Ruiz-Arias, A., C. Juarez-Ramirez, D. de los Cobos-Vasconcelos, N. Ruiz-Ordaz, A. Salmeron-Alcocer, D. Ahuatzi-Chacon and J. Galindez-Mayer. 2010. Aerobic biodegradation of a sulfonated phenylazonaphthol dye by a bacterial community immobilized in a multistage packed-bed BAC reactor. Appl. Biochem. Biotech. 162: 1689–1707.

[SFWMD] (South Florida WATER Management District) Lake Okeechobee: Algal Bloom Monitoring Program. www.sfwmd.gov. http://www.sfwmd.gov/portal/page/portal/pg_grp_sfwmd_watershed/pg_ sfwmd_watershed_inlake?_piref2294_4946961_2294_4946960_4946960.tabstring=tab20940638&_ piref 294_4946961_2294_4946960_49469602294_20940817_2294_4946514_20940639. tabstring=tab209 40808. Accessed on June 23. 2014.

Steidinger, K., J.H. Landsburg, C.R. Omas and J.W. Burns. 1999. Harmful algal blooms in Florida. The Harmful Algal Bloom Task Force Technical Advisory Group. Review Discussion. http://myfwc.com/ media/202228/HAB_whitepaper2006_UPDATE.pdf. Accessed on June 23, 2014.

Tindall, B.J., R. Rossello-Mora, H.J. Busse, W. Ludwig and P. Kampfer. 2010. Notes on the characterization of prokaryote strains for taxonomic purposes. Int. J. Syst. Evol. Microbiol. 60: 249–266.

Tsuji, K., M. Asakawa, Y. Anzai, T. Sumino and K. Harada. 2006. Degradation of microcystin using immobilized microorganisms isolated in a eutrophic lake. Chemosphere 65: 1117–124.

Valeria, A.M., E.J. Ricardo, P. Stephan and W.D. Alberto. 2006. Degradation of microcystin-RR by *Sphingomonas* sp. CBA4 isolated from San Roque reservoir (Co´rdoba—Argentina). Biodegrad. 17: 447–455.

Wei, G., J. Yua, Y. Zhua, W. Chena and L. Wanga. 2008. Characterization of phenol degradation by *Rhizobium* sp. CCNWTB 701 isolated from *Astragalus chrysopteru* in mining tailing region. J. Hazard. Mater. 151: 111–117.

Williams, C.D., M.T. Aubel, A.D. Chapman and P.E. Aiuto. 2007. Identification of cyanobacterial toxins in Florida's Freshwater systems. Lakes Reservoir Manage. 23: 144–152.

Woo, P.C.Y., J.L.L. Teng, J.K.L Wu, F.P.S. Leung, H. Herman Tse, A.M.Y. Fung, S.K.P. Lau and K. Yuen. 2009. Guidelines for interpretation of 16S rRNA gene sequence-based results for identification of medically important aerobic Gram-positive bacteria. J. Med. Microbiol. 58: 1030–1036.

World Health Organization, Geneva. 1998. Guidelines for Drinking-Water Quality. 2nd edition. Addendum to Vol. 2. Health Criteria and Other Supporting Information 95–110.

Closing Thoughts on the Role of Microbial Ecology in Management and Monitoring of the Greater Everglades Ecosystem

Andrew Ogram,[1,a,*] *James A. Entry,*[2] *Andrew Gottlieb,*[3,b,*] *K. Ramesh Reddy*[1] and *Krish Jayachandran*[4]

Introduction

Microbial communities underpin the biogeochemical cycles that sustain the Everglades ecosystem and play critical roles in processing and regulating various ecosystem services that are important to the health and economic status of the populations of South Florida. Microbial communities, along with their interactions with periphyton and vegetation communities, serve as major drivers regulating macro-elemental cycles in the Everglades. Their structure, function, and productivity are strongly affected by the anticipated changes in temperatures, hydrology, and salinity that are likely to accompany global climate change. Examples of potential changes in wetland ecosystem processes include altered microbial communities which may decrease or increase

[1] Soil and Water Science Department, University of Florida, Gainesville, Florida 32611-0290.
[a] Email: aogram@ufl.edu
[b] Email: krr@ufl.edu
[2] Nutrigrown LLC., 7389 Washington Boulevard, Suite 102, Elkridge Maryland 21075.
   Email: jim.entry@nutrigrown.com
[3] 141 Harvard Drive, Lake Worth, FL 33460-6332.
   Email: adgottlieb71@gmail.com
[4] Earth and Environment Department, Florida International University, Miami, FL 33199.
   Email: jayachan@fiu.edu
* Corresponding author

relatives rates of biogeochemical processes thus affecting nutrient regeneration and availability to periphyton and vegetation (Reddy and Delaune 2008). The intention of this volume was to provide the reader with a greater understanding of the current research used to characterize the structure and function of microbial communities in the Everglades ecosystem. It was also our hope that this volume identified gaps in our understanding of the microbial ecology of the Everglades that must be addressed to properly restore and manage this critical ecosystem.

Abiotic Factors Influencing the Microbial Ecology of Waters and Soils

The activities of microorganisms are affected by their environment and in turn affect many properties of that environment. For wetlands such as the Everglades, water is the dominant driver of landscape organization. The hydrology of the Everglades has changed dramatically over the last 100 years, with over half of the original area of the wetland drained. The construction of over 2500 km of canals (and water control structures), conversion of 700,000 acres of land immediately south of Lake Okeechobee to farmland for the Everglades Agricultural Area (EAA), as well as large increases in the human population and related infrastructure of South Florida contribute to the interruption of the natural flow of water, generally sheet flow, from the banks of the Lake south into Florida Bay and the Gulf of Mexico (Larsen et al. 2011; Harvey et al. 2009).

In addition to decreases in the spatial extent and quantity of water and changes to flow patterns, the quality of the water has declined significantly due to runoff from the EAA and from agricultural chemicals input to Lake Okeechobee. Many of the ecosystem properties of the historic Everglades were shaped by the very low available phosphorus (P) concentrations, such that the ecosystem was well-adapted to life in which P availability controlled growth rates and community structures across trophic levels. Runoff from the agricultural operations of the EAA into the adjacent areas of the northern Everglades previously contained even higher concentrations of P than today, and to a lesser degree, sulfate (SO_4). This input of P into extremely P-limited waters resulted in dramatic shifts in plant communities and in a wide range of microbially mediated processes, including the rates of peat accumulation and methane production.

The continued accumulation of elevated concentrations of P, S, and N in soils delivered by runoff from the EAA and urban sources will result in a significant legacy of these nutrients that will be released by microbial activity for years to come. Management strategies must consider that these nutrients, and mercury, will be transported downstream from the impacted soils to relatively unimpacted areas. Sulfate and dissolved organic carbon drive SO_4 reduction, the primary process responsible for Hg methylation in these areas, increasing the urgency for development of comprehensive management strategies for this area.

Complicating future management strategies is the continued salt water intrusion into both groundwater and surface water from the Gulf of Mexico and Florida Bay into the Everglades associated with sea level rise. Much of the South Florida coast is susceptible to sea level rise, from major flooding to salt intrusion into fresh water resources (Guha and Panday 2012; Zhang et al. 2010).

Substantial tracts of the Everglades and other low-lying coastal areas will be submerged or will be susceptible to storm surges, flooding, erosion, and other problems associated with 1 m of sea level rise (Noss 2011; IPCC 2007). Rising sea levels will cause increased salinity in fresh and brackish water areas of the Florida coastal zone and will increase the risk of storm surge-induced flooding and saltwater exposure in oligohaline areas of the Everglades near the top of the estuarine ecotone (Teh et al. 2008; Pearlstine et al. 2010). Saha et al. (2011) predict that, based on tolerance to drought and salinity, sea level rise will change species composition of coastal hardwood hammocks and buttonwood (*Conocarpus erectus* L.) forests in Everglades National Park, and hence affect the quality and quantity of carbon sources available to Everglades' microbial communities.

The Everglades is a highly modified and managed system with a large number of canals, ditches, and levies used to divert water for flood control, agricultural water supply, and human consumption. The changes in the quantity, timing, and distribution of freshwater delivery to the coastal zone caused by hydrologic alterations throughout the Everglades are believed to be amplifying saltwater intrusion and the rate of its landward movement (Nuttle et al. 2000). Although Florida's climate is predicted to warm less than northern regions of the US (IPCC 2007; Von Holle and Nickerson 2010), a climate inventory over the past 35 to 108 years indicated Florida will experience greater climatic extremes, with trends of increased summer and fall maximum temperatures and decreased winter and spring minimum temperatures (Von Holle and Nickerson 2010). The intensity of tropical storms is also predicted to increase, although frequency may decrease (Misra et al. 2011; IPCC 2007).

Not only will increased salinity affect the structure of plant communities and associated carbon (C), but salt water intrusion is likely to increase sulfate (SO_4) concentrations. Increased concentrations of sulfate (SO_4) in the water column will increase the activities of sulfate (SO_4) reducing bacteria, which may increase the rates of peat decomposition and potentially increase rates of Hg methylation in affected areas. Additional factors associated with increased salinity may impact the growth rates of plants and peat accretion rates, which may further alter flow patterns. Research to promote the understanding of peat deposition and decomposition in estuarine and freshwater environments is warranted, along with gaining a better understanding of marine effects on freshwater derived peats and associated microbial communities.

The Impact of Microorganisms on Water Quality

Water quality has been improving in the Everglades Protection Area (EPA) areas since 1990 (Entry and Gottlieb 2015, Chapter 2; Entry and Gottlieb 2013; Entry 2012a). However, Everglades soils continue to accumulate P (Osborne et al. 2015, Chapter 3) and large areas originally dominated by sawgrass continue to be converted to cattail. When nutrients, including P, are added to an oligotrophic ecosystem, periphyton and plants immediately begin to take up these nutrients by removing them from the water column (Noe et al. 2003). In addition to the myriad of design problems (Entry et al. 2012b; 2013), water quality monitoring in the Everglades is further complicated by microorganisms. The elevated TP values reported by water quality monitoring in the

Everglades ecosystem are the result of nutrient input into marsh faster than the algae and higher plants' ability to assimilate nutrients.

The reduction of Everglades water TP values may mask the fact that although nutrient load into the Everglades has decreased, large amounts of nutrients continue to be delivered to the EPA (Entry and Gottlieb 2014). These nutrients are being taken up, cycled through microbes, periphyton, and vegetation, deposited in flocculent organic matter, and ultimately incorporated into the soil (Reddy et al. 1999; Noe et al. 2003). The rate at which the soils are accumulating P and vegetative change has most likely decreased, however the change is still occurring and will continue to occur until water being delivered to the EPA is at 10 µg TP L^{-1} or less. Traditional water quality monitoring in the Everglades is generally less effective because measures of surface water TP concentration do not account for the fact that periphyton removes substantial amounts of P from the water column. Additionally, water quality monitoring in the Everglades does not take into account daily variance and hence may mask extreme event measures. If the monitoring of marsh water quality in the Everglades ecosystem is to be effective in helping prevent P accumulation in soils and vegetation change, algal P uptake rates should be incorporated into Everglades nutrient monitoring.

Implications for Periphyton Monitoring, Assessment and Forecasting

Periphyton is a critical component of the greater Everglades ecosystem, serving as the base for aquatic food webs (Trexler et al. 2015, Chapter 8), is an important regulator of N and P biogeochemistry (Inglett 2015, Chapter 9; Komárek et al. 2015, Chapter 10), and is considered to be a sensitive indicator of nutrient status (Gaiser et al. 2015, Chapter 6; Gottlieb et al. 2015, Chapter 7), hydrology (Gottlieb et al., Chapter 7), and salinity (Wachnicka and Wingard 2015, Chapter 11; Frankovich and Wingard 2015, Chapter 12). Periphyton mats are composed primarily of cyanobacteria, green algae, and diatoms imbedded in a complex matrix of polysaccharides, detritus and inorganic material. Depending on nutrient status and hydrology, periphyton may take a variety of forms, ranging from thick benthic and floating mats to the epiphytic coatings on plants.

As a consequence of its importance in the landscape and its utility as an ecological indicator (Gaiser 2009), research to understand the effects of sea level rise, nutrient enrichment and changes in conductivity on the structure and function of Everglades periphyton mats is critical. As the estuarine ecotone shifts inland, changes in the structure of periphyton communities have implications for nutrient cycling and food web dynamics. Studies that characterize P assimilation and N fixation rates associated with varying landscapes, and salinity and nutrient regimes will help us understand the implications of potential future changes in algal community structure. Changes at the species and community level occur rapidly with changes in water quality and are therefore important as early indicators of change (Gaiser et al. 2015, Chapter 6). Other methods, such as remote sensing (Gann et al. 2015, Chapter 14) and chemotaxonomy (Louda et al. 2015, Chapter 13) can eventually aid in mapping landscape scale changes in structure and related function. Continued work to compare ongoing changes in Everglades periphyton communities to those that occur in both space and time provides us with a better understanding of potential variance and related biogeochemical implications. That is, studying changes in historical assemblages in South Florida using soil cores, and comparative studies between south Florida assemblages and those in other systems helps us understand how the local changes in structure and function we observe fit within a more global context.

Importance of Microbial Ecology for Management

Prokaryotes do not possess homeostatic mechanisms as sophisticated as those of more complex multicellular organisms, such that they may respond rapidly to changes in their environment through changes in activities and numbers. Microorganisms, either regarded as individual species or as part of complex communities, have therefore been used as sensitive indicators of environmental change and the nutrient status of their environment. Development and evaluation of microbial communities as indicators of ecosystem state can be complex, however, due to complex interactions that microbial communities have with each other and with the environment. Development of management strategies should include microbial ecological components both from a mechanistic perspective and from their potential development and use as indicators of ecosystem status.

Microbial communities in anoxic systems such as wetland soils form networks of interacting trophic levels that transfer C from plant material such as cellulose to methane and CO_2. Wetlands such as the Everglades are a primary natural source of CH_4, such that nutrient input and management strategies may impact global methane budgets. In nutrient-impacted soils of the Water Conservation Areas (WCAs), microbial activities, including methanogenesis and soil respiration, are higher than in low-nutrient areas. Predictably, the numbers of fermenters, methanogenic archaea and SO_4 reducing prokaryotes are higher in nutrient impacted soils than in the unimpacted soils, and the structures of the respective guilds are dependent on nutrient status, indicating differences in the fundamental pathways through which these groups respond to the changes in nutrient limitations. As described by Chauhan et al. (2015) in Chapter 17, the relationships between methanogens and SO_4 reducing bacteria can be very complex: many of the SO_4 reducers they describe also function cooperatively with methanogens as syntrophs. When SO_4 is limiting, these versatile prokaryotes shift to a fermentative metabolism and supply H^+ and acetate to the appropriate methanogens. Not surprisingly, the activities and structures of methane oxidizing bacteria also depend on the amount of CH_4 production, and hence the site. In addition to differences in CH_4 production, these changes in microbial community structure may also impact corollary processes such as SO_4 reduction and mercury methylation/demethylation, and CH_4 oxidation.

Studies such as those presented in this section that describe the phylogeny of the different critical groups in CH_4 production, and their occasional competitors and cooperators, the SO_4 reducing prokaryotes, are critical to developing a long term strategies for management. This information may seem esoteric; however, knowledge of the impacts of different environmental conditions on the distribution of the specific phylogenetic groups of prokaryotes will provide insights into the pathways through which CH_4 is produced and consumed, which in turn will provide greater insight into how CH_4 production and consumption may be managed. Potentially more important is the insight into environmental controls on microbial groups responsible for related processes such as Hg methylation and demethylation (Weidow and Ogram 2015, Chapter 15). For example, a recent report (Bae et al. 2014) indicated that the methanogen-dependent syntrophs may be the dominant Hg methylators in much of the Everglades, while iron reducers may be important in certain regions.

In a narrow region of the WCAs prolonged input of P due to agricultural runoff resulted in a shift from a historically P limited ecosystem to a system that is nitrogen limited (White and Reddy 2000; Inglett et al. 2004). Nitrogen limitation typically translates into increased nitrogen fixation rates to satisfy the increased N demand, such that N fixation rates, as reflected in nitrogenase activity, can be a sensitive indicator of P impact in Everglades soils and periphyton (Inglett et al. 2004; 2008). Nitrogen fixation rates are typically assessed via measurement of activities of the nitrogenase enzyme complex, and can be estimated through direct measurement of the concentrations of microbial genes involved in N fixation, such as *nifH* (encoding dinitrogenase reductase) or expression of that gene. As Inglett (2015) describes in Chapter 9, nitrogenase activity can be a sensitive indicator of nutrient status in both soil and periphyton; however, the response can be complicated due to issues related to hydrology. As with all microbial processes, N fixation is intimately tied to management decisions, such that water flow and depth will impact both the species distribution of nitrogen fixers and their activities.

In addition to their use as indicators of ecosystem status, microorganisms can be employed for practical means, such as their use in decontaminating water and in control of invasive species. One potential application is in treating waters contaminated with a potent class of hepatotoxins produced by cyanobacteria in natural waters. These toxins are produced by a range of common algal genera including *Microsystis*, *Anabaena*, *Oscillatoria*, and *Nostoc*, among others. These toxins are produced during algal blooms that result from increases in nutrients, and have been associated with blooms in Lake Okeechobee. Jayachandran and Shetty (2015, Chapter 19) have described the isolation and characterization of bacterial strains that are capable of detoxification of microcystins. This study and others like it lay the groundwork for a greater understanding of the dynamics of microcystin concentrations during algal blooms, and may eventually form the basis of a treatment system to decontaminate water intended for human consumption.

The Everglades have been plagued for many years with exotic species of plants and animals that have escaped their native enemies and were able to establish at the expense of native species. The greater Everglades ecosystem is very complex, such that different invasive plants are of concern in different regions: Brazilian pepper and *Maleleuca* are of great concern below Lake Okeechobee, whereas hydrilla (*Hydrilla verticillata* L.f.) Royle), water hyacinth (*Eichornia crassipes* L.), and water grass (*Luziola fluitans* (syn. *Hydrochloa caroliniensis* L.) are of great concern above Lake Okeechobee. Native fungi that could be developed into biocontrol agents against these exotic plants show promise (Shetty and Jayachandran 2015, Chapter 18), but require more research and development.

Implications of Environmental Restoration on the Microbial Ecology of the Greater Everglades Ecosystem

Everglades restoration projects are expected to affect the quantity, quality, timing and distribution of water throughout south Florida. These changes will be gradual as projects are implemented but will have profound implications on landscape and community structure in the Everglades. The goal to provide additional flow to hydrate chronically dry regions of the Everglades will shift many highly oxidized regions of the EPA back to a more reduced state. It is anticipated that this change in flow will result in shifts in microbial communities and processes associated with anaerobic conditions, including increases in organic matter accumulation. In addition to hydrating regions impacted by drying, incoming water quality will likely continue to improve through ongoing implementation of stormwater treatment areas and agricultural best management practices. As water quality improves, it is likely that shifts in related structure and function will be observed. These improvements will affect the northern boundary of most of the WCAs. Additionally, any improvements to the distribution of water will not only improve hydrology and reduce fire and oxidation, but will also reduce localized nutrient loading associated with flow at structures (i.e., bridges and culverts) where increased loading occurs (Gottlieb et al. 2015, Chapter 7).

Everglades restoration is aimed at providing additional freshwater flow with improved timing to estuarine end members of Florida Bay and the Gulf of Mexico, yet this is overlaid upon a backdrop of climate change mediated sea-level rise. Not only will direct changes in salinity associated with sea level rise affect upstream communities, but changes in water level elevation will have landscape scale changes in flow discharge from upstream freshwater marshes. Further study to understand the effects of estuarine and marine end member encroachment on freshwater systems is needed.

While understanding of Everglades microbial ecology is growing, further work to characterize the interactions between tightly coupled algal and bacterial communities is needed. Research suggests that communities interact and develop emergent properties that are greater than the sum of their parts, and/or develop physiochemical feedback mechanisms that regulate growth and structure. Varying forms of periphyton contain different distributions of bacteria and algae, thereby affecting fish and invertebrate communities differently. Further research to understand these biogeochemical pathways and related functions in periphyton mats is warranted. This can aid achievement of water quality and water quantity goals while promoting beneficial habitats.

The preceding chapters presented research on a broad range of issues related to environmental quality, yet there is still much to be learned. Microbial ecology monitoring and research are critical to our understanding of the functioning of the Everglades and wetlands globally. Additionally, given the current observations regarding climate change and sea level rise, understanding changes in community structure in large landscapes and related functionality is a priority for human health. Microbial communities not only affect C cycling and climate, but also provide the earliest indicators of environmental change and hence are important monitoring tools, particularly for tracking migration in internal soil P load and related effects.

Although the chapters within Microbiology of the Everglades Ecosystem provide a balanced overview of the current theoretical models and Everglades microbial ecology research, significant gaps in understanding remain. We hope that this book provides a useful snapshot of current research and helps provide direction for future research and applied needs. Gaps noted above are not all inclusive. Further work to understand detrital pathways and the role of fungi in the Everglades landscape is also important, as fungi often control C form and availability, making organic material available to plants and animals in the system.

Conclusions

The Everglades ecosystem has been negatively impacted by changes to water quantity, timing and distribution, as well as water quality for decades. Although restoration efforts have reduced nutrient input to the Everglades and started to improve water timing and distribution more intensive efforts are necessary if the Everglades is to be restored. The more recently identified threat in the form of salt water intrusion due to sea level rise is predicted to change freshwater marsh to brackish mangrove or in some cases to saltwater habitat. Microbial communities serve as early warning indicators of nutrient loading impacts in the Everglades, as compared with vegetation and other ecosystem components. Also, microbial communities can serve as an indicator of ecosystem recovery, once external nutrient loads are curtailed. Continued research is needed to better characterize the distribution and function of Everglades microbial communities and associated biogeochemical processes occurring across the south Florida landscape to understand the implications of these changes to habitat, climate, and human health. Only a limited fraction of Everglades microbial communities have been characterized using molecular, microscopic, and other traditional methods, and in most cases these communities have only been sporadically studied. The structure and function of Everglades algal, bacterial and fungal communities should be characterized to determine the future effects of improved water quality, increased flow volume (with improved distribution) and the competing effects of marine end-member encroachment and related saltwater intrusion on the Everglades landscape.

References

Bae, H.-S., F.E. Dierberg and A. Ogram. 2014. Syntrophs dominate sequences associated with the mercury-methylating gene *hgcA* in the Water Conservation Areas of the Florida Everglades. Appl. Environ. Microbiol. 20: 6517–6526.

Benscoter, A.M., J.S. Reece, R.F. Noss, L.A. Brandt and F.J. Mazzotti. 2013. Threatened and endangered subspecies with vulnerable ecological traits also have high susceptibility to sea level rise and habitat fragmentation. PLoS ONE 8: e70647, doi:10.1371/journal.pone.0070647.

Chauhan, A., A. Pathak and A. Ogram. 2015. Ecological perspective on the associations of syntrophic bacteria, methanogens and methanotrophs in the Florida Everglades WCA-2A soils. *In*: J.A. Entry, A.D. Gottlieb, K. Jayachandran and A.V. Ogram (eds.). Microbiology of the Everglades Ecosystem. CRC Press, Boca Raton (this book).

Christensen, J.H., B. Hewitson, A. Busuioc, A. Chen and X. Gao. 2007. 2007: regional climate projections. pp. 847–940. *In*: S. Solomon, D. Qin, M. Manning, Z. Chen and M. Marquis (eds.). Climate Change 2007: The Physical Science Basis. Contribution of Working Group I to the Fourth Assessment Report of the Inter-governmental Panel on Climate Change. Cambridge University Press, Cambridge and New York.

Entry, J.A. 2012a. Water quality characterization in the northern Florida Everglades. Wat. Air Soil Pollut. 223: 3237–3247.

Entry, J.A. 2012b. The efficacy of the four-part test network to monitor water quality in the Loxahatchee National Wildlife Refuge. Wat. Air Soil Pollut. 223: 4999–5015.

Entry, J.A. 2013. The impact of station location on water quality characterization in the Loxahatchee National Wildlife Refuge using four different monitoring networks. Environ. Monit. Assess. 185: 7605–7615.

Entry, J.A. and A.D. Gottlieb. 2015. Water quality in the Everglades protection area. *In*: J.A. Entry, A.D. Gottlieb, K. Jayachandran and A.V. Ogram (eds.). Microbiology of the Everglades Ecosystem. CRC Press, Boca Raton (this book).

Frankovich, T.A. and A. Wachincka. 2015. Epiphytic diatoms along phosphorus and salinity gradients in Florida Bay (Florida, USA), an illustrated guide and annotated checklist. *In*: J.A. Entry, A.D. Gottlieb, K. Jayachandran and A.V. Ogram (eds.). Microbiology of the Everglades Ecosystem. CRC Press, Boca Raton (this book).

Gaiser, E. 2009. Periphyton as an early indicator of restoration in the Florida Everglades. Ecol. Indicators 6: S37–S45.

Gaiser, E., A.D. Gottleib, S. Lee and J. Trexler. 2015. The importance of species-based microbial assessment of water quality in freshwater everglades. Wetlands. *In*: J.A. Entry, A.D. Gottlieb, K. Jayachandran and A.V. Ogram (eds.). Microbiology of the Everglades Ecosystem. CRC Press, Boca Raton (this book).

Gann, D., J. Richards, S. Lee and E. Gaiser. 2015. Detecting calcareous periphyton mats in the Greater Everglades using passive remote sensing methods. *In*: J.A. Entry, A.D. Gottlieb, K. Jayachandran and A.V. Ogram (eds.). Microbiology of the Everglades Ecosystem. CRC Press, Boca Raton (this book).

Gottleib, A.D., E. Gaiser and S. Lee. 2015. Changes in hydrology, nutrient loading and conductivity in the Florida Everglades, and concurrent effects on periphyton community structure and function. *In*: J.A. Entry, A.D. Gottlieb, K. Jayachandran and A.V. Ogram (eds.). Microbiology of the Everglades Ecosystem. CRC Press, Boca Raton (this book).

Guha, H. and S. Panday. 2012. Impact of sea level rise on groundwater salinity in acoastal community of South Florida. J. Am. Wat. Res. Assoc. 48: 510–529.

Harvey, J.W., R.W. Schaffranek, G.B. Noe, L.G. Larsen, D.J. Nowacki and B.L. O'Connor. 2009. Hydroecological factors governing surface water flow on a low-gradient floodplain. Water Res. 45, DOI: 10.1029/2008WR007129.

Inglett, P. 2015. Nitrogenase activity in everglades periphyton: patterns, regulators and use as an indicator of system change. *In*: J.A. Entry, A.D. Gottlieb, K. Jayachandran and A.V. Ogram (eds.). Microbiology of the Everglades Ecosystem. CRC Press, Boca Raton (this book).

Inglett, P.W., K.R. Reddy and P.V. McCormick. 2004. Periphyton chemistry and nitrogenase activity in a northern Everglades ecosystem. Biogeochemistry 67: 213–233.

Inglett, P.W., E.M. D'Angelo, K.R. Reddy, P.V. McCormick and S.E. Hagerthey. 2008. Periphyton nitrogenase activity as an indicator of wetland eutrophication: spatial patterns and response to phosphorus dosing in a northern Everglades ecosystem. Wetlands Ecol. Manag. 17: 131–144.

[IPCC 2007]. Intergovernmental Panel on Climate Change. 2007. Climate Change 2007: Synthesis Report. Contribution of Working Groups I, II and III to the Fourth Assessment Report of the Intergovernmental Panel on Climate Change. Core Writing Team: R.K. Pachauri and A. Reisinger (eds.). Intergovernmental Panel on Climate Change, Geneva.

Jayachandran, K. and K. Shetty. 2015. Algal toxin degradation by indigenous bacterial communities in the Everglades region. *In*: J.A. Entry, A.D. Gottlieb, K. Jayachandran and A.V. Ogram (eds.). Microbiology of the Everglades Ecosystem. CRC Press, Boca Raton (this book).

Komárek, J., D. Sirová, J. Komárková and E. Rejmánková. 2015. Structure and function of cyanobacterial mats in the wetlands of Belize. *In*: J.A. Entry, A.D. Gottlieb, K. Jayachandran and A.V. Ogram (eds.). Microbiology of the Everglades Ecosystem. CRC Press, Boca Raton (this book).

Larsen, L., N. Aumen, C. Bernhardt, V. Engel, T. Givnish and S. Hagerthey. 2011. Recent and historic drivers of landscape change in the Everglades ridge, slough, and tree island mosaic. Crit. Rev. Environ. Sci. Technol. 4: 344–381.

Larsen, L.G. and J.W. Harvey. 2010. How vegetation and sediment transport feedbacks drive landscape change in the everglades and wetlands worldwide. Am. Nat. 176: E66–E79.

Louda, J., C. Grant, J. Browne and S. Hagerthey. 2015. Pigment-based chemotaxonomy and its application to Everglades periphyton. *In*: J.A. Entry, A.D. Gottlieb, K. Jayachandran and A.V. Ogram (eds.). Microbiology of the Everglades Ecosystem. CRC Press, Boca Raton (this book).

Misra, V., E. Carlson, R.K. Craig, D. Enfield and B. Kirtman. 2011. Climate scenarios: a Florida-centric view. Florida Climate Change Task Force. Center for Ocean-Atmospheric Prediction Studies 14: 1–61.

Noe, G.B., L.J. Scinto, J. Taylor, D.L. Childers and R.D. Jones. 2003. Phosphorus cycling and partitioning in an oligotrophic Everglades wetland ecosystem: a radioisotope tracing study. Freshwater Biol. 48: 1993–2008.

Noss, R.F. 2011. Between the devil and the deep blue sea: Florida's unenviable position with respect to sea level rise. Clim. Change 107: 1–16.

Osborne, T., S. Newman, K.R. Reddy, L.R. Ellis and M.S. Ross. 2015. Spatial distribution of soil nutrients in the Everglades protection area. *In*: J.A. Entry, A.D. Gottlieb, K. Jayachandran and A.V. Ogram (eds.). Microbiology of the Everglades Ecosystem. CRC Press, Boca Raton (this book).

Reddy, K.R. and R.D. Delaune. 2008. Biogeochemistry of Wetlands: Science and Applications. CRC Press, Boca Raton, Florida, pp. 774.

Ross, M.S., J.J. O'Brien, R.G. Ford, K. Zhang and A. Morkill. 2009. Disturbance and the rising tide: the challenge of biodiversity management on low-island ecosystems. Front. Ecol. Environ. 7: 471–478.

Saha, A.K., S. Saha, J. Sadle, J. Jiang, M.S. Ross, R.M. Price, L.S.L. Sternberg and K.S. Wendelberger. 2011. Sea level rise and South Florida coastal forests. Clim. Change 107: 81–108.

Shetty, K.G. and K. Jayachandran. 2015. Potential for biological control of invasive plants in the everglades ecosystem using native microorganisms. *In*: J.A. Entry, A.D. Gottlieb, K. Jayachandran and A.V. Ogram (eds.). Microbiology of the Everglades Ecosystem. CRC Press, Boca Raton (this book).

Trexler, J., E. Gaiser, J. Kominoski and J. Sanchez. 2015. The role of periphyton mats in consumer community structure and function in calcareous wetlands: lessons from the everglades. *In*: J.A. Entry, A.D. Gottlieb, K. Jayachandran and A.V. Ogram (eds.). Microbiology of the Everglades Ecosystem. CRC Press, Boca Raton (this book).

Von Holle, B., Y. Wei and D. Nickerson. 2010. Climatic variability leads to later seasonal flowering of Floridian plants. PLoS ONE 5: 1–9.

Wachnicka, A.H. and L.G. Wingard. 2015. Biological indicators of changes in water quality and habitats of the coastal and estuarine areas of the Greater Everglades ecosystem. *In*: J.A. Entry, A.D. Gottlieb, K. Jayachandran and A.V. Ogram (eds.). Microbiology of the Everglades Ecosystem. CRC Press, Boca Raton (this book).

Weidow, C. and A. Ogram. 2015. The microbial ecology of mercury methylation and demethylation in the Florida Everglades. *In*: J.A. Entry, A.D. Gottlieb, K. Jayachandran and A.V. Ogram (eds.). Microbiology of the Everglades Ecosystem. CRC Press, Boca Raton (this book).

White, J.R. and K.R. Reddy. 2000. The effects of phosphorus loading on organic nitrogen mineralization of soils and detritus along a nutrient gradient in the northern Everglades, Florida. Soil Sci. Soc. Am. J. 64: 1525–1534.

Zhang, K. 2010. Analysis of non-linear inundation from sea-level rise using LIDAR data: a case study for South Florida. Clim. Change. 106: 537–565.

APPENDICES TO CHAPTER 13

APPENDIX A: Plant/algal pigments, selected terminology (**abbreviation**) and pigment implications (**comments**) to Chemotaxonomy and CERP-RECOVER/MAP activities with Everglades periphyton communities. Alternate names given in italics (e.g., *2-hydroxy-zeaxanthin*) are meant only to impart an understanding of structural relationships are certainly not legitimate nomenclature!*

Astacene ASTA An oxidation product (3,3',4,4'-tetraketo-b-carotene: aka 3,3',4,4'-tetraketo-β-carotene; aka β,β-carotene-3,3',4,4'-tetrone) of astaxanthin, the major carotenoid in Crustacea, notably zooplankton such as copepods.

Allomer allo-Literally, "other form". A term originally applied to chlorophylls and chlorophyll derivatives that exhibited identical UV/Vis spectra as the parent compound but slightly different (more polar) chromatographic behavior. Strutural studies revealed that the "allomer" form of a chlorophyll or 'pheopigment' was the 13^2-oxy (=hydroxy)-derivative. As the allomer coexists in structures with intact 'carbomethoxy' groups, 2 allomers, the native and the epimer, also exist.

Alloxanthin ALLO The biomarker pigment ("ALLO") of choice for cryptophytes (Chryptophyceae). Presently, The FAU-OGG lab uses a CHLa/allo ratio of 3.8:1 for the back estimation of cryptophyte contributed CHLa to a natural microalgal community. Literally, 'different form yellow pigment' (7',8'didehydro-monadoxanthin).

Antheraxanthin ANTHERA Antheraxathin is most commonly known as one of the main components of the "xanthophyll cycle" in green algae (Chlorophyta) and higher plants. Literally, 'yellow pigment of anthers' (5,6-epoxy-zeaxanthin). Found in very high abundance as antheraxanthin di-(fatty acid) esters in the anthers of Tiger Lilies [(3S,5S,6R,3'R)-5,6-epoxy-5,6-dihydro-bb-carotene-3,3'-diol].

Aphanizophyll APHAN An indicator of N_2-Fixing (diazotrophic) colonial cyanobacteria. Originally isolated from *Aphanizomenon flos-aquae* [2'-(L-Rhamnosyloxy)-3',4'-didehydro-1',2'-dihydro-b,y-carotene-3,4,1'-triol].

Astaxanthin ASTAX Literally, 'the yellow pigment {however it is reddish orange} of *Asctacus*, a lobster genus (3,3'-dihydroxy-4,4'-diketo-β-carotene). Commonly found bound to a protein in a complex called crustacyanin (literally—crustacean blue pigment). Found in many micro-/macro-zooplankton as well as avian feathers (Flamingo, Roseate Spoonbill, etc.). Oxidation leads to astacene.

Bacteriochlorophyll-*a* BCHL*a* Indicator of reducing (anoxic) conditions and linked to potential sites of mercury methylation (see Cleckner et al. 1999). Existence of BCHL*a* indicates anoxia and the activities of sulfate reducing and potential coexistence of the methanogenic corsotium. Present in purple-sulfur bacteria (Thiorhodaceae) as the major photosynthetic pigment but also co-occurs bacteriochlorophylls-*c* (aka Chlorobium chlorophyll-660) and -*d* (aka Chlorobium chlorophyll-650) in species of the green and brown sulfur bacteria (*Chlorobium* sp.) (Chapters in Grimm et al. 2006; Scheer 1991).

Bacteriochlorophyll-*b* **BCHL*b*** Found in *Rhodopseudomonas viridis* and *R. sulfoviridis* (Chapters in Grimm et al. 2006; Scheer 1991). Regrouped into *Bastochloris* gen. nov. by Hiraishi 1997). Bacteriochlorophyll-b now reported in 13 species, in a tight clade, of *Bastochloris viridis* (Hoogewerf et al. 2003) To date, not identified in Everglades periphyton.

Bacteriochlorophyll-*c* **BCHL*c*** aka *Chlorobium* chlorophyll-660. Found in green sulfur bacteria (Chlorobiaceae). A pseudohomologous series with a C20 (δ-methine bridge) methyl. (Smith et al. 1980: Chapters in Grimm et al. 2006; Scheer 1991).

Bacteriochlorophyll-*d* **BCHL*d*** aka *Chlorobium* chlorophyll-650. Found in green sulfur bacteria (Chlorobiaceae). A pseudohomologous series lacking methine bridge alkyl substitution (Chapters in Grimm et al. 2006; Scheer 1991).

Bacteriochlorophyll-*e* **BCHL*e*** Minor bacteriochlorophylls found in *Chlorobium phaeobacteriodes* (Chapters in Grimm et al. 2006; Scheer 1991). Not found in CERP-MAP samples as of this writing (06/07).

Bacteriochlorophyll-*f* **BCHL*f*** Aka 20-desmethyl-bacteriochlorophyll-*e*. Minor bacteriochlorophyll. Not found in nature, including CERP-MAP samples as of this writing (see Vogl et al. 2012; cf. Blakenship 2004: Chapters in Grimm et al. 2006; Scheer 1991).

Bacteriochlorophyll-*g* **BCHL*g*** Found in anaerobic Heliobacteria, brownish-green N_2-fixing phototrophs. Not identified in CERP-MAP samples as of this writing (Blakenship 2004: Chapters in Grimm et al. 2006; Scheer 1991).

Bilin: bilin A straight chain tetrapyrrole pigment. The prosthetic group of phycobiliproteins (see below).

α-carotene αCAR or α [β,ε-carotene; aka (6'R)-β,ε-carotene]. A hydrocarbon (carotene) present in Cryptophyte algae. Differs from β-carotene only in the position of 1 double bond (1 C=C is out of conjugation resulting in a UV/Vis spectrum with blue shifted {shorter wavelengths} absorption maxima) relative to β-carotene. Easily separated from the β-analog. NOT presently used in CHEMOTAXNOMIC estimations of Cryptophytes. Useful only in aiding confirmation. Alloxanthin is used for quantitative estimation of Cryptophytes.

β-carotene βCAR or β The most widespread natural carotenoid [β,β-carotene]. Pro-vitamin-A.

Caloxanthin: CALO (2,3,3'-trihydroxy-β-carotene; aka [2R,3R,3'R-β,β-carotene-trione]). Minor carotenoid sometimes isolated from cyanobacteria. A zeaxanthin derivative (~ *2-hydroxy-zeaxanthin*).

Canthaxanthin CANTH (aka 4,4'-diketo-b-carotene) Present, with myxoxanthophyll and echinenone, in (non-N_2-fixing) colonial cyanobacteria (Grant and Louda 2010). Also found in lower quantities accompanying (as a precursor to) astaxanthin in zooplankton and birds.

Chlorin CHLORIN A generic term used to describe a (17, 18-dihydroporphyrin and/or an UV/Vis spectrum with a band order of I>II>III>IV (Baker and Louda 1986, 2002).

Chlorophyll-*a* CHL*a* Most abundant and widespread chlorophyll. Routinely used as a biomass proxy for oxygenic photoautotrophs.

Chlorophyll-*a*-epimer CHL*a*-epi or CHL*a*' A derivative of chlorophyll-a in which the C13^2-H and -COOCH$_3$ groups are in the 'flipped' (epimerized) configuration relative to the normal enzymatically determined configuration (see Appendix B1 CHL*a*-allo and the epimer of the allomer for configurational change). The 13^{2}S-derivative (H is down and carbomethoxy is up, relative to the plane of the chlorophyll nucleus) of CHL-a as conventionally drawn. "Normal" or biologic CHL-a is 13^{2}R (H-up, carbomethoxy-down) (see Baker and Louda 1986, 2002; Louda 2008; Jeffrey et al. 1997; Scheer 1991).

Chlorophyll-*a*-allomer CHL*a*-allo An oxidized derivative of chlorophyll-a in which the C13^2-H is replaced with an -OH group (see Appendix B1). Can occur as both the 'normal' and epimeric forms, eluting prior to CHL*a*, in reversed phase HPLC chromatograms (see Baker and Louda 1986, 2002; Louda 2008; Jeffrey et al. 1997; Scheer 1991).

Chlorophyllide-*a* CHLide-*a* Chlorohyllides are chlorophylls which have had the phytol ester removed by hydrolysis leaving a free propionic acid group. This polar compound is prevalent in samples high in diatoms due to their high chlorophyllase activity (Baker and Louda 1986, 2002; Jeffrey et al. 1997; Scheer 1991; Louda et al. 1998, 2002).

Chlorophyll-*b* CHL*b* Aka 7-formyl-7-desmethyl-chlorophyll-*a*. The accessory chlorophyll in Chlorophytes (green algae) and higher plants. Structurally identical to CHL-a except for the conversion of the C7 methyl to a formyl moiety.

Chlorophylls-c_1/-c_2 CHLS-c_1/-c_2 Major accessory chlorophylls in diatoms (Bacillariophyceae) and other members of the Chrysophyta. Difficult to impossible to separate using conventioanl reversed phase C2, C8 or C18 HPLC. Can be separated over TLC using powedered polyethylene (Jeffrey and Wright 1987; Stauber and Jeffrey 1988) or specific RP-HPLC methods (Kraay et al. 1992). The nucleus is that of a true porphyrin (fully aromatized) rather than that of a 17,18-dihydropporphyrin (aka 'chlorin'). Structurally-c_1 is similar to CHL-a except for (a) 17,18-didehydro, (b) the propionic acid group at C17 being converted to an acrylic acid group, and (c) lacking estrification to the 20 carbon isoprenoid alcohol, phytol. More properly termed a Mg-protochlorophyllide. CHL-c_2 differs from CHL-c_1 in that the C8 ethyl moiety is a vinyl group (Bidigare et al. 1990; Dougherty et al. 1970; Jeffrey 1989: Chapters in Jeffrey et al. 1997; Roy et al. 2011).

Chlorophyll-c_3 CHLc_3 An accessory chlorophyll found in some prymnesiophtes, notably *Emiliania huxleyi*, and certain diatoms (Jeffrey and Wright 1987; Stauber and Jeffrey 1988: Chapters in Jeffrey et al. 1997; Roy et al. 2011). Not found to date, nor expected, in CERP-MAP samples.

Crocoxanthin CROCO Minor allenic carotenoid found in some cryptomonads (aka 3R, 6'R-7,8-didehydro-β,ε-caroten-3-ol: Chapters in Jeffrey et al. 1997; Roy et al. 2011). Not found to date in Everglades periphyton or water samples.

Cryptoxanthin CRYPTO Minor carotenoid found in some cryptomonads (aka β, β-caroten-3-ol: Chapters in Jeffrey et al. 1997; Roy et al. 2011). Only traces to date in Everglades periphyton.

Diadinoxanthin DIADINO A major carotenoid in diatoms, prymnesiophytes and some crysophytes and dinoflagellates (3S,5R,6S,3'R-5,6-epoxy-7',8'-didehydro-5,6-dihydro-β,β-carotene-3,3'diol) (Chapters in Jeffrey et al. 1997; Roy et al. 2011) Part of the "xanthophyll cycle" in diatoms (Hager 1980). Prevalent in CERP-MAP samples when diatoms (fucoxanthin as marker) are present.

Diatoxanthin DIATO The non-epoxy precursor of diadinoxanthin (aka 3R,3'R-7,8-didehydro-β,β-carotene-3,3'diol) (Chapters in Jeffrey et al. 1997; Roy et al. 2011). Usually a trace to minor pigment in diatoms, etc. (see DIADINO). Part of the "xanthophyll cycle" in diatoms (Hager, 1980). Present in CERP-MAP samples, at lower levels relative to DIADINO, when diatoms (fucoxanthin as marker) are present.

Dinoxanthin DINO [aka (3S,5R,6S,3'S,5'R,6'R)-5,6-epoxy-3'-ethanoyloxy-6',7'-6',7'-didehydro-5,6,5',6'-tetrahydro-β,β-carotene-3,5'diol] (Chapters in Jeffrey et al. 1997; Roy et al. 2011). A minor pigment in dinoflagellates (Chapters in Jeffrey et al. 1997; Roy et al. 2011). Seldom found in CERP-MAP samples, even when the dinoflagellate marker is present.

Echinenone ECHIN (aka β,β-caroten-4-one) (Chapters in Jeffrey et al. 1997; Roy et al. 2011). The main biomarker pigment for filamentous cyanobacteria. Quite prevalent in CERP-MAP samples (β,β-caroten-4-one). A keto-carotenoid often accompanied by canthaxanthin in higher light conditions (cf. Grant and Louda 2010).

Epimer epi-Derivatives of non-pyro (i.e., native) pheopigments in which the natural orientation (down) of the 13^2-carbomethoxy moiety is reversed ("flipped", UP) relative to the plane of the phorbide macrocycle. A derivative of chlorophylls and their derivatives in which the $C13^2$-H and –COOCH$_3$ groups are in the 'flipped' (epimerized) configuration relative to the normal enzymatically determined configuration (see Appendix B1 CHLa-allo and the epimer of the allomer for configurational change). The 13^2S-derivative (H is down and carbomethoxy is up, relative to the plane of the chlorophyll nucleus) of CHL-a as conventionally drawn. "Normal" or biologic CHL-a is 13^2R (H-up, carbomethoxy-down) (see Baker and Louda 1986, 2002; Louda 2008; Jeffrey et al. 1997; Scheer 1991).

Epoxide(s) epoxy-An oxygen functional group formed by the insertion of an atom of O across a double bond (~ a 3 membered cyclic ether: see Appendix B3, e.g., Antheraxanthin, Diadinoxanthin and others).

Fucoxanthin FUCO The main biomarker pigment for diatoms. Also found in other chrysophytes and certain marine dinoflagellates. Many derivatives exist in marine chrysophytes (Chapters in Jeffrey et al. 1997; Roy et al. 2011). Essentially only fucoxanthin and its breakdown product fucoxanthinol in fresh waters such as Everglades periphyton and phytoplankton.

Fucoxanthinol FUCOL A common derivative of FUCO formed by the hydrolysis of the 3'-acetate functional group leaving an alcohol (hydroxyl). Added to FUCO when determining the total amount of fucoxanthin for purposes of estimating diatom derived CHLa in chemotaxonomy. Common in CERP-MAP samples.

Glycoside(s) glycol Indicates the presence of a sugar (~glyco). Sugar is bound to a pigment (carotenoid) via a glycosidic bond (Dehydration, removal an -OH on the sugar and H on the pigment giving an ether-like bond {R-O-R', where R or R' is a sugar} equaling the 'glycosidic' linkage (see Britton et al. 2004).

Lutein LUT A major carotenoid in green algae (Chlorophyta) and higher plants (3,3'-dihydroxy-α-carotene; aka3R,3'R,6'R-β,ε-carotene-3,3'-diol). Thought at one time to be a photosynthetic accessory pigment but most likely a sunscreen (cf. Grant and Louda 2010) and antioxidant. Chlorophyll-b, not lutein, is used for the quantitative estimation of chlorophyte CHLa.

Meso- meso-A modifier used with chlorophylls and their derivatives to indicate that the 3-vinyl moiety has been reduced (hydrogenated) to an ethyl. Meso can also indicate a "meso" position in the tetrapyrrole macrocycle (viz. an α, β, γ or δ carbon in the Fischer nomenclature or carbon numbers 5, 10, 15 or 20 in the revised system (see Baker and Louda 2002; Scheer 1991).

Monadoxanthin MONAD Basically, 7,8-didehydro-lutein. Found in cryptomonads as a trace to minor component (Chapters in Jeffrey et al. 1997; Roy et al. 2011). Also found as a trace component in a marine *Picocystis saliinariun* and lichen *Coccomyxa* sp. chlorophyte (see Jeffrey et al. 2011). Sporadic occurrence in Everglades periphyton in conjunction with ALLO and α–CAR, indicating a cryptophyte origin.

Myxol MYXOL The aglycone of myxoxanthophyll (see Cottingham et al. 2000; Jeffrey et al. 1997; Britton et al. 2004). Added with MYXO to determine the total MYXO in a sample.

Myxoxanthophyll MYXO (aka myxol-quinovoside) A carotenoid gylcoside with quinovose (6-deoxy-D-glucopyranose) appears to be the most common MYXOL glycoside [aka 3R,2'S-2'(α-L-chinovosyloxy)-3',4'-didehydro-1',2'-dihydro-β,ψ-carotene-3,1'diol: see Jeffrey et al. 1997; Britton et al. 2004]. However, "myxoxanthin" is also know to be the rhamnose (Hertzberg and Liaaen-Jensen 1969; Mochimura et al. 2008), fucose (Takaichi et al. 2001, 2010) and several other glycosides with and without modifications such as ketone functionalities and/or additional sites of dehydrogenation (see Mochimura et al. 2008; Takaichi et al. 2010). The 'myxoxanthophylls', as well as MYXOL, all share a common visible spectral fingerprint (451, 476, 507+/– 2 nm) and the glycosides exhibit slightly different retention times during RP-HPLC. All are added with MYXOL to determine the total MYXO in a sample.

Common in cyanobacteria where it appears to be required for normal cell wall construction (Mohamed et al. 2005) and increases in concert with lowered pH (West and Louda 2011). Very common in CERP-MAP samples.

Neochrome NEOCHR ([9'-cis-(3S, 5R, 6R, 3'S,5'R,8'RS)-5',8'-Epoxy-6,7-didehydro-5,6,5',6'-tetrahydro-β,β-3,5,3'-triol]). A mixture of the two (8'R and 8'S) optical isomers (see Roy et al. 2011). The furanoid rearranged derivative of 9'-cis-neoxanthin. Not noticed in Everglades periphyton to date.

Neoxanthin NEO (9'-cis-neoxanthin; aka [9'-cis-(3S, 5R, 6R, 3'S,5'R,6'S)-5',6'-Epoxy-6,7-didehydro-5,6,5',6'-tetrahydro-β,β-3,5,3'-triol]). The rearranged derivative of violaxanthin in which 2 additional H atoms are lost forming an allene (-C=C-C=C-) subunit and a trans-to-cis rearrangement has occurred. Often thought of as a 'storage' form of violaxathin in the xanthophyll cycle (cf. Demmig-Adams 1990; Hager 1980). Common in Everglades periphyton when a minor to significant abundance of chlorophytes are present.

Nostoxanthin: NOSTO (2,3,2',3'-tetrahydroxy-b-carotene; aka [(2R,3R,2'R,3'R-β,β-carotene-2,3,2',3'-tetrol]. Minor carotenoid often isolated from unicellular cyanobacteria. A zeaxanthin derivative (~*2,2'-dihydroxy-zeaxanthin*).

Oscillaxanthin OSCIL [aka Oscillol diquinovoside: 2,2'-Di-(6-deoxy-α-L-glucopyranosyloxy)-3,4,3',4'-tetradehydro-1,2,1',2'-tetrahydro-ψ,ψ-carotene-1,1'-diol]. Restricted to *Oscillatoria* and related species (see Roy et al. 2011 and references given). Not found in Everglades periphyton to date.

Pheophorbide-*a* PBID*a* A common breakdown product of CHL*a* in which the Mg of CHL*a* is replaced by 2 H atoms and the phytol group has been hydrolysed leaving a free propioninc acid group. Also equates to a derivative of CHLide*a* involving Mg loss (Baker and Louda 1986, 2002; Louda et al. 1998, 2002; Roy et al. 2011; Scheer 1991; Jeffrey et al. 1997).

Pheophytin-*a* PTIN*a* A common breakdown product of CHL*a* in which the Mg of CHL*a* is replaced by 2 H atoms (Baker and Louda 1986, 2002; Louda et al. 1998, 2002; Roy et al. 2011; Scheer 1991; Jeffrey et al. 1997).

Pheophytin-*b* PTIN*b* A common breakdown product of CHL*b* in which the Mg of CHL*b* is replaced by 2 H atoms (Baker and Louda 1986, 2002; Louda et al. 1998, 2002; Roy et al. 2011; Scheer 1991; Jeffrey et al. 1997).

Pheopigment(s) pheopigments(s) A general or "catch-all" term applied to olive-brown to olive-green derivatives (breakdown products) of chlorophylls. Phorbide structures that includes pheophytins and pheophorbides and their isomers. Viz. Mg-free derivatives of the chlorophylls in which the "isocyclic ring". A dihydropheoporphyrin. (13,15-cycloethano moiety) remain (Baker and Louda 1986, 2002 and references therein).

Pheoporphyrin pheophorphyrin A pheopigment that is in the fully aromatized (porphyrin) oxidation state (Baker and Louda 1986, 2002; Scheer 1991).

Phycobilins, phycobiliproteins: Photosynthetic accessory pigments in which a straight-chain tetrapyrrole pigment [aka a bilin: Phycoerytrobilin (red), phycocyanobilin (cyan/blue), etc.] are bound to an apo-protein (see Rowan 1989; Falkowski and Raven 2007 and references in each).

Porphyrin porphyrin A fully aromatized macrocycle tetrapyrrole.

Pyro-pyro-(p) A term used to describe a chlorophyll, a phorbide or a pheoporphrin in which the 13^2 "carbomethoxy (aka methylformate, aka methoxycarbonyl) moiety has been lost. Historically derived from the *in vitro* method of partial synthesis, namely heating (pyro ~ fire) (Pennington et al. 1964; Baker and Louda 1986, 2002).

Pyrochlorophyllide-*a* **pCHLide*a*** The "pyro" derivative of chlorophyllide-*a*.

Pyropheophorbide-*a* (-*b*) **pPBID*a* (*b*)** The "pyro" derivative of pheophorbide-*a* (-*b*).

Pyropheophytin-*a* (-*b*) **pPTIN*a* (*b*)** The "pyro" derivative of pheophytin-*a* (-*b*).

Scytonemin **SCYTO** A dimeric indole-phenol pigment common as an UV sunscreen in cyanobacteria (e.g., *Scytonema hoffmanii, Schizothrix calcicola*, etc.). Very prevalent in CERP-MAP samples. Used as an indicator of high light/low hydroperiod conditions (Garchia-Pichel and Castenholz 1991; Proteau et al. 1993; West and Louda 2011).

Scytonemin-imine **SCYTO-imine** (scytonemin-3a-imine) An imine containing derivative of scytonemin with strong visible absorption (Grant and Louda 2013). This pigment is quite prevalent in low hydroperiod intense light areas of the Everglades such as the red to mahogany colored epilithic/epipelic periphyton of the Shark River Slough (*Scytonema* sp.).

Tetrapyrrole tetrapyrrole A general term to describe molecules containing 4 pyrrole units. These can be macrocyclic, such as the porphyrins and phorbides, or linear such a the straight-chain bilin pigments.

Tetraterpenoid tetraterpenoid isoprenoids with 4 'terpenoid' (viz. C10) subunits. Actually comprised of 8 isopentanyl equivalents. A forty carbon isoprenoid. The carotenoids.

UNKNOWN **UNKN** Refers to a pigment, tetrapyrrole (~chlorophyll + derivatives) or tetraterpenoid (carotenoid), that does not match existing physico-chemical data (specifically coincidence of a chromophore {~ spectrum} and retention time. Usually, we make an educated guess of the pigment class or tentative {tent.} identification: Example, quite often we encounter lutein epoxides eluting prior to lutein. The numerous isomers and the non-inclusion of such pigments in chemotaxonomic estimations makes additional study for the purposes of CERP-MAP not worth the effort, except academically.

Violaxanthin **VIOLA** The diepoxide derived from zeaxanthin in the green algal/ plant "xanthophyll cycle (5, 6, 5', 6'-diepoxy-zeaxanthin or 5, 6, 5',6' -β,β-carotene-3,3'-diol). Part of the 'xanthophyll cycle' (cf. Demmig-Adams 1990; Hager 1980).

Zeaxanthin **ZEA** (3,-3'dihydroxy-β-carotene; aka [(3R,3'R)-β,β-carotene-3,3'-diol]. The major biomarker pigment (carotenoid) for coccoidal cyanobacteria. Usually, if not always, found in CERP-MAP samples. Always factored into chemotaxonomic regression formulae. Used alone for coccoidal cyanobacteria after subtraction of ZEA contributed by MYXO and ECHIN containing filamentous cyanobacteria. Also present in the 'xanthophyll cycle' of chlorophytes (cf. Demmig-Adams 1990; Hager 1980). but 'usually' present in very low to trace amounts in those cases.

**For general references see: Chapters in Goodwin 1965, 1988; Isler 1971; Jeffrey et al. 1997; Roy et al. 2011; Scheer 1991). For chlorophyll breakdown products see Baker and Louda 1986, 2002; Louda 2008; Louda et al. 1998, 2002 and references in each.

Appendix B1: Structures of chlorophylls and their derivatives.

Chlorophyll-*a* (CHL*a*)

Chlorophyll-*b* (CHL*b*)
(**Note**: *7-formyl-7-desmethyl* CHL*a*)

Bacteriochlorophyll-*a* (BCHL*a*)
Bactetriochlorophyll-*b* (BCHL*b*)

Chlorophyll-c_1 (CHL*c1*)

Chlorophyll-c_2 (CHL*c2*)

Chlorophyll-c_3 (CHL*c3*)

Appendix B1 cont: Structures of chlorophylls and their derivatives.

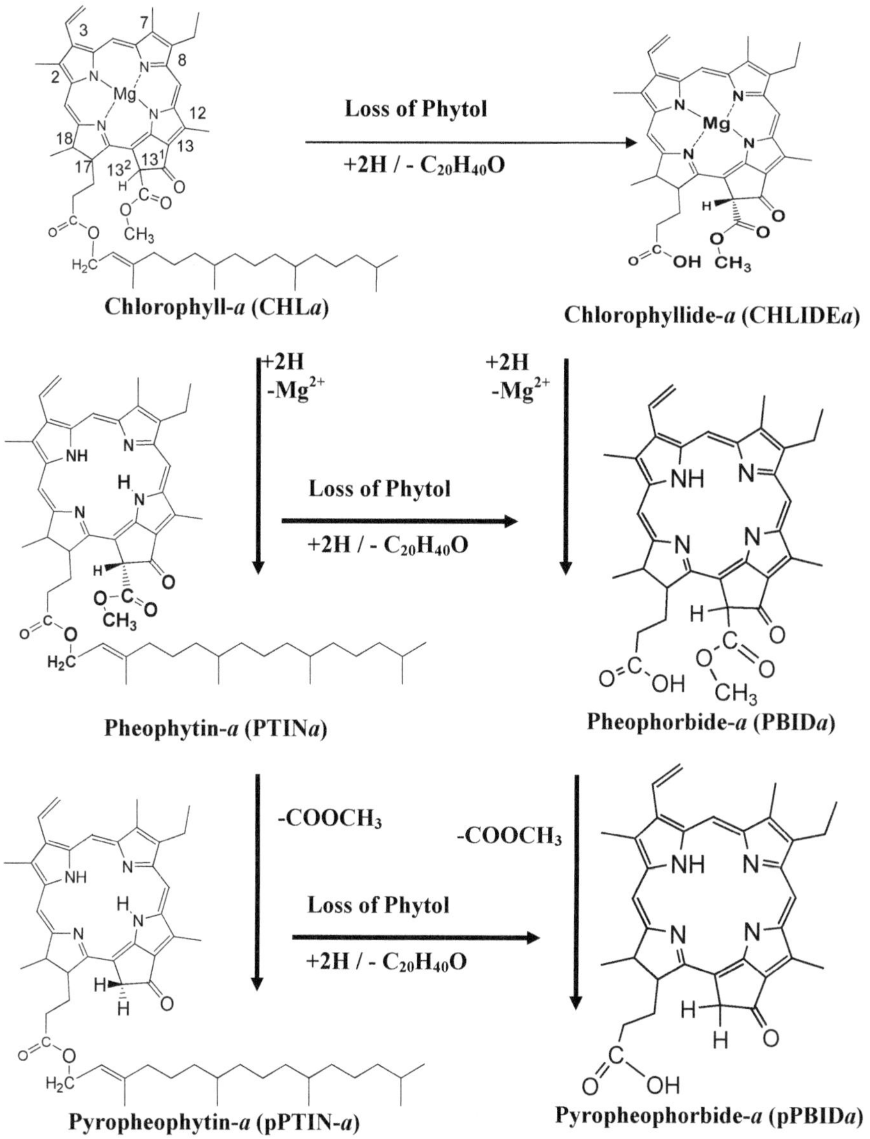

Loss of Phytol

$+2H / - C_{20}H_{40}O$

Chlorophyll-*a* (CHL*a*)

Chlorophyllide-*a* (CHLIDE*a*)

$+2H$
$-Mg^{2+}$

$+2H$
$-Mg^{2+}$

Loss of Phytol

$+2H / - C_{20}H_{40}O$

Pheophytin-*a* (PTIN*a*)

Pheophorbide-*a* (PBID*a*)

$-COOCH_3$

$-COOCH_3$

Loss of Phytol

$+2H / - C_{20}H_{40}O$

Pyropheophytin-*a* (pPTIN-*a*)

Pyropheophorbide-*a* (pPBID*a*)

<u>**Appendix B1 cont**</u>: Structures of chlorophylls and their derivatives.

Chlorophyll-*a*-allomer (CHL*a*-allo)

Chlorophyll-*a*-allomer-epimer (CHL*a*'-allo)

Pheophorbide-*a*-allomer (PBID*a*-allo)

Purpurin-18 (PUR-18)

Chlorin-e₆ (Chl-e6)

Chlorin-p₆ (Chl-p6)

Appendix B1 cont: Structures of chlorophylls and their derivatives.

Bacteriochlorophylls-c and -d
BCHL*c* , BCHL*d*
(Many isomers)

BCHL*c* Series: R3 = -CH$_3$
(aka Chlorobium Chlorophylls-660)

BCHL*d* Series: R3 = -H
(aka Chlorobium Chlorophylls-650)

R1 = -C$_2$H$_5$, -C$_3$H$_7$, or – C$_4$H$_9$
R2 = -CH$_3$ or –C$_2$H$_5$

Bacteriochlorophylls-e and -f
BCHL*e* , BCHL*f*
(Many isomers)

BCHL*e*Series: R3 = -H

BCHL*f* Series: R3 = -CH$_3$

R1 = -C$_2$H$_5$, -C$_3$H$_7$, or – C$_4$H$_9$
R2 = -CH$_3$ or –C$_2$H$_5$

Appendix B2: Structures of carotenoids.

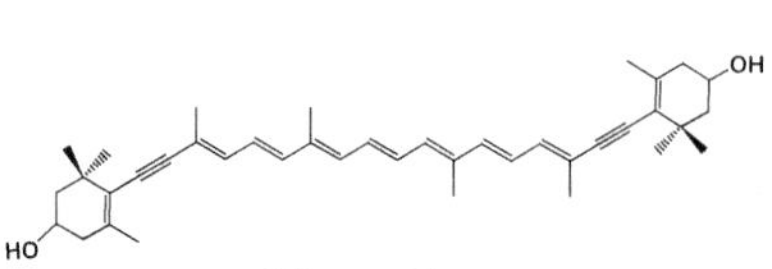

Alloxanthin

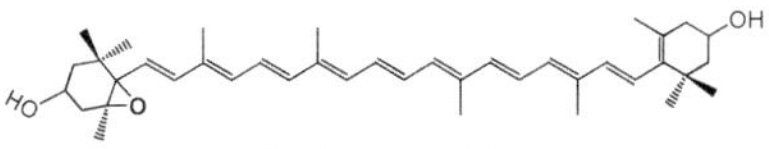

Antheraxanthin

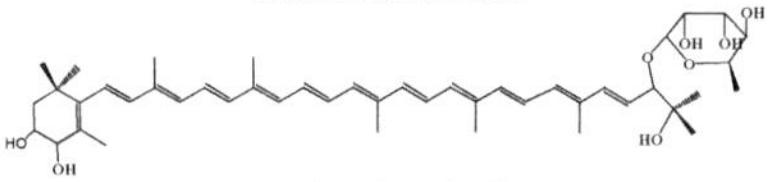

Aphanizophyll

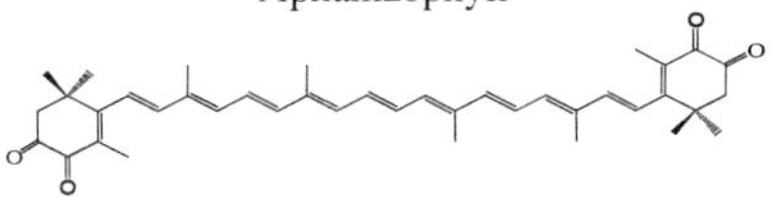

Astacene

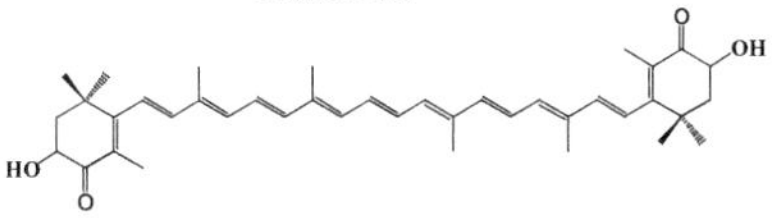

Astaxanthin

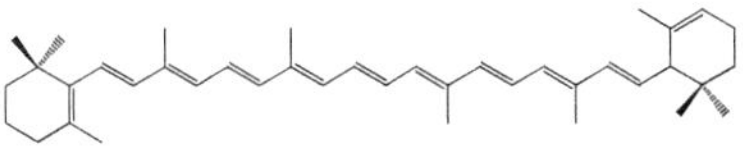

α-Carotene (β,α-carotene)

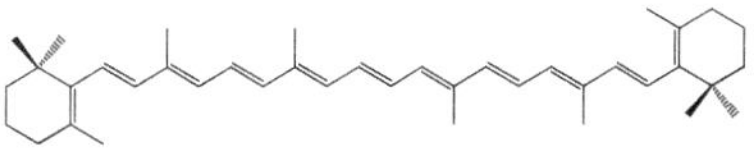

β-Carotene (β,β-carotene)

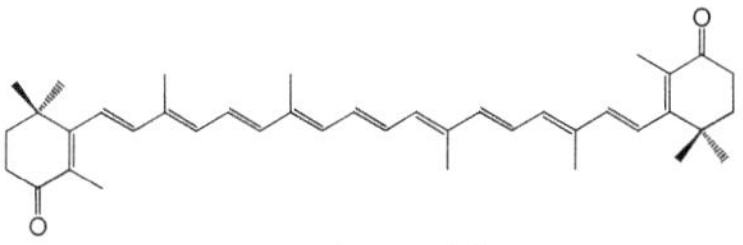

Canthaxanthin
(4,4'-diketo-β-Carotene)

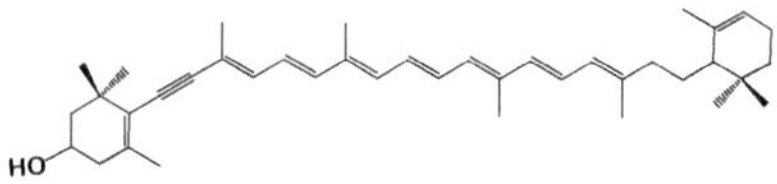

Crocoxanthin

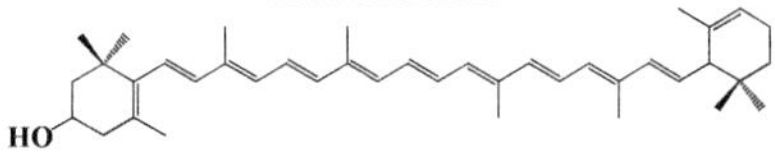

α-Cryptoxanthin

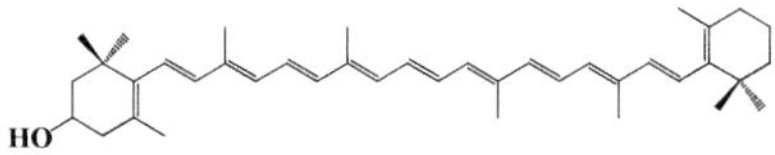

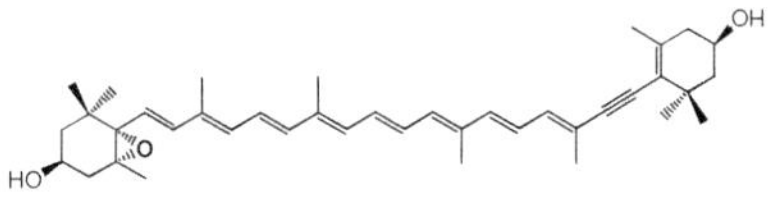

β-Cryptoxanthin

Diadinoxanthin

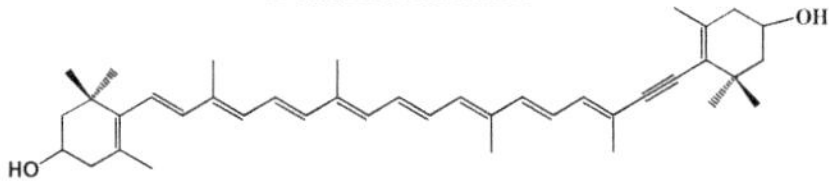

Diatoxanthin

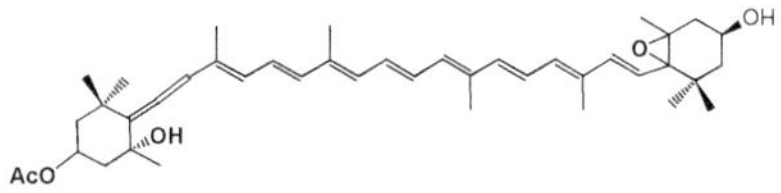

Dinoxanthin

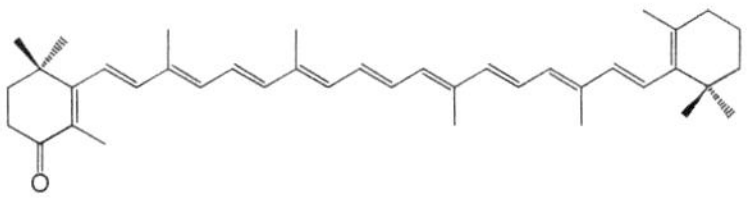

Echinenone
(4-keto-β-Carotene)
(aka β,β-caroten-4-one)

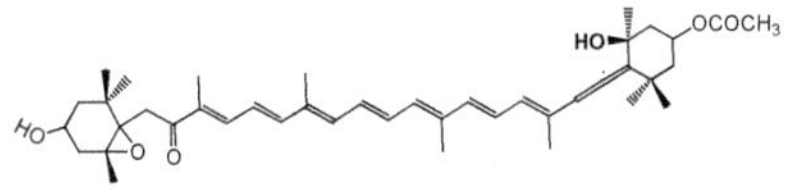

Fucoxanthin

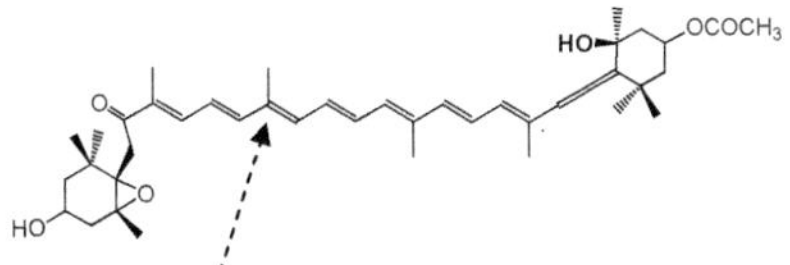

9-cis-Fucoxanthin
(*13-cis-fucoxanthin* also known)

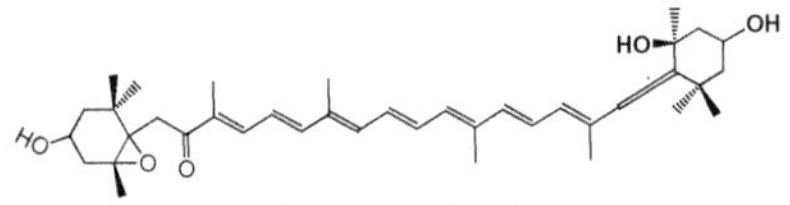

Fucoxanthinol

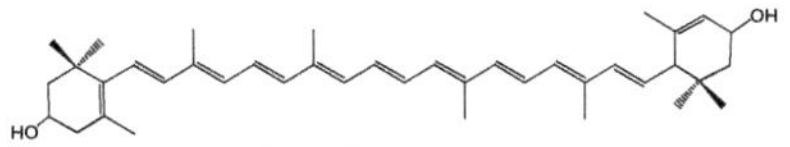

Lutein

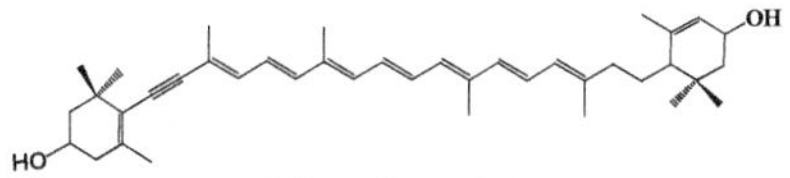

Monadoxanthin
(7,8-didehydro-lutein)

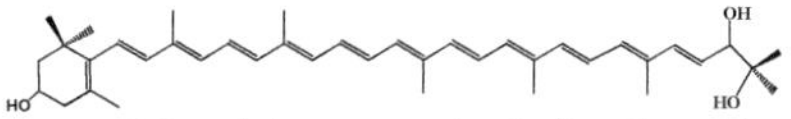

Myxol (myxoxanthophyll aglycone)

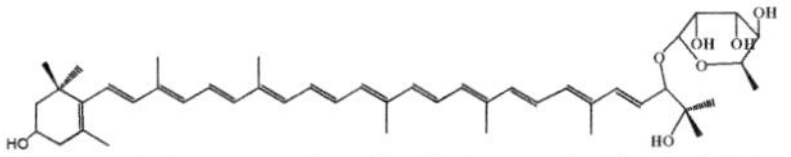

Myxoxanthophyll (myxol glycoside)

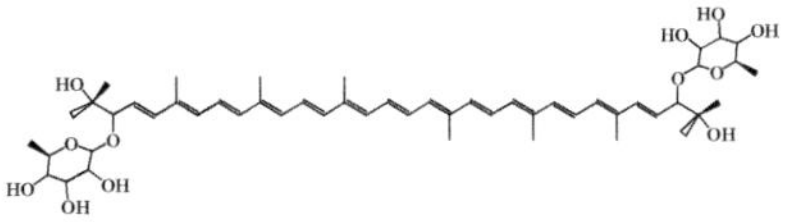

Oscillaxanthin

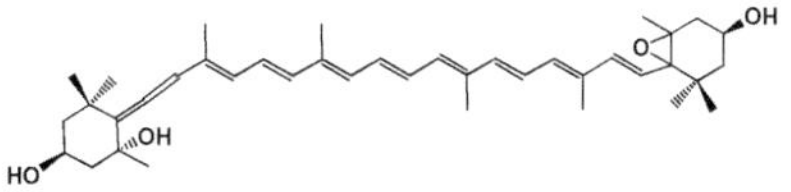

Neoxanthin

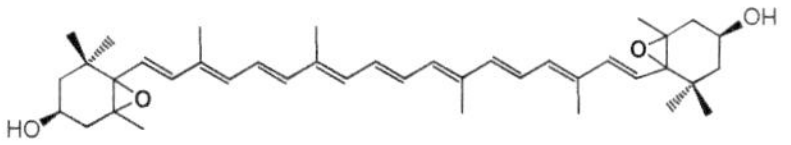

Violaxanthin

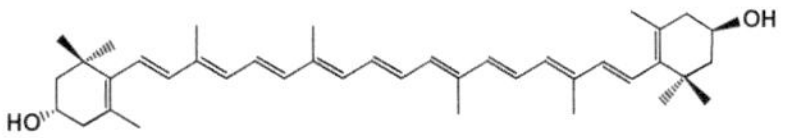

Zeaxanthin

Appendix B3: Structures of scytonemin and derivatives.

Scytonemin (oxidized form)

Dimethoxyscytonemin

Scytonemin (reduced form)

Tetramethoxyscytonemin

Scytonine

Scytonemin-3a-imine
(Grant and Louda 2013)

Index

Color Plate Section

Chapter 2

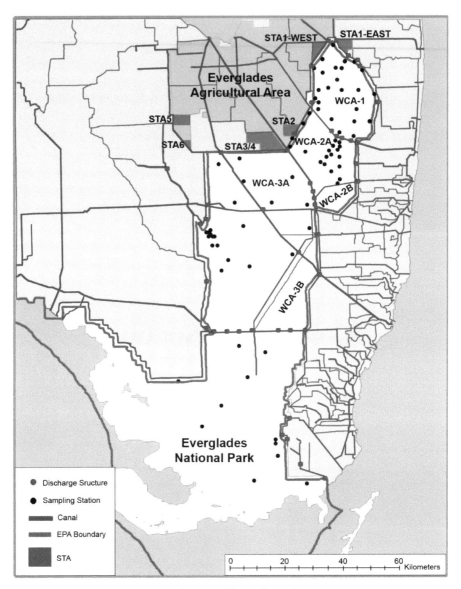

Figure 1. Map of the Everglades Protection Area with sampling stations.

Chapter 3

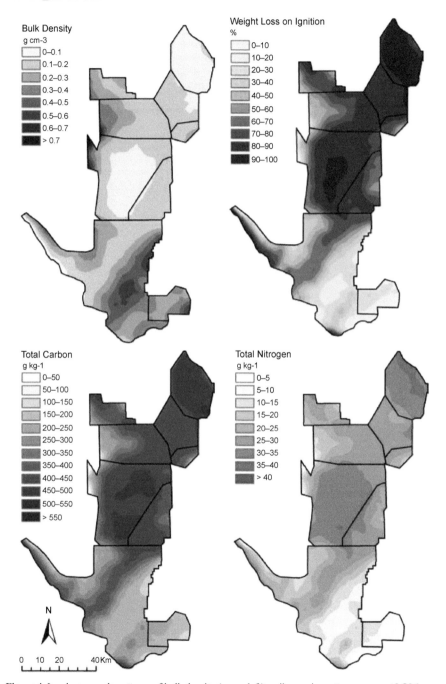

Figure 4. Landscape scale patterns of bulk density (upper left), soil organic matter as percent LOI (upper right), total carbon (lower left) and total nitrogen (lower right) in the Everglades Protection Area. Figure adapted from Reddy et al. 2005.

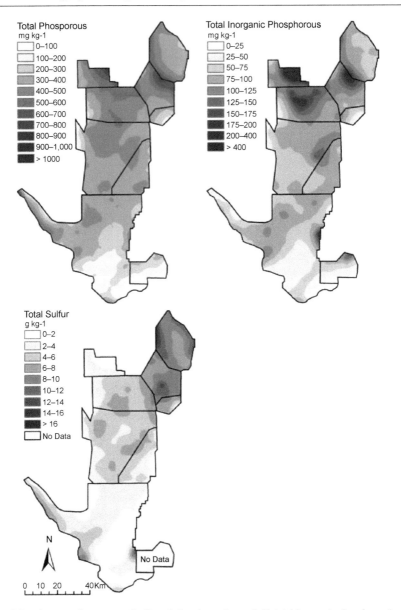

Figure 5. Landscape scale patterns of soil total phosphorus (upper left), total inorganic phosphorus (upper right), and total sulfur (lower left) in the Everglades Protection Area. Figure adapted from Reddy et al. 2005.

Chapter 6

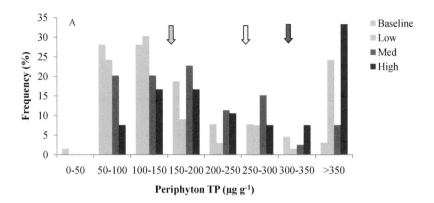

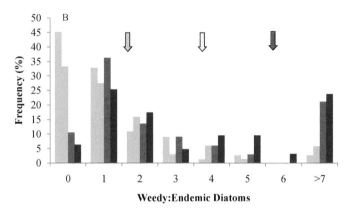

Figure 3. Frequency histogram of (A) periphyton TP values, and (B) ratio of weedy to endemic diatoms, in baseline (control) conditions, and in low (+5 µg P L⁻¹), medium (+15 µg P L⁻¹) and high (+30 µg P L⁻¹) enrichment channels over the first two years of continuous P dosing in three experimental flumes in central Shark River Slough in Everglades National Park. Green, yellow and red arrows represent the mean, and one and two standard deviations from the mean, respectively.

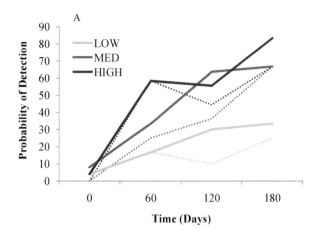

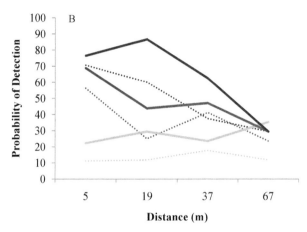

Figure 4. Probability that P-enrichment will be detected by periphyton TP indicator (dotted lines) and by the combined TP and weedy:endemic diatom indicator (solid lines) in low (+5 µg P L⁻¹), medium (+15 µg P L⁻¹) and high (+30 µg P L⁻¹) enrichment channels in the second year of the dosing study, using the one standard deviation criterion. Plot also shows how probability increases with (A) increasing duration of exposure, and (B) decreasing distance, to inflows of a constant concentration.

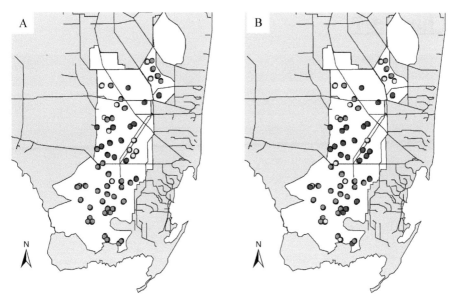

Figure 5. Assessment of phosphorus enrichment across the Everglades watershed, based on (A) periphyton TP concentrations, and (B) a combination of periphyton TP and weedy:endemic diatoms averaged over 2006–2011. Sites are coded green, yellow and red if they are less than one, greater than one and greater than two standard deviations, respectively, from the mean TP concentration (A) or either mean TP or weedy:endemic diatom ratio (B).

Chapter 7

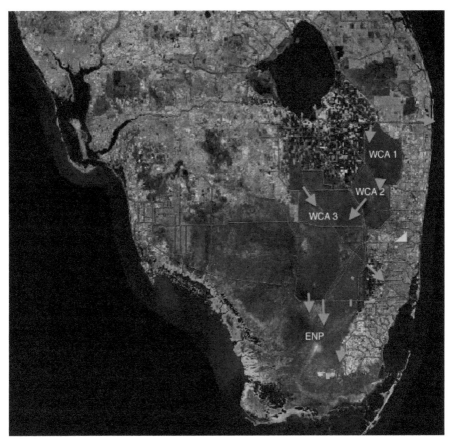

Figure 1. Map of the Florida Everglades including Water Conservation Areas (WCAs) and Everglades National Park (ENP). Primary canals and structure flows noted in red and blue, respectively (Landsat 2004; SFWMD canals 1997). Arrows indicating Lake flow to the Caloosahatchee and St. Lucie not indicated.

Chapter 8

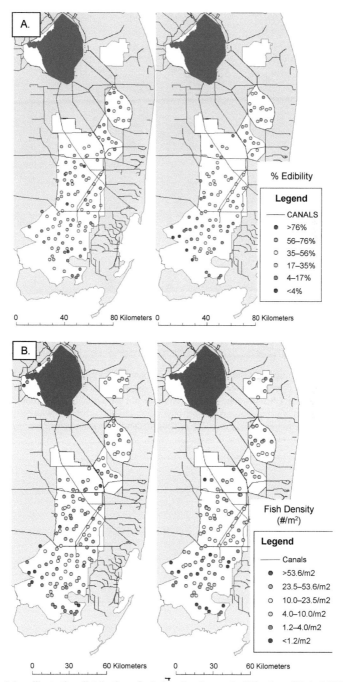

Figure 4. A. Maps illustrating distribution of relative abundance of edible algae (%) in 2005 and 2006. B. Map illustrating distribution of fish density (#/m²) in 2005 and 2006.

Chapter 9

Figure 1. Photos depicting the major types of periphyton assayed for nitrogen fixation activity in various Everglades systems as discussed in the text. Included are: epiphytic and floating mat periphyton from low nutrient areas (upper left), benthic mats lifting off from the soil surface and cross section of a floating mat showing thickness and layering (upper right), periphytic communities associated with detritus in eutrophic areas (lower left), and periphyton/soil crust during drought conditions in short hydroperiod marshes of the southern Everglades (lower right).

Chapter 10

Figure 2. Examples of various types of cyanobacterial mats from marshes of Belize. 2A Aerial view of Belizean marshes; 2B Benthic mat (epipelon); 2C Floating mat (metaphyton); 2D Mat forming on plant stems (periphyton); 2E Dry mat; 2F Vertical cross-section of a benthic mat.

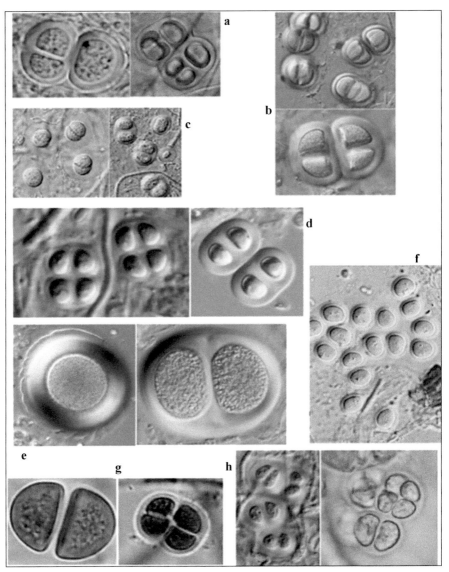

Figure 5. Examples of morphological diversity of *Chroococcus* types from Belizean marshes: a. *Chroococcus maior*, b. *C. mediocris*, c. *C. minutus*, d. *C. mipitanensis*, e. *C. occidentalis* f. *C. cf. occidentalis*, g. *C. pulcherrimus*, h. *C. subsphaericus*. From Komárek et al. 2005.

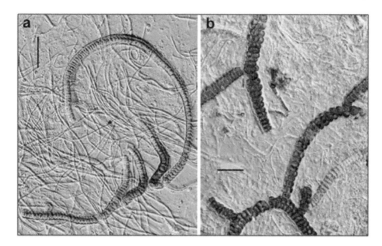

Figure 7. Examples of two heterocytous, probably endemic species from cyanobacterial mats from marshes of N Belize, a. *Chakia ciliosa* (with numerous thin filaments of *Leptolyngbya*), b. *Stigonema eliskae*; bars = 50 μm.

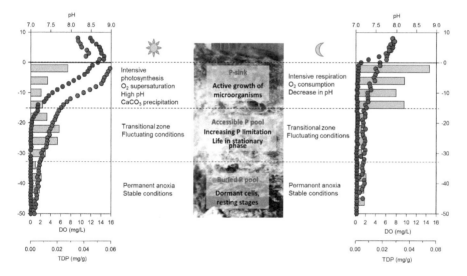

Figure 9. Diurnal changes in selected physico-chemical parameters of studied CBM at 5:00 pm (day) and 5:00 am (night).

Chapter 11

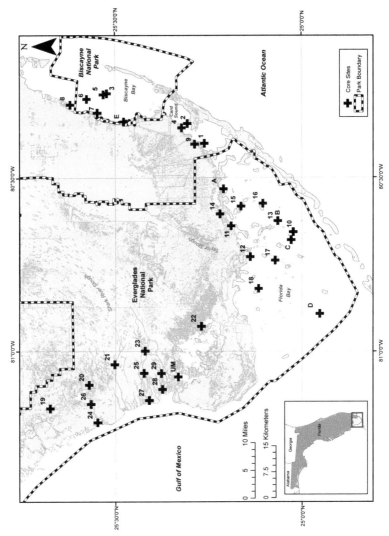

Figure 1. Map showing USGS, FIU and UM coring locations in Florida Bay, Biscayne Bay and along the southwest coast of the Everglades (Florida, U.S.A.). See Table 1 for detailed information about the core locations.

Chapter 13

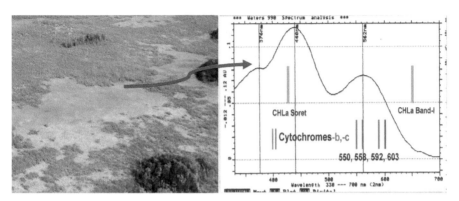

Figure 38. (a) An aerial view of a low hydroperiod periphyton mat in the Shark River Slough. (SFWMD) (b) UV/Vis of scytonemin-imine isolated from Shark River Slough samples and cultures of *Scytonema hofmannii* (Grant and Louda 2013).

Chapter 14

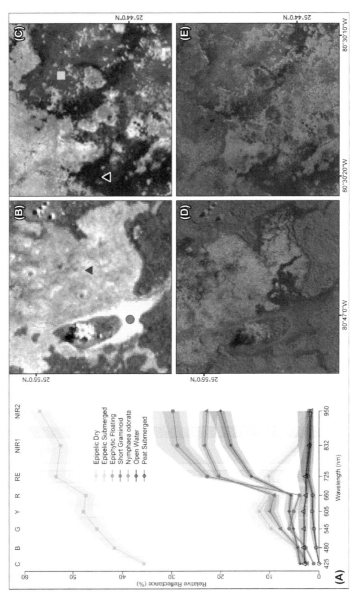

Figure 5. (A) Selected spectral signatures of slough and wet prairie classes derived from WV2 data: dark shades of confidence bands are 25th to 75th percentile; lighter shades are below 5th and above 95th percentiles. Background shading and band abbreviations as in Fig. 1. (B, C) WV2 imagery showing extraction locations of signatures for epiphytic floating periphyton (triangle) vs. *Nymphaea odorata* (circle) in slough within WCA3A (B) and submerged epipelic periphyton (square) vs. submerged peat (triangle) in wet prairie within ENP (C). (D, E) Corresponding aerial photography for B (D) and C (E). Sample locations given in Fig. 4.

Chapter 18

Plate 1. Severe outbreak of leaf spot on *Lygodium microphyllum*, background shows dead plants with disease symptoms.

Plate 2. Severe outbreak of leaf spot and blight on leaves and fertile fronds of *Lygodium microphyllum*.

Plate 3. Close-up of *Lygodium microphyllum* rachis showing brown soft scale insect and tissue necrosis.